# IFIP Advances in Information and Communication Technology 499

## IFIP – The International Federation for Information Processing

IFIP was founded in 1960 under the auspices of UNESCO, following the first World Computer Congress held in Paris the previous year. A federation for societies working in information processing, IFIP's aim is two-fold: to support information processing in the countries of its members and to encourage technology transfer to developing nations. As its mission statement clearly states:

> *IFIP is the global non-profit federation of societies of ICT professionals that aims at achieving a worldwide professional and socially responsible development and application of information and communication technologies.*

IFIP is a non-profit-making organization, run almost solely by 2500 volunteers. It operates through a number of technical committees and working groups, which organize events and publications. IFIP's events range from large international open conferences to working conferences and local seminars.

The flagship event is the IFIP World Computer Congress, at which both invited and contributed papers are presented. Contributed papers are rigorously refereed and the rejection rate is high.

As with the Congress, participation in the open conferences is open to all and papers may be invited or submitted. Again, submitted papers are stringently refereed.

The working conferences are structured differently. They are usually run by a working group and attendance is generally smaller and occasionally by invitation only. Their purpose is to create an atmosphere conducive to innovation and development. Refereeing is also rigorous and papers are subjected to extensive group discussion.

Publications arising from IFIP events vary. The papers presented at the IFIP World Computer Congress and at open conferences are published as conference proceedings, while the results of the working conferences are often published as collections of selected and edited papers.

IFIP distinguishes three types of institutional membership: Country Representative Members, Members at Large, and Associate Members. The type of organization that can apply for membership is a wide variety and includes national or international societies of individual computer scientists/ICT professionals, associations or federations of such societies, government institutions/government related organizations, national or international research institutes or consortia, universities, academies of sciences, companies, national or international associations or federations of companies.

More information about this series at http://www.springer.com/series/6102

Luis M. Camarinha-Matos · Mafalda Parreira-Rocha
Javaneh Ramezani (Eds.)

# Technological Innovation for Smart Systems

## 8th IFIP WG 5.5/SOCOLNET Advanced Doctoral Conference on Computing, Electrical and Industrial Systems, DoCEIS 2017 Costa de Caparica, Portugal, May 3–5, 2017 Proceedings

*Editors*
Luis M. Camarinha-Matos
FCT- Department of Electrical Engineering
Universidade Nova de Lisboa
Monte da Caparica
Portugal

Javaneh Ramezani
FCT- Department of Electrical Engineering
Universidade Nova de Lisboa
Monte da Caparica
Portugal

Mafalda Parreira-Rocha
FCT- Department of Electrical Engineering
Universidade Nova de Lisboa
Monte da Caparica
Portugal

ISSN 1868-4238 ISSN 1868-422X (electronic)
IFIP Advances in Information and Communication Technology
ISBN 978-3-319-85820-3 ISBN 978-3-319-56077-9 (eBook)
DOI 10.1007/978-3-319-56077-9



Printed on acid-free paper

This Springer imprint is published by Springer Nature
The registered company is Springer International Publishing AG
The registered company address is: Gewerbestrasse 11, 6330 Cham, Switzerland

# Preface

This volume of the proceedings of the **Advanced Doctoral Conference on Computing, Electrical and Industrial Systems** (DoCEIS) 2017 presents a series of selected articles produced in the context of engineering doctoral programs. The theme was "Technological Innovation for Smart Systems" and contributions reflect the growing interests in research, development, and application of smart systems. The rapid evolution in smart sensors, actuators, and embedded intelligence technology and its seamless integration into multiple system architecture and platforms have revolutionized the technological world, and even the way we live, since the last decade. The pervasive nature of this technology has enabled rapid permeation into all facets and levels of engineering disciplines and has earned immense attention and focus not only within academic circles and research communities worldwide, but also in the practical applications development.

Potential benefits can be found in all engineering fields and at all levels, e.g., supporting systems-of-systems, facilitating the industrial Internet and networked enterprises, enabling effective smart energy grids, creating the basis for smart environments, etc. A "smart systems" approach can change the way engineering systems are designed and operated while leading to exciting challenges for researchers and industrial practitioners. Smart systems are undeniably the technology of the future, with unparalleled possibilities, hence the need to further explore and exploit its prospects.

DoCEIS is aimed as an international forum providing a platform for the presentation of research results generated in PhD works, and a space for discussion of post-graduate studies, PhD thesis plans, and practical aspects of a PhD work and results from doctoral research in these inter-related areas of engineering, while promoting a strong multi-disciplinary dialog. As such, participants were challenged to look beyond their specific research question and relate their work to the selected theme of the conference, namely, to identify in which ways their research topics can benefit from, or contribute to, smart systems-based solutions.

A basis for innovation nowadays is to embrace the application of multi-disciplinary and interdisciplinary approaches in the context of research. In fact, more and more funding agencies are including this element as a key requirement in their funded programs. As such, the challenge put forward by the DoCEIS series of conferences to its authors can be seen as a contribution to the process of acquiring such skills, which are mandatory in the profession of a PhD.

This eighth edition of DoCEIS, which was sponsored by SOCOLNET, IFIP WG5.5, and IEEE IES, attracted a considerable number of paper submissions from a large number of PhD students and their supervisors from 23 countries. This book comprises the works selected by the international Program Committee for inclusion in the main program and covers a wide spectrum of application domains. As such, research results and on-going work are presented, illustrated, and discussed in areas such as:

- Collaborative networks
- Computational intelligence
- Systems analysis
- Smart manufacturing systems
- Smart sensorial systems
- Embedded and real-time systems
- Energy management
- Energy optimization
- Distributed infrastructure
- Solar energy
- Electrical machines
- Power electronics
- Electronics

As anticipated, and confirmed by the submissions, it is shown that virtually any research topic in this broad engineering area can either benefit from a smart systems perspective, or be a direct contributor with models, approaches, and technologies for further development of such systems.

We expect that this book will provide readers with an inspiring set of promising ideas and new challenges, presented in a multi-disciplinary context, and that by their diversity these results can trigger and motivate richer research and development directions.

We would like to thank all the authors for their contributions. We also appreciate the efforts and dedication of the DoCEIS international Program Committee members, who both helped with the selection of articles and contributed with valuable comments to improve their quality.

February 2017

Luis M. Camarinha-Matos
Mafalda Parreira-Rocha
Javaneh Ramezani

# Organization

**The 8th IFIP/SOCOLNET Advanced Doctoral Conference on Computing, Electrical, and Industrial Systems was held in Costa de Caparica, Portugal, during May 3–5, 2017.**

## Conference and Program Chair

Luis M. Camarinha-Matos, Portugal

## Organizing Committee Co-chairs

Luis Gomes, Portugal
João Goes, Portugal
Pedro Pereira, Portugal

## International Program Committee

Adriana Giret, Spain
Ahmed F. Zobaa, UK
Alok Choudhary, UK
Amir Assadi, USA
Andrea Bottino, Italy
Andrew Adamatzky, UK
Angel Ortiz, Spain
Antoni Grau, Spain
Antonios Tsourdos, UK
Armando Pires, Portugal
Asal Kiazadeh, Portugal
Barbora Buhnova, Czech Republic
Bernadetta Kwintiana, Germany
Constantin Filote, Romania
Diego Gachet, Spain
Dirk Lehmhus, Germany
Duc Pham, UK
Eduard Shevtshenko, Estonia
Enrique Romero-Cadaval, Spain
Erik Bruun, Denmark
Eugénio Oliveira, Portugal
Ezio Bartocci, Austria
Fausto Pedro Garcia Marquez, Spain
Florin G. Filip, Romania
Ghazanfar Safdar, UK
Giuseppe Buja, Italy
Gordana Ostojic, Serbia
Hans-Jörg Kreowski, Germany
Horacio Neto, Portugal
Igor Kuzle, Croatia
Ip-Shing Fan, UK
João Goes, Portugal
João Martins, Portugal
João Paulo Pimentão, Portugal
Jose M. De La Rosa, Spain
José Igreja, Portugal
José M. Fonseca, Portugal
Juan Jose Rodriguez Andina, Spain
Klaus-Dieter Thoben, Germany
Kleanthis Thramboulidis, Greece

Laura Carnevali, Italy
Luigi Piegari, Italy
Luis Bernardo, Portugal
Luis Gomes, Portugal
Luis M. Correia, Portugal
Luis M. Camarinha-Matos, Portugal (Chair)
Luis Oliveira, Portugal
Manuela Vieira, Portugal
Marcin Paprzycki, Poland
Maria Helena Fino, Portugal
Marin Lujak, Spain
Marko Beko, Portugal
Michael Huebner, Germany
Niels Lohse, UK
Nik Bessis, UK
Noelia Correia, Portugal
Nuno Paulino, Portugal
Olga Battaia, France
Paul Grefen, The Netherlands
Paulo Miyagi, Brazil
Paulo Pinto, Portugal
Pedro Pereira, Portugal
Peter Marwedel, Germany
Peter Palensky, Austria
Pierluigi Siano, Italy
Ricardo Gonçalves, Portugal
Ricardo Rabelo, Brazil
Rita Ribeiro, Portugal
Roberto Canonico, Italy
Rolf Drechsler, Germany
Ruggero Donida Labati, Italy
Rui Melicio, Portugal
Simon Pietro Romano, Italy
Stefano Di Carlo, Italy
Sven-Volker Rehm, Germany
Thilo Sauter, Austria
Thomas Strasser, Austria
Vladimir Katic, Serbia
Wojciech Cellary, Poland
Youcef Soufi, Algeria
Zbigniew Leonowicz, Poland

## Organizing Committee (PhD Students)

Adriano Ferreira, Portugal
Fernando Monteiro, Portugal
Javaneh Ramezani, Iran
José Gonçalves, Portugal
Kankam Adu, Ghana
Luis Irio, Portugal
Mafalda Rocha, Portugal
Miguel Teixeira, Portugal
Pedro Monteiro, Portugal
Ricardo Peres, Portugal
Shabnam Pasandideh, Iran
Majid Zamiri, Iran

## Technical Sponsors

Society of Collaborative Networks

IFIP WG 5.5 COVE
Co-Operation infrastructure for Virtual Enterprises and electronic business

IEEE–Industrial Electronics Society

## Organizational Sponsors

**Organized by:** PhD Program on Electrical and Computer Engineering FCT-UNL.

# Contents

## Collaborative Networks



## Computational Intelligence



## Systems Analysis

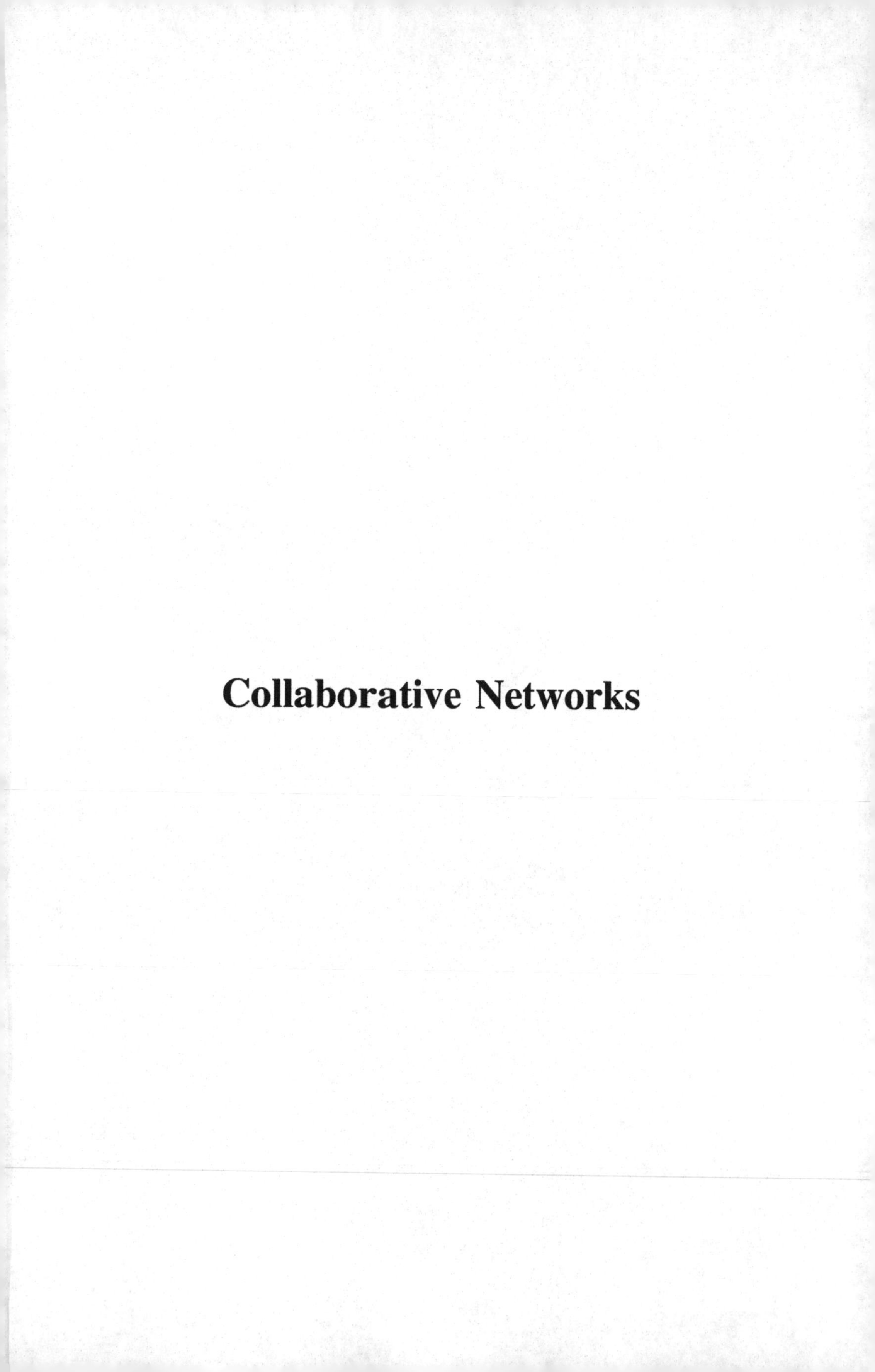

# Collaborative Networks

# Supporting the Strategies Alignment Process in Collaborative Networks

Beatriz Andres(✉) and Raul Poler

Research Centre on Production Management and Engineering (CIGIP), Universitat Politècnica de València (UPV), Calle Alarcón, 03801 Alcoy, Spain
{bandres,rpoler}@cigip.upv.es

**Abstract.** The establishment of collaborative relationships with the network partners provides them important advantages, such as competitiveness and agility, when responding to the current rapid market evolutions. Nonetheless, the participation in collaborative networks becomes a complex process that starts with the alignment of all the enterprises' objectives and strategies. Smart systems and approaches are needed in order support collaborative partners to deal with the strategies alignment challenge. The lack of alignment emerges because each enterprise defines its own objectives and strategies, to perform their business, and it could happen that non-compatible strategies are activated, involving the appearance of conflicts between strategies of different enterprises. To this regard, a decision support system is proposed, consisting of a mathematical model, a system dynamics method, a simulation tool and a guideline, with the main aim of supporting the process of identifying aligned strategies, among the enterprises of the collaborative network.

**Keywords:** Alignment · Strategy · Objective collaborative network · Performance indicator · System dynamics · Simulation

## 1 Introduction

The concept of collaborative networks (CN) has been widely studied over the last years due to the positive effects undergone by the enterprises that collaborate [1]. In the work of consolidating a new discipline in CN, Camarinha-Matos and Afsarmanesh [2] define CN as a network consisting of a variety of autonomous and heterogeneous entities that collaborate to better achieve common or compatible goals, to jointly generate value, and whose interactions are computer network supported. Collaborative processes have been widely studied over the last years due to their decisive contribution in the proper operation of the CN. With the aim of consolidating the wealth of knowledge in the research area of collaborative processes Andres and Poler [3] perform a deep analysis that has allowed to (i) classify the most relevant collaborative processes according to the decision making level: strategic, tactical, and operational, and (ii) analyse for each process the models, guidelines and tools proposed in the literature to address them. The authors conclude that amongst all the collaborative processes studied, the ones that need to be


L.M. Camarinha-Matos et al. (Eds.): DoCEIS 2017, IFIP AICT 499, pp. 3–19, 2017.
DOI: 10.1007/978-3-319-56077-9_1

addressed from the collaborative perspective, through proposing new contributions to fill the decentralized and collaborative features, are (i) at the strategic decision-making level: the strategies alignment process; (ii) at the tactical level: share costs and profits, and uncertainty management; and (iii) at the operational level: collaborative lotsizing.

Enterprises willing to collaborate must overcome a set of barriers not only associated with the establishment of collaborative processes identified by [3, 4] (e.g. products design, demand forecasting, operations planning, replenishment, uncertainty management, share costs and profits, scheduling, information exchange, interoperability, etc.), but also when defining compatible goals, activating complementary strategies [5] or aligning their core values [6, 7]. Focusing on the strategies alignment process, the mere consideration of all the enterprises' objectives when deciding which strategies are the best ones to carry out will allow achieving higher levels of adaptability, agility, and competitiveness [1], strengths that are specially valued in current turbulent contexts and dynamic markets. Considering this, the strategies alignment process is hereafter addressed; with the main aim of dealing with the conflicts appearing when strategies misalignments emerge, in the CN context. Intuitively, as the activation of strategies has a direct influence on the objectives achievement, it can be understood that the strategies will be characterised by being aligned when each activated strategy not only promotes the achievement of the objectives defined by the enterprise that formulates such strategy, but also when positively influences the accomplishment of the objectives defined by the rest of the network partners.

Considering the importance of aligning strategies, among the enterprises of the same network, in terms of improving the CN relationships, there is a lack of an integrated approach to support enterprises on the modelling, assessment and solution of the strategies alignment process from a collaborative and integrated perspective. In the light of this, the following research questions are raised to support the strategies alignment process, in order to solve them as the objective of this research.

**RQ1.** *How to model the impact that each strategy, formulated by one enterprise, has on the objectives defined by the other network enterprises? That is, how to model the impact of the strategies at the inter-enterprise level?*

**RQ2.** What would be an adequate model to support the process of identification of aligned strategies, through modelling the strategies impact in the objectives, in CN context?

**RQ3.** What would be an adequate method to support the process of identification of aligned strategies, and to represent causal relationships (impacts) between the strategies and the objectives, in CN context?

**RQ4.** What would be an adequate tool to support the process of identification and assessment of aligned strategies, and to compute the strategies impact on the objectives performance at enterprise and network level, in CN context?

**RQ5.** What would be an adequate guideline to support the process of identification and assessment of aligned strategies, and to analyse the strategies impact on the objectives and identify misalignments, in CN context?

In order to give response to the raised research questions, the paper is organized as follows: Sect. 2 discusses the relationship between the approach proposed in this paper, to deal with the strategies alignment process, and the Smart Systems; Sect. 3 summarizes literature review performed in the research area of strategies alignment; Sect. 4 introduces the decision support system, consisting of a mathematical model, a system dynamics method, a simulation tool and a guideline, with the main aim of supporting the process of identifying aligned strategies among the enterprises of the CN; Sect. 5 the proposed approach is validated in a real use case from food industry; finally, Sect. 6 provides the conclusions and the paper discussion.

## 2 Relationship with Smart Systems

Smart Systems (SS) have multi-disciplinary applications in different areas of research, such as the social, economic, healthcare, energy, safety and security, logistics, ICT, and manufacturing. The area under study of this paper is focused on the CN operation, which includes service and manufacturing sectors. The novelty in SS is the integration of different components, regardless the technologies and materials in which are created [29]. The proposed strategies alignment approach is applied in different industries and sectors that have in common the CN to which they belong; in this regard, SS can support the diversity associated to this approach. The decision support system proposed could be used as part of a SS that allows, through using real time information of other interoperable components, identifying in each enterprise the most appropriate aligned business strategies. The negotiation process, for identifying the most appropriate strategies, could be also included in a SS in order to allow enterprises make smart decisions with regards the strategies to activate during their participation in the CN. In the light of this, the strategies alignment approach proposed could benefit from the real-time information, response capability, tracking and monitoring features that provide the SS. The integration of different systems, for the implementation of the strategies alignment approach in SS, is a key question to answer, bringing together interdisciplinary technological approaches and solutions for overcoming potential limitations in the establishment of collaborative process and the stable and sustainable operation of the CN.

## 3 Problem Definition and Conceptualization

The literature review carried out in [3] has allowed identifying, firstly, the most important processes to perform in a CN, and secondly, amongst all these processes, those that have a lack of contributions from the CN context. As stated in Sect. 1, the strategies alignment process is included in the group of potential processes to propose solutions in collaborative decentralised scenarios. According to the analysis carried out in [3] it can be concluded that, to the best of our knowledge, the strategies alignment process is a collaborative process that requires to be studied, so that models, guidelines and tools for its analysis, assessment and resolution, in the CN context, have to be proposed.

In a network of enterprises, the alignment can be defined as a proper or desirable coordination or relationship of the components of this network. More concretely, in the management field, the concept of alignment can be considered as a situation in which the strategies, formulated by the entities belonging to the network, are strictly combined under a set of functions to achieve the objectives [8]. CNs consist of autonomous and heterogeneous enterprises [9] each one defining its own objectives. The formulation of the strategies answers the question: How to reach the objectives? Once the planner decides the scope, situation or problem that aims to modify, a goal is drawn to guide the processes of change and then to trace the trajectory of necessary events over time to achieve that purpose. Strategy is the way forward to achieve the objectives. The business strategies are the set of actions raised to achieve the defined objectives; therefore, each enterprise of the CN formulates its own strategies with the main aim of achieving the defined objectives. There will be times in which all the strategies formulated are activated. Nevertheless, sometimes only a few of the formulated strategies will be activated, due to, for example, a restriction associated with the budged. Lets consider two enterprises (E) in a CN, each one defines two objectives ($O$) and formulates two strategies (*str*). Each objective has associated a KPI to measure its achievement. In this regard, $E_1$ acquires the role of the distributor in the CN and defines $O_{11}$: *Increase the product sales by a 10%*, and $O_{12}$: *Reduce the product costs by a 30%*; and formulates $str_{11}$: *Invest 0,5 m.u on marketing activities*, and $str_{12}$: *Conduct negotiations with other manufacturers to reduce the purchasing costs*. $E_2$ acquires the role of the manufacturer in the CN and defines $O_{21}$: *Increase the profit by a 15%*, and $O_{22}$: *Reduce the quantity of product that cannot be sold by 100%*; and formulates $str_{21}$: *Use different distribution channels to sell the product in other markets*, and $str_{22}$: *Buy one machine to make derivative products, reprocessing the product that cannot be sold (i.e. low cost product)*. With this example it can be observed that the $str_{12}$ is not compatible with the $str_{21}$, because $str_{12}$ is devoted to establish new relations with other manufacturers, which will involve the reduction of the profit defined in $O_{21}$. Moreover, if $E_1$ conducts negotiations with other manufacturers ($str_{12}$), the $O_{22}$ will be negatively influenced. Focusing on $E_2$, the $str_{21}$ focuses on the generation of alliances with distribution channels different from the one provided by $E_1$. The activation of $str_{21}$ negatively influences $O_{11}$, defined for reducing the product sales; and consequently the $O_{12}$ that leads to reduce the product costs. Considering the aforementioned, the $str_{12}$ and $str_{21}$ are considered misaligned, if activated at the same time. On the other hand, $str_{11}$ and $str_{22}$ are considered to be aligned because the two formulated strategies positively influence the achievement of the objectives defined. Assuming that, the strategies alignment concept is defined next as: "the set of strategies, formulated by the enterprises belonging to the CN, whose activation positively influence, on the whole, the objectives achievement of the majority of the enterprises participating in the CN; obtaining the best performance at the network level, although small number of the strategies negatively influence any of the defined objectives" [10]. It must be considered that (i) individual enterprises take part in several networks, so that it is likely that some of the enterprises taking part in these networks have contradictory objectives and consequently contradictory strategies; and (ii) the enterprises belonging to one specific CN are heterogeneous and contradictory objectives and strategies might arise. Therefore, for enterprises belonging to a CN, the defined objectives and the strategies formulated by one enterprise could favour, or not, the objectives defined by other enterprises. In order to achieve the ideal situation, enterprises belonging to a CN should be able to identify those aligned strategies, whose activation promotes the improvement of the objectives defined by the majority of the networked enterprises, or at least the activated strategies do not negatively influence on the objectives attainment [5].

# 4 Literature Review

A summary of the review performed to analyse how the strategies alignment process has been treated in the literature is presented. In the light of this, some models guidelines and tools are identified and briefly described. The gaps and trends related to the strategies alignment process from a collaborative perspective are identified, as a result of the analysis performed. The initial round of search was based on a broad meaning of keywords and contexts (enterprise and network level), to ensure that papers adopting an alternative nomenclature, were identified. *Alignment of strategies*, *alignment of actions*, *alignment of decisions*, *collaborative decisions design*, *collective decisions* and *alignment in supply chain* where the keywords used. The found works proposed models, guidelines and tools to deal with the alignment of decisions from different decision making levels and different perspectives of application (i) one in which the decisions are collaboratively made and from the beginning of the decision making the decisions are aligned, and (ii) another one in which the each partner defines its own decisions and then these are pooled in order to identify those that are more aligned with the decisions of other network partners. Considering the reviewed works, Table 1 is generated, listing and briefly describing the works. Due to space restrictions the table presents works from 2012. The selected contributions are analysed considering if the proposed approaches are designed when (i) the decisions are collaboratively and centralised (C) made or, unlike, (ii) the decisions are decentralised (D) made by each CN partner and after that the decisions are aligned. A set of models, guidelines and tools are proposed in the literature with the main aim of aligning decisions among the enterprises of the network. Some of them can be highlighted: classified as models it can be found the multi-criteria methods such as FMCDS or MCDM; fuzzy approaches deal with uncertain information. As regards the guidelines, collaborative strategies or negotiation-based schemes such as S-DSP are found. Considering the methods, TOPSIS, MCOGA, GA, ANP, causal maps can be emphasised. Concerning tools MECDSS is found. Despite of the importance of aligning strategies, in terms of avoiding partnership conflicts, to the best of our knowledge, there are some gaps in the literature as regards contributions that provide a holistic approach that allows considering all the strategies formulated by all the partners in the CN context. The performed review has allowed identifying possible trends, gaps and actions in the topic under study. This actions are summarised as follows: (i) propose a complete approach to deal with the strategies alignment process by considering all the strategies formulated by all the enterprises of the CN; (ii) identify the aligned strategies from an holistic perspective regardless of their nature and type, taking into account the CN context; (iii) model the strategies alignment process considering the intra-enterprise strategies alignment (alignment of the strategies defined in the same enterprise), and inter-enterprise strategies alignment (alignment among the strategies defined by different enterprises of the network); (iv) consider the performance approach to measure the strategies influence, that when activated, it will be measured considering the increase and decrease of the KPIs defined in each enterprise. Definitely, from the performed literature review, there is a need to propose a framework consisting of a model method, tools and guidelines to address the strategies alignment process from a holistic perspective by equally considering all the network partners.

**Table 1.** Contributions dealing with the research topic decisions alignment

| Author | Proposal | C | D |
|---|---|---|---|
| [11] | An optimisation model to configure supply chain network, by combining optimisation at the strategic and tactical level | X | - |
| [12] | A set of decisions are previously identified and with IMGP (Interactive Meta-Goal Programming) approach is used to analyse a decision under consideration and identify how it favours other enterprises performance | - | X |
| [13] | Sustainability – Decision Support Protocol (S-DSP) is a collaborative centralised solution that enables to maximise the sustainability of supply network in production and scheduling planning | X | - |
| [14] | Multi-enterprise collaborative decision support system (MECDSS) that allows each enterprise to propose its own decisions and dispose of all the data as regards the decision-making components of its partners. Enterprises identify how its decisions affect itself and its partners | - | X |
| [15] | A multi-criteria decision making (MCDM) technique, is proposed to support decision making in B2B collaboration | X | - |
| [16] | Analytic Network Process (ANP) used to determine how the decisions defined by the CN enterprises influence on the strategies achievement of the network | - | X |
| [17] | Fuzzy analytic hierarchy process (FAHP) that uses weightings to support the identificaiton of the importance of the dimensions that influence the decisions | X | - |
| [18] | Fuzzy causal maps and qualitative assessment methods designed to asses the values alignment among the CN partners | - | X |
| [19] | Fuzzy TOPSIS (Technique for order preference by similarity to ideal solution) and multi-criteria decision making (MCDM) are used to make collective decisions to solve a number of problems, which are characterised by various quantitative and qualitative criteria | X | - |
| [20] | TOPSIS combined with a multi-criteria genetic optimisation feature. Modified multi-criteria genetic optimisation feature (MCOGA) are proposed to create centralised collaborative decision | X | - |
| [21] | Optimisation model to integrate the decision of partners' selection and collaborative transportation scheduling. A novel Genetic Algorithm (GA) is proposed | X | - |
| [22] | An approach based on subjective probabilities is proposed to evaluate (i) the probability that a decision of one partner is optimal for itself and (ii) the probability that a decision for the first partner is optimal for other partners | - | X |

# 5 Approach to Support the Strategies Alignment Process

An approach that consists of a model, method, tool and guideline is proposed, to deal with the strategies alignment process, in the CN context.

## 5.1 Mathematical Model

The proposed model allows to formally represent, in a mathematical notation, the influences that the strategies activated in one enterprise have on the performance indicators (KPI) defined to measure the achievement of the objectives, both in the same enterprise

**Table 2.** Index and model parameters

| Index | |
|---|---|
| $net$ | set of networks, $net = (1, \ldots, N)$ |
| $i$ | set of enterprises, $i = (1, \ldots, I)$ |
| $x$ | set of objectives, $x = (1, \ldots, X)$ |
| $k$ | set of key performance indicators, $k = (1, \ldots, K)$ |
| $s$ | set of strategies where $s = (1, \ldots, S)$ |
| **Model Parameters** | |
| $n$ | number of enterprises belonging to the network |
| $o_{ix}$ | objective $x$ defined in enterprise $i$ |
| $b_i$ | budget owned by the enterprise $i$ to invest in the activation of the strategies $str_{is}$, in monetary units *[m.u.]* |
| $str_{is}$ | strategy $s$ defined by enterprise $i$ |
| $kpi_{ixk}$ | key performance indicator (KPI) $k$ used to measure the objective $o_{ix}$ |
| $\Delta kpi_{ixk}$ | increase observed in the $kpi_{ixk}$ when the $str_{is}$ is activated. It can be decomposed in:<br>• $\Delta kpi_{ixk}^{intra}$ $\Delta kpi_{ixk}^{intra}$ increase of the $kpi_{ixk}$ when the $str_{is}$ of the same enterprise $i$ ($e_i$) is activated<br>• $\Delta kpi_{ixk}^{inter}$ $\Delta kpi_{ixk}^{inter}$ increase of the $kpi_{jxk}$ when the $str_{js}$ of a different enterprise $j$ ($e_j$) is activated |
| $\Delta kpi_{ixk}^{max}$ | maximum increase of $kpi_{ixk}$ estimated by the *enterprise* $i$ (used to homogenise all the KPIs) |
| $Threshold\_kpi_{ixk}$ | value from which the associated $kpi_{ixk}$ is affected by the activation of a strategy $str_{is}$. Below $Threshold\_kpi_{ixk}$ the influence of $str_{is}$ is not observed, from $Threshold\_kpi_{ixk}$ the influence exerted by $str_{is}$ is considered |
| $\Delta kpi_{ixk}\_T$ | increase experienced by the $kpi_{ixk}$ once the $Threshold\_kpi_{ixk}$ is computed |
| $\Delta kpi_{ixk}\_min$ | minimum increase that the enterprise estimates for the $kpi_{ixk}$ , once the $Threshold\_kpi_{ixk}$ is computed |
| $fulfillment\_kpi_{ixk}$ | binary parameter that indicates that increase experienced by the $kpi_{ixk}$ is higher than the minimum increase that the enterprise estimates for the $kpi_{ixk}$, once the $Threshold\_kpi_{ixk}$ is computed |
| $w_{ixk}$ | weight of $kpi_{ixk}$, determines the relevance that the $kpi_{ixk}$ has for enterprise $i$ |
| $\Delta kpi_i$ | increase experienced by the KPI defined at enterprise $i$ level |
| $\Delta kpi_{net}$ | increase experienced KPI defined at network $net$ level |
| $f\_inf\_str_{is}\_kpi_{ixk}(t)$ | function that models the behaviour of the $kpi_{ixk}$ $kpi_{ixk}$ when $str_{is}$ is activated |
| $f\_kpi_{ixk}(t)$ | function that models the overall behaviour of the $kpi_{ixk}$ considering all the activated strategies |
| $f\_kpi_{ixk}\_T(t)$ | function that models the behaviour of the $kpi_{ixk}$ when the $Threshold\_kpi_{ixk}$ value is computed |
| $c\_str_{is}$ | cost of activating one unit of strategy $str_{is}$ [m.u.] |
| $str_{is}\_mu$ | monetary units invested in the activation of $str_{is}$ [m.u.] |
| $val\_str_{is}\_kpi_{ixk}$ | numerical value estimated by the enterprise $e_i$, that registers the increase or decrease of the $kpi_{ixk}$ when one unit of $str_{is}$ is activated ($u\_str_{is}$) |
| $inf\_str_{is}\_kpi_{ixk}$ | maximum level of influence on the $kpi_{ixk}$ when certain number of units of strategy ($u\_str_{is}$) are activated |
| $slope\_str_{is}\_kpi_{ixk}$ | slope of the ramp in represented in $f\_inf\_str_{is}\_kpi_{ixk}(t)$ $f\_inf\_str_{is}\_kpi_{ixk}(t)$ |
| $H$ | horizon, time units [t.u.], period of time in which the set of strategies are to be activated. Normalised to the unit, $H = 1$. |
| $d_1\_str_{is}$ | delay, time period between the initial time of activation of $str_{is}$ ($ti\_str_{is}$) and the time when the $kpi_{ixk}$ is started to be influenced by the activated $str_{is}$ [t.u.] |
| $d_2\_str_{is}$ | time period between the $str_{is}$ starts to influence the $kpi_{ixk}$ until the maximum level of influence in is achieved ($inf\_str_{is}\_kpi_{ixk}$), [t.u.] |
| $d_3\_str_{is}$ | time period in which $str_{is}$ is exerting the highest influence ($inf\_str_{is}\_kpi_{ixk}$) on the $kpi_{ixk}$ [t.u.] |
| $d_4\_str_{is}$ | total duration of $str_{is}$ [t.u.] |
| $tf\_str_{is}$ | time unit when $str_{is}$ is finished [t.u.] |
| **Decision Variables** | |
| $u\_str_{is}$ | units of strategy [u.s] $str_{is}$ to be activated |
| $ti\_str_{is}$ | initial time of activation of $str_{is}$ [t.u.] |

and in other CN enterprises [10]. In order to represent the influences and relations between the KPIs and the strategies a mathematical notation model is proposed: the Strategies Alignment Model (SAM). First of all, the set of parameters and decision variables, used to model the SAM, are defined in Table 2.

The SAM is hereafter developed, consisting of an objective function and the associated restrictions, representing the relations amongst all the defined variables and parameters. The main aim is to identify, amongst all the strategies defined, those strategies that have higher level of alignment. The activation of the aligned strategies positively influences the majority of the objectives defined by the networked partners, maximising the performance at network level. The SAM computes the KPIs improvement or worsening when a strategy is activated. Thus, the developed model supports enterprises on the decision making as regards the number of units of strategy ($u\_str_{is}$) to be activated and the time in which the strategies have to be activated ($ti\_str_{is}$) with the objective of maximising the network performance, given by $kpi'_{net}$ as the homogenised version of the $kpi_{net}$. Therefore, the objective function of the SAM is mathematically represented by the following Eq. (1):

$$max. \quad \Delta kpi'_{net} \tag{1}$$

The homogenised version of $\Delta kpi_{net}$ is obtained through the homogenisation of parameters related to KPIs ($\Delta kpi'_{ixk}$) (2); and the normalisation of the parameters related to durations and time (3), based on the horizon ($H$) of time in which the strategies alignment process is modelled.

$$\Delta kpi'_{ixk} = \frac{\Delta kpi_{ixk}}{\Delta kpi^{max}_{ixk}} \tag{2}$$

$$\begin{gathered} H' = \frac{H}{H} = 1;\ d'_1\_str_{is} = \frac{d_1\_str_{is}}{H};\ d'_2\_str_{is} = \frac{d_2\_str}{H};\ d'_4\_str_{is} = \frac{d_4\_str}{H}; \\ d'_3\_str_{is} = d'_4\_str_{is} - 2 \cdot d'_{2_{str_{is}}} - d'_1\_str_{is}; \\ t'_i\_str_{is} = \frac{t_i\_str_{is}}{H};\ t'_f\_str_{is} = \frac{t_f\_str_{is}}{H} \end{gathered} \tag{3}$$

Two *decision variables*, $u\_str_{is}$ and $t\_str_{is}$, are defined in order to maximise the parameter $\Delta kpi'_{net}$. The decision variable $u\_str_{is}$ decomposes the strategy ($str_{is}$) in units of strategy, allowing representing the "intensity" in which each strategy $str_{is}$ is activated. One unit of strategy has an associated a cost ($c\_str_{is}$). Therefore, depending on the parameter $c\_str_{is}$, the enterprise' budget ($b_i$) will be reduced in a lesser or larger extent (4). The budget, $b_i$, owned by each company defines the monetary capacity constraint (5). In order to identify the influence that one unit of strategy ($u\_str_{is} = 1$) has over the $\Delta kpi'_{ixk}$, parameter $val\_str_{is}\_kpi'_{ixk}$ is used (6).

$$str_{is}\_mu = u\_str_{is} \cdot c\_str_{is} \tag{4}$$

$$b_i \geq \sum_s str_{is}\_mu \quad \forall s \tag{5}$$

$$inf\_str_{is}\_kpi'_{ixk} = u\_str_{is} \cdot val\_str_{is}\_kpi'_{ixk} \tag{6}$$

The influence that one strategy $str_{is}$ has on a particular $\Delta kpi'_{ixk}$ is modelled through the function $f\_inf\_str_{is}\_kpi'_{ixk}$. This function, $f\_inf\_str_{is}\_kpi'_{ixk}$ (8), is a piecewise function that depends on the time $[f_1(t)]$, that is, the duration parameters ($d_1\_str_{is}$, $d_2\_str_{is}$, $d_3\_str_{is}$ and $d_4\_str_{is}$) and the decision variable $ti\_str_{is}$. Besides, $f\_inf\_str_{is}\_kpi'_{ixk}$ is modelled according to a ramp shape ($slope\_str_{is}\_kpi'_{ixk}$) (7). The representation of the ramp allows modelling that, after the delay time ($d_1\_str_{is}$), the $str_{is}$ progressively influences the $kpi'_{ixk}$.

$$slope\_str_{is}\_kpi'_{ixk} = \frac{inf\_str_{is}\_kpi'_{ixk}}{d'_2\_str_{is}} \tag{7}$$

$$f\_inf\_str_{is}\_kpi'_{ixk}(t) = \\ = \begin{cases} 0 \rightarrow t \leq t'_i\_str_{is} + d'_1\_str_{is} \wedge t \geq t'_i\_str_{is} + d'_4\_str_{is} \\ slope\_str_{is}\_kpi'_{ixk} \rightarrow t'_i\_str_{is} + d'_1\_str_{is} < t < t'_i\_str_{is} + d'_1\_str_{is} + d'_2\_str_{is} \\ inf\_str_{is}\_kpi'_{ixk} \rightarrow t'_i\_str_{is} + d'_1\_str_{is} + d'_2\_str_{is} \leq t \leq t'_i\_str_{is} + d'_1\_str_{is} + d'_2\_str_{is} + d'_3\_str_{is} \\ -slope\_str_{is}\_kpi'_{ixk} \rightarrow t'_i\_str_{is} + d'_1\_str_{is} + d'_2\_str_{is} + d'_3\_str_{is} < t < t'_f\_str_{is} \end{cases} \tag{8}$$

The influence received by the KPIs defined in one enterprise $i$ (11) is caused by both intra-enterprise influence, $\Delta^{intra} kpi'_{ixk}$, (9) and inter-enterprise influences, $\Delta^{inter} kpi'_{ixk}$, (10).

$$\Delta^{intra} kpi'_{ixk} = \int_{t'_i\_str_{is}+d'_1\_str_{is}}^{H'} f\_inf\_str_{is}\_kpi'_{ixk}(t) \cdot dt \tag{9}$$

$$\Delta^{inter} kpi'_{ixk} = \int_{t'_i\_str_{js}+d'_1\_str_{js}}^{H'} f\_inf\_str_{js}\_kpi'_{ixk}(t) \cdot dt \tag{10}$$

$$\Delta kpi'_{ixk} = \Delta^{intra} kpi'_{ixk} + \Delta^{inter} kpi'_{ixk}; \; \Delta kpi'_{ixk} = \int_0^{H'} f\_kpi'_{ixk}(t) \cdot dt \tag{11}$$

After being depicted the function $f\_kpi'_{ixk}$ and computed the $\Delta kpi'_{ixk}$, the value estimated by the threshold ($Threshold\_kpi'_{ixk}$) must be considered (12). At enterprise and network level the parameters $\Delta kpi'_i$ and $\Delta kpi'_{net}$ are defined as (13).

$$\Delta kpi'_{ixk}\_T = \int_a^b f\_kpi'_{ixk}(t) \cdot dt - \int_0^{H'} Th\_kpi'_{ixk} \cdot dt \tag{12}$$

$$\Delta kpi'_i = \frac{\sum_{x,k} \Delta kpi'_{ixk}\_T \cdot w_{ixk}}{\sum_{x,k} w_{ixk}}; \quad \Delta kpi'_{net} = \frac{\sum_i \Delta kpi'_i}{n} \tag{13}$$

## 5.2 System Dynamics Method

The method used is based on system dynamics (SD), and will allow to graphically represent and solve the proposed mathematical model, from a CN perspective. SD will enable to characterise the causal relationships between the strategies and the objectives; modelling the influences that the objectives experience when certain set of strategies are activated. Moreover, SD will favour to understand the structure and dynamics of complex systems, such as the CN [10, 23]. The causal loop diagram is the graphical description that represents the system in SD. It includes all the system elements and

**Table 3.** Equations of the flow diagram

| |
|---|
| dimension_KPIixk, representing the indexes of the KPIs defined in the model index_KPIixk<br>dimension_Sis, representing the indexes of the strategies defined in the model index_Sis |
| bi - ΣSis_mu |
| Sis_mu = u_Sis · c_Sis.get(index_Sis) |
| tf_Sis = ti_Sis + d4_Sis.get(index_Sis) |
| d3_Sis = d4_Sis.get(index_Sis) - d1_Sis.get(index_Sis) - (2·d2_Sis.get(index_Sis)) |
| slope_Sis_KPIixk = (u_Sis · val_Sis_KPIixk [dimension_KPIixk])/ d2_Sis.get(index_Sis) |
| Inf_Sis_KPIixk = delay (ramp (slope_Sis_KPIixk[dimension_KPIixk], ti_Sis, ti_Sis + d2_Sis.get(index_Sis)) - ramp (slope_Sis_KPIixk[dimension_KPIixk], ti_Sis + d2_Sis.get(index_Sis) + d3_Sis, ti_Sis + 2 · d2_Sis.get(index_Sis) + d3_Sis) , d1_Sis.get(index_Sis)) |
| curve_KPIixk = ΣInf_S11_KPIixk[dimension_KPIixk] |
| KPIixk = ∫ curve_KPIixk[dimension_KPIixk] |
| Curve_KPIixk_T = IF ((curve_KPIixk[dimension_KPIixk] >= Threshold_KPIixk[dimension_KPIixk]) THEN (curve_KPIixk[dimension_KPIixk] - Threshold_KPIixk[dimension_KPIixk]) ELSE (IF (curve_KPIixk[dimension_KPIixk]<0) THEN curve_KPIixk[dimension_KPIixk] ELSE 0)) |
| KPIixk_T = ∫ curve_KPIixk_T[ dimension_KPIixk ] |
| fulfill_KPIixk_min = IF ((KPIixk_T[ dimension_KPIixk ] >= KPIixk_min[ dimension_KPIixk ]) THEN 1 ELSE 0) |
| KPI_i = ΣKPIixk_T.get(index_KPixk) · Wixk[dimension_KPIixk] |
| KPI_GLOBAL = Σ KPI_i / n |

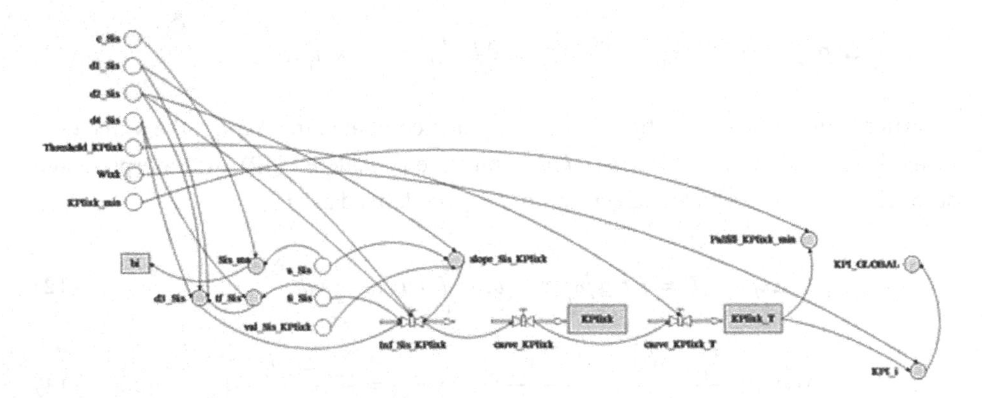


**Fig. 1.** Flow diagram in SD of the strategies alignment model.

represents the relationships among them. The causal diagram allows to qualitatively represent the behaviour of the modelled system. In order to carry out a quantitative analysis the flow diagram is constructed. The flow diagram interprets the causal loop diagram (the information and the casual relationships depicted) into a terminology that allows transcribing the equations within a simulation software. The parameters modelled in the SAM are translated for its use in the SD simulation software. Moreover the equations remain as shown in Table 3. The flow diagram of the SD SAM is represented in Fig. 1, and will be extended according to the number of enterprises, the number of KPIs defined and the strategies formulated.

### 5.3 Simulation Tool

The proposed simulation software tool is used to solve and represent the strategies alignment model, based on SD rigorous method. The use of computational tools allows automatically solving the strategies alignment process. System-dynamic's simulation based models supports on the process of computing the strategies to activate and the time slot in which activate them, optimising the global performance of the CN. Considering the SAM developed and the SD resolution method described, three tools used to address the strategies alignment process, from a CN context, are described: (i) *AnyLogic* simulation software is selected to support the system dynamics (SD) method, in which the SAM is solved; (ii) a Database Management System (DMS) is proposed to store all the information required in the SAM. The parameters required to feed the SAM are gathered in a *Microsoft Access Database* specifically designed; and (iii) the *Strategies Alignment GENerator* (SAGEN) is designed as an application to automatically generate the SAM in SD simulation software. In this regard, SAGEN contains the set of procedures that allow generating the required structure, in XML language, to create the strategies alignment simulation model in the SD simulation software selected (AnyLogic). The procedures are created according to the requirements of the XML schema, for its reading in AnyLogic. The programing language used to build SAGEN is Pascal. Lazarus [24] is used as an *Integrated Development Environment* (IDE) that uses Free Pascal compiler. In order to have a deeper insight of SAGEN programming and the procedures creation, we refer readers to [25]. To automatically generate the SAM in SD simulation software, the user firstly introduces the information required to solve the SAM in the DMS, through SAGEN user interface (SAGEN UI) (see Fig. 2). SAGEN UI is connected with *Microsoft Access Database 2010* through an *OCDBConnection*. The information stored in *Microsoft Access Database 2010* contains all the tables and fields necessary to create the XML file that contains the SAM to be simulated in AnyLogic simulation software. In a second step, the user creates the XML file, which results from the execution of the procedures programmed in SAGEN. The XML file automatically created in SAGEN contains the strategies alignment simulation model, which can be loaded in AnyLogic simulation software. The SAM is automatically created containing the flow diagram, as well as the simulation and the optimisation experiments. *AnyLogic* simulation software is selected due to brings together the most common modelling methods: System Dynamics (SD), Discrete Events (DE), and Agent Based (AB). *AnyLogic* integrates both simulation and optimisation experiments. Accordingly, in the optimization experiments, *AnyLogic* searches the values of the model parameters that lead to obtain greater performance levels of the model, given an objective function and the set constraints and requirements. *OptQuest* is the engine used by AnyLogic to carry out the optimisation of the represented simulation model [26].

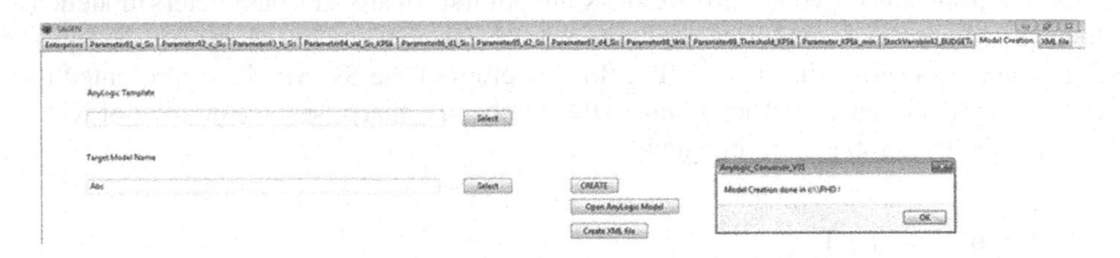


**Fig. 2.** SAGEN user interface

## 5.4 Guideline

A guideline is proposed as a complementary mechanism to the model, method and tool, with the main aim of supporting the enterprises, which belong to a CN, on addressing, assessing and solving the strategies alignment process [27]. The guideline consists of twelve phases, hereafter briefly described. Phase 1 starts with the identification of the CN partners, willing to align their strategies. Phase 2 focuses on the enterprises' roles definition. Phase 3 continues with the collection of the data required as an input of the SAM related with the KPIs and the parameters associated ($kpi_{ixk}$, $\Delta kpi_{ixk}$, $\Delta kpi_{ixk}^{max}$, $Threshold\_kpi_{ixk}$, $w_{ixk}$). Phase 4 is devoted to the collection of data, from the CN enterprises, related with the strategies and the parameters associated ($str_{is}$, $c\_str_{is}$, $d_1\_str_{is}$, $d_2\_str_{is}$, $d_4\_str_{is}$). In Phase 5 the collaborative partners agree the type of collaboration to carry out in the CN. Three collaboration levels (CL) are defined depending on the data exchanged: (i) CL1, enterprises only exchange information as regards the KPIs defined and enumerated $kpi_{ik}$; (ii) CL2, enterprises exchange information about the KPIs and the parameters that characterize them, and the number of strategies (only the IDs of the strategies, not the definition) and the parameters that characterize them; and (iii) CL3, enterprises exchange information as regards the KPIs defined and the parameters that characterize them, and the definition of the strategies formulated and the parameters that characterize them. In Phase 6, the values of influence are estimated by each enterprise, $val\_str_{is}\_kpi_{ixk}$. The data retrieved in Phase 4, 5 and 6 is gathered in Phase 7, by using a template. In Phase 8 the gathered data is introduced in the DMS. SAGEM allows automatically creating, in Phase 9, the SAM in the simulation software selected, AnyLogic. The resolution of the model is performed in Phase 10, and the SAM solutions are generated. The negotiation of the SAM results is performed in Phase 11, which depends on the collaboration type previously agreed. When negotiating, each enterprise selects the alternative of solution, that best fits to its requirements. The alternative of solution is exchanged with the other partners of the CN, and a negotiation process is started until the CN partners agree on the alternative of solution selected, which generates the closest performance to the optimum for each partner. In order to give the reader a better insight of the negotiation process, a scheme of the Negotiation Process for the Level 1 of Collaboration is described in [27]. Finally, Phase 12 allows, after carrying out the negotiation, identifying potential appearing conflicts when activating certain strategies. In Phase 12, possible misalignments and negative-influences appearing in the alternative of solution selected are to be identified, analysed and solved.

## 6 Validation of the Proposal

The stage of verification and validation aim to assess, give credibility and accredit the proposed original work [28]. In order to show the relevance of the model, method, tool and guideline proposed to deal with the strategies alignment problem a three validation elements are considered: (i) validation of the research by peer reviewed publications; (ii) development of empirical experiments; and (iii) real application of the complete approach in two networks belonging to the food (Pilot 1) and automotive industry (Pilot 2). The implementation of the proposed contribution allows identifying critical points of application; and the pilots allow showing the use that the enterprises give to the proposed contribution, as well as determining the practical relevance when applying the strategies alignment model in the CN. For the validation of the proposed approach, a real simplified use case from food industry is presented. The simulated CN consists of two enterprises, the distributor ($E_1$) and the manufacturer ($E_2$), each one defining two objectives ($o_{ixk}$) whose achievement is measured through the KPIs ($kpi_{ixk}$): $E_1$ defines $kpi_{111}$ and $kpi_{121}$; $E_2$ defines $kpi_{211}$ and $kpi_{221}$. In order to achieve the objectives defined, each enterprise formulates two strategies ($str_{is}$): $E_1$ formulates $str_{11}$ and $str_{12}$; $E_2$ formulates $str_{21}$ and $str_{22}$. Each enterprise also defines the data related to the strategies (durations and costs) and the associated to the corresponding KPIs (minimum values, threshold and weights). The objectives and the strategies are described in Sect. 3, in order not to repeat we refer the reader to that section. Moreover, the enterprises have a certain budget ($b_i$) to carry on the formulated strategies. The values of influence that each strategy has on the defined KPIs are given by the parameter $val\_str_{is}\_kpi_{ixk}$. All the data related with the objectives and strategies defined in the food industry use case are shown in Table 4. The data depicted on the cells in dark grey correspond to the values of influence that the strategies defined in one enterprise have on the KPIs defined in the same enterprise (intra-enterprise values of influence). While the white coloured cells represent the values related to the inter-enterprise influences. In the non-collaborative scenario only the inter-enterprise values of influence will be used. Whilst in the collaborative scenario will take into consideration both intra and inter-enterprise values of influence.

In the collaborative scenario the enterprises participating take into account the influences of all the strategies formulated by the enterprises. The optimisation experiment carried out in the simulation software used (AnyLogic) generates a set of solutions, as regards the units of strategies to activate and the time in which to activate them. The values concerning the enterprise performance indicators (kpi'i) and the network performance indicator (kpi'net) are computed in the simulation experiment. The experiments have been also performed in the non-collaborative scenario, in which the decision-making is made from an isolate perspective without considering how the strategies

**Table 4.** Real simplified from food industry use case: data

| Distributor ($E_1$) $b_1 = 4$ | | | | | | | | | $kpi'_{111}$ | | $kpi'_{121}$ | |
|---|---|---|---|---|---|---|---|---|---|---|---|---|
| | | | | | | | | | $w_{111}$ | 0,5 | $w_{121}$ | 0,5 |
| | | | | | | | | | $Threshold\_kpi'_{111}$ | 0,05 | $Threshold\_kpi'_{121}$ | 0,01 |
| $str_{11}$ | $c\_str_{11}$ | 0,5 | $d'_1\_str_{11}$ | 0,05 | $d'_2\_str_{11}$ | 0,01 | $d'_4\_str_{21}$ | 0,3 | $val\_str_{11}\_kpi'_{111}$ | 0,4 | $val\_str_{11}\_kpi'_{121}$ | 0 |
| $str_{12}$ | $c\_str_{12}$ | 4 | $d'_1\_str_{12}$ | 0,2 | $d'_2\_str_{12}$ | 0,03 | $d'_4\_str_{21}$ | 0,7 | $val\_str_{12}\_kpi'_{111}$ | 0,2 | $val\_str_{12}\_kpi'_{121}$ | 0,9 |
| | | | | | | | | | $val\_str_{21}\_kpi'_{111}$ | -0,6 | $val\_str_{21}\_kpi'_{121}$ | -0,6 |
| | | | | | | | | | $val\_str_{22}\_kpi'_{111}$ | 0,3 | $val\_str_{22}\_kpi'_{121}$ | 0,4 |
| **Manufacturer ($E_2$) $b_2 = 9$** | | | | | | | | | $kpi'_{211}$ | | $kpi'_{221}$ | |
| | | | | | | | | | $w_{211}$ | 0,5 | $w_{221}$ | 0,5 |
| | | | | | | | | | $Threshold\_kpi'_{211}$ | 0,2 | $Threshold\_kpi'_{221}$ | 0,1 |
| $str_{21}$ | $c\_str_{21}$ | 7 | $d'_1\_str_{21}$ | 0,1 | $d'_2\_str_{21}$ | 0,02 | $d'_4\_str_{21}$ | 0,75 | $val\_str_{21}\_kpi'_{211}$ | 1 | $val\_str_{21}\_kpi'_{221}$ | 0,7 |
| $str_{22}$ | $c\_str_{22}$ | 3 | $d'_1\_str_{21}$ | 0,05 | $d'_2\_str_{21}$ | 0,01 | $d'_4\_str_{21}$ | 0,5 | $val\_str_{22}\_kpi'_{211}$ | 0,2 | $val\_str_{22}\_kpi'_{221}$ | 0,8 |
| | | | | | | | | | $val\_str_{11}\_kpi'_{211}$ | 0,8 | $val\_str_{11}\_kpi'_{221}$ | 0,6 |
| | | | | | | | | | $val\_str_{12}\_kpi'_{211}$ | -0,5 | $val\_str_{12}\_kpi'_{221}$ | -0,5 |

formulated by other network enterprises affect the achievement of its objectives (performance). In Table 5 the results of both scenarios, non-collaborative (NC) and collaborative (C) are compared. The optimised solution of the collaborative scenario (using the SAM) generates, at network level, a performance significantly higher than the performance resulting from the solution obtained in the non-collaborative scenario. Moreover, the solution obtained in the non-collaborative scenario breaches the restriction of non-negativity of all the KPIs of the network (fulfilment_kpi'ixk > 0). Whereas that the solution of the collaborative scenario complies with the restriction of non-negativity being the fulfilment of all the KPIs 1.

**Table 5.** Collaborative scenario vs. non-collaborative scenarios: optimization results

| | NC | C | Performance Improvement |
|---|---|---|---|
| $u\_str_{111}$ | 0 | 4,5 | |
| $ti\_str_{111}$ | 0,105 | 0,013 | |
| $u\_str_{121}$ | 1 | 0 | |
| $ti\_str_{121}$ | 0 | 0,017 | |
| $u\_str_{211}$ | 1 | 0 | |
| $ti\_str_{211}$ | 0,1 | 0,01 | |
| $u\_str_{221}$ | 0 | 2 | |
| $ti\_str_{221}$ | 0,3 | 0 | |
| $\nabla kpi'_{11}$ | -0,288 | 0,674 | 334,03% |
| $fulfilment\ kpi'_{111}$ | 0 | 1 | |
| $\nabla kpi'_{121}$ | 0,04 | 0,348 | 770,00% |
| $fulfilment\ kpi'_{121}$ | 1 | 1 | |
| $\nabla kpi'_{211}$ | 0,226 | 0,951 | 320,80% |
| $fulfilment\ kpi'_{211}$ | 1 | 1 | |
| $\nabla kpi'_{221}$ | 0,141 | 1,307 | 826,95% |
| $fulfilment\ kpi'_{221}$ | 1 | 1 | |
| $kpi'_1$ *(Distributor)* | -0,122 | 0,511 | 518,85% |
| $kpi'_2$ *(Manufacturer)* | 0,204 | 1,129 | 453,43% |
| $kpi'_{net}$ | 0,041 | 0,82 | 1900,00% |

## 7 Conclusions

The developed research aims to provide a better understanding on the ways of establishing sustainable collaborative relationships within the partners of a CN. In this regard, a complete approach consisting of a model, a method a tool and a guideline is proposed, to support the strategies alignment process, in the CN context. The complete approach allows to automatically identifying the set of strategies to be activated, and the time in which to activate them, in order to obtain maximum levels of network performance. SD method is proposed to solve the SAM, and three tools support the computation of the SAM: (i) simulation software; (ii) DMS; and (iii) SAGEN tool, that automatically builds the SAM in the simulation software. Finally, a guideline is proposed, to give the CN partners a vision of how to perform the strategies alignment process. Despite the advantages of the application of the strategies alignment approach, there is a main drawback related with the information gathering as regards the value $val\_str_{is}\_kpi'_{ixk}$, especially if the strategy $str_{is}$ has never been activated before, this parameter it is very difficult to estimate. In the light of this, network enterprises can opt for (i) estimating the parameter $val\_str_{is}\_kpi'_{ixk}$ or (ii) waiting until the strategy ($str_{is}$) is activated and measure the real value of $val\_str_{is}\_kpi'_{ixk}$. If the enterprise has stored the increase of the KPIs when a strategy specific strategy was activated in the past ($\Delta kpi'_{ixk} | str_{is}$), the enterprise can objectively compute $val\_str_{is}\_kpi'_{ixk}$, for strategies activated in the same enterprise; and $val\_str_{is}\_kpi'_{ixk}$ for strategies active in different network enterprises.

Future research work leads to deal with the collection of the data required in an accurate way. For doing this, complementary sensitivity analysis is to be proposed in order to identify the robustness of the optimised solution obtained, resulting from the implementation of the SAM in the simulation software AnyLogic. A second future line of research leads to design a distributed multi-agent system model so that each network node is represented as an agent and simulates in its own hardware and software one part of the strategies alignment model (its own part). Moreover, other applications can be identified to the proposed work, such as supporting the partners' selection process from a collaborative perspective.

**Acknowledgments.** This work has been funded in part by Programa Val i+d para investigadores en formación (ACIF 2012).

## References

1. Poler, R., Carneiro, L.M., Jasinski, T., Zolghadri, M., Pedrazzoli, P.: Intelligent Non-hierarchical Manufacturing Networks. Networks and Telecommunications Series. iSTE Wiley, Hoboken (2012)
2. Camarinha-Matos, L.M., Afsarmanesh, H.: Collaborative Networks: Reference Modelling. Springer International Publishing, Heidelberg (2008)
3. Andres, B., Poler, R.: Models, guidelines and tools for the integration of collaborative processes in non-hierarchical manufacturing networks: a review. Int. J. Comput. Integr. Manuf. **2**(29), 166–201 (2016)

4. Andres, B., Poler, R.: Relevant problems in collaborative processes of non-hierarchical manufacturing networks. J. Ind. Eng. Manag. **6**(3), 723–731 (2013)
5. Andrés, B., Poler, R.: Computing the strategies alignment in collaborative networks. In: Mertins, K., Bénaben, F., Poler, R., Bourrières, J.-P. (eds.) Enterprise Interoperability VI. PIC, vol. 7, pp. 29–40. Springer, Cham (2014). doi:10.1007/978-3-319-04948-9_3
6. Bititci, U., Turnera, T., Mackaya, D., Kearneyc, D., Parunga, J., Waltersb, D.: Managing synergy in collaborative enterprises. Prod. Plan. Control Manag. Oper. **18**(6), 454–465 (2007)
7. Macedo, P., Abreu, A., Camarinha-Matos, L.M.: A method to analyse the alignment of core values in collaborative networked organisations. Prod. Plan. Control **21**(2), 145–159 (2010)
8. da Piedade Francisco, R., Azevedo, A., Bastos, J.: Managing performance to align the participants of collaborative networks: case studies results. In: Camarinha-Matos, L.M., Boucher, X., Afsarmanesh, H. (eds.) PRO-VE 2010. IAICT, vol. 336, pp. 545–552. Springer, Heidelberg (2010). doi:10.1007/978-3-642-15961-9_65
9. Camarinha-Matos, L.M., Afsarmanesh, H.: Collaborative networks: a new scientific discipline. J. Intell. Manuf. **16**(4–5), 439–452 (2005)
10. Andres, B., Poler, R.: A decision support system for the collaborative selection of strategies in enterprise networks. Decis. Support Syst. **91**, 113–123 (2016). doi:10.1016/j.dss.2016.08.005
11. Kristianto, Y., Gunasekaran, A., Helo, P., Sandhu, M.: A decision support system for integrating manufacturing and product design into the reconfiguration of the supply chain networks. Decis. Support Syst. **52**(4), 790–801 (2012)
12. Lin, H.W., Nagalingam, S.V., Kuik, S.S., Murata, T.: Design of a global decision support system for a manufacturing SME: towards participating in collaborative manufacturing. Int. J. Prod. Econ. **136**(1), 1–12 (2012)
13. Seok, H., Nof, S.Y., Filip, F.G.: Sustainability decision support system based on collaborative control theory. Annu. Rev. Control **36**(1), 85–100 (2012)
14. Shafiei, F., Sundaram, D., Piramuthu, S.: Multi-enterprise collaborative decision support system. Expert Syst. Appl. **39**(9), 7637–7651 (2012)
15. Tan, P.S., Lee, S.S.G., Goh, A.E.S.: Multi-criteria decision techniques for context-aware B2B collaboration in supply chains. Decis. Support Syst. **52**(4), 779–789 (2012)
16. Verdecho, M.J., Alfaro, J.J., Rodriguez, R.: Prioritization and management of inter-enterprise collaborative performance. Decis. Support Syst. **53**(1), 142–153 (2012)
17. Lu, T.P., Trappey, A.J.C., Chen, Y.K., Chang, Y.D.: Collaborative design and analysis of supply chain network management key processes model. J. Netw. Comput. Appl. **36**(6), 1503–1511 (2013)
18. Macedo, P., Camarinha-Matos, L.M.: A qualitative approach to assess the alignment of value systems in collaborative enterprises networks. Comput. Ind. Eng. **64**(1), 412–424 (2013)
19. Singh, R.K., Benyoucef, L.: A consensus based group decision making methodology for strategic selection problems of supply chain coordination. Eng. Appl. Artif. Intell. **26**(1), 122–134 (2013)
20. Zhang, H., Deng, Y., Chan, F.T.S., Zhang, X.: A modified multi-criterion optimization genetic algorithm for order distribution in collaborative supply chain. Appl. Math. Model. **37**(14–15), 7855–7864 (2013)
21. Dao, S.D., Abhary, K., Marian, R.: Optimisation of partner selection and collaborative transportation scheduling in virtual enterprises using GA. Expert Syst. Appl. **41**(15), 6701–6717 (2014)
22. Guillaume, R., Marques, G., Thierry, C., Dubois, D.: Decision support with ill-known criteria in the collaborative supply chain context. Eng. Appl. Artif. Intell. **36**, 1–11 (2014)

23. Andres, B., Poler, R., Sanchis, R.: Collaborative strategies alignment to enhance the collaborative network agility and resilience. IFIP Adv. Inf. Commun. Technol. **463**, 88–99 (2015)
24. Lazarus Free Pascal: Lazarus (2016)
25. Andres, B., Poler, R., Rosas, J., Camarinha-Matos, L.: Technological Innovation for Cyber-Physical Systems. IFIP Advances in Information and Communication Technology, vol. 470, pp. 3–10. Springer, Heidelberg (2016)
26. Kleijnen, J.P.C., Wan, J.: Optimization of simulated systems: OptQuest and alternatives. Simul. Model. Pract. Theory **15**(3), 354–362 (2007)
27. Andres, B., Poler, R.: Towards a methodology to support the strategies alignment process in collaborative networks. In: Mertins, K., Jardim-Gonçalves, R., Popplewell, K., Mendonça, J.P. (eds.) Enterprise Interoperability VII. PIC, vol. 8, pp. 295–305. Springer, Cham (2016). doi: 10.1007/978-3-319-30957-6_24
28. Kleijnen, J.P.C.: Verification and validation of simulation models. Eur. J. Oper. Res. **82**, 145–162 (1995)
29. Akhras, G.: Smart materials and smart systems for the future. Can. Mil. J. **1**(3), 25–31 (2000)

# Service Personalization Requirements for Elderly Care in a Collaborative Environment

Thais Andrea Baldissera(✉), Luis M. Camarinha-Matos, and Cristiano De Faveri

Centre Technology and Systems, UNINOVA, NOVA University of Lisbon, Caparica, Portugal
{tab,cam}@uninova.pt, c.faveri@campus.fct.unl.pt

**Abstract.** In diverse sectors companies collaborate to offer integrated user-centric services to obtain competitive advantage. However, regarding elderly care, services are typically isolated, and mostly provided by a single provider. Moreover, there is a need to personalize services in respect of the individuality of each senior and making these services evolve according to the lifestyle and necessities of the person. In this context, the concept of Elderly Care Ecosystem (ECE) emerges as a computer-based collaborative environment that promotes the integration of distinct services and providers. ECEs are complex systems that demand a clear description of their requirements, especially of those related to personalization of services. In this paper, we describe a set of core requirements and challenges for designing personalized services in an elderly care ecosystem. In particular, our contributions are: the identification of the main stakeholders of an ECE, the requirements to build a persona profile in ECE, and a reference requirements model based on i* for service personalization in ECE.

**Keywords:** Collaborative business service · Service personalization · Requirements · Elderly care · Aging

## 1 Introduction

Aging of the global population represents one of the most significant demographic changes in the humanity history. The increase of the average life expectancy associated to the declination of fertility led to the growing of the median age of the population, being expected that those over 80 years will exceed the young in 2050 [1]. This aging process is responsible for changes at various levels, which may impact a number of different aspects of older adults' lives and limit the extent to which they are able to perform certain activities. This reality reflects the need for adaptations in the society to provide better healthcare and inclusion solutions for elderly people.

Research and development as well as industry practice in elderly care services has conventionally dedicated on the development of isolated services, contemplating only a single service provider and often showing an extreme focused on technology. Also services delivery often does not fully cover customers care needs and remain static until there is a customer request for change. As it is well-known in other sectors, to answer current demanding market challenges, organizations must collaborate to overcome their


L.M. Camarinha-Matos et al. (Eds.): DoCEIS 2017, IFIP AICT 499, pp. 20–28, 2017.
DOI: 10.1007/978-3-319-56077-9_2

weaknesses and strengthen their expertise, to offer better integrated services, focusing on user-centric services and gaining competitive advantage [2]. Furthermore, in the elderly care context, there is a need for personalized services that respect the individuality of each senior and cope with the evolution of needs, since the senior's life context changes along the aging process [3].

In order to emphasize this perspective, the term Elderly Care Ecosystem (ECE) was proposed in [1]. In particular, the ECE is aimed at enabling the integration of different services from distinct providers. In this way, care and assistance services for elderly can be seen as "the result of collaboration among various stakeholders, including local communities, governmental institutions, professionals, family and caregivers, and thus require a supporting collaboration environment" [1].

As such, this research aims at contributing to a more effective provision of services for elderly care in the context of collaborative networks. The main research question addressed by this work is: "*How to provide personalized care services for elderly?*" The pursued hypothesis is: "*Effective and reliable personalized services for elderly care can be provided if a suitable set of multi-provider business services is composed and integrated in the context of a collaborative network environment*".

The first step to design such a system is to understand the domain of the problem we want to solve. This is covered during the first stages of requirements engineering, where domain analysis and elicitation are typical activities. These steps are highly intertwined and seek at grasping the system as-is, identifying problems and opportunities, discovering the real needs of the stakeholders with respect to the ecosystem, and exploring alternative ways to address those needs [4]. In this paper, we highlight requirements for elderly care ecosystems, including the set of stakeholders and their interactions.

The relationship to smart system is presented in Sect. 2. Background and approach paradigm (Sect. 3) support the provision of integrated services and help service providers to acquire agility and better survive in market. The Elderly Care Ecosystem Description (Sect. 4) introduces the environment and potential challenges in this domain. In Sect. 4, the ecosystem stakeholders' identification and relations between them are described as well as the elderly care ecosystem characterization and business service targeting. Finally, we conclude in Sect. 5 identifying the future work.

## 2 Relationship to Smart Systems

Recent trends in the development of smart environments and services for human-centered applications point to personalized care and improved quality of life [5]. The elderly community represents a relevant portion of the world population and tends to be more active and independent, assisted by emergent technologies, such as those based on the Internet of Things, embedded and wearable systems as well as cloud computing. The inclusion of progressive levels of intelligence in these systems is likely to provide better support to these users and increase their quality of life.

A practical application of smart system for elderly care is exemplified by the Susan's illustrative scenario in Fig. 1. Susan is 82 years old, lives alone, but has a very active life. She likes to go out with her friends, attends social events, and does social work.

She takes care of her health and preserves natural resources as much as she can. She has a SPA device (Smart Personal Assistant) which interacts with her in different moments of Susan's life and guides her steps in order to respect her lifestyle and intentions.

Hi, I'm Susan
Nutrition and exercise habits
My SPA prepared me a daily diet with calorie control and exercises planning. I need balanced meals between proteins, liquids, nutrients, calcium, etc.
Sustainability
The fridge reads the RF-ID tags of the several items inside, and identify types of food, expiration date and keeps my consume behavior. With this information my SPA suggests recipes based on my health condition and adds items to the shopping list on my smart phone.
Mobility
Next time, I go shopping, a price reporting and an itinerary including the location of the shopping centers containing the required items is prepared by my SPA.
Relationships
I could shop online , but my SPA suggests that I walk to the mall, since I did not burn the expected amount of calories yet for today. In addition, my SPA informed me that my friend Dorothy would also be in the mall. We can take a coffee together and plan a dinner for the weekend.


**Fig. 1.** Susan's illustrative scenario

## 3 Trends in Elderly Care Ecosystem Design

Our focus is to describe requirements for the design and construction of an elderly care ecosystem that considers two main aspects: personalization and evolution of services. Both aspects depend on a set of data about customers, providers, regulations and the network itself, to determine a ranking of potential services for each customer. This ranking considers individual customer's needs and a set of criteria that suggests the best services in a given moment.

Domain understanding and requirements elicitation are typically performed by studying key documents, investigating similar ecosystems and interviewing or observing the identified stakeholders. In the elderly care domain, potential ECE stakeholders include professional entities and care people, suppliers and support entities, and customers. We analyzed a number of research projects on ICT and aging as part of the elicitation process. These projects have been developed since 2001 as part of ICT and Aging agenda of European framework programs. For instance, AMIGO[1] Project – "Ambient Intelligence for the Networked Home Environment" - was one of the first addressing collaborative providers with integrated services in 2004. EMOTIONAAL[2] project effectively focused on services delivered through various providers addressing the healthcare-concept for elderly persons in rural areas in Europe. ePAL project (Extending Professional Active Life) developed a strategic research roadmap focused on innovative collaborative solutions and ensuring a balanced postretirement life-style.

BRAID project – "Bridging Research in Ageing and ICT Development" [6] introduced four perspectives or life settings (adopted in our work): (i) independent living: how technology can assist in normal daily life activities, e.g., tasks at home, mobility,

[1] ftp://cordis.europa.eu/pub/ist/docs/directorate_d/st-ds/amigo-project-story_en.pdf.
[2] http://www.aal-europe.eu/projects/emotionaal/.

safety, agenda management (memory help), etc.; (ii) health and care in life: how technology can assist in health monitoring, disease prevention, and compensation for disabilities; (iii) occupation in life: how technology can support the continuation of professional activities along the ageing process; (iv) recreation in life: how technology can facilitate socialization and participation in leisure activities. Since 2010, 21 other relevant projects were identified, which focused on some form of collaborative providers. Out of these projects, 18 delivered atomics services and 3 developed integrated services, all with larger focus on independent living. In particular, Ambient Assistance Living for All - AAL4ALL[3] project conducted a field survey aimed at both characterizing current users of AAL technologies and understanding how potential future users feel about and are willing to use this type of technology.

# 4 Requirements of ECE

## 4.1 ECE Stakeholders

An ECE model include many components, including customers (seniors) and their wants and limitations, care services, and service provision entities [1]. The organization of this ecosystem model is proposed per four environments: customer, care needs, service, and service providers. *Customer* relates to relevant information about elderly as an entity within the ECE. *Care need* represents the main necessities of individuals within the elderly domain. *Service* is provided by service providers, and it represents a care service available within the ecosystem.

A detailed customer's profile, including historical record, is fundamental to identify the best option (ranking) of care services that attends his/her individual and specific needs. In addition, context analysis, considering available stakeholders and devices, can provide a valuable contribution for potential service rating. We identified the following environments in ECE:

A. *Customer Environment*, representing the elderly customers and their ambient and context. It includes:
   - Elderly. In ECE, a customer is characterized by typical personal attributes such as name, surname, nickname, gender, birth date, contact, marital status, etc. Geographical location is highlighted for guiding regional preferences and service coverage.
   - Elderly Family, Caregivers and Friends. Many people live with their family when they begin to lose capabilities, representing the traditional way to deal with aging but that requires intense care from family. Therefore, other relevant information is about caregivers, family members and relatives who are responsible for them. Relatives need to actively participate in the aging process and support the elderly.
   - Applications and Devices. Software and hardware which interact with the elderly leading to context-aware systems. Sensorial device information contributes to

[3] http://www.aal4all.org/.

frequent contextual analysis and to input for efficient service adaptation to current elderly context.

B. *Intermediaries Environment,* representing people or organizations whose main function is to provide opportunities for the relation between customers and service providers. It includes:
   - Broker. An entity responsible to receive customer's request and identify the services available to attend the request.
   - Professional stakeholders. Representing medical professionals, care homes, housing associations, local communities, etc. These stakeholders are particularly associated with those individuals who live by themselves but require professional care to assist them.
   - Suppliers. They are enterprises with a business in tele-medicine or tele-care, providers of the ICT infrastructures – networks and applications, enterprises of hardware and software and/or service provision, etc.
   - Regulation and Support entities. Examples include policy-makers, social insurance companies, public administrations, government roles, standardization organizations, civil society organizations, media services providers, etc.

C. *Organization Environment*, representing the care provision stakeholders with the goal to provide care and assistance services. It includes:
   - Single Entity, representing an individual service provider.
   - Virtual Organization, representing a set of collaborative services providers working together for a common goal. A single provider probably cannot properly fulfill all needs of a customer, if it operates alone. When a situation of tailored care arises, service providers can organize themselves in a partnership and deliver an integrated service.

### 4.2 ECE Characterization

In this section, we characterize the envisioned Elderly Collaborative Ecosystem and describe the requirements of the personalization process. In the elderly care domain, personalization involves the analysis of the senior's life context and can be achieved by composing a set of specific business services. We assume a data-rich environment in which data can be acquired via specific sensors and devices. In an automated environment (fully or semi), the characteristics and composition of the services set can evolve. As a result, a new service can start, a current service may evolve, or a service can be disabled or excluded from the customer's portfolio. For each new context change, the responsible care and assistance provider organization shall analyze the situation (in collaboration with all relevant stakeholders), and evolve the service to fit that context.

### 4.3 Service Personalization and Evolution

The main requirements for a service personalization and evolution are synthetized in Fig. 2. A goal-oriented approach is used to express a strategic dependency model using i* language [7], which allows to describe goals, softgoals, tasks, resources, and dependencies among multiple actors. The main goal of the model is to support a middle layer for service provision between the Service Provider Environment (actor Service Providers) and the Elderly Living Environment (actor Customer).

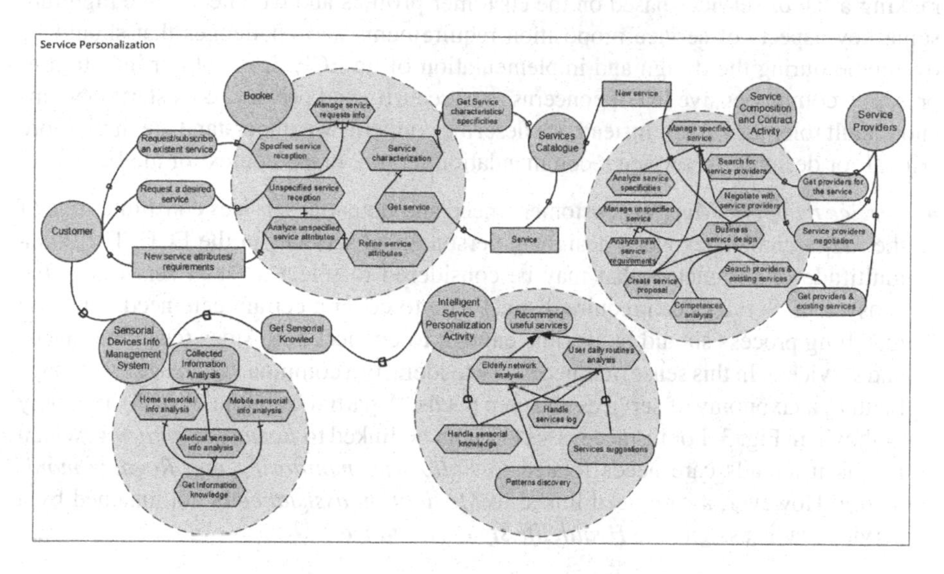


**Fig. 2.** i* Rationale strategic model for the service personalization

The Broker is responsible for collecting information about ECE costumer's care and assistance service requirements (through questionnaires, identification of similarities with other customers, etc.) and verifying if an adequate service description is already available in the service catalogue, as expressed in the tasks within the Broker boundary (Manage service requests info and its decompositions). If the service is available in the ECE, the goal *Request/subscribe an existing service* can be achieved. If it is an undefined service, the Broker is responsible for launching a new care and assistance service design, involving the relevant stakeholders along with the *Service Composition* activity [8]. This is expressed in the goal *Request a desired service*.

The Sensorial Devices Information Management System gets information from the multiple sources and analyzes these data. This result is sent to the *Intelligent Service Personalization Activity* which, by using relevant information, is able to recommend and/or suggest changes in current services for each ECE customer. Customer profile is required to be constantly updated to reflect most fresh information about a person. Types of input information include: sensorial information, essentially for context-awareness; relevant information from the ECE customer, to verify if additional services are necessary; analysis of the ECE customer's daily habits to recommend or suggest a care and assistance services adapted to the corresponding routines; and analysis of ECE to discover new patterns in ECE customer's behavior. A challenge here is to guarantee privacy of users and non-intrusiveness.

## 4.4 Business Service Targeting

Service proposition is a key component of the ECE. It is responsible for selecting and ranking a list of services based on the customer profiles and care needs. We highlight some key aspects of service proposition requirements and challenges that should be overcome during the design and implementation of an ECE. It is not our intention to present a comprehensive list of concerns (e.g. security and privacy, cold start phenomenon, fault tolerance, etc.). Instead, we describe concerns we understand are those more critical for designing a service recommendation feature in the context of the ECE.

- *Service mapping*. Mapping customer's need and potential services constitutes one of the major challenges when designing personalized services in the ECE. There is a multitude of parameters that may be considered to select a set of services in the ecosystem. Services can require distinct data to cover a certain care need, thus, the matching process should be flexible enough to connect both sides (customer needs and services). In this sense, it is necessary to identify a common "language" that maps them to a taxonomy of services and care needs. A partial example of such taxonomy is shown in Fig. 3. For instance, a service can be linked to *Remote monitoring*, which means it attends care needs related to *Follow-up monitoring* and *Regular monitoring*. However, a care need linked to *Medication assistance* is not matched by a service that is assigned to *Health life style assistance*.

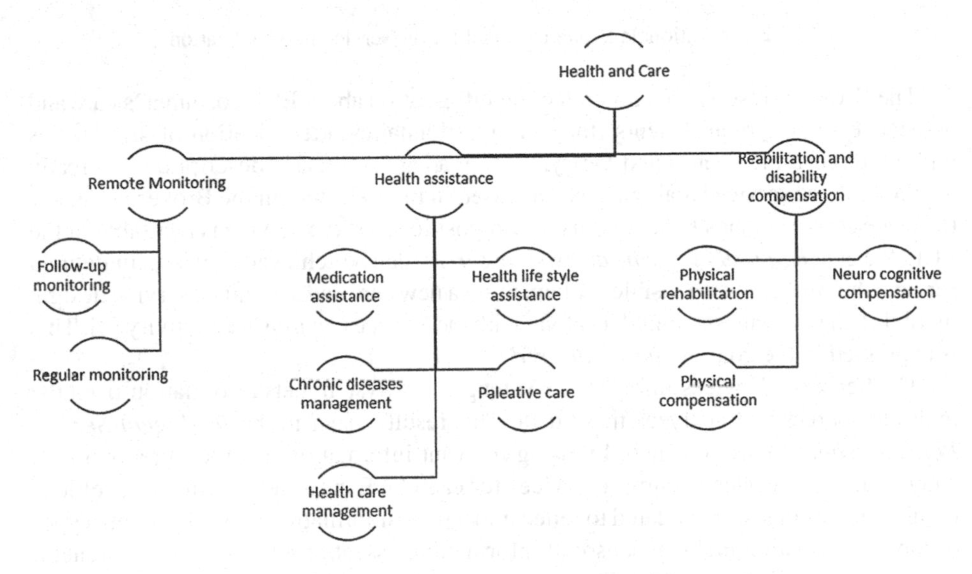


**Fig. 3.** Health and care in life taxonomy partial example.

- *Ranking*. Once a list of potential services is built, there is a need to rank relevant services to be suggested to the customer. This is a classical problem in recommendation systems that depend on a number of parameters to determine the most relevant outcome for a particular customer.

- *Scalability*. As the amount of providers and customers grow, selecting services in response to new customer's state can be computationally infeasible. There is a need to optimize the searching for potential services, while keeping the recommendation list relevant for a particular customer.
- *Timing*. As a customer may require several services that require multiple sensors to acquire proper data, determining the time interval of sensing remains a challenge. If this time is too short, the ECE can suffer for a lack of resources. On the other hand, if this time is too long, the ECE may be inefficient to determine a service evolution.
- *Accuracy prediction*. Sensors data can be incomplete, due to failures on sensing devices, noise in the communication, among other factors. Under this situation, a recommendation function demands predictive mechanisms to keep ECE operational.

A preliminary approach to service recommendation and composition, following a set of service adherence criteria based on a fuzzy causal network for rating care services, was presented in [9]. This system is now being refined, taking into account the identified requirements. An overview of its main elements is shown in Fig. 4.

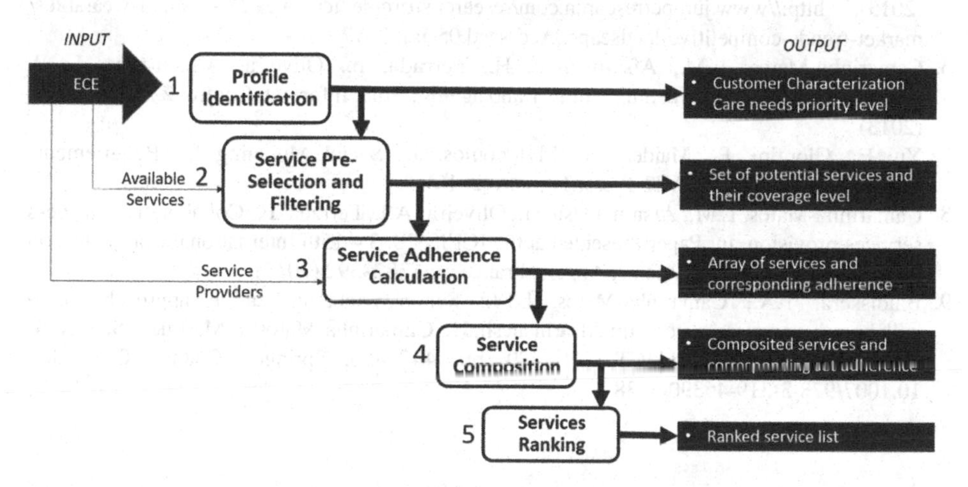


**Fig. 4.** Service recommendation and composition process on ECE

## 5 Conclusions and Future Work

Elderly Care Ecosystems can play an important role in delivering collaborative and personalized services for the elderly community. ECE represents a complex system that demands a careful identification and analysis of requirements, such that more useful systems can be built. With this purpose, we presented a set of core requirements for service personalization in ECE and a set of relevant challenges regarding service recommendation, as part of a collaborative environment for elderly care. As part of our future work, we are refining the design of a multi-criteria mechanism to support the selection and ranking process of personalized services.

**Acknowledgment.** This work has been funded in part by the Center of Technology and Systems and the Portuguese FCT-PEST program UID/EEA/00066/2013 (Impactor project), and by the Ciência Sem Fronteiras program and Erasmus Mundi project (Brazil and European Commission).

## References

1. Baldissera, T.A., Camarinha-Matos, L.M.: Towards a collaborative business ecosystem for elderly care. In: Camarinha-Matos, L.M., Falcão, A.J., Vafaei, N., Najdi, S. (eds.) DoCEIS 2016. IAICT, vol. 470, pp. 24–34. Springer, Cham (2016). doi:10.1007/978-3-319-31165-4_3
2. Camarinha-Matos, L.M., Rosas, J., Oliveira, A.I., Ferrada, F.: Care services ecosystem for ambient assisted living. Enterp. Inf. Syst. **9**(5–6), 607–633 (2015)
3. O'Grady, M.J., Muldoon, C., Dragone, M., Tynan, R., O'Hare, G.M.: Towards evolutionary ambient assisted living systems. J. Ambient Intell. Hum. Comput. **1**(1), 15–29 (2010)
4. Van Lamsweerde, A.: Requirements Engineering: From System Goals to UML Models to Software Specifications. Wiley Publishing, New York (2009)
5. Moar, J.: Smart Wearable Devices: Market Trends and Competitive Landscape 2015–2020 (2015). http://www.juniperresearch.com/researchstore/devices-wearables/smart-wearables/market-trends-competitive-landscape. Accessed 05 Jan 2017
6. Camarinha-Matos, L.M., Afsarmanesh, H., Ferrada, F., Oliveira, A.I., Rosas, J.: A comprehensive research roadmap for ICT and ageing. Stud. Informat. Control **22**(3), 233–254 (2013)
7. Yu, E., Giorgini, P., Maiden, N., Mylopoulos, J.: Social Modeling for Requirements Engineering, vol. 1. The MIT Press, Toronto (2010)
8. Camarinha-Matos, L.M., Afsarmanesh, H., Oliveira, A.I., Ferrada, F.: Collaborative business services provision. In: Paper Presented at the ICEIS 2013 – 15th International Conference on Enterprise Information Systems, Angers, France, pp. 382–392 (2013)
9. Baldissera, T.A., Camarinha-Matos, L.M.: Services personalization approach for a collaborative care ecosystem. In: Afsarmanesh, H., Camarinha-Matos, L.M., Lucas Soares, A. (eds.) PRO-VE 2016. IAICT, vol. 480, pp. 443–456. Springer, Cham (2016). doi: 10.1007/978-3-319-45390-3_38

# A System Dynamics and Agent-Based Approach to Model Emotions in Collaborative Networks

Filipa Ferrada[1(✉)] and Luis M. Camarinha-Matos[1,2]

[1] CTS – Center of Technology and Systems, Uninova, NOVA University of Lisbon, 2829-518 Caparica, Portugal
{faf,cam}@uninova.pt

[2] Department of Electrical Engineering, Faculty of Sciences and Technology, NOVA University of Lisbon, 2829-518 Caparica, Portugal

**Abstract.** A good amount of research within the last few decades has been focusing on computational models of emotion and the relationships they have with human emotional processes and how they affect the surrounding environments. The study of emotions is interdisciplinary and ranges from basic human emotion research, like in psychology, to the social sciences studies present in sociology. The interactions between those and the computational sciences are becoming a challenge. One particular challenge that is presented in this paper is the study of collaborative emotions within a Collaborative Network (CN) environment. A CN is composed of different participants with different interaction characteristics such as, expectations, will to cooperate and share, leadership, communication, and organizational abilities, among others. This paper presents an approach, based on system dynamics and agent-based modeling, to model the emotional state of an individual member of the network (via a non-intrusive way). Some simulation results illustrate the approach.

**Keywords:** Emotions · Collaborative emotions · Collaborative networks · System dynamics · Agent-based modeling

## 1 Introduction

Nowadays the area of Collaborative Networks (CNs) is being challenged by the necessity of improvements not only in technical terms but also in relation to social interactions among their participating members. According to some research in socio-technical systems [1, 2], the failure of large complex systems, such as CNs, is not directly related to the technology neither to the operational systems that compose them. Rather, they fail because they do not recognize the social and organizational complexity of the environment in which these systems are deployed. A survey conducted by Morris et al. [2] highlights that neglecting social and organizational complexity can cause large, and often serious, technological failures and also recognize that there is a need to provide "human-tech" friendly systems with cognitive models of human factors like stress, emotion, trust, leadership, expertise, or decision-making ability.


L.M. Camarinha-Matos et al. (Eds.): DoCEIS 2017, IFIP AICT 499, pp. 29–43, 2017.
DOI: 10.1007/978-3-319-56077-9_3

Emotion is an important factor in human cognition and social communication [3] and has been used as a mean of interaction in several fields of science like psychology, sociology, AI, and HCI with the use of emotional agents. In this context, a large amount of research within the last few decades has been focusing on computational models of emotion and the relationship they have with human emotional processes and how they affect the surrounding environments.

A new approach that is expected to improve the performance of existing CNs, namely the collaboration sustainability and interactions, is introduced here by adopting some of the models developed in the psychology, sociology and affective computing areas. The idea is to "borrow" the concept of *human emotion* and apply it within the context of a CN environment, turning it into a more "human-tech" friendly system without being intrusive, i.e. without violating the intimacy of each member.

When thinking about complex systems such as CNs that are composed of several nodes representing organizations, SMEs, large companies, among others, collaborating with the aim to achieve a purpose, it is reasonable to imagine that all of these entities interacting might also generate "emotions" that would be affected by the dynamics of the collaborative environment. Thus, the emotional state of each participating organization (CN Member) would contribute to the assessment of the collective emotional state of the CN and in this way contribute for its well-functioning. The individual emotional state of a member would affect its performance and relationships within the CN [4]. In this paper, it is assumed that the modeling of individual emotions will allow a CN Administrator to have a better understanding of each member's reality within the network environment.

In this context, the main research question guiding this work is: *What could be a suitable modeling framework to support the concept of collaborative emotions and assist on simulation experiments in order to understand the behavioral dynamics within a collaborative networked environment?*

The modeling framework developed to give an answer to this research question is based on System Dynamics and Agent-based Modeling and Simulation. System Dynamics allow the understanding of the behavior of each member over time. Agent-based models and simulation offer a good representation of the real-world environments with appropriate level of complexity and dynamism and allow to explain (simulate) a variety of situations and behaviors through selection of scenarios, which are difficult to analyze with the traditional approaches.

The remainder of this paper is organized as follows: Sect. 2 identifies the relationship of this work to smart systems; Sect. 3 gives a brief overview of the psychological and computational models of emotion; Sect. 4 presents the proposed modeling framework; Sect. 5 presents the model implementation and the simulation results; and finally Sect. 6 concludes and identifies the future work.

## 2 Relationship to Smart Systems

The current increase of systems complexity and the fast growing hyper connectivity raises the necessity to have mechanisms to assess the behavior of such systems. Their complex dynamism not only incorporates technological and operational mechanisms but also integrates social and organizational constructs [1, 2], becoming socio-technical systems. These socio-technical systems are composed of complex intelligent subsystems with a vast degree of autonomy and tendentiously configured as collaborative systems.

In this direction, the work presented in this paper proposes an approach to bridge the gap between the technological and social aspects. The introduction of the concept of collaborative emotions to complex collaborative systems, contributes to a new generation of hyper-connected smart socio-technical systems [5].

# 3 Emotions and Computational Models of Emotion

Emotions are unique to each human being and deviations from person to person are the result of each person's genes and the involved environment in conjunction. Emotions can be distinguished from feelings, affects, moods, and sentiments. *Emotions* are driven by specific events, actions or objects. They are more dynamic and episodic processes than *moods*, which are generally less intense [6, 7], longer lasting [8] and not directed at specific stimuli or event [9], although this distinction is more often made theoretically than empirically [10]. *Affect* is a broader term and can be defined as a valence evaluation in reference to the self [11]. Put simply, affect is an umbrella concept that covers a broad range of feelings, indicating if something is good or bad for oneself. Affect is often used as the denominator for both emotion and mood [12]. *Sentiments* are, according to Gordon: "*socially constructed pattern[s] of sensations, expressive gestures, and cultural meanings organized around a relationship to a social object, usually another person… or group such as a family*" [13].

Numerous theories involving the origins, mechanisms and nature of emotions have been generated over the years. This is a challenge since emotions can be analyzed from many different perspectives. All of the classic theories of emotion have fallen under criticism at various times, though many modern theorists still use them as a basis to work from. The most known theoretical models of emotion are [14]:

1. **Physiological or Somatic Emotion Theories** [15–17]: concede that emotions are primary to cognitive processes [17]. Prior to analyzing a perceived object, and even before recording any impressions, the (human) brain is able to immediately invoke an emotion associated with this object.
2. **Basic Emotion Theories** [3, 18–22]: adopt a certain number of basic emotions. The fundamental assumption is that a specific event triggers a specific affect corresponding to one of the basic emotions producing physiological response mostly through facial expressions.
3. **Appraisal Theories of Emotion** [23–27]: suggest that before the occurrence of emotion, there are certain cognitive processes that analyze stimuli [25, 28]. In such a way, the emotions are related to a certain history of a human (agent or robot). The relation to the history should follow the process of recognition (since the objects and their relations to the agent's emotion should be first recognized). Thus, the appraisal theory postulates a certain priority of cognitive processes over emotions.

4. **Dimensional Emotion Theories:** providing a suitable framework for representing emotions from a structural perspective. These theories established that emotions can be differentiated on the basis of dimensional parameters, such as *arousal* and *valence.* Russell [29, 30] proposes a two-dimensional framework consisting of pleasantness and activation to characterize a variety of affective phenomena such as emotions, mood and feelings. Another approach is the three-dimensional framework proposed by Russell and Mehrabian [31], which describes emotions based on their level of pleasantness, arousal, and dominance. This model is known as the PAD model [32].

Furthermore, there are also a number of psychological theories of emotion that do not fit exactly into the ones outlined above, focusing on a **specific aspect** or **component of emotion**, such as *motivation* or *action preparation*, or **combined features from the major theoretical orientations** (see [33]).

*Computational Models of Emotion* are complex software systems conceived to embrace design decisions and assumptions, inherited from the psychological and computational traditions from where they emerged, and synthesize the operations and architectures of some components that constitute the process of human emotions [34]. In general, computational models of emotion include mechanisms for the evaluation of emotional stimuli, the elicitation of emotions, and the generation of emotional responses, creating means for the recognition of emotions from human users and artificial agents, the simulation and expression of emotional feelings and the execution of emotional responses (or behaviors) [35]. Table 1 illustrates some computational models of emotion according to some theoretical models of emotion. For other reviews of computational models of emotion consult [34, 36, 37].

**Table 1.** Some computational models of emotion

| Computational model | Theoretical model | Computational techniques |
|---|---|---|
| CATHEXIS [38] | Physiological theory by Damasio [3] and appraisal theories by Roseman et al. [39]. Minsky's paradigm [40] | Synthetic agents |
| FLAME [41] | Combination of OCC [27] and Roseman et al.'s [39] appraisal theories | Software agent (does not incorporate group behavior); fuzzy logic |
| ALMA [42] | OCC appraisal theory [27] and PAD dimensional theory [32] | Virtual character or agent |
| WASABI [43] | OCC [27] and Scherer [44] appraisal theory and PAD dimensional theory [32] | Software agents; BDI |
| KISMET [45] | Physiological theories of Tinbergen and Lorenz and PAD dimensional theory | Robotic agent |

The work presented in this paper is based on a combination of the WASABI and KISMET's computational models, applied to organizations in a collaborative environment, as presented in the following sections.

## 4 Proposed Modeling Framework

In this work, the C-EMO modeling framework is proposed, which was developed to simulate dynamically changing *emotions* in virtual agents representing members of a CN. As known, members of a CN are organizations that might be dispersed geographically with different purposes and competences, and not human beings, yet they are managed by humans. Emotions are without any doubt related to humans and it is unquestionable that organizations cannot feel emotions in the same way humans do. Nevertheless, the authors believe that a kind of *individual emotional state* of an organization can be appraised once it makes part of a virtual environment that presupposes interaction and collaboration among its members.

In this context, the main challenges herein are twofold: first what aspects of the emotion theories should be applicable to organizations and second how to capture the stimulus, the evidences and the concepts that should be used in order to mount a model based on human-related emotional theories but applied to organizations. Furthermore we should take into consideration that the applied methods should be non-intrusive, i.e., preserving the privacy of each organization. Another aspect that is also developed, but out of this paper's scope, is the concept of *collaborative emotional state* that represents the CN's emotional state as a whole.

In order to accomplish the proposed challenges, the C-EMO Modeling Framework illustrated in Fig. 1 was designed, comprising four main systems:

- ***Perception System***: Collects the external events, environmental states and stimuli from the CN environment in conjunction with the internal state of the member agent in order to prepare an emotional evidences vector of the individual agent.
- ***Internal System***: Maintains the information about the individual agent updated.

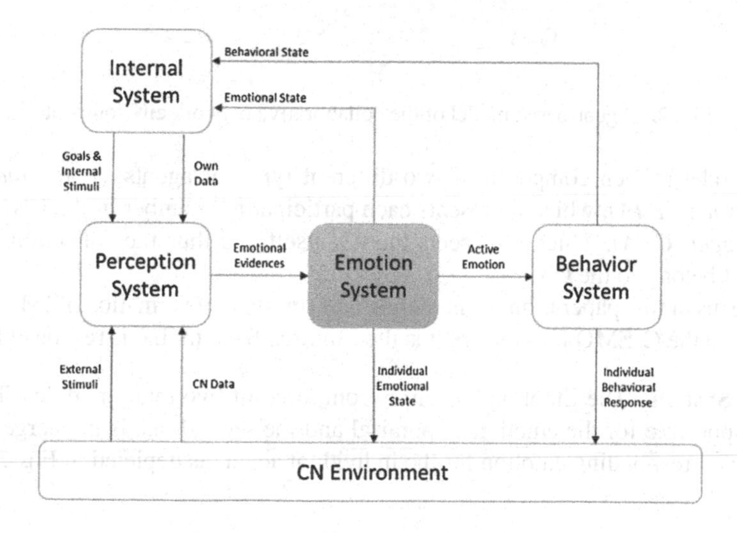


**Fig. 1.** C-EMO modeling framework

- ***Emotion System***: Responsible for assessing the emotional evidences, activating the corresponding agent's emotion and making the emotion manifest to the CN environment.
- ***Behavior System***: According to the activated emotion, a behavioral response(s) is designated to the agent and made available to the CN environment.

Due to the complexity and dynamic nature of emotions, the C-EMO framework was modeled using the Agent-Based and System Dynamics methodologies. In *Agent-Based Modeling (ABM)*, a complex system is modeled as a collection of autonomous decision-making entities called agents (either individual or collective entities such as organizations or groups). Each agent individually evaluates its situation and makes decisions on the basis of a set of rules. According to Siebers et al. [46], an Agent-Based Modeling system should be used when the problem has a natural representation of agents, i.e., when the goal is modeling the behavior and interactions of individual entities in a diverse population in the form of a range of alternatives or futures. In this line, the individual entities are the CN members and the diverse population is the collection of individual members that belong to the collaborative network. Thus, each CN individual member is represented by an agent and the collection of members are represented by a population of agents that "live" inside the agent that represents the Collaborative Network as illustrated below (Fig. 2).

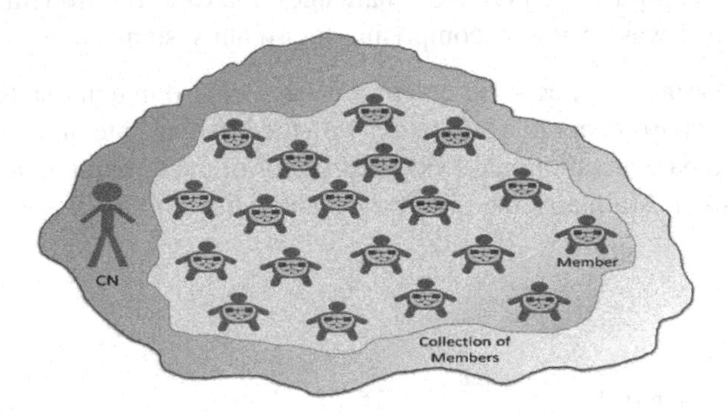


**Fig. 2.** Agent-based model of the collaborative network environment

The model is then composed of two different types of agents; *(i)* the *Individual Member Agent (IMA)*, which represents each participating member of the CN, and *(ii)* the *CN Agent (CNA)*, which represents the CN itself and thus the collection of IMA agents that belong to the CN.

The focus of this paper is on the modeling and simulation/estimation of IMA agent's emotions. In the C-EMO Framework it is the Emotion System that is responsible for it.

**Emotion System.** The Emotion System is composed of two main modules. The first one is responsible for the emotional appraisal and the second one is in charge of activating the corresponding emotion for the individual agent, as depicted in Fig. 3.

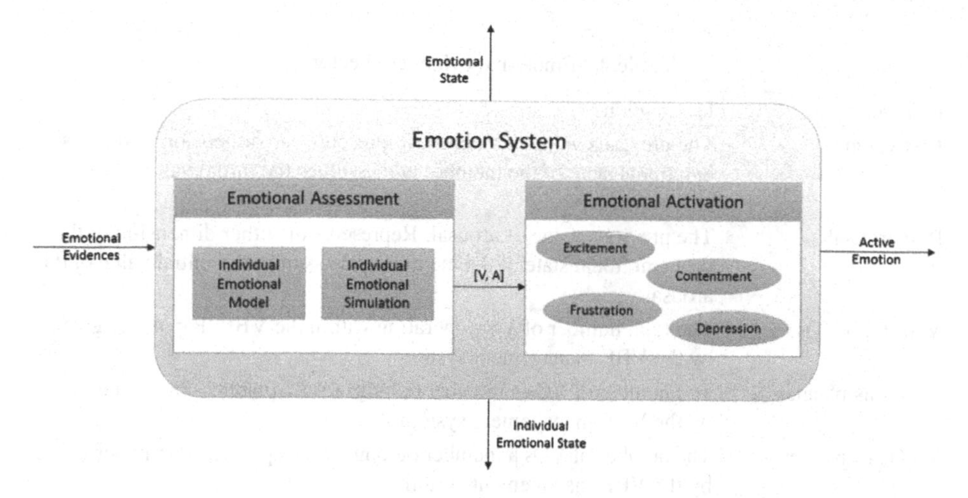


**Fig. 3.** Emotion system

The Emotional Assessment module receives, from the Perception System, an emotional evidences vector that is composed of different parameters relative to the IMA agent own data and also with information about the CN environment, through the CNA agent. For this paper purposes, the interactions of the IMA agent and the CNA agent are not taken into consideration. Table 2 describes the parameters composing the emotional evidences.

The two modules composing the Emotional Assessment are the Individual Emotional Model and the Individual Emotional Simulation. Due to uncertainty of how emotions emerge in a virtual collaborative network and how they influence the collaboration itself, a qualitative approach for modeling emotions is proposed using the methodology of System Dynamics.

*System Dynamics Modeling (SDM)*, initially proposed by Forrester [47], is a methodology and set of modeling tools that allows the understanding of the behavior of complex systems over time. It deals with internal feedback loops and time delays that affect the behavior of the entire system. It has two model representations: *(i) causal loop diagrams (qualitative)*, which are used to depict the basic cause-effect mechanisms of the system and also the circular chains of those mechanisms that form a feedback or closed loop; and *(ii) stock and flow diagrams (quantitative)* that show relationships between variables that have the potential to change over time and distinguishes between different types of variables. The resulting structure of the system, built with stocks and flows, determines the behavior of the system. The corresponding emotion causal loop diagram and stock and flow diagram of the IMA agent is illustrated in Figs. 4 and 5 respectively.

**Table 2.** Emotional evidences vector

| Parameters | Description |
|---|---|
| Past valence | The previous value of valence. Represents one dimension of the past emotional state of the member and assumes the initial value of the valence variable |
| Past arousal | The previous value of arousal. Represents the other dimension of the past emotional state of the member and assumes the initial value of the arousal variable |
| VBE total VOs | The total number of VOs operating within the VBE. Parameter given by the VBE management system |
| # VOs as planner | The number of VOs a member belongs to as a planner. Parameter given by the VBE management system |
| # VOs as partner | The number of VOs a member belongs to as a partner. Parameter given by the VBE management system |
| Net income value | The total earnings or profit of a member. This is the result of the difference between the total *revenue* and the total *expenses*. Parameter calculated by the VBE management system |
| Member satisfaction | Represents the level of satisfaction of the member. This parameter is calculated through a questionnaire that is sent to the VBE members periodically |
| Member needs and expectations | Represents the level of expectancy a member has achieved/met regarding its involvement in the VBE. This parameter is calculated through a questionnaire that is sent to the member when it joins the VBE and whenever the member wishes, during the VBE lifecycle |
| Performance evaluation | The performance evaluation value of the member. This parameter is given by the VBE management system |
| Belonging groups | The percentage of groups a member belongs to in a certain time. This parameter is calculated by the VBE management system and is the difference between the number of groups joined and the number of groups left by the member in relation to the total number of existing groups within the VBE |
| Shared knowledge and resources | The percentage of knowledge and resources a member shares within the VBE. This parameter is estimated by the VBE management system and represents the relation between the knowledge and resources a member shares and the generation of knowledge that result from it |
| Communication frequency | The rate at which the member communicates with others within the VBE. This parameter is based on a social network analysis of the VBE environment and is given by the VBE management system |
| Communication intensity | The measure of the effectiveness of communication within the collaborative VBE environment. This parameter, which is also based on the theory and analysis of social networks, is delivered by the VBE management system |
| Invitations to form VOs | The percentage of invitations to form VOs a member has, in relation to the total of existent VOs in preparation phase within the VBE |

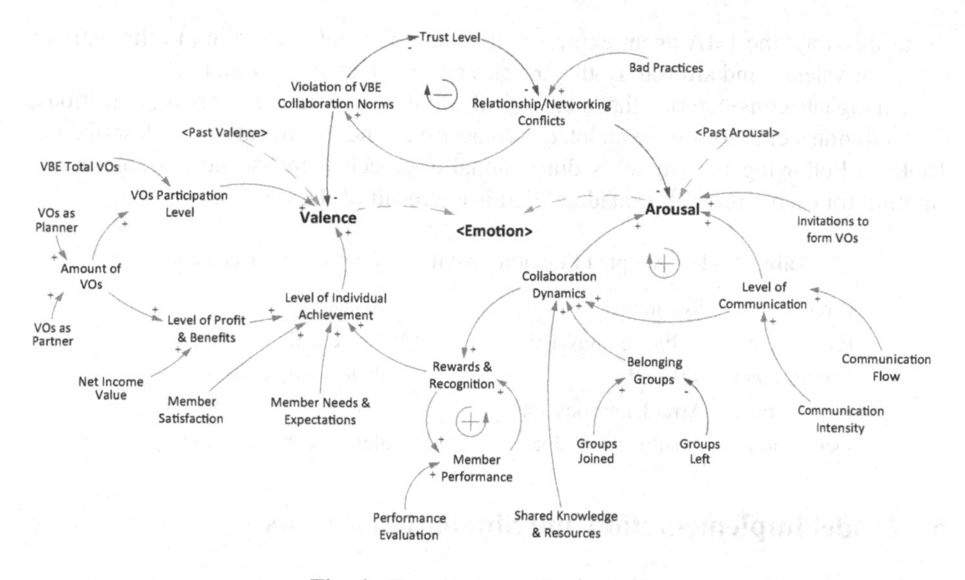


**Fig. 4.** Emotion causal loop model

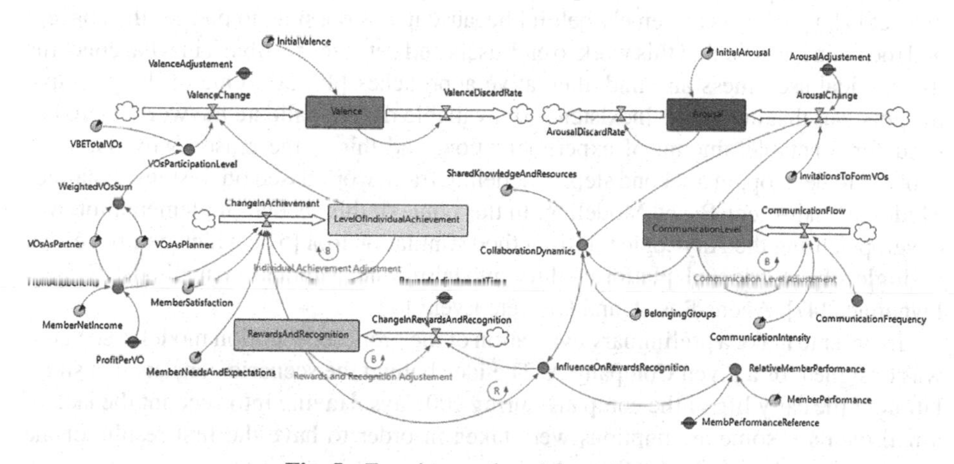


**Fig. 5.** Emotion stock and flows model

The dynamic model proposed for the representation of individual emotion in CNs is based on the Russell's pair of variables and follows the agent dynamics of Garcia's baseline model for emotional dynamics [48]. Russell's circumplex model [29], states that all affective states arise from two fundamental variables: *Valence*, a pleasure-displeasure continuum, and *Arousal* that is related to the level of activation, uncertainty, novelty, expectation and complexity of the stimuli. Therefore, each individual emotion can be understood as a linear combination of these two dimensions and the IMA agent emotional state described as:

$$e_i(t) = <V_i(t), A_i(t)> \tag{1}$$

In this way, the IMA agent expresses its emotional state according to the tuple of values of valence and arousal. Both variables ranging between −1 and 1.

Taking into consideration that members of collaborative networks are organizations, four individual emotions were adopted, two positives and two negatives as described in Table 3. Following the Russell's dimensional approach there is one possible active emotion for each circumplex quadrant that is the result of the Eq. (1).

**Table 3.** The Adopted Emotions and their dimensional placement

| Emotions | Synonyms | Dimensions |
|---|---|---|
| Excitement | Active, enthusiastic | Valence > 0; arousal > 0 |
| Contentment | Relaxed | Valence > 0; arousal < 0 |
| Frustration | Afraid, nervous, angry | Valence < 0; arousal > 0 |
| Depression | Apathy, miserable | Valence < 0; arousal < 0 |

## 5 Model Implementation and Simulation Results

The first developments of the Emotional System were conducted within the GloNet project [4], which was extremely helpful because it was possible to present the concept and the main ideas behind this work to end-users and get some feedback in what concerns its practical usefulness and find alternative approaches [49, 50]. One of the principal insights was that due to the inexistence of available data to validate the work, there is a need for a notable amount of experimentation, and this is the reason why the work evolved to develop, in a second step, a modeling framework based on System Dynamics Modeling and Agent-Based Modeling. In this context, this second implementation was developed using the AnyLogic multi-method simulation tool [51], which comprises, on a single object-oriented platform, three modeling and simulation methods: System Dynamics [47], Agent-Based, and Discrete Event [52].

In order to make a preliminary evaluation of the proposed emotion model, a scenario was designed for a given Company A. The idea behind the scenario consists in a simulation of the daily life of the company during 100 days. Having into account the lack of a real data set, some assumptions were taken in order to have the first results of the proposed model, as follows (Table 4).

**Table 4.** Simulation assumptions

| Assumptions | |
|---|---|
| $A_1$ | A company when invited to form a VO gets enthusiastic |
| $A_2$ | A company seeing its expectations met tends to stay comfortable |
| $A_3$ | A company whose performance evaluation is high tends to be pleased |
| $A_4$ | A company that shares its knowledge and resources and maintains working groups is more energetic |
| $A_5$ | A company that is satisfied keeps engaged |

During the simulation period, several parameters of the emotional evidences vector change according to the previous assumptions, as shown in Table 5. The initial emotional state of the member company is neutral. The values in bold are the ones that change in order to cope with the assumptions.

**Table 5.** Adopted simulation scenario for a Company A

| Parameters | Day 0–15 ($A_2 + A_3$) | Day 15–30 (all) | Day 30–50 ($A_2 + A_3 + A_5$) | Day 50–70 (none) | Day 70 ($A_4$) |
|---|---|---|---|---|---|
| Past valence | 0 | 0.4 | 0.7 | −0.3 | −0.3 |
| Past arousal | 0 | −0.3 | 0.2 | −0.1 | 0.3 |
| VBE total VOs | 3 | 3 | 3 | 3 | 3 |
| VOs as planner | 1 | 1 | 1 | 1 | 1 |
| VOs as partner | 1 | 1 | 1 | 1 | 1 |
| Net income value | 100 000 | 100 000 | 100 000 | 100 000 | 100 000 |
| Member satisfaction | 50% | **100%** | 100% | **20%** | 20% |
| Member needs and expect | **100%** | 100% | **20%** | 20% | 20% |
| Performance evaluation | **5** | 5 | **2** | **0.5** | 0.5 |
| Belonging groups | 50% | **100%** | 100% | **20%** | **100%** |
| Shared knowledge and Resources | 50% | **100%** | 100% | **20%** | **100%** |
| Communication frequency | 0.2 | 0.2 | 0.2 | 0.2 | 0.2 |
| Communication intensity | 2.5 | 2.5 | 2.5 | 2.5 | 2.5 |
| Invitations to form VOs | 0 | **1** | 1 | **0** | 0 |
| Emotion activation | Contentment | Excitement ↑ | Excitement ↔ | Depression | Frustration |

The implementation of the model and the simulation results are presented in Fig. 6. On the left hand side the evolution of the tuple (Valence, Arousal) is presented. On the right hand side, a representation of the circumplex model.

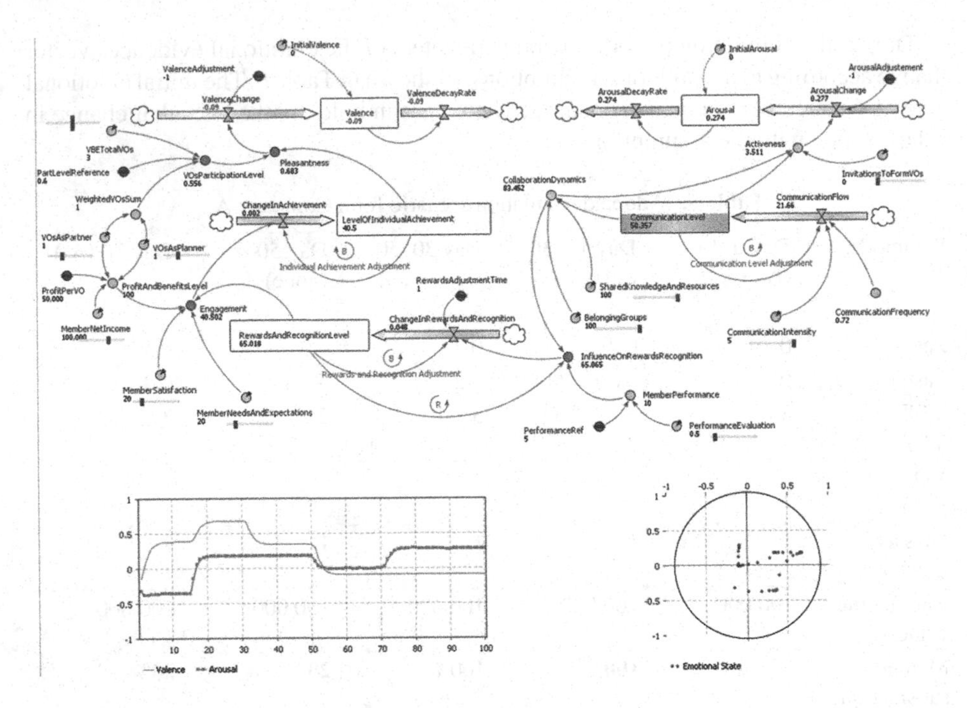


**Fig. 6.** Simulation runs and results

In a first overall analysis, the proposed model provided promising indications. Nevertheless, it was found that some adjustments are needed on two main aspects: (a) the weight of some of the parameters and, (b) the time of the emotion decay.

## 6 Conclusions and Future Work

The C-EMO Modeling Framework is introduced with the aim to conceptualize the notion of Collaborative Emotions within the context of a collaborative environment. A model based on Agent-Based and System Dynamics Modeling and Simulation is presented in order to model and simulate individual emotions generated by IMA agents. At the end some simulations results for a partial validation of the model are shown.

Future work relies on developing the *Behavior System* of the C-EMO Modeling Framework and developing a model to estimate the Collective Emotional State of the CN.

**Acknowledgments.** This work has been funded in part by the Center of Technology and Systems and the Portuguese FCT-PEST program UID/EEA/00066/2013 (Impactor project), and partly by the GloNet project funded by the European Commission.

## References

1. Baxter, G., Sommerville, I.: Socio-technical systems: from design methods to systems engineering. Interact. Comput. **23**(1), 4–17 (2011)
2. Morris, A., Ross, W., Ulieru, M.: A system dynamics view of stress: towards human-factor modeling with computer agents. In: IEEE International Conference on Systems, Man and Cybernetics, Istanbul, Turkey. IEEE (2010)
3. Damasio, A.R.: Descarte's Error: Emotion, Reason, and the Human Brain. Gosset/Putnam Press, New York (1994)
4. Ferrada, F., Camarinha-Matos, L.M.: An emotional support system for collaborative networks. In: Camarinha-Matos, L.M., Baldissera, T.A., Di Orio, G., Marques, F. (eds.) DoCEIS 2015. IAICT, vol. 450, pp. 42–53. Springer, Heidelberg (2015). doi: 10.1007/978-3-319-16766-4_5
5. EPoSS: Smart Systems in the Multi-Annual Strategic Research and Innovation Agenda of the JTI ECSEL (2014)
6. Mandler, G.: The nature of emotion. In: Miller, J. (ed.) States of Mind: Conversations with Psychological Investigators, pp. 136–153. BBC, London (1983)
7. Siegert, I., Böck, R., Wendemuth, A.: Modeling users' mood state to improve human-machine-interaction. In: Esposito, A., Esposito, A.M., Vinciarelli, A., Hoffmann, R., Müller, V.C. (eds.). LNCS, vol. 7403, pp. 273–279Springer, Heidelberg (2012). doi:10.1007/978-3-642-34584-5_23
8. Ekman, P.: Expression and the nature of emotion. In: Scherer, K.R., Ekman, P. (eds.) Approaches to Emotion, pp. 319–344. Erlbaum, Hillsdale (1984)
9. Parrott, W.: Emotions in Social Psychology. Psychology Press, Philadelphia (2001)
10. Fredrickson, B.L.: The broaden-and-build theory of positive emotions. Philos. Trans. R. Soc. Lond. **359**(Series B), 1367–1377 (2001)
11. Baumeister, R.F., et al.: How emotion shapes behavior: feedback, anticipation, and reflection, rather than direct causation. Pers. Soc. Psychol. Rev. **11**, 167–203 (2007)
12. Robbins, S.P., Judge, T.A.: Organizational Behavior, 15th edn, pp. 258–297. Pearson, Hoboken (2013)
13. Gordon, S.L.: The sociology of sentiments and emotion. In: Rosenberg, M., Turner, R.H. (eds.) Social Psychology: Sociological Perspectives, pp. 562–592. Basic Books, New York (1981)
14. Scherer, K.R.: Emotions are emergent processes: they require a dynamic computational architecture. Philos. Trans. R. Soc. B: Biol. Sci. **364**(1535), 3459–3474 (2009)
15. James, W.: What is an emotion? Mind **9**(34), 188–205 (1884)
16. Schachter, S., Singer, J.E.: Cognitive, social and physiological determinants of emotional state. Psychol. Rev. **69**, 379–399 (1962)
17. Zajonc, R.B.: On the primacy of affect. Am. Psychol. **39**, 117–123 (1984)
18. Darwin, C.: Expression of the Emotions in Man and Animals. J. Friedmann, London (1979)
19. Izard, C.E.: Human Emotions. Plenum Press, New York (1977)
20. Izard, C.E.: Basic emotions, relations among emotions, and emotion–cognition relations. Psychol. Rev. **99**, 561–565 (1992)
21. Tomkins, S.S.: Affect as amplification: some modification in theory. In: Plutchik, R., Kellerman, H. (eds.) Emotion: Theory, Research, and Experience, pp. 3–33. Academic Press Inc., New York (1980)
22. Ekman, P.: An argument for basic emotions. Cogn. Emot. **6**, 169–200 (1992)
23. Arnold, M.B.: Emotion and Personality. Columbia University Press, New York (1960)
24. Lazarus, R.S.: Thoughts on the relations between emotions and cognition. Am. Psychol. **37**, 1019–1024 (1982)

25. Lazarus, R.S.: Emotion and Adaptation. Oxford University Press, New York (1991)
26. Scherer, K.R., Schorr, A., Johnstone, T.: Appraisal Processes in Emotion: Theory, Methods, Research. Oxford University Press, New York and Oxford (2001)
27. Ortony, A., Clore, G.L., Collins, A.: The Cognitive Structure of Emotions. Cambridge University Press, Cambridge (1988)
28. Frijda, N.H.: The Emotions. Cambridge University Press, Cambridge (1986)
29. Russell, J.A.: A circumplex model of affect. J. Pers. Soc. Psychol. **39**, 1161–1178 (1980)
30. Russell, J.A.: Core affect and the psychological construction of emotion. Psychol. Rev. **110**, 145–172 (2003)
31. Russell, J.A., Mehrabian, A.: Evidence for a three-factor theory of emotions. J. Res. Pers. **11**(3), 273–294 (1977)
32. Mehrabian, A.: Pleasure-arousal-dominance: a general framework for describing and measuring individual differences in temperament. Curr. Psychol. **14**(4), 261–292 (1996)
33. Moors, A.: Theories of emotion causation: a review. Cogn. Emot. **23**(4), 625–662 (2009)
34. Marsella, S., Gratch, J., Petta, P.: Computational models of emotion. In: Scherer, K.R., Bänziger, T., Roesch, E.B. (eds.) Blueprint for Affective Computing: A Source Book. Oxford University Press, Oxford (2010)
35. Rodríguez, L.-F., Ramos, F.: Development of computational models of emotions for autonomous agents: a review. Cogn. Comput. **6**(3), 351–375 (2014)
36. Handayani, D., et al.: Systematic review of computational modeling of mood and emotion. In: Information and Communication Technology for the Muslim World (ICT4 M) (2014)
37. Kowalczuk, Z., Czubenko, M.: Computational approaches to modeling artificial emotion – an overview of the proposed solutions. Front. Robot. AI **3**(21) (2016). doi:10.3389/frobt.2016.00021
38. Velásquez, J.D.: Cathexis: a computational model for the generation of emotions and their influence in the behavior of autonomous agents. In: Department of Electrical Engineering and Computer Science, Massachusetts Institute of Technology (MIT) (1996)
39. Roseman, I.J., Spindel, M.S., Jose, P.E.: Appraisals of emotion-eliciting events: testing a theory of discrete emotions. J. Pers. Soc. Psychol. **9**(5), 899–915 (1990)
40. Minsky, M.: The Society of Mind. Simon & Schuster, New York (1986)
41. El-Nasr, M.S., Yen, J., Ioerger, T.R.: FLAME-Fuzzy logic adaptive model of emotions. Auton. Agents Multi-agent Syst. **3**, 219–257 (2000)
42. Gebhard, P.: ALMA - a layered model of affect. In: Proceedings of the Fourth International Joint Conference on Autonomous Agents and Multiagent Systems (AAMAS 2005), Utrecht, pp. 29–36 (2005)
43. Becker-Asano, C.: WASABI: Affect Simulation for Agents with Believable Interactivity. University of Bielefeld, Bielefeld (2008)
44. Scherer, K.R.: What are emotions? And how can they be measured? Soc. Sci. Inf. **44**(4), 695–729 (2005)
45. Breazeal, C.: Emotion and sociable humanoid robots. Int. J. Hum Comput Stud. **59**(1–2), 119–155 (2003)
46. Siebers, P.-O., et al.: Discrete-event simulation is dead, long live agent-based simulation! J. Simul. **4**(3), 204–210 (2010)
47. Forrester, J.W.: Industrial Dynamics. The M.I.T. Press, Cambridge (1961)
48. Garcia, D., Schweitzer, F.: Emotions in product reviews - empirics and models. In: Proceedings of 2011 IEEE International Conference on Social Computing, SocialCom, pp. 483–488 (2011)
49. Camarinha-Matos, L.M., et al.: Collaborative enterprise networks for solar energy. In: ICCCT 2015 - IEEE International Conference on Computing and Communications Technologies, Chennai, India. IEEE (2015)

50. Camarinha-Matos, L.M., Oliveira, A.I., Ferrada, F.: Supporting collaborative networks for complex service-enhanced products. In: Camarinha-Matos, L.M., Bénaben, F., Picard, W. (eds.) PRO-VE 2015. IAICT, vol. 463, pp. 181–192. Springer, Heidelberg (2015). doi: 10.1007/978-3-319-24141-8_16
51. AnyLogic: Multimethod Simulation Software (2000). http://www.anylogic.com/. Accessed 2015
52. Gordon, G.: A general purpose systems simulation program. In: Proceedings of EJCC: Computers - A Key to Total Systems Control, pp. 87–104. McMillan, Washington D.C. (1961)

# Computational Intelligence

# Efficient Fuzzy Controller to Increase Soybean Productivity

Bruno S. Miranda[1(✉)], Gian M. Meira[1], Sidney J. Montebeller[1], Edinei P. Legaspe[1], Joel R. Pinto[1], Diolino J. Santos Filho[2], and Paulo E. Miyagi[2]

[1] Faculdade de Engenharia de Sorocaba, Alto da Boa Vista, Sorocaba, Brazil
miranda.bsm@live.com, gian.meira@hotmail.com, {sidney.montebeller,joel.rocha}@facens.br, edineilegaspe@hotmail.com

[2] Universidade de São Paulo, Butantã, São Paulo, Brazil
{diolino,pemiyagi}@usp.br

**Abstract.** Soybean production has expanded intensively in South America over the last decades. As the second leading worldwide soybean producer, Brazil has in prospect to increase market share through production growth in soybean areas. Therefore, soybean fields ought to encompass a sizeable region among different types of soil and climate conditions, and so advanced irrigation methods should be advantageous. This present work aims to optimize soybean production through an irrigation system control based on environmental and plant requirements. It was developed an embedded fuzzy controller that is able to process soil moisture, air moisture, temperature, soil type and soybean growth stages and it returns the ideal amount of water. Also, to accomplish the data acquisition, a multiparameter sensor device establishes remote connection and provides continuous in-field measurement. The efficiency of the fuzzy controller and the monitoring unit was verified through simulations, and so, results reached the expected model.

**Keywords:** Soybean · Fuzzy · Multiparameter sensor

## 1 Introduction

Brazilian soybean fields have gradually increased production over the last three decades. Consequently, Brazil has firmly established as the second leading worldwide soybean producer. Although agriculture technology has increased average yield per acre significantly, soybean frontier's expansion in Brazilian agriculture has been especially held out as the main responsible for high production levels [1, 2].

Eventually, expanding soybean fields over a huge country such as Brazil ought to bring unprecedented challenges to the producers. In addition, the diversity of climates and soil types characteristics have led farmers to search for new technologies [3, 4].

Therefore, potential scenario begins to emerge as soybean farmers increase investments, and modern irrigation techniques appear to be suitable in soybean fields, but, in generally, irrigation systems are quite inefficient, spend water resources and electricity


L.M. Camarinha-Matos et al. (Eds.): DoCEIS 2017, IFIP AICT 499, pp. 47–54, 2017.
DOI: 10.1007/978-3-319-56077-9_4

poorly and do not goal production results. Although Brazil has a privileged stand before many countries containing large part of available fresh water in the world, global trends defend the conscious consumption of resources, promoting sustainability in agriculture [5].

This paper presents the development of a control system that applies to irrigation devices in soybean fields. It was developed an embedded fuzzy controller, a supervisory control software and continuous monitoring sensor units. The control decision was settled according to soybean characteristics such as growth stages and water demand and also soil type, temperature and relative air moisture. The main objective of the system is to automate soybean irrigation mechanisms in order that production could increase based on fuzzy control and agricultural knowledges.

## 2 Relationship to Smart Systems

Smart system describes embedded and cyber-physical solutions consisting of diverse sub-systems and smart embedded applications are expected to deliver miscellaneous functionalities and services [6]. Although the development of precision agriculture claims to achieve a high-quality management of cultivable land and environment resources, wireless sensor network and advanced control techniques have been studied apart on agricultural developments and so, combined solutions are hardly available among farm applications [7, 8].

The proposed irrigation system control could increase efficiency in soybean fields crossing precision wireless monitoring and specialist control techniques, which base on environment characteristics and soybean water requirements. As shown in Fig. 1, the proposed system integrates in-field measurement and supervisory control software.

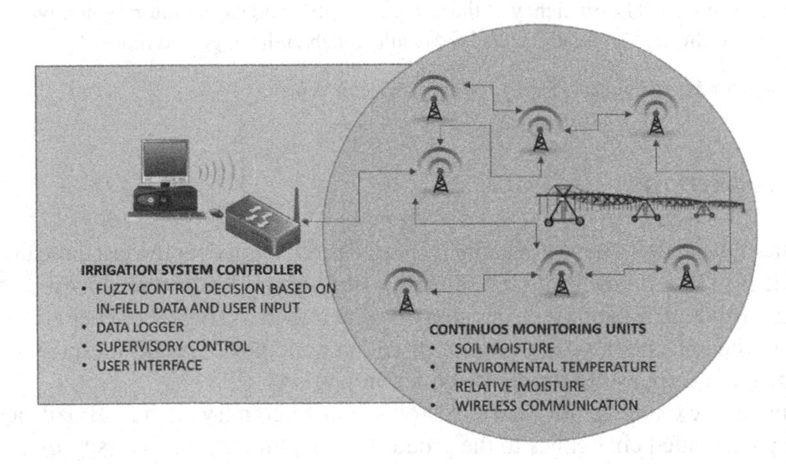


**Fig. 1.** Architecture of the irrigation control system. It represents continues monitoring units (*right*) which operate in-field and transmit data over wireless network and the irrigation system controller (*left*) that processes received data and communicates to supervisory control software.

## 3 Design of Fuzzy Logic Controller

Fuzzy logic has been used in many segments such as industrial controllers, artificial intelligence applications, medical researches and robotics. Fuzzy logic approaches real world attributes/linguistics and computational capabilities. As a result, the amount of time spent by developers decreases as well as precision, optimization and efficiency increase slightly [9]. A fuzzy control model was appropriate for the present work due to the complexity of the input variables and long cycle operation required, which has many variations throughout each stage, besides fuzzy logic easily approaches the specialist knowledge and could be comprehend and accepted by farmers.

The control model bases on two fuzzy logics: main control algorithm and equalizer algorithm. The selection of input variables based on soybean water requirements, which relate to soil water status and soybean growth stage, and environmental conditions, such as temperature and relative moisture. The following topics present input variables.

**Soybean Growth Stage.** Soybean growth stages divide in vegetative stages and reproductive stages. A slight shortage of water during the vegetative stage do not affect production drastically. However, water deficits over reproductive stages could decrease soybean yield severely. In addition, the required amount of water depends on stage of growth due to the requirement of maintaining adequate water supply till the bottom of the root system [1, 4, 5].

**Soil Type.** The identification of soil type is important due to the distinct physical and chemical properties which take action on the movement and storage of water and nutrients. The soil texture is based on the relative amount of three main components: sand, silt and clay. According to the soil texture, each type of soil has a limited capacity of water storage known as Field Capacity. This typical feature is fundamental to the irrigation process and it should be a limit on water distribution. As a consequence of exceeding field capacity, there will be infiltration and nutrient loss eventually. Then soil type must be identified and based on each kind, the proposed system can adjust the amount of water required [10, 11].

**Soil Moisture.** Monitoring soil moisture over irrigation fields is fundamental on crop management practices applied to boost yields. Water availability is a control factor over soybean growth and development and water stress may reduce nutrient absorption because it increases the difficulty of nutrient uptake. However, the management of irrigation fields should be more reliable combining soil moisture data and climate monitoring information in irrigation fields.

**Air Temperature and Relative Air Moisture.** In order to attach environmental conditions to the control model, air temperature and relative air moisture balance the controller output. These input variables have strong action under harsh environmental conditions such as high temperature or poor air moisture levels.

The equalizer algorithm was planned according to the temperature and air moisture classification. The design of equalizer model rules bases on the statement that water demand slightly increases on a hot day and decreases on cold day. The same strategy embraces air moisture. The developed membership functions were shown in Fig. 2.

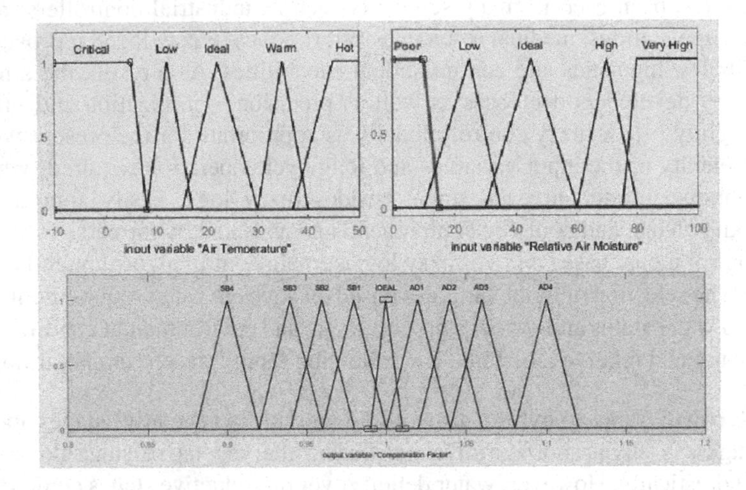


**Fig. 2.** This figure shows the membership functions of air temperature (*top and left*) and relative air moisture (*top and right*). The membership function of compensation factor (*bottom*) presents the response of equalizer algorithm.

The rule design in the main control algorithm considers the type of soil, soybean growth stages and soil moisture and the output was defined as the ideal amount of water to reach field capacity according to each soil, which increases production because then the fuzzy controller could assure that water supply have been provided to the entire soybean root system. Also, the main algorithm was developed to be in control of the whole plant cycle. The output variable was divided in many classes to achieve precision

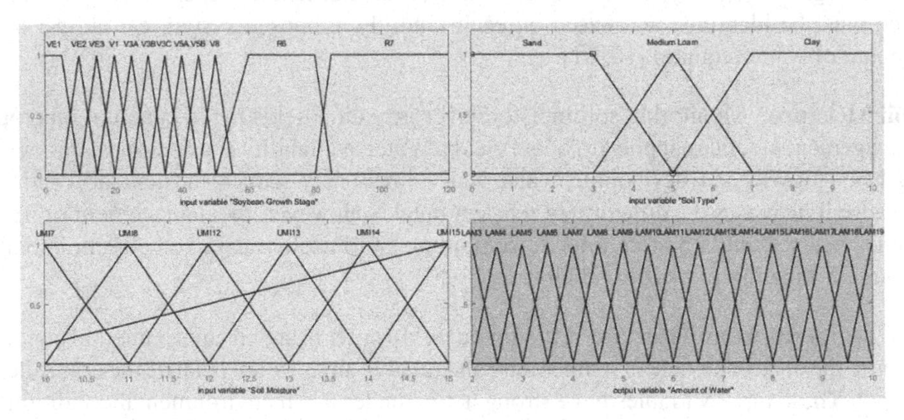


**Fig. 3.** As shown in figure, the membership functions are soybean growth stage (*top and left*), soil type (*top and left*), soil moisture (*bottom and left*) and amount of water (*bottom and right*). The last two membership functions are partial views because there is a large number of linguistic variables in soil moisture and amount of water.

and a large number of rules were requested to approach the ideal amount of water over soybean stages. The fuzzy control output represents the amount of water which should be applied based on agricultural knowledge and ought to be compensated due to environmental conditions. The membership functions are presented in Fig. 3.

According to Fig. 4, the control model was developed on two fuzzy algorithms due to computational resource limitations. The divided solution reduces computational time and simplifies code compilation. The amount of water (*SHEET OF WATER*) can be obtained through the main algorithm (*CONTROLLER*) without compensation factor. The environmental compensation relays on temperature and air moisture combined through compensation algorithm (*EQUALIZER*). The large number of divisions in the output membership function simplifies defuzzification process because the linguistic terms from amount of water corresponds directly to the control variable. This application encompasses 268 rules in main control algorithm and 25 rules in equalizer algorithm.

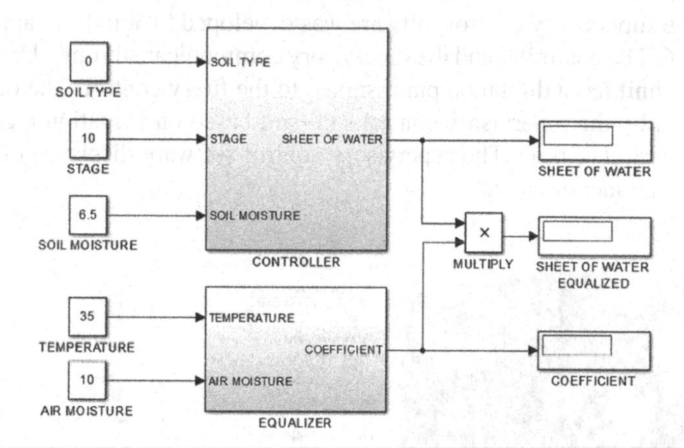


**Fig. 4.** The architecture of the developed fuzzy control model has based on two individual fuzzy algorithms.

## 4 Design of Continuous Monitoring Sensor Unit

The developed sensor bases on a capacitive element and the calibration method provides a soil moisture relationship. The sensor unit has an electronic circuit that is able to measure capacitance precisely. Also, the sensor unit integrates wireless connection, air temperature and relative air moisture elements.

### 4.1 Procedure of Calibration

The developed capacitive probe was calibrated in the Laboratory of Chemical Engineering at Faculdade de Engenharia de Sorocaba, where it was possible to detect a variation of the capacitance according to the mass quantity of water applied in each sample.

Two soil samples were used: clay soil and medium loam soil. These samples were dried and sifted to accomplish low moisture level. Each dried sample was weighed and had capacitance measured, and so, an acknowledged mass of water was applied on the soil sample and capacitance measured again.

Through collected data on capacitive probe calibration, it was highlighted the relationship between the percentage of moisture and measured capacitance, therefore it was possible to reach the exponential equation of calibration for clay soil and medium loam soil that works on developed sensor.

## 5 Design of Controller Unit and Supervisory Control Software

As shown in Fig. 5, it presents the developed electronic controller device and the capacitive sensor element. The wireless communication network operates between sensor and controller. The supervisory control software was developed to Windows application as shown in Fig. 6. The controller and the supervisory communicate through USB protocol. The controller unit has a digital output assigned to the fuzzy control. The output value can be adjusted by the supervisory control software based on irrigation mechanism to correspond a period of time. The supervisory control software displays collected data and storages each measurement.

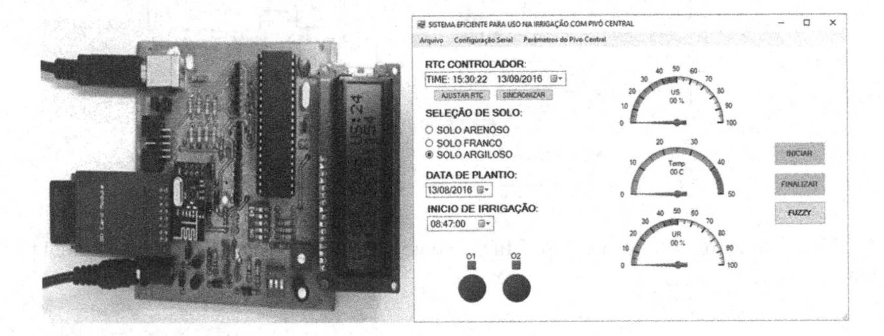


**Fig. 5.** Irrigation system controller (*left*) and supervisory control software (*right*). The operator can adjust parameters from the irrigation mechanism on software.

## 6 Results

A simulation was built in the software Matlab Simulink version 2015a. Adopted parameters have been used to better analyze the developed control logic. As shown in Fig. 6, it has demonstrated that the created system was able to process the input variables and the project met expectations for the proposed model. In addition, the surface view on the right of the picture presents the output response without equalizer action based on clay soil characteristics. Also, these surfaces present fixed soil type due to the number of available axes. The surface on the left shows the equalizer output, which adjusts system output based on air temperature (*TEMPERATURE*) and relative air moisture

(*AIR-MOIS*TURE). The designed system has been able to operate on the complete soybean cycle and the control output has reached the ideal amount of water, according to the proposed objective.

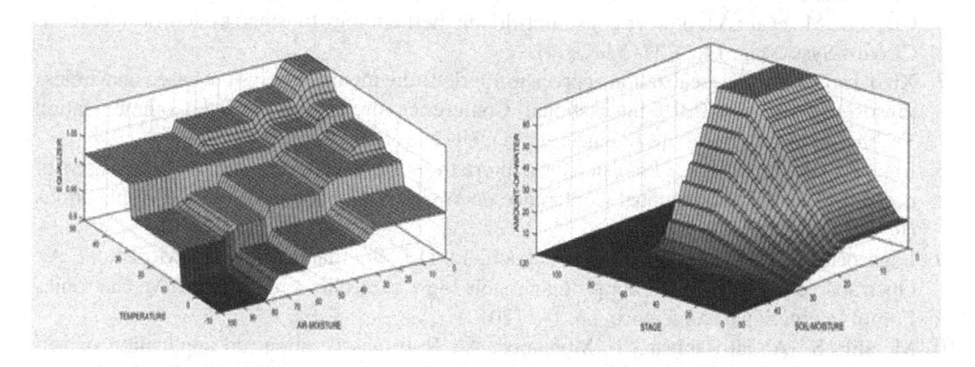


**Fig. 6.** Surface view model of the equalizer block (*left*) and surface view model of the controller block for the soil type clay (*right*).

## 7 Conclusion

The proposed control system provides resources for continuous data acquisition of particular field variables and the developed control algorithm analyses these aspects successfully.

In order to evaluate the performance of the developed devices, it was monitored the system operation and it was realized that the control module and the sensor unit have been acting satisfactory, according to the design specifications.

The present solution can further be applied to others cultivable plants. In addition, the proposed system might integrate a meteorological station to accomplish prediction capability. Furthermore, the fuzzy architecture could be easily modified to aggregate other unexplored features, which could improve overall performance.

## References


1. Sutton, M.C., Klein, N., Taylor, G.: A comparative analysis of soybean production between the United States, Brazil, and Argentina. J. Am. Soc. Farm Managers Rural Appraisers 33–41 (2005)
2. Meade, B., Puricelli, E., McBride, W., Valdes, C., Hoffman, L., Foreman, L., Dohlman, E.: Corn and soybean production costs and export. U.S. Department of Agriculture, Economic Research Service (2016)
3. Salin, D.L., Ladd, J.E.: Soybean Transportation Guide: Brazil. United States Department of Agriculture, Washington (2007)
4. Formaggio, A.R., Vieira, M.A., Rennó, C.D.: Object Based Image Analysis (OBIA) and Data Mining (DM) in Landsat time series for mapping soybean in intensive agricultural regions. In: 2012 IEEE International Geoscience and Remote Sensing Symposium, pp. 2257–2260 (2012)

5. Guimaraes, R.F., Andrade, L.R., Cavalho, O.A.: Identification of erosion susceptible areas in Grande river basin (Brazil). In: Proceedings of 2003 IEEE International Geoscience and Remote Sensing Symposium, GARSS 2003 (IEEE Cat. No. 03CH37477), vol. 4, pp. 2442–2443 (2003)
6. Crepaldi, M., et al.: A top-down constraint-driven methodology for smart system design. IEEE Circuits Syst. Mag. **14**(1), 37–57 (2014)
7. Xiao, L., Guo, L.: The realization of precision agriculture monitoring system based on wireless sensor network. In: 2010 International Conference on Computer and Communication Technologies in Agriculture Engineering, pp. 89–92 (2010)
8. Liao, Z., Dai, S., Shen, C.: Precision agriculture monitoring system based on wireless sensor networks. In: IET International Conference on Wireless Communications and Applications (ICWCA 2012), pp. 1–5 (2012)
9. Legaspe, E.P., Dias, E.M., Dias, M., Santos Filho, D.J., Squillante Júnior, R., Moscato, L.A.: Open source fuzzy controller for programmable logic controllers. In: The 13th Mechatronics Forum International Conference, pp. 1–7 (2012)
10. Mutalib, S., Abdul-Rahman, S., Mohamed, A.: Soil classification: an application of self organizing map and k-means. In: 10th International Conference on Intelligent Systems Design and Applications, pp. 439–444 (2010)
11. Yang, H., Kuang, B., Mouazen, A.M.: Effect of different preprocessing methods on principal component analysis for soil classification. In: 2011 Third International Conference on Measuring Technology and Mechatronics Automation, pp. 355–358 (2011)

# A Hybrid Expert Decision Support System Based on Artificial Neural Networks in Process Control of Plaster Production – An Industry 4.0 Perspective

Javaneh Ramezani[1(✉)] and Javad Jassbi[2]

[1] Faculty of Sciences and Technology, NOVA University of Lisbon, Campus da Caparica, 2829-516 Monte Caparica, Portugal
m.ramezani@fct.unl.pt
[2] UNINOVA-CTS, FCT-UNL, Caparica, Portugal
j.jassbi@uninova.pt

**Abstract.** Emerging technologies could affect future of factories and smartness is the main trend to receive that points. Quality was important and will be crucial in future but the question is how to build Smart Systems to guaranty quality in workshop level. This is an important challenge in Industry 4.0 paradigm. In this paper the main objective is to present practical solution under the light of Industry 4.0. The aim of this study is to presents propose a Hybrid Expert Decision Support System (EDSS) model, which integrates Neural Network (NN) and Expert System (ES) to detect unnatural CCPs and to estimate the corresponding parameters and starting point of the detected CCP. For this purpose, Learning Vector Quantization (LVQ) and Multi-Layer Perceptron (MLP) networks architecture have been designed to identify unnatural CCPs. Moreover, a rule based ES has been developed for diagnosing causes of process variations and subsequently recommending corrective action. The proposed model was successfully implemented in Construction Plaster producing company to demonstrate the capabilities and applicability of the model.

**Keywords:** Expert Decision Support System · Neural network · Statistical process control · FMEA · Control chart pattern

## 1 Introduction

Minute variations or/and differences always exist in any production process, regardless of the quality of design and maintenance. SPC is a powerful and capable set of solving problem tools, which is very useful in stabilizing the process and improving its efficiency by reducing variability [1]. One of the main objectives of SPC is quickly realizing the existence of assignable causes and any change in the process to investigate the reason of the diversions and applying corrective actions to avoid more defective products in lines. The most essential process of implementing SPC is control charting. Control charts are useful in determining the process behaviour [2]. Usually existence of the non-randomness trend in the control chart has significant impact on process performance. In


L.M. Camarinha-Matos et al. (Eds.): DoCEIS 2017, IFIP AICT 499, pp. 55–71, 2017.
DOI: 10.1007/978-3-319-56077-9_5

the meantime, control chart patterns such as Cycle, Trend, Shift, and Systematic as the basic patterns, have the process roots and usually appear in the most control charts of qualitative characteristics. As control charts consider only the current sample data, are not capable of presenting any pattern-related information. On the other hand, applying the accessible run rules can lead to increase false alarms [1]. Since the analysis of control charts is difficult, because it requires statistical knowledge and also experience of process, the main motivation of this study is to deal with the challenge of developing intelligent system which could identify defects, detect resources of deviations and recommend corrective actions automatically. Our investigation shows there is no existing model capable of handling mentioned challenge. To address this specific challenge, the main research question would be:

*What would be a suitable intelligent methodology to describe and analyze control charts for the fault diagnosis of the process in order to advise decision makers?*

Considering the fact that neural networks are well used alternatives for pattern recognition, and expert systems are effective in identifying causes of deviations and also recommending corrective action, we will try to answer the research question merging mentioned models using current successful experiences in both areas. The rest of paper is organized as follows: Sect. 2 is devoted to the literature review. Research methodology is discussed in Sect. 3. In Sect. 4 EDSS model is presented and in Sect. 5 conclusions are provided.

## 2 Contribution to Smart Systems

Smartness is the heart of Industry 4.0. Smart systems refer to diverse range of technological systems that can perform autonomously or in collaboration with other systems. These systems are capable to combine functionalities including sensing and controlling a particular situation in order to describe and analyze it. Smart systems have the ability to predict, decide and communicate with the user through a user interfaces. Smart systems have been also used in different areas, such as energy, transportation, security, ICT, industrial manufacturing, control, etc. [3]. The intelligence of smart systems is associated whit autonomous operation based on adaptability and learn-ability. This idea carries a sense of evolution and refers to process of modification and improvement over time. Neural networks with the ability of learning and expert systems with the command and rule-based features are perfect examples of smart systems [4]. In control process area, the main purpose of utilizing smart systems is to develop an intelligent system for real time control and monitoring the process. This work intends to contribute with the issues related to supervisory aspects of smart systems, considering the data acquisition, information transmitting, command-and-control and cognitive features of NNs and ESs. The main objective is to provide a model to support more intelligent and adaptive monitoring of smart system.

## 3 State of the Art

In this section related works that have used ESs and NNs to interpret SPC charts, were reviewed.

### 3.1 Application of ESs in SPC

In order to be competitive in the global market, more attention should be paid to extracting quality engineering knowledge in a systematic manner. Availability, consistency, extensibility and testability are the major advantages of ESs for SPC users [5]. In the following application of ES in SPC shall be briefly described. Evans [6] designed an ES for interpretation of x-bar and R charts using three sets of rules. [7] proposed a knowledge-based SPC system based on general knowledge of the process, which is capable of monitoring variations in a process. In another study [8] they proposed a hybrid system for SPC implementation. This system used ANN models to analyze control charts and an ES to diagnose the plausible causes. Their system can only recognize patterns that have full features. Reference [9] developed a knowledge based assistance for monitoring the process. Chung [10] integrated a model of decision support system (DSS), ANN, SPC and ICT to facilitate making decisions in production line. Reference [11] employed image processing and multivariate SPC to develop a visual detection ES. [12] has focused on ES and SPC, for selecting collaborative commerce system. And [13] developed an ES, based on multivariate control charts, to fault detection in induction motors.

### 3.2 Application of ANNs in SPC

The principle reason for applying NNs in SPC is to automate SPC chart interpretation [14]. This author has divided the literature of this area into structural change identification (changes in the average or variance of process) and pattern recognition. Most of the early research focused on detecting mean and variance shifts using similar approaches to [15] including [16–18]. Cheng [18, 19] designed multi-layer networks to simulate variance change. [24] presents a NN-based approach for detecting bivariate process variance shifts. NNs in the issues of pattern recognition are used to determinate random and non-random patterns. For example, [20] used LVQ to detect normal patterns, trends, sudden shifts and cycles. [21] proposed a hybrid-learning model using back propagation networks (BPN) and decision trees. [22] proposed a selective NN ensemble approach for CCP identifications. [23] proposed a NN to address the problem of monitoring a multivariate–multistage process. [24] using the method of Fourier descriptors and NNs identifies the CCPs. In the [25] the hybrid model of Recurrent Neural Network (RNN) and regression were utilized to recognize the CCPs. [26] developed a NN classifier for CCPs by Generalized Autoregressive Conditional Heteroskedasticity (GARH) Model. [27] Applies multivariate exponentially weighted moving average (MEWMA) and NNs for identifying the start point of the variations. [28] presents combination of intelligent model of ANN and support vector machine (SVM) learning methods for CCP recognition. [29] is an attempt to separate basic and mixture CCPs using NN and Independent

Component Analysis (ICA). [30] is an example of using Paraconsistent ANN (PANnet) and SPC in electrical power System.

## 4 Research Methodology

As the conceptual model is observed in the Fig. 1, in this study the structure of EDSS model based on ANNs has been presented to control the plaster manufacturing process. In order to describe the implementation steps of the model and validate it, the case study has been done in Semnan Noor plaster factory, which is producer of construction plaster and micronized plaster, according to Iran National Standard No. 1-12015 in the field of construction plaster production. The investigated product in this study is POP (Plaster of Paris) or construction plaster that is produced through calcinations process at the temperature of 150 °C in a closed reactor system called baking kiln while gypsum loses 1.5 mol of molecule water and converts to the construction plaster. In this study, order to improve the quality of the process, after identifying all parameters of each control station in the plaster production process, the interview was done with the experts focusing on the importance of each variable in the process. Then, according to the existing records, the causes of defect occurrence in the construction plaster production that were associated with quality features of the plaster were examined. Finally, initial setting time of the plaster was diagnosed as the critical parameter of the process that

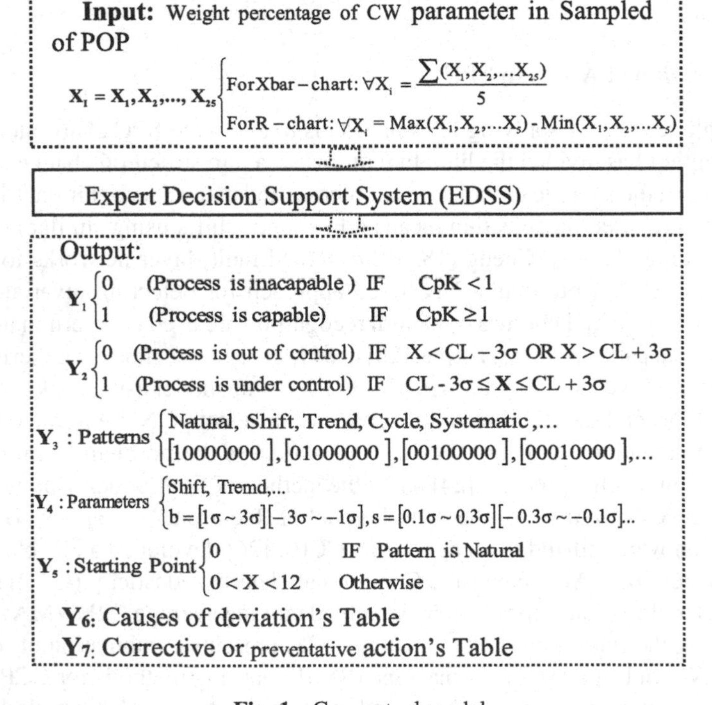


**Fig. 1.** Conceptual model

should take at least 7 min and at most 15 min. This parameter is dependent on the plaster's crystal of water after baking while the acceptance range in the investigated factory is between LSL = 5.0 to USL = 5.08 weight percentage and the process works under control with the basic limits of about LCL = 5.26 to UCL = 5.56. This parameter as a CW index has been analyzed by "Failure Modes and Effects Analysis" (FMEA) and "SPC" methodologies. The statistical population of this study included baked plaster of Low Burn kiln in the plaster production process that has been produced in a specified time period by stratified random sampling method from the baking tail silo, at random times, different shifts and operators. According to previous research on quality control issues, 25 subgroups of 5 samples (n = 125) were found suitable. Due to the nature of plaster production process, 8 times per shift and each time, five 25 grams' sample from a random point of the silo was taken to measure.

# 5 Model Design Construction

EDSS is composed of 3 subsystems. The first subsystem using statistical formulas, determines control chart limits for sampled data (weight percentage of CW), calculates process capability (Cpk) and gives warning if any of the points is out of control limits. The second subsystem is responsible for identifying unnatural patterns and the purpose of third subsystem is interpreting the causes of deviations and recommending preventative or corrective action.

## 5.1 Developing an ANN

In this section first the procedure of pattern simulation and then the development steps of NN will be explained.

**Simulating Different Patterns of the Process Control Charts.** In statistical issues any natural deviations could be determined according to probability distribution function of the corresponding random variable. This is the base of natural behaviors simulation in different charts of process control [31]. In this study, due to lack of large volume of data for unnatural patterns, it is tried to simulate the data, according to the original data of process.

**Designing the Structure of NN Model.** Overall structure of NN model is composed of, two separated sets, module I and module II (Fig. 2).

**Topology of the LVQ Network Designed in Module I.** First part of the model has been designed with general aim of identifying and classifying the input patterns. For this purpose, using competitive algorithms instruction, a LVQ network with two half-connected layers has been designed. The recognition window has 25 components (input layer neurons). In the first layer of network, 175 neurons and in the second layer 8 neurons are considered. The main reason in determining the required number of first layer neurons, were significantly identification incorrectly patterns. In

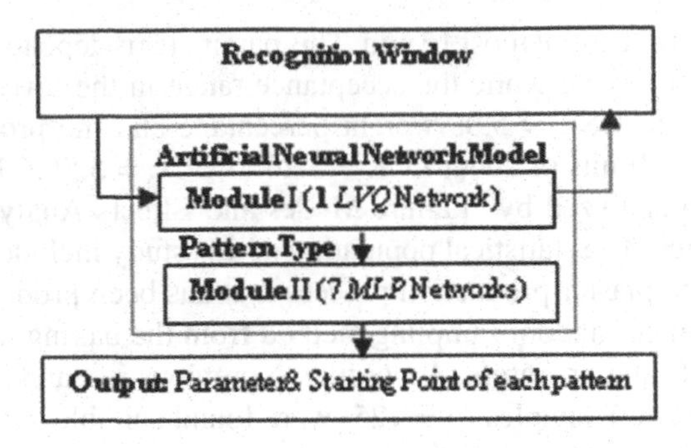


**Fig. 2.** Structure of NN mode

addition, it has been tried to assign almost equal neurons to patterns with equal number of parameters (Table 1).

**Table 1.** Neurons of the first layer

| Class | Subgroups | Class | Subgroups |
|---|---|---|---|
| 1. Natural | 1 | 5. Systematic | 4 |
| 2. Upward shift | 12 | 6. U.Shift + U.Trend | 18 |
| 2. Downward shift | 12 | 6. D.Shift + D.Trend | 18 |
| 3. Upward trend | 18 | 7. Shift + Cycles | 27 |
| 3. Downward trend | 18 | 8. Trend + Cycles | 27 |
| 4. Cycles | 20 | Total | 175 |

**Topology of MLP Networks Designed in Module II.** The second part of the model has been developed with the aim of estimating the parameters of each unnatural CCP, based on determined definitions (Table 2), and also estimating the starting point of the unnatural patterns. For this purpose, 7 two-layer perceptron networks, analyze the basis and simultaneous patterns. Because different directions of changes in process (upward or downward pattern) should be well recognized, so due to considering the output values in the range $[-1, 1]$ the activation function has been considered bipolar sigmoid with a constant $A = 0.1$. Total input of all MLP networks in module II is 26 that one of them is for bias and is considered equivalent to one. In all MLP networks for a special error value of the network, the number of required iterations to reach the desired error is calculated and the lowest repeated number until reaching to the considered error, is selected as the number of hidden layer neurons (Table 3). In output layer of each MLP, the number of neuron is selected according to the number of corresponding parameters of each pattern.

**Table 2.** Range of changes in unnatural patterns parameters

| Pattern type | X(t) | Range of parameters changes | Descriptions |
|---|---|---|---|
| Shift (Sh.) | x(t) = n(t) + u.**b** | $\mathbf{b} = [1\sigma \sim 3\sigma] \Rightarrow$ [0.1,0.3]<br>$\mathbf{b} = [-3\sigma \sim -1\sigma] \Rightarrow$ [−0.3,−0.1] | Parameter of Sh. (ascending)<br>Parameter of Sh. (descending) |
| Trend (Tr.) | x(t) = n(t) + **s**.t | $\mathbf{s} = [0.1\sigma \sim 0.3\sigma] \Rightarrow$ [0.01, 0.03]<br>$\mathbf{s} = [-0.3\sigma \sim -0.1\sigma] \Rightarrow$ [−0.03,−0.01] | Parameter of Tr. (ascending)<br>Parameter of Tr. (descending) |
| Cycles (Cyc.) | $x(t) = n(t) + \mathbf{l} * \sin(\frac{2\pi.t}{\mathbf{T}})$ | $\mathbf{l} = [1\sigma \sim 3\sigma] \Rightarrow$ [0.1,0.3]<br>**T** = 8,12, … | T is considered variable |
| Systematic (Sys.) | $x(t) = n(t) + \mathbf{g} * \cos(\pi.t)$ | $\mathbf{g} = [1\sigma \sim 3\sigma] \Rightarrow$ [0.1,0.3] | Concurrent combination Sh. (ascending) and Tr. (ascending) |
| Shift + Trend (Sh. + Tr.) | x(t) = n(t)) + u.**b** + **s**.t | $\mathbf{b} = [1\sigma \sim 3\sigma] \Rightarrow$ [0.1,0.3]<br>$\mathbf{s} = [0.1\sigma \sim 0.3\sigma] \Rightarrow$ [0.01, 0.03]<br>$\mathbf{b} = [-3\sigma \sim -1\sigma] \Rightarrow$ [−0.3,−0.1]<br>$\mathbf{s} = [-0.3\sigma \sim -0.1\sigma] \Rightarrow$ [−0.03,−0.01] | Concurrent combination Sh. (descending) and Tr. (descending) |
| Shift + Cycle (Sh. +Cyc.) | $x(t) = n(t) + u.\mathbf{b} + \mathbf{l} * \sin(\frac{2\pi.t}{\mathbf{T}})$ | $\mathbf{b} = [1\sigma \sim 3\sigma] \Rightarrow$ [0.1,0.3]<br>$\mathbf{l} = [1\sigma \sim 3\sigma] \Rightarrow$ [0.1,0.3]<br>**T** = 8,12,… | Concurrent combination Sh. and Cyc ascending |
| Trend + Cycle (Tr. + Cyc.) | $x(t) = n(t) + \mathbf{s}.t + \mathbf{l} * \sin(\frac{2\pi.t}{\mathbf{T}})$ | $\mathbf{s} = [0.1\sigma \sim 0.3\sigma] \Rightarrow$ [0.01, 0.03]<br>$\mathbf{l} = [1\sigma \sim 3\sigma] \Rightarrow$ [0.1,0.3]<br>**T** = 8,12,… | Concurrent combination Tr. and Cyc. ascending |

**Table 3.** Optimum number of hidden layer neurons, iteration trainings and maximum cumulative error in module II.

| Network name | Maximum cumulative error | Number of training repeat | Minimum number in each training | Neuron of hidden layer | Output number | Error of MLPs |
|---|---|---|---|---|---|---|
| 1. Natural | 10 | 10 | 10537 | 14 | 2 | 0.01 |
| 2. Upward shift | 18 | 13 | 16423 | 15 | 2 | 0.018 |
| 2. Downward shift | 25 | 10 | 35360 | 22 | 2 | 0.016 |
| 3. Upward trend | 12 | 10 | 48343 | 17 | 2 | 0.012 |
| 3. Downward trend | 18 | 10 | 48343 | 25 | 3 | 0.012 |
| 4. Cycles | 25 | 11 | 46703 | 23 | 4 | 0.012 |

**NNs Training.** The LVQ network in module I with learning rate λ = 0.01 is trained by enabling competition to take a place among the Kohonen neurons. The competition is based on the Euclidean distances between the weight vectors of these neurons and the input vector. The training method used in MLP networks is back propagation with adaptive learning rate, where the weight of each layer, by the output and output derivative is corrected until the network is fully trained [31]. The training data set is applied to corresponding networks in a category form and error is calculated at each step until the learning process performed. Network's error which is a cumulative error is defined as below in which: p stands for pattern number, o output neurons and $d_{ij}$ desired value for the j output of i pattern.

$$E = \frac{1}{2}\sum\nolimits_{i=1}^{p}\sum\nolimits_{j=1}^{o}\left(d_{ij} - o_{ij}\right)^{2}. \tag{1}$$

In this study, there were total of 11,000 training samples. 4000 samples as training data set in the LVQ network which for each pattern equally 500 data have been assigned and 7000 samples in the MLPs which for each of 7 networks equally 1000 samples were produced. In order to producing training set, first maximum and minimum values of input parameters were defined and then desired data were scaled and mapped considering the type of used transfer functions.

$$Ascale = \min + \frac{\max - \min}{\mathrm{Amax} - \mathrm{Amin}}(\mathrm{A} - \mathrm{Amin}). \tag{2}$$

The scaling data range in LVQ network between [−5, +5] and in MLP networks [−1, +1] is considered. But before that the inputs of MLPs, between [3.4, 7.4] had been scaled and the outputs with separate max and min values scaled (Table 4). Because each pattern has two orders of changes (upward and downward), the desired output is set in form 1 or −1. For example, the output vectors of Natural would be [10000000] and Downward Shift [0–1000000]. Maximum cumulative error (MCE) for training the LVQ is calculated 0.047 (188 in 4000 training data), and for testing 0.0525. MCE and training iteration of each MLP can be seen in Table 3.

**Table 4.** Scaled values range corresponding output of MLP networks

| Pattern | Output1 | | Output2 | | Output3 | | Output4 | |
|---|---|---|---|---|---|---|---|---|
| | min | max | min | max | min | max | min | max |
| Shift | −0.4 | 0.4 | 0 | 12 | | | | |
| Trend | −0.04 | 0.04 | 0 | 12 | | | | |
| Cycle | 0.01 | 0.4 | 6 | 14 | 0 | 12 | | |
| Systematic | 0.01 | 0.4 | 0 | 12 | | | | |
| ShiftTrend | −0.4 | 0.4 | −0.04 | 0.04 | 0 | 12 | | |
| ShiftCycle | 0.01 | 0.4 | 0.01 | 0.4 | 6 | 14 | 0 | 0 |
| TrendCycle | 0.001 | 0.04 | 0.01 | 0.4 | 6 | 14 | 0 | 0 |

**Test the NN Model.** In the training stage, the network's efficiency increases by minimizing the error between real outputs and in the testing step only the input data are given to the network, which are validated through the response prediction, between input and output variables.

**Evaluation of LVQ Network in Module I.** One of the major criteria in the development of LVQ network in purposed model has been the reduction of possibility of incorrect identification. Considering the diversity of trained patterns, the model shows a good response in this problem and also in indirect detection of individual patterns and incomplete diagnosis of mixed behaviors (Table 5). Moreover, for each identification window (input vector) network makes a decision on the process status. Hence the occurrence possibility of errors related to the decision making topics will be discussed. If the network detects by mistake the behavior of the under control process as unnatural, type I error and if the unnatural pattern in the process could not be detect, type II error has taken placed. Performance of module I and LVQ network by 400 test vectors were measured. Each of 400 test vectors is a sample of 25 Pieces that represents one of the 8 identified patterns. Amount of LVQ error is calculated 0.052 and the results of 400 experimental vectors is integrated in Table 5.

**Table 5.** Evaluation results of module I

| Pattern | Direct identification | Identification (incomplete/indirect) | Wrong identification | Type I error | Type II error |
|---|---|---|---|---|---|
| Shift | 41 | 6/50 = 0.12 | 3/50 = 0.06 | 0.00 | – |
| Trend | 48 | 2/50 = 0.04 | 0.00 | 0.00 | 1/50 = 0.02 |
| Cycles | 48 | 1/50 = 0.02 | 1/50 = 0.02 | 0.00 | 1/50 = 0.02 |
| Systematic | 50 | 0.00 | 0.00 | 0.00 | 0.00 |
| Shift + Trend | 50 | 0.00 | 0.00 | 0.00 | 0.00 |
| Shift + Cycles | 50 | 0.00 | 0.00 | 0.00 | 0.00 |
| Trend + Cycles | 44 | 0.00 | 6/50 = 0.00 | 0.00 | 0.00 |
| Total | 21/400 = 0.052 | 9/400 = 0.022 | 10/400 = 0.025 | 0.00 | 2/400 = 0.005 |

**Test and Evaluation the MLP Networks in Module II.** One of the critical problems in NNs training is the over- adaption problem of training data. Topics "generalization" and "training" in NNs have the same significance [32]. For solving this problem, part of the training vectors is considered as validation data. In this way training data follow the flow of parameters modification and validity data will follow error during the learning process. The validity data set error such as training data set error naturally decreases but as soon as there is over learning in the network, despite fixing or decreasing error of validity data set, the amount of training data error will increase. In this time the training process stops and the parameters according to the minimum errors of validity data, will be considered as an algorithm ultimate answer. After the training process, performance of the network was tested by several examples. Results show, module II in identification and analysis of defined parameters, acts successfully and efficiently (Table 3).

**Verification of the NN Model.** After evaluating the NN model by the test data set, in order to determine the accuracy, repeatability and the amount of results stability in the test repetition, the verification of the model by comparing the amount of NN error and discriminant analysis (DA) error has been done. It should be noted that DA is a similar statistical method in classification and in this study the statistical software SAS was used for this purpose. Figure 3 shows related errors to any classes in the Rate row, and in the Priors row the weight of each class is shown. 0.3325 represents the total error in using DA for the test data set. As it is seen in Fig. 4, neural networks compared to DA method, has much better performance and precision.

| Error Count Estimates for target | | | | | | | | | |
|---|---|---|---|---|---|---|---|---|---|
| | 1 | 2 | 3 | 4 | 5 | 6 | 7 | 8 | Total |
| Rate | 0.6458 | 0.5763 | 0.5577 | 0.0943 | 0.0000 | 0.5870 | 0.0000 | 0.1321 | 0.3325 |
| Priors | 0.1200 | 0.1475 | 0.1300 | 0.1325 | 0.1250 | 0.1150 | 0.0975 | 0.1325 | |

**Fig. 3.** DA error in the test data set classification

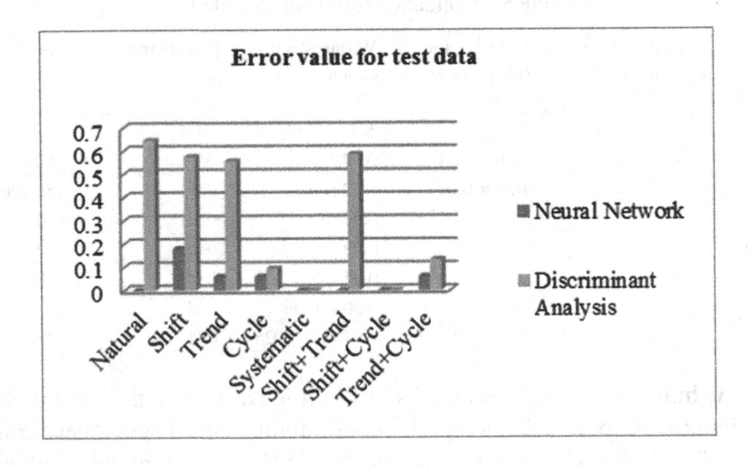


**Fig. 4.** Comparing the error of NN and DA for each pattern

## 5.2 Developing an ES

Steps discussed in this section include knowledge acquisition, knowledge representation and implementation:

**Knowledge Acquisition.** In this study following the knowledge management system (KMS) approach, to help the diagnostic process, Western Electric tests are used as general knowledge and in order to obtain specific knowledge of the process, first Cause and Effect diagrams were prepared to determine the most problem causes in the process (Fig. 5). After several reviews, effective variables on rate of crystal water were categorized in two categories of controllable and uncontrollable variables. Controllable variables include: kiln's negative pressure (filter), temperature of kiln body, flame's profile, fuel pressure adjustment, kiln blades and humidity in raw material. Uncontrollable variables include: Air temperature and change of trend in raw material because of mine. Next the FMEA methodology were used to identify and prioritize failure modes in the plaster production process and then necessary actions to eliminate or reduce the occurrence of failure modes were taken and finally the results of the analysis with the aim of creating the full reference for future problems were recorded in the Knowledge base of the designed system.

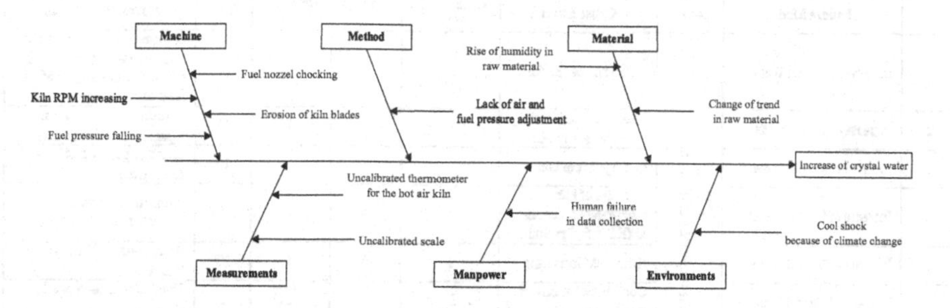


**Fig. 5.** Cause and effect diagram for increasing of crystal water

**Knowledge Representation.** In this study the rule-based approach has used as a general framework for the knowledge representation. In this research experts' problem-solving knowledge were encoded in the form of IF <Situation> THEN <Action> rules. This set of rules has formed the knowledge base of ES. The knowledge rules have obtained from manuals, procedures, technical documentations and forms of interviews with experienced engineers and technicians. During the interview process, conversations recorded in detail and then transferred to the FMEA worksheet. Finally, a total of 60 rules applied as specific and general knowledge for interpreting X-bar and R control charts.

**Implementation.** The proposed hybrid model (EDSS) has designed using three main modules. The first module is related to knowledge base, the second module is related to designing the interface and the third module is related to running the system and dialogue to the user. EDSS which can be called intelligent statistical process control (ISPC), its components are as follows:

(1) Knowledge base, which is composed of three main parts below:
   (a) Events, which extracts from the records of ISO, preventive maintenance, calibration, brainstorming sessions and FMEA forms.
   (b) Procedures, which Includes technical instructions, plaster production standard, ISPC manual and etc.
   (c) Rules, which consists of 60 rules extracting from experts and documents such as Western Electric tests, statistical formulas and NN's analysis, has presented in the form of IF-THEN statement.

Western Electric tests which are using general knowledge of process to control charts interpretation is including: points out of control limits (1 point more than $+3\sigma$ or $-3\sigma$), gradual changes in levels (9 points larger or smaller than CL), trends (6 points in a row steadily increasing or decreasing), Systematic variations (14 points in a row alternating up and down), cycles (4 of 5 points more than $+2\sigma$ or $-2\sigma$), and mixtures (8 points in a row more than $+1\sigma$ or $-1\sigma$) (Table 6).

**Table 6.** FMEA form for the critical parameter (crystal water)

| Failure Modes and Effects Analysis (FMEA) | | | | | | | | |
|---|---|---|---|---|---|---|---|---|
| | Failure Mode | index | Cause Effect | Occurrence (O) | Severity (S) | Detection (D) | RPN | Corrective Actions |
| 1 | Increase of Crystal Water | Decreasing of middle temperature in kiln body | Fuel nozzle chocking | 3 | 5 | 5 | 75 | Cleaning the fuel nozzle and establishment of PM for burner and installation of filter for fuel. |
| 2 | Increase of Crystal Water | | Lack of air and fuel pressure adjustment | 3 | 4 | 4 | 48 | Adjustment of fuel and air regulator in defined period |
| 3 | Increase of Crystal Water | | Leakage of pre kiln | 2 | 5 | 1 | 10 | Chang or patching of pre-kiln and thickness monitoring |
| 4 | Increase of Crystal Water | | Decrease of kiln temperature according to falling fuel pressure | 2 | 3 | 1 | 6 | Installation a shutdown sensor for fuel pressure. |
| 5 | Increase of Crystal Water | | Kiln RPM Increasing | 1 | 5 | 4 | 20 | Adjustment of RPM via frequency. |
| 6 | Increase of Crystal Water | | Cool shock because of weather temperature | 1 | 2 | 1 | 2 | Thermal insulation of kiln shell. |
| 7 | Increase of Crystal Water | | Erosion in kiln blades | 1 | 5 | 5 | 25 | Change of runner blade and periodically monitoring |
| 8 | Increase of Crystal Water | | Pollution of air nozzles of burner | 4 | 2 | 3 | 24 | Cleaning the air nozzles and instating a multilayer filter fair. |
| 9 | Increase of Crystal Water | | Heat transition between kiln and environment because of lake of refractory | 5 | 5 | 5 | 125 | Establishment of PM and thickness monitoring of refractory. |
| 10 | Increase of Crystal Water | | Negative pressure of kiln (filter) rising | 5 | 5 | 1 | 25 | P Meter Installation.Δ |
| 11 | Decrease of Crystal Water | Increasing of middle temperature | Rising rate fuel pressure | 1 | 2 | 1 | 2 | Pressure monitoring and CO controlling unit. |
| 12 | Decrease of Crystal Water | | Decrease of negative pressure of exhaust fan | 5 | 5 | 3 | 75 | P Meter Installation.Δ |
| 13 | Decrease of Crystal Water | | Decrease of RPM of kiln gear box | 1 | 5 | 4 | 20 | Establishment of preventive maintenance for gear box and inverter. |
| 14 | Decrease of Crystal Water | | Heat shock because of water temperature | 1 | 2 | 1 | 2 | Thermal insulation of kiln shell. |
| 15 | Variation in crystal water | - | Change of trend in raw material because of mine | 4 | 5 | 4 | 80 | XRF test for raw material. |
| 16 | Increase of Crystal Water | - | Rise of humid in raw material | 3 | 5 | 1 | 15 | Setup a raw material storage salon. |
| 17 | Variation in crystal water | - | Change of raw material spec | 3 | 5 | 1 | 15 | XRF test for raw material. |

**Example.** The Following example represents a kind of typical rule, based on specific knowledge of process:

IF interpretation is "Upward Trend"
AND Failure Mode is "Increase of Crystal Water"
AND Process Index is "Decrease of kiln's temperature"
THEN Special Cause can be "Fuel nozzle Chocking"
AND Corrective Actions can be either "Cleaning the fuel nozzle, establishment of PM for the burner or installing fuel filter".

(2) Inference Engine, which backward-chain inference engine has used in this study in order to troubleshooting. This means, the written program has performed with two

sets of rules. The first group define goals for the properties and if it is unable to determine the value of the properties with the existing rules, the user is asked to identify them. The second group updates actions like modifying the rules and transmitting the satisfied goals.

(3) Working Memory, which Consists of events and facts to be applied by the rules. This memory is used to store temporary data that is provided during problem-solving process. This data includes the user's answers to the system's questions and also deriving facts from reasoning process like unnatural patterns, the starting point and corresponding parameters of patterns that are identified by the NN.

(4) User Interface, which works with the inference engine and the knowledge base in order to two-way communication between user and the EDSS. Users can answer a question by selecting Yes or No or select one item from a menu on the screen.

## 5.3 Integration of NN and ES

To develop ISPC, capabilities of NNs and ES are used in automatic interpretation of control charts. The purposed system is able to do most of traditional SPC operations including calculation of control limits (LCL, CL and UCL), calculation of Cpk, data normalization test and checking for X-bar and R charts being under control. In SPC first the based charts must be prepared. For this purpose, after initial sampling, control limits and Cpk must be calculated in control mode. In this study according to the plaster factory experts' opinion if "Cpk > 1" the chart is accepted as the basis (Fig. 6).

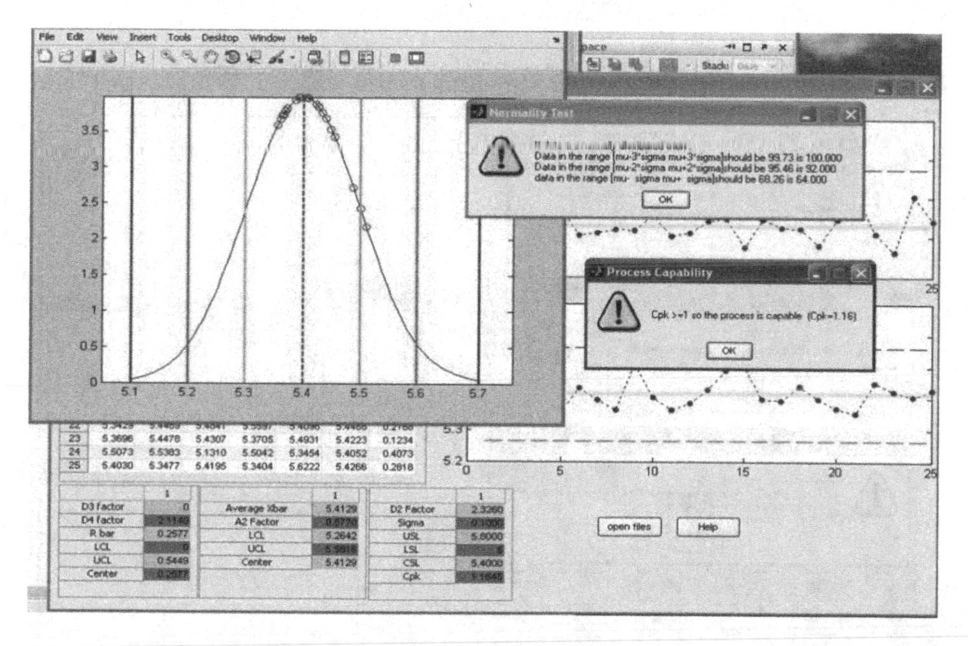


**Fig. 6.** SPC operations

**Experimental Result of EDSS Model.** Using real data from the current process of plaster producing and interface software, a practical example is presented (Fig. 7). As shown in Fig. 6 although there is no any out of control mode in the R-chart but the process is not capable of meeting specifications because the "Cpk < 1" and equal to 0.81. On the other hand, by selecting x-bar chart, the user has faced with the message of "X-bar chart is out of control". Then, the ES using Western Electric tests has announced that points falling outside the control limits may be the result of: carelessness in the measurement, machinery stop or off spec materials. Later, ES suggests the user to check the unnatural patterns by NN. As you can see, the NN has not only has identified downward shift pattern in X-bar chart but it has also estimated the starting point of unnatural pattern (point 6) and the displacement parameter (−0.161). After that, ES asks the user to explain the defect to find causes of potential failure modes in the process. If the answer is not in the options, ES says: "refer to the experts!". In this example according to the appearance of downward shift pattern and as the user observation was "kiln body scarlet", the cause of defect has announced "temperature exchange of kiln with environment due to the loss refractory and thickness" and "establishment of maintenance and inspection of refractory" have proposed as corrective or preventive actions. In this practical example, after doing corrective action and re-sampling the process, the control charts did not show any out of control mode and also process capability improved from Cpk = 0.81 to Cpk = 1.15. Ultimately, the designed system examined by several examples and the results were considered satisfactory. However, the system performance needs to be improved more so it is important that the long-term use of the model

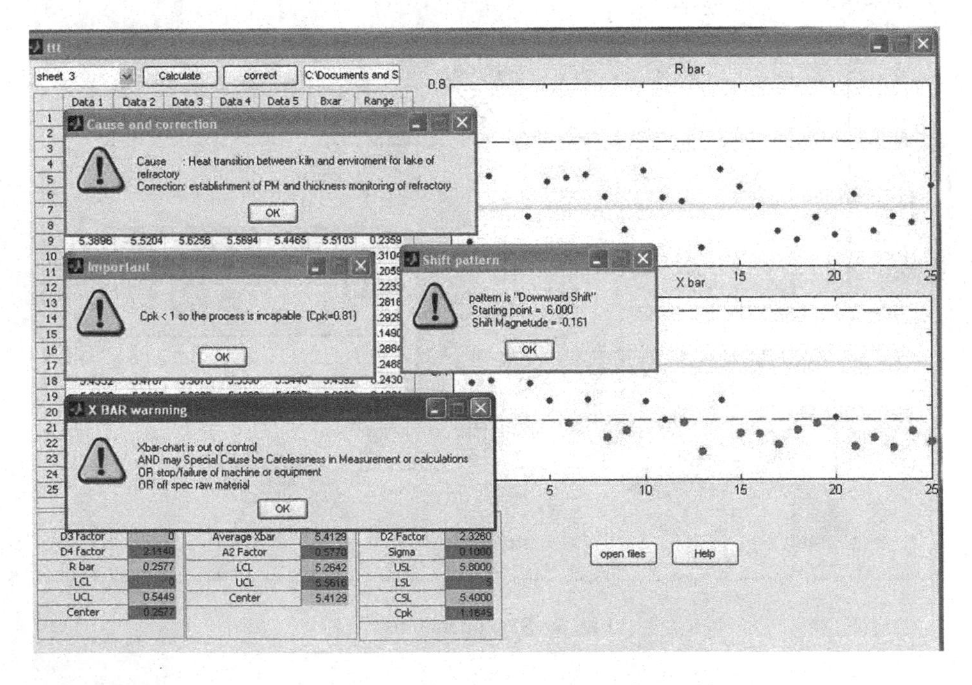


**Fig. 7.** Results of EDSS for a practical example

be identified as part of the objectives of the study because the model design is highly influenced by it.

## 6 Conclusions

The main goal of this work was to present the capabilities of emerging technologies and algorithms, dealing with quality issues in shop floor. In this work a hybrid EDSS was designed to support the operators in troubleshooting of plaster production process. For this purpose, ES and NN were integrated via designed interface software for reasoning of deviation sources and recommending corrective actions. The ES tries to determine the fault area as far as possible until the serviceman focuses on the point. On the other hand, in the structure of current model, features of LVQ and MLP networks are used in two modules. Therefore, the competitive power of LVQ nctwork in pattern classification and the interoperability of multi-layer perceptron networks for parameter estimation of abnormal patterns in different process control chart were used simultaneously. This integrated approach has provided an appropriate condition for implementing desired thoughts. Considering the training of the network for identifying basic and mixed significant behaviors, the numerical results in Table 5 shows output of module I in diagnosis of behavioral patterns, which is acceptable. The results also show that module II who is estimating the parameters of corresponding patterns, is effective and reliable (Table 3). The work has many contributions and was covering multi tasks including: production, delivery and encryption of NN input, using a wide range of data in training the NNs, and doing most of SPC required via integrated system such as drawing of basis chart, checking X-bar and R charts for being under control and also calculating Cpk. Smartness is key criteria in Industry 4.0 and the case study shows the capability of proposed model to bring more intelligence to the production lines.

## References


1. Western Electric Company: Statistical Quality Control Handbook. Western Electric Company, New Jersey (1956)
2. Montgomery, D.C.: Introduction to Statistical Quality Control. Wiley, New York (1996)
3. CORDIS Archive: European Commission: Smart components and smart systems integration. EXTRACT FROM WORK PROGRAMME (2011). http://cordis.europa.eu/fp7/ict/micro-nanosystems/calls_en.html
4. Moshnyaga, V.: Guidelines for developers of smart systems. In: 8th International Conference on Intelligent Systems (IS), pp. 455–460. IEEE (2016)
5. Feigenbaum, E.A.: The Rise of the Expert Company: How Visionary Companies are Using Artificial Intelligence to Achieve Higher Productivity and Profits. Macmillan, London (1983)
6. Evans, J.R., Lindsay, W.M.: A framework for expert system development in statistical quality control. Comput. Ind. Eng. **14**, 335–343 (1988)
7. Pham, D.T., Oztemel, E.: Control chart pattern recognition using multi-layer perceptrons and learning vector quantization networks. J. Syst. Control Eng. **207**, 113–118 (1993)

8. Pham, D.T., Oztemel, E.: Control chart pattern recognition using learning vector quantization networks. Int. J. Prod. Res. **32**, 721–729 (1994)
9. Masud, A.S.M.: A knowledge-based advisory system for statistical quality control. Int. J. Prod. Res. **31**, 1891–1900 (1993)
10. Chung, W.C., Wong, K.C.M.: An ANN-based DSS system for quality assurance in production network. J. Manuf. Technol. Manag. **18**, 836–857 (2007)
11. Lyu, J., Chen, M.: Automated visual inspection expert system for multivariate statistical process control chart. Expert Syst. Appl. **36**, 5113–5118 (2009)
12. Chang, D.S., Liao, C.C.: Expert system for multi-criteria decision making to assess the collaboration commerce systems. In: The 2nd International Conference on Next Generation Information Technology, pp. 68–71. IEEE (2011)
13. Garcia-Escudero, L., Duque-Perez, O., Morinigo-Sotelo, D., Perez-Alonso, M.: Robust multivariate control charts for early detection of broken rotor bars in an induction motors fed by a voltage source inverter. In: International Conference on Power Engineering, Energy and Electrical Drives, pp. 1–6. IEEE (2011)
14. Jacob, D.A.: Training artificial neural networks for statistical process control. In: The 10th Biennial University/Government/ Industry Microelectronics Symposium, pp. 235–239. IEEE (1993)
15. Pugh, G.A.: Synthetic neural networks for process control. Comput. Ind. Eng. **17**, 24–26 (1989)
16. Pugh, G.A.: A comparison of neural networks to SPC charts. Comput. Ind. Eng. **21**, 253–255 (1991)
17. Guo, Y.: Identification of change structure in statistical process control. J. Prod. Res. **7**, 1655–1669 (1992)
18. Cheng, C.S.: A multi-layer network model for detecting changes in the process. Comput. Ind. Eng. **28**, 51–61 (1995)
19. Cheng, C.S.: A neural network model approach for the analysis of control chart patterns. Int. J. Prod. Res. **35**, 667–697 (1997)
20. Cheng, C.H.: Using neural networks to detect the bivariate process variance shifts pattern. Comput. Ind. Eng. **60**, 269–278 (2011)
21. Guh, R.S., Hsieh, Y.: Neural network based model for abnormal pattern recognition of control charts. Comput. Ind. Eng. **36**, 97–108 (1999)
22. Guh, R.S.: Optimizing feed forward neural networks for control chart pattern recognition through genetic algorithms. Int. J. Pattern Recogn. **18**, 75–99 (2005)
23. Yu, J., Wu, B.: A neural network ensemble approach for recognition of SPC patterns. In: 3rd International Conference on Natural Computation, vol. 2, pp. 575–579. IEEE (2007)
24. Niaki, S.: Designing a multivariate–multistage quality control system using artificial neural networks. J. Prod. Res. **47**, 251–271 (2009)
25. Phokharatkul, P., Phaiboon, S.: Control chart pattern classification using Fourier descriptors and neural networks. In: 2nd International Conference on Artificial Intelligence, Management Science and Electronic Commerce (AIMSEC), pp. 4587–4590. IEEE (2011)
26. Behmanesh, R., Rahimi, I.: Control chart forecasting: a hybrid model using recurrent neural network, design of experiments and regression. In: IEEE Business, Engineering and Industrial Applications Colloquium (BEIAC), pp. 435–439. IEEE (2012)
27. Lavangnananda, K., Sawasdimongkol, P.: Neural network classifier of time series: a case study of symbolic representation preprocessing for Control Chart Patterns. In: 8th International Conference on Natural Computation, pp. 344–349. IEEE (2012)
28. Ebrahimzadeh, A., Addeh, J., Ranaee, V.: Recognition of control chart patterns using an intelligent technique. Appl. Soft Comput. **13**, 2970–2980 (2013)

29. Thaiupathump, T., Chompu-inwai, R.: Impact of kurtosis on performance of mixture control chart patterns recognition using independent component analysis and neural networks. In: 4th International Conference on Advanced Logistics and Transport (ICALT), pp. 94–99. IEEE (2015)
30. da Cruz, C.M., Rocco, A., Mario, M.C., Garcia, D.V., Lambert-Torres, G., Abe, J.M., Torres, C.R.: Application of paraconsistent artificial neural network in statistical process control acting on voltage level monitoring in electrical power system. In: 18th International Conference on Intelligent System Application to Power Systems (ISAP), pp. 1–6. IEEE (2015)
31. Fatemi Ghomi, M., Lesani, A.: Design a Model Based on Neural Networks to Pattern Identification. Tehran University, Tehran (1385)
32. Menhaj, M.: Computational Intelligence. Amirkabir University, Tehran (1387)

# Flexibilizing Distribution Network Systems via Dynamic Reconfiguration to Support Large-Scale Integration of Variable Energy Sources Using a Genetic Algorithm

Marco R.M. Cruz[1], Desta Z. Fitiwi[1], Sérgio F. Santos[1], and João P.S. Catalão[1,2,3](✉)

[1] C-MAST, University of Beira Interior, 6201-001 Covilhã, Portugal
marco.r.m.cruz@gmail.com, destinzed@gmail.com, sdfsantos@gmail.com, catalao@fe.up.pt

[2] INESC TEC and the Faculty of Engineering, University of Porto, 4200-465 Porto, Portugal

[3] INESC-ID, Instituto Superior Técnico, University of Lisbon, 1049-001 Lisbon, Portugal

**Abstract.** In recent years, the level of variable Renewable Energy Sources (vRESs) integrated in power systems has been increasing steadily. This is driven by a multitude of global and local concerns related to energy security and dependence, climate change, etc. The integration of such energy sources is expected to continue growing in the coming years. Despite their multifaceted benefits, variable energy sources introduce technical challenges mainly because of their intermittent nature, particularly at distribution levels. The flexibility of existing distribution systems should be significantly enhanced to partially reduce the side effects of vRESs. One way to do this is using a dynamic network reconfiguration. Framed in this context, this work presents an optimization problem to investigate the impacts of grid reconfiguration on the level of integration and utilization of vRES power in the system. The developed combinatorial model is solved using a genetic algorithm. A standard IEEE 33-node distribution system is employed in the analysis. Simulation results show the capability of network switching in supporting large-scale integration of vRESs in the system while alleviating their side effects. Moreover, the simultaneous consideration of vRES integration and network reconfiguration lead to a better voltage profile, reduced costs and losses in the system.

**Keywords:** Distributed generation · Network reconfiguration · RESs · Genetic algorithm · Variable energy resources

## 1 Introduction

In recent years, integration of Distributed Generations (DGs) in electric distribution systems has attracted more attention due to various reasons [1]. The purpose of DG is to install small generation plants (mainly of non-conventional energy sources) close to end-users. Because of the growing concerns on climate changes, increased demand for electricity and other reasons, several states are adopting new policies to support the development of "clean" energy resources and address these concerns [2]. Nevertheless,


L.M. Camarinha-Matos et al. (Eds.): DoCEIS 2017, IFIP AICT 499, pp. 72–80, 2017.
DOI: 10.1007/978-3-319-56077-9_6

dramatically increasing the level of variable Renewable Energy Sources (vRESs) often poses some technical challenges that often undermine the stability and reliability of the system as well as the quality of power delivered to consumers [3]. This is because such intermittent energy sources substantially increase the level of operational variability and unpredictability in distribution network systems, which further complicate the "operation, planning and control" of such systems [1]. This can be addressed by enhancing the flexibility of existing systems. One way to do this is using dynamic distribution system reconfiguration (DSR). A dynamic DSR can considerably enhance the flexibility of the system and improve voltage profiles, thereby, increasing chances of accommodating large-scale vRES power. The DG placement and sizing problem has been widely studied especially in recent years. In relation to this, authors in [4] present a detailed review of existing literature in DG allocation and associated issues. Depending on the size and type of DGs, their integration in distribution network systems may have positive or negative effects. Authors in [4, 5] employ several techniques to analyze the impact of DGs in electrical systems as a whole. Some researchers resort to ways that can be used to develop appropriate rules for allocating DGs in distribution network systems while minimizing their possible side effects [6–10]. In this regard, the most relevant issues that need to be accounted for in DG integration include the network topology, DG capacity and suitable location among others. In the absence of advanced enabling mechanisms such as smart-grids, each connection point in distribution network systems has a maximum DG penetration limit beyond which some of the system performance indicators such as losses and technical issues are undermined [11, 12]. One of the main aims of this work is to develop appropriate tools for optimally integrating variable energy sources so that the quality of power delivered to end-users is enhanced at a minimal possible cost. To this end, an optimization problem is proposed to optimally deploy such resources in distribution systems, and investigate the impacts of grid reconfiguration on the penetration level of vRESs in such systems. The developed model is solved using a genetic algorithm. A standard IEEE 33-node distribution system is employed in the analysis. Hence, one of the main contributions of this work is an improved mathematical model for jointly optimizing distribution network reconfiguration and allocation of RES-based DGs. Another contribution is the quantitative and qualitative analysis made to assess the impacts of network reconfiguration on the penetration and utilization level of intermittent power, and relevant system variables in the distribution system.

## 2 Relationship to Smart Systems

The integration of DGs in electrical distribution network systems (renewable types, in particular) has been gaining momentum since recently. This is driven by a number of factors of technical, economic and environmental nature. Growing concerns related to the increasing demand for electricity, energy security and climate change will further lead to a continuously increasing share of vRESs (such as wind and solar) in the final energy consumption. However, a very high penetration level of such resources results in some technical challenges in the system that may undermine the integrity, stability, reliability and quality of power delivered to consumers [1]. This is mainly due to the intermittent nature of such energy sources that substantially increase operational variability and unpredictability in the power production. To deal with such challenges, existing systems should be equipped with new flexibility mechanisms. For example, the deployments of smart-grid technologies are expected to adequately overcome the technical challenges posed by RES-based DGs as a result of the dramatically enhanced system flexibility [1, 13]. In connection to this, a dynamic (and possibly automated) distribution network reconfiguration is one of these technologies that are expected to be rolled out in distribution systems for the same purpose. Distribution System Operators (DSOs) can reap the benefits of automated DSR that along with other smart systems helps to address the challenges of DG sizing and allocation problem [14]. In general, the contribution of the current work to smart systems can be summarized with the following points:

1. DSOs can invest in smart-grid systems that have the capability to automatically switch branches in order to operate the system with reduced costs and improved system stability.
2. After the integration of vRES based DGs, smart systems that have the capability to switch branches can efficiently route RES power generated elsewhere where the demand is low to locations where demand is relatively higher, hence reducing curtailment. All this leads to an enhanced voltage profile and reduced energy losses in the system.
3. Smart grids and isolated systems are often based on vRES technologies. Innovation in smart systems will influence control and operation of these systems, reducing the impact of the associated variability and uncertainty of such DGs. This in turn implies better performances in terms of stability, reliability and quality of the systems. In the near future, isolated systems may operate automatically, without DSO's intervention. And, this can bring fewer concerns for remotely located systems.

## 3 Model Formulation

### 3.1 Objective Function

The main objective of the problem is to minimize the total costs of operation in the system, which is given by (1).

$$F = \min C_i \tag{1}$$

where Ci is the cost of each population, contained in the fitness vector. The formulation of the problem is based on the AC network model. In order to solve the AC optimal power flow problem, MatPower is used. The combinatorics of DSR is handled via a genetic algorithm. Details of the algorithm are presented below.

(a) *Initialization:* In order to set the connected branches, a binary population is assumed. First, populations whose number is equal to the number of links are generated. The generated populations should keep the network radial, given by the number of buses minus the number of generators. The DG sizing problem assumes integer values. Hence, a DG chromosome generates integer numbers, representing the size of a DG and with a length equivalent to the number of buses to simultaneously handle the allocation problem. This means that each population has the size and the location information of each DG type.

(b) *Mutation:* Uniform crossover, Boolean mutation for the connected branch is used. At the same time, integer mutation for the DG placement and size is used for new populations based on the best populations.

(c) *Selection:* Tournament selection is used to select the best minimal cost. In tournament selection, $n$ individuals of a population are selected randomly with the same probability. The individual with the greatest fitness among them is selected and passed to an intermediate population. The process ends when the intermediate population is fulfilled. Crossover uses the information in two or more individuals to generate one or more individuals. Mutation can diversify when new information is introduced in the individual, and consequently to the population.

### 3.2 Restrictions

In order to keep the radiality of the network system, one needs to check if the generated population fulfils the following criterion:

$$N_{closed\ branches} = N_{buses} - N_{generators} \tag{2}$$

This condition is applied to check if the criterion is satisfied. This is a necessary condition but not sufficient to obtain a radial topology. It is therefore necessary to verify if all buses in the system are connected. This condition is expressed as follows [15]:

$$\forall_{X_i X_j} \in N, \ni \{C_i \cup C_j\} \left( \prod_k^n Z_{i_k} \prod_k^n Z_{j_k} = 1 \right) \tag{3}$$

where $N$ represents the set of nodes in the system; $C_i \cup C_j$ is the unique path connecting nodes $i$ and $j$; $Z_{i_k}$ and $Z_{j_k}$ refer to the statuses (0 or 1) of the $n$ branches in the system that constitute the path $C_i \cup C_j$; and $k$ is an index of the branches that constitute each path. In addition to the above constraints, the lower and upper limits of system variables such as voltage, current, active and reactive power generation are included.

### 3.3 Active and Reactive Power Balances

The constraint related to the active power balance is enforced using Eq. (4):

$$gP(\theta;V_m;P_g) = P_{bus}(\theta;V_m) + P_d - C_g P_g = 0 \tag{4}$$

Similarly, reactive power balance should be ensured, which is given by (5):

$$gQ(\theta;V_m;Q_g) = Q_{bus}(\theta;V_m) + Q_d - C_g Q_g = 0 \tag{5}$$

where $\theta$ and $V_m$ refer to the voltage angle and magnitude; $P_g$ and $Q_g$ are generator injections; $P_d$ and $Q_d$: load injections; $C_g$: a matrix whose $(i,j)^{th}$ element is 1 if generator $j$ is located at bus $i$ and 0 otherwise.

## 4 Tests and Results

To perform the analysis in this work, a standard 33-bus radial distribution network, taken from [16], is employed. The rated voltage, active and reactive power demand in this system are 12.66 kV, 3.715 MW and 2.3 MVAr, respectively. Further information about this system can be found in [16, 17]. The voltage at any node in the system is allowed to deviate up to a maximum value of ±5%. All DGs are assumed to have a power factor of 0.95. The costs of the generators at the feeders are given by polynomial functions $C(P) = 150 + 20P + 0.01P^2$ €/h. For integrating the DG as a PV bus and add to the cost of the system given by the OPF, one polynomial function was taken into consideration $C(P) = 8P$€/h. For the sake of simplicity, the availability of wind and solar power sources is assumed to be uniform across the nodes in the system. Furthermore, Node 1 is considered as the reference node, and a base power of 100 MVA is used. A total of six cases, designated as Case $i$, are considered in the analysis. Case 1 is the base case; Case 2 considers reconfiguration; Case 3 is a scenario where minimizes only losses. Cases 4, 5 and 6 all handle reconfiguration along with DG integration but they differ in that, in Case 4, the considered DGs are capable of producing active power as well as injecting and absorbing reactive power, Case 5 considers DGs that can only produce active power, and the DGs considered in Case 6 are capable of only producing or consuming reactive power. The variations of system parameters when considering different are analyzed as shown in Table 1. Comparing Case 1 with Case 2, we see that reconfiguration slightly lowers the total costs and losses. The total cost reduction is about 0.54%. The active and reactive power losses are also reduced by 61.59% and 17.38%, respectively. In Fig. 1, the voltage profiles of the system after reconfiguration and the base case are shown. Clearly, the positive contribution of reconfiguration to the voltage profiles can be observed. In addition, in Table 1, there is little difference between minimization of losses and minimization of costs, the difference is approximatcly 0.0057% for total costs, 0.5147% for active power losses and 0.5695% for reactive power losses. In Table 2, the only difference lies between the opened branches that is 9–11 in Case 2, and 10–11 in Case 3. Only one branch is different, and almost leads to similar fitness function values. There is a small difference and we can conclude that these configurations are minimized but may not be the global optima. Analyzing the results in Table 1 further, there is a significant difference in total costs and in total losses in Cases 4 and 5 compared to the first three cases. The effects of having DGs capable of producing only active or reactive power are also seen in Fig. 2. With active-power-only DGs, we can also have a better voltage profile, not as linear as in Case 4 but significantly better than in the base case. Deploying reactive-power-only DGs also has some noticeable impact in systems losses, and voltage profiles. In Case 5, the reduction in total costs is 20.65% compared with Case 2 and 21.08% when compared with Case 1. Compared with Case 2, active and reactive power losses are slashed by 76.22% and 66.68%, respectively. As in Case

6, there is no big impact in total costs, only 0.87% when compared with Case 1 but, there is a huge difference in terms of losses. Compared with Case 1, the active and reactive power losses are reduced by 53.04% and 32.69%, respectively. Although the costs are slightly increased, the benefits of having DGs with this technology are evident with the substantial reduction of losses and improvement in voltage profile. The placement and sizing of DGs may not be optimal because of the solution method. However, there are small differences from one generation type to another, probably indicating the closeness of the solution to the optimal one.

**Table 1.** Values of relevant system variables corresponding to the different cases.

| Cases | Total Cost [€/h] | Total active power losses [MW] | Total reactive power losses [Mvar] | Total installed DG size [MVA] | Computation time [s] |
|---|---|---|---|---|---|
| 1 | 228.1816 | 0.1865 | 0.0999 | 0 | – |
| 2 | 226.9463 | 0.1249 | 0.0825 | 0 | 24.423398 |
| 3 | 226.9593 | 0.1256 | 0.0830 | 0 | 27.102581 |
| 4 | 179.7385 | 0.0023 | 0.0020 | 23 | 32.926209 |
| 5 | 180.0747 | 0.0443 | 0.0333 | 17 | 39.599780 |
| 6 | 226.1954 | 0.0876 | 0.0672 | 15 | 34.074488 |

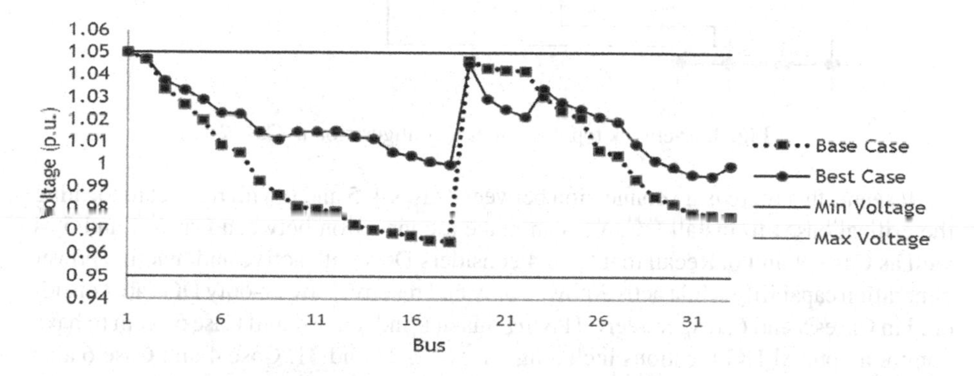


**Fig. 1.** Voltage profiles corresponding to Cases 1 and 2.

**Table 2.** Branches opened and DG location in the 33-bus distribution system

| Cases | Opened branches | DG bus location |
|---|---|---|
| 1 | 21-8; 9-15; 12-22;18-33;25-29 | – |
| 2 | 7-8; 9-10; 14-15; 32-33;25-29 | – |
| 3 | 7-8; 10-11; 14-15; 32-33; 25-29 | – |
| 4 | 7-8; 11-12; 15-16; 21-22; 28-29 | 4; 9; 16; 17; 20; 22; 23; 24; 26; 30; 31; 32 |
| 5 | 6-7; 11-12; 14-15; 26-27; 32-33 | 5; 8; 12; 13; 14; 17; 23; 25; 28; 31; 33 |
| 6 | 7-8; 8-9; 14-15; 28-29; 32-33 | 5; 6; 13; 15; 22; 24; 30; 32 |

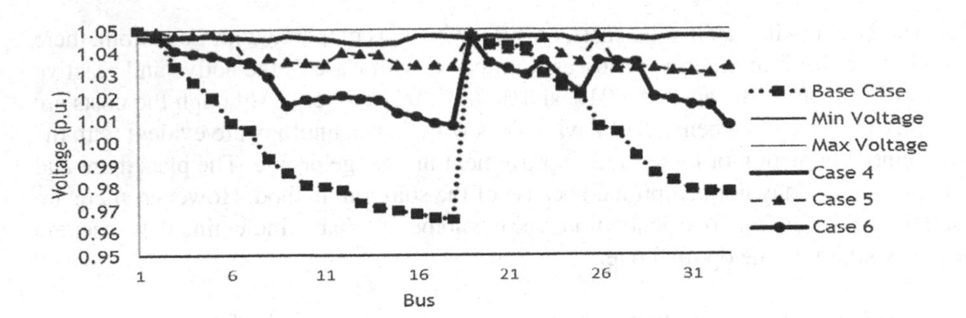


**Fig. 2.** Voltage profiles corresponding to Cases 1, 4, 5 and 6.

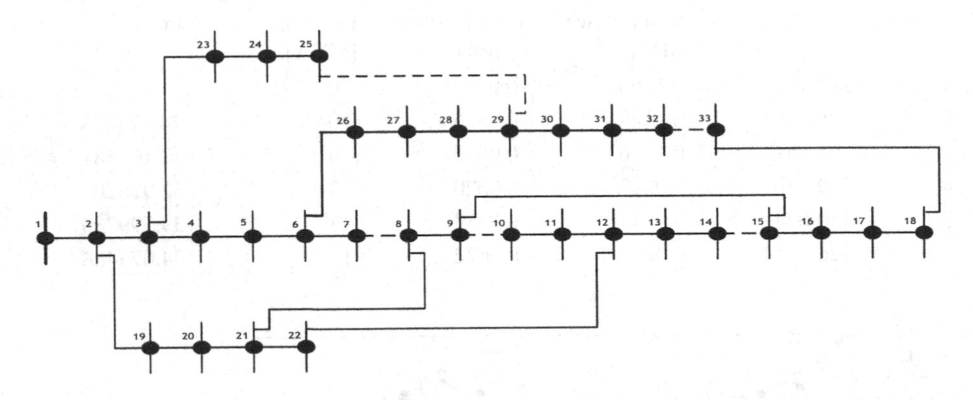


**Fig. 3.** Network topology after reconfiguration in Case 2

It seems that there is no connection between Cases 4, 5 and 6 with respect to locating the critical buses to install DG. We can make a connection between Cases 4 and 5 as well as Cases 4 and 6. Recall that Case 4 considers DGs with active and reactive power generation capability while active-power-only and reactive-power-only DGs are considered in Cases 5 and 6, respectively. Having this in mind, Case 4 and Case 5 seem to have common optimal DG locations including buses 17, 23 and 31. Case 4 and Case 6 also have common "optimal" DG locations such as buses 22, 24, 30 and 32. When we look at the demand and at the total installed size of DG, there seems to be a lot of discrepancies among the different cases. The configuration outcome of Case 2 is shown in Fig. 3.

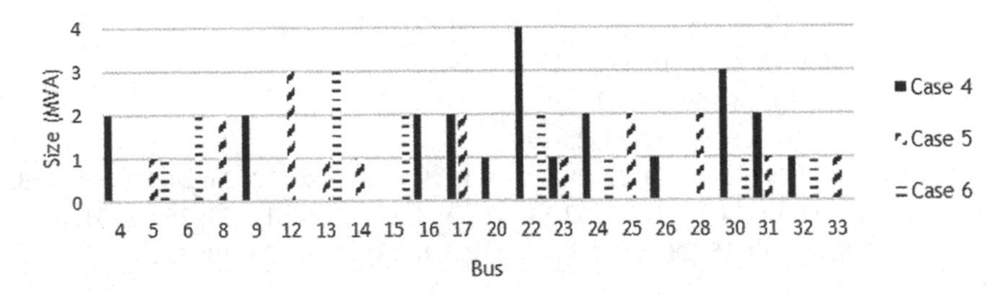


**Fig. 4.** Location and size of DGs in Cases 4, 5 and 6.

Figure 4 shows the DG location and size corresponding to Cases 4, 5 and 6. Figure 5 shows the configuration and DG placement outcomes for Case 4.

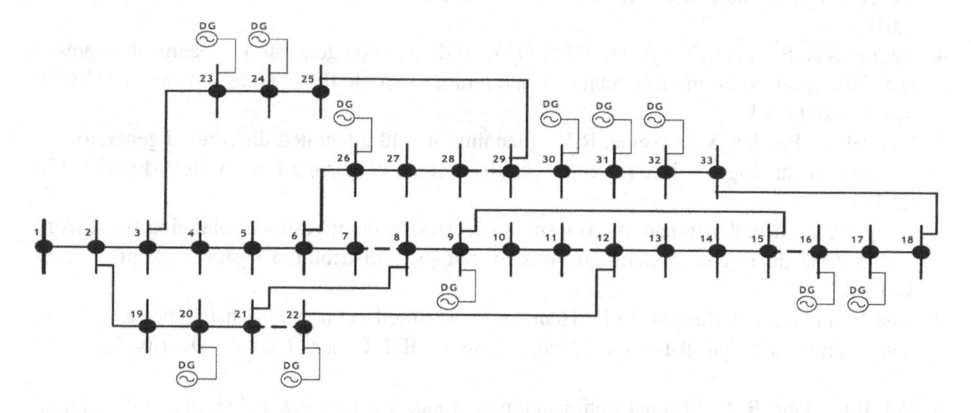


**Fig. 5.** Network topology and DG placement in Case 4.

## 5 Conclusions

In this work, an improved mathematical model has been developed with the aim of performing quantitative and qualitative analysis with regards to the impacts of dynamic network reconfiguration on the penetration and utilization level of variable renewable power and overall system performance. The developed model has the overall cost minimization as an objective, and is based on the most accurate AC network model. A genetic algorithm has been used to efficiently handle the combinatorial nature of the reconfiguration problem. Generally, the outcome of the analysis shows the capability of dynamic network reconfiguration in increasing the penetration and utilization level of vRES as well as improving the overall system performance in terms of enhanced voltage profile, reduced costs and losses.

**Acknowledgment.** This work was supported by FEDER funds through COMPETE 2020 and by Portuguese funds through FCT, under Projects SAICT-PAC/0004/2015 - POCI-01-0145-FEDER-016434, POCI-01-0145-FEDER-006961, UID/EEA/50014/2013, UID/CEC/50021/2013, and UID/EMS/00151/2013. The research leading to these results has received funding from the EU Seventh Framework Programme FP7/2007-2013 under grant agreement no. 309048.

## References

1. Santos, S.F., Fitiwi, D.Z., Shafie-Khah, M., Bizuayehu, A.W., Cabrita, C.M.P., Catalao, J.P.S.: New multistage and stochastic mathematical model for maximizing RES hosting capacity—part I: problem formulation. IEEE Trans. Sustain. Energy **8**(1), 304–319 (2017)
2. Paliwal, P., Patidar, N.P., Nema, R.K.: Planning of grid integrated distributed generators: a review of technology, objectives and techniques. Renew. Sustain. Energy Rev. **40**, 557–570 (2014)

3. Colmenar-Santos, A., Reino-Rio, C., Borge-Diez, D., Collado-Fernández, E.: Distributed generation: a review of factors that can contribute most to achieve a scenario of DG units embedded in the new distribution networks. Renew. Sustain. Energy Rev. **59**, 1130–1148 (2016)
4. Georgilakis, P.S., Hatziargyriou, N.D.: Optimal distributed generation placement in power distribution networks: models, methods, and future research. IEEE Trans. Power Syst. **28**(3), 3420–3428 (2013)
5. Paliwal, P., Patidar, N.P., Nema, R.K.: Planning of grid integrated distributed generators: a review of technology, objectives and techniques. Renew. Sustain. Energy Rev. **40**, 557–570 (2014)
6. Hung, D.Q., Mithulananthan, N., Bansal, R.C.: An optimal investment planning framework for multiple distributed generation units in industrial distribution systems. Appl. Energy **124**, 62–72 (2014)
7. Abu-Mouti, F.S., El-Hawary, M.E.: Heuristic curve-fitted technique for distributed generation optimisation in radial distribution feeder systems. IET Gener. Transm. Distrib. **5**(2), 172 (2011)
8. Pal, B.C., Jabr, R.A.: Ordinal optimisation approach for locating and sizing of distributed generation. IET Gener. Transm. Distrib. **3**(8), 713–723 (2009)
9. Maciel, R.S., Rosa, M., Miranda, V., Padilha-Feltrin, A.: Multi-objective evolutionary particle swarm optimization in the assessment of the impact of distributed generation. Electr. Power Syst. Res. **89**, 100–108 (2012)
10. Vinothkumar, K., Selvan, M.P.: Fuzzy embedded genetic algorithm method for distributed generation planning. Electr. Power Compon. Syst. **39**(4), 346–366 (2011)
11. Ugranlı, F., Karatepe, E.: Convergence of rule-of-thumb sizing and allocating rules of distributed generation in meshed power networks. Renew. Sustain. Energy Rev. **16**(1), 582–590 (2012)
12. Murty, V.V.S.N., Kumar, A.: Optimal placement of DG in radial distribution systems based on new voltage stability index under load growth. Int. J. Electr. Power Energy Syst. **69**, 246–256 (2015)
13. Ma, Y., Tong, X., Zhou, X., Gao, Z.: The review of smart distribution grid. In: 2016 IEEE International Conference on Mechatronics and Automation (ICMA), pp. 154–158 (2016)
14. Zhou, X., Cui, H., Ma, Y., Gao, Z.: Research review on smart distribution grid. In: 2016 IEEE International Conference on Mechatronics and Automation (ICMA), pp. 575–580 (2016)
15. Hayfa, S., Omar, K., Hsan, H.A.: Optimal power distribution system reconfiguration using genetic algorithm. Presented at the 16th International Conference on Sciences and Techniques of Automatic Control and Computer Engineering, Monastir, Tunisia (2015)
16. Cruz, M.R.M.: Benefits of coordinating distribution network reconfiguration with distributed generation and energy storage systems. Universidade do Porto (2016)
17. Cruz, M.R.M., Fitiwi, D.Z., Santos, S.F., Catalão, J.P.S.: Influence of distributed storage systems and network switching/reinforcement on RES-based DG integration level. Presented at the 2016 13th International Conference on the European Energy Market (EEM), OPorto (2016)

# Data Fusion of Georeferenced Events for Detection of Hazardous Areas

Sérgio Onofre[1(✉)], João Gomes[2], João Paulo Pimentão[2], and Pedro Alexandre Sousa[2]

[1] Holos SA, Caparica, Portugal
onofre@holos.pt

[2] Faculty of Sciences and Technology, NOVA University of Lisbon, Monte de Caparica, 2829-516 Almada, Portugal
jfn.gomes@campus.fct.unl.pt, {pim,pas}@fct.unl.pt

**Abstract.** When dealing with events in moving vehicles, which can occur over widespread areas, it is difficult to identify sources that do not derive from material fatigue, but from situations that occur in specific spots. Considering a railway system, problems could occur in trains, not because of train's equipment failure, but because the train is crossing a specific location. This paper presents a new smart system being developed that is able to generate geo-located sensor-data; transmit it for smart processing and fusing to the inference engine being built to correlate the data, and drill-down the information. Using a statistical approach within the inference engine, it is possible to combine results collected over long periods of time in a "heat-map" of frequent fault areas, mapping faulty events to detect hazardous locations using georeferenced sensor data, collected from several trains that will be integrated in these maps to infer high probability risk areas.

**Keywords:** Data-fusing · Smart system · Industry 4.0 · IoT · Event geographic position · Data correlation · Forecast

## 1 Introduction

The association of faults to geographic locations can be used to alert and provide support to incident intervention crews to deal with current issues and, especially to prevent forecast events in a specific location in a nation-wide (or trans-national) railway network.

Deciding on whether a location is faulty or not depends on the quotient between the number of trains that crossed the railway and how much breakdowns occurred around that location. A statistical approach is used where data is be auto correlated based on past events. To get the information needed to feed the system's databases it uses a sensor fusing system, where it combines and study the data collected from the trains that populate the railway network.


L.M. Camarinha-Matos et al. (Eds.): DoCEIS 2017, IFIP AICT 499, pp. 81–89, 2017.
DOI: 10.1007/978-3-319-56077-9_7

This system will be integrated in a multi-agent architecture [1] used in a distributed surveillance system (DVA[1]), where events, sensors and human resources are georeferenced. New features will be added to the system such as reasoning in terms of what should be considered a correlated event (geographically speaking) and forecast of events using the knowledge acquired from previous events and data.

This paper starts by presenting a state of the art in data fusion architectures, Geolocation sensing, data correlation, event forecasting and finally presenting the foreseen method to integrate this system in the DVA for providing data correlation and forecasting functionalities.

## 2 Contribution to Smart Systems

The work described in this paper will encompass the collection of real-time raw data from sensors, which are then sent to a central processing unit that will perform reasoning and prediction based on the knowledge acquired data along a large period of time. In this paper a third-generation smart system [2], is presented, that performs human-like perception and acts autonomously, i.e. without any human decision. Moreover, it is equipped with the capability to predict and adapt according to this readings.

In practice, one could argue that the current system encompasses all the components that are the base for the concept of "Industry 4.0", although applied to mobility: the cyber-physical systems that is achieved by the monitoring of the trains and the availability of the data through Internet protocols; the "IoTization" of trains allowing them to be remotely sensed and monitored, and finally the application of Cloud Computing and Big Data technologies and methodologies to both gather and process the high volumes of information that each train produces every second.

## 3 State of the Art

### 3.1 Data Fusion Architectures

Nowadays sensor nodes integrate multiple capabilities that include sensing, global positioning system (GPS), computing and communication. Fusing multiple simple sensor nodes, one can deliver scalable, reliable and more complex sensor based systems.

Multiple Architectures have been proposed for sensor fusing. First example is the one proposed in Fig. 1(a) where static sensors nodes are wired to a central gateway that handles all the requests and processing. An implementation of this type is [3] where all the equipment in an office (printers, thermometers, ventilation, etc.) are connected to a central gateway. All the clients that require information connect to the gateway that proceeds to service the requests. Figure 1(b) is used when the data collected volume is large and therefore needs some local processing before sending the compressed/processed version to the server. Finally, when different layers of processing are responsible for different aspects of data manipulation there is a hierarchical processing as

[1] Advanced Surveillance System, for more information see: dva.holos.pt.

presented in Fig. 1(c), in which the collected data passes through several layers of data processing and manipulation before reaching the central server.

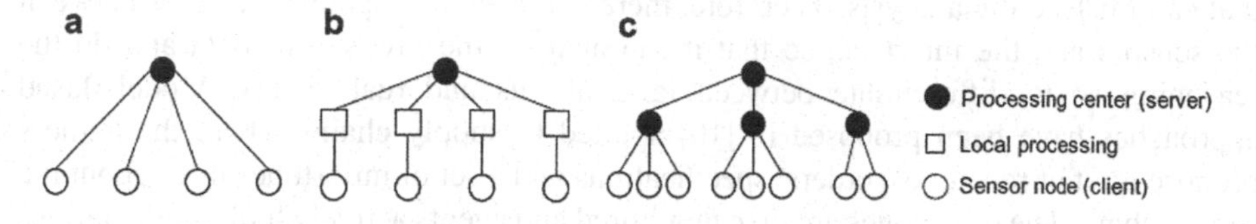


**Fig. 1.** Typical sensor fusing architectures [4].

Recently there has been a more hybrid approach where mobile agents are used in combination with static nodes [5]. An example of this is the Interoperable-agent-mode for sensor network [4] that is FIPA standards compliant [6, 7], which considers static wireless nodes distributed in an area and a mobile agent that goes physically to the sensor nodes to collect the data. The mobile agent is responsible for making his own itinerary around the sensor nodes and when the battery runs low he returns to the central station.

As another example, the DVA surveillance system [8] is a human-machine collaborative distributed system where static nodes are in place and in case of alert the mobile agents (be it policeman, security, firefighter or civil protection) are called in to deal with the occurrence.

### 3.2 Geolocation Relation

The next question that arises is how one should relate events, since the goal is to associate locations to breakdowns. It makes sense to define a radius around said events so that it is possible to search for previous nearby occurrences (since location is not 100% accurate). Speed (Velocity) of the train is an important aspect to consider in defining said radius since the faster the train goes the more difficult it is to track down the exact location of the breakdown, therefore, the radius should be wider.

Several approaches have been developed in terms of geolocation radius but in different contexts such as [9] where automatic safety envelopes (that can be both dynamic or static) are defined around different construction equipment based on their width, length and velocity, the goal here is to alert the workers that can, without their knowledge, be putting their safety in jeopardy by going too near dangerous equipment. These envelopes radius are created according to the Eq. (1).

$$R = \sqrt{(\frac{L}{2} + r + \frac{vt}{2})^2 + (r + \frac{w}{2})^2} \text{ where } \quad v = |\bar{v}| \tag{1}$$

Where L, W, v are the length, width and velocity (both in X and Y axis) respectively of the moving equipment. The r represents the Safe distance of the equipment (which means the distance it will go before completely stopping), t represents the time it will take to break.

### 3.3 Event Processing

To make the system reliable and robust it is very important not to have false positives that can mislead the analysis. Therefore, there is a need to implement a layer between the sensors and the interface, so that it can analyze the raw sensor data and do the reasoning so to differentiate between false alarms and real events. Model-Based approaches have been proposed in [10] applied to supply chains where the authors propose a set of resources, orders, specifications and a set of milestones throughout the supply chain. The milestones are then monitored and agents actively look for deviations of the expected models. Fuzzy logic has also been widely used in the literature to solve these kinds of problems.

In [11] the authors define multiple arrays of input variables (Vi), their domains (DDV), and their constraints (Φ) the values that the system receives from each individual set of input variables are classified in terms of "normality" (Ncn) which is a scalar between [0,1] being 1 the most normal situation and the 0 the most abnormal. Several solutions have been proposed to make decisions towards this "normality factor". One simple solution as presented in [11] is finding the minimum value of normality (Ncn) and consider only this value to classify the situation, however, this is limiting because it is only considered one set of input variables while the others are discarded, which means that there is the possibility that values with information important for the classification are being ignored. Choquet Integral, as defined in [11] proposed, to define specific weights (**μ(Ai)**) to normality concepts and correlate them. Equation (2) it is shows how Choquet Integral correlates the normality factors.

$$C_\mu(N^j_{c_1}, N^j_{c_2}, \ldots N^j_{c_n}) = \sum\nolimits_{i=1}^{n} (N_{(i)} - N_{(i-1)}) \times \mu(A_i) \quad (2)$$

This means that once the normality concepts are defined (Nc1, Nc2, … Ncn), it is needed to relate them by defining weights. The conclusions are made based on the final value of the equation. The OWA aggregation operators presented in [12], use the normality factors times a weight this makes a junction between different criteria (normality factors) using weights. In (3) is presented how the OWA aggregators are calculated.

$$OWA(a_1, a_2, \ldots a_n) = \sum\nolimits_{j=1}^{n} w_j.b_j \quad (3)$$

Where (a1, a2, …, an) represents the values that are collected to make the decision, wj represents the weights and bj is the ordered set of (a1, a2, … an).

### 3.4 Event Prediction (Forecasting)

In [12] it is presented a way to predict events in real time streams of data. The authors look for event signatures (what happened before the event) and make sequences of readings that lead to the event. The readings are then compared with the sequences in the database that are known to cause breakdown events. Since there can be a lot of events signatures in the database it is not feasible to compare every sensor reading with the values in the database, the authors, propose to look for triggers. These triggers are then used in forward rules which means, for example, that If Trigger 1 and Trigger 2 are found THEN event C will happen. In the paper it is also proposed to once a new signature is found to generate readings (real or unreal) that lead to this event. The idea is to have a model of the event.

# 4 Architecture

The DVA system as explained in [13] is "is a Geo-referenced multi-agent surveillance system, composed by several agents: Sensor agent – provides sensor information; Processor agent – transforms sensor information into parameters; Inference agent – uses parameters in rules for event detection; Action agent – executes predefined actions for each event; Backup agent - stores all the system information; Interface agent – shows (in maps) the values of the sensors, events, actions and system status; Mobile agent – Associated with a human, equipped with a mobile device who is responsible to perform events' actions, such as confirming the event or handling the event; Monitor agent – monitors all system's agents, ensuring correct system performance." In this paper, two new agents to be added to this architecture are presented: "**correlate agent**" (that makes links between events) and the "**forecast agent**" (that looks for current sensor readings and predicts problems that might arise in the near future).

## 4.1 Proposed Architecture

The Fig. 2 represents the proposed changes (in red) to the DVA architecture. The "Correlate" Agent will be activated every time a new event is detected. It will look for previously events nearby (stored in the Backup Agent) and if it detects a similar event in the area, it will create a new type of event (Correlated Event) that will be shown on

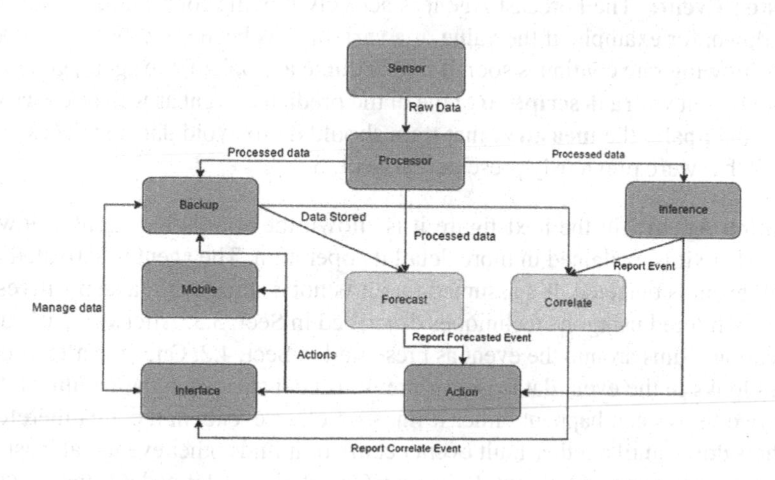


**Fig. 2.** Architecture proposed. (Color figure online)

the interface. The Forecast Agent is receiving the data from the processor and previously readings stored in the Backup Agent to make predictions about potential problems.

It will also generate a new event type (Forecasted Event) in the system so it can make decisions to prevent damage.

## 4.2 Proposed Changes to the Architecture

**Correlated Event:** The correlated event, links separate events to geographical areas. It is described by: location of the **center of the event** (since the goal is to map these events, it is need a geographical center given by the center of mass of the composing events), a **geographical radius** (distance between the center of mass and the further away event), a **time of occurrence** (when it was detected), **number of occurrences** (how many events it is linking) and a **vector with the events correlated**.

To link the geographical events an adaptation is made from the Eq. (1) considering that the width as well as the breaking time (time to full stop) of the train are irrelevant since the events will be detected while the train is still moving.

$$R = \frac{L}{2} + \frac{v}{\delta} \text{ where } \quad v = |\bar{v}| \tag{4}$$

Where L represents the length of the train, and the v (composed by its X and Y component) represents the velocity of the train at the time of the fault detection. Delta ($\delta$) is a parameter, between 0 and 1 that allows the adjustment of the radius according to what is verified in practice. For the time being it is considered to be 0.25 s which at a standard train speed of 100 km/h will produce a length of roughly 110 m which is half the usual length of a train. Events are therefore correlated if the circumferences centered at the location of each event with their respective radius (R) intersect within a timeframe (the timeframe being a parameter that will be adjusted as more and more information is collected).

**Forecasted Event:** The Forecast Agent is actively looking for readings that can lead to breakdown, for example, if the value of a variable has been steadily rising to a point that if its growing rate continues soon it will become a problem, the agent, generates an Forecasted event with a **description** of the of the predicted event as well as the **readings relevant** and finally the **measures** that train should do to avoid damage. These are the "triggers" that were previously presented in Sect. 3.4.

**Correlation Agent:** In the next figure it is shown the correlation agent's flow chart (Fig. 3.) where it is explained in more detail its operation. The agent is activated once a new fault event is detected. It's assumed that it is not dealing with false positives since it is already filtered using the techniques described in Sect. 3.3. Afterwards it's defined a geolocation radius around the event as presented in Sect. 4.2. Once the area is defined the agent looks in the event database for previous occurrences in the vicinities. Subsequently two things can happen: either it finds no other events in the area therefore the agent shuts down until another fault event occurs or it finds other events (at least 2) and generates a new Correlation Event. If the new Correlation is defined it is then sent to the

Interface Agent that flags this area in a form of heat map that has the warm colors to show the areas of found malfunctioning.

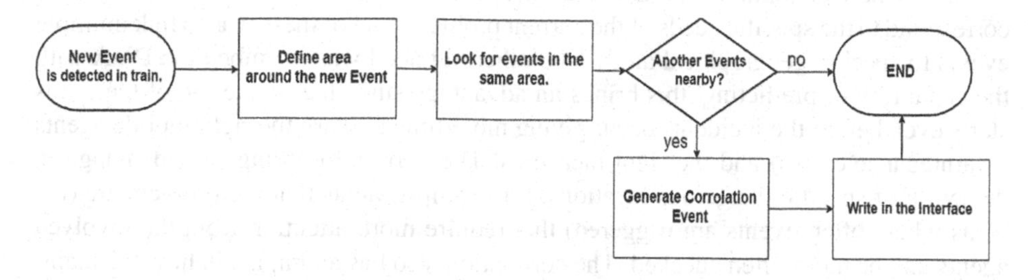


**Fig. 3.** Flow chart correlation agent.

**Forecast Agent:** The Forecast Agent looks for patterns in the readings before breakdown events (called event signatures), in brief, every time a fault is detected it will make a history of readings just before the event. These readings will then be used as comparison for the active trains and if a pattern that previously led to breakdown is a found in an active train this agent will generate Forecasted Events.

There are critical points in the railway grid where power sub-stations are changed, these points usually are associated with high voltage differences (more than 5 kV). These variations tend to cause peeks of currents/voltage in the trains' electrical systems, this can eventually become a source of malfunctioning or breakdown.

The forecasting needs to be done on multiple active trains in the grid, this means that it needs threads to take care of these multiple tasks. Due to the high volume of data that it is collected and sent (at a rate of 3000bps), the Apache Hadoop's Map-Reduce Paradigm (from big data analysis) is used. This gives the ability to separate trains into different parallel tasks. The Hadoop then processes all the information sent by the trains and looks for patters that can be known to be troublesome. The results are shown in the interface with a delay associated with the receiving the information, the processing and the actual writing in the interface, however, this delay is not critical since the goal of this system its not to control the train in real time, but to give insight to the maintenance crews of potential problems within the electrical grid and/or with the trains. Every time there is a forecasted event of breakdown this agent produces a report of how many critical points did the train endure before. The idea is to give an average of the number of how many critical points the trains can endure before needing a maintenance team to check its electrical protections. More detailed information about Apache Hadoop can be found in [14]. To extract knowledge the data mining techniques presented in Sect. 3.4 are used.

## 5 Conclusion

DVA is a developed and tested system that protects both people and goods from all sorts of dangers (both natural and human caused) using statically nodes spread out across vast areas. In the presented approach, some of its behavior had to be adapted because we are now considering moving trains, which makes the sensors' positions change over time.

So, no static location relations to identify events are now available. And this was a cornerstone for the inference engine of DVA.

In this new adaptation, the DVA architecture had to be adjusted so that it could correspond to the specific needs of the current problem, one of these was to link multiple events to specific geographical areas. The other big need was to embed the DVA with the capability of predicting, this brings an advantage since the protection of the trains starts even before the incidents occur giving more time to warn the right mobile agents (Maintenance crews) and the right measures. The correlation brings an advantage in detection of specific dangerous locations, for example, areas that are problematic (i.e. areas where often events are triggered) that require more attention from the involved agents and be more often checked. The correlation also has an impact in how the maintenance crews are spread across their working, for example, after a careful study of the common problem areas of the agents (maintenance crews) should be closer to where the problems usually appear. This architecture is currently under implementation and results are expected to be presented soon.

## References

1. Jennings, N.R.: An agent-based approach for complex software systems. Commun. ACM **44**(4), 35–41 (2001)
2. Gessner, T., Messe, M., Gmbh, F.: Smart systems integration and simulation. Integr. VLSI J. 5–22 (2016)
3. Suarez, J., Quevedo, J., Vidal, I., Corujo, D., Garcia-Reinoso, J., Aguiar, R.L.: A secure IoT management architecture based on information-centric networking. J. Netw. Comput. Appl. **63**, 190–204 (2016)
4. Biswas, P.K., Qi, H., Xu, Y.: Mobile-agent-based collaborative sensor fusion. Inf. Fusion **9**(3), 399–411 (2008)
5. Xu, Y., Qi, H.: Distributed computing paradigms for collaborative signal and information processing in sensor networks. J. Parallel Distrib. Comput. **64**(8), 945–959 (2004)
6. FIPA97 Foundation for intel. Physical agents spec. Part 1 Agent Management. Quality (1997)
7. FIPA 97 Specification Part 2: Agent Communication Language (1997)
8. Onofre, S., Sousa, P., Pimentao, J.P.: Multi-sensor geo-referenced surveillance system. In: 10th International Symposium on Mechatronics and its Applications (ISMA), pp. 1–6 (2015)
9. Li, Z., Zhou, S., Choubey, S., Sievenpiper, C.: Failure event prediction using the Cox proportional hazard model driven by frequent failure signatures. IIE Trans. **39**(3), 303–315 (2007)
10. Fernández, E., Toledo, C.M., Galli, M.R., Salomone, E., Chiotti, O.: Agent-based monitoring service for management of disruptive events in supply chains. Comput. Ind. **70**, 89–101 (2015)
11. Albusac, J., Vallejo, D., Castro-Schez, J.J., Glez-Morcillo, C., Jiménez, L.: Dynamic weighted aggregation for normality analysis in intelligent surveillance systems. Expert Syst. Appl. **41**(4), 2008–2022 (2014). PART 2
12. Yager, R.R.: On ordered weighted averaging aggregation operators in multicriteria decisionmaking. **18**(1), 183–190 (1988)

13. Dias, B., Rodrigues, B., Claro, J., Pimentão, J.P., Sousa, P., Onofre, S.: Architecture and message protocol proposal for robot's integration in multi-agent surveillance system. In: Cornelis, C., Kryszkiewicz, M., Ślęzak, D., Ruiz, E.M., Bello, R., Shang, L. (eds.) RSCTC 2014. LNCS (LNAI), vol. 8536, pp. 366–373. Springer, Heidelberg (2014). doi: 10.1007/978-3-319-08644-6_38
14. Liroz-Gistau, M., Akbarinia, R., Agrawal, D., Valduriez, P.: FP-Hadoop: efficient processing of skewed MapReduce jobs. Inf. Syst. **60**, 69–84 (2016)

# Systems Analysis

# Student's Attention Improvement Supported by Physiological Measurements Analysis

Andreia Artífice(✉), Fernando Ferreira, Elsa Marcelino-Jesus, João Sarraipa, and Ricardo Jardim-Gonçalves

CTS, UNINOVA, DEE/FCT, Universidade Nova de Lisboa, Lisbon, Portugal
{afva,flf,ej,jfss,rg}@uninova.pt

**Abstract.** The focus of the most recent theories of emotional state analysis is the Autonomic Nervous System. Those theories propose that sympathetic and parasympathetic nervous systems interact antagonistically accordingly to each emotional state implying variations of interbeat intervals of consecutive heart beats. Emotional arousal and attention can be inferred based on the electrocardiogram (ECG) specifically through Heart Rate Variability (HRV) analysis, including the Low Frequency (LF), High Frequency (HF), and ratio LF/HF. The aim of this study is to analyze the impact of classic background music, in students' emotional arousal and attention, and performance in the context of e-Learning training courses. As a result, it is foreseen the development of a system integrating wearables to smoothly gather the mentioned biosignals, which will be able to sense user's emotions to further automatically propose recommendations for better learning approaches and contents, aiming student's attention improvement.

**Keywords:** eLearning · Heart rate variability · Emotional arousal · Attention

## 1 Introduction

The present research work analyses the role of classic background music in the students' attention, emotional arousal, and performance in the e-Learning environment. The proposed approach is based on ECG readings with the aim of feeding a smart system able to perform analysis and propose recommendations for improved learning approaches.

In the context of Internet based teaching, E-learning is becoming an innovative approach for delivering well-designed, learner-centered, interactive methodology. In that context E-learning is providing a learning environment to anyone, anyplace, anytime by utilizing the attributes and resources of various digital technologies, and for that it is growing its importance in the market [1]. The near future of E-learning will, most probably, be supported by the integration of sensing, diagnose, describe and qualify a given situation. Such exercises will be related with the specification of its related Smart Systems. The aimed characteristic of that Smart Systems would be the potential to sense using biosensors able to measure various physiological parameters such as heart rate (HR), galvanic skin response (GSR), electroencephalogram (EEG) and electrocardiogram (ECG) [2, 3].


L.M. Camarinha-Matos et al. (Eds.): DoCEIS 2017, IFIP AICT 499, pp. 93–102, 2017.
DOI: 10.1007/978-3-319-56077-9_8

The hereby-presented research work follows the traditional research method and encloses the following research question: Does eLearning training courses with classic background music influence emotional states and will it have an impact in attention and performance?

Concerning the proposed research question, aforementioned, we argue, by hypothesis that if we define a prototype, able to conduct analysis of arousal and attention based on biosignals, then the process to determine the emotional arousal and attention of learning students could be seamlessly executed. Such development would be able to proof the usefulness of background music in eLearning materials.

This paper is organized as follows: next section is dedicated to Smart Systems and e-Learning. The state of the art, concerning background music and Emotional State and Attention detection using Electrocardiogram is presented in Sect. 3, followed by the experiment in Sect. 4. Section 5 reveals the results and finally Sect. 6 presents the conclusions of the study.

## 2 Smart Systems and E-Learning

Smart systems are understood as systems able to sense, diagnose, describe and qualify a given situation as well as be capable to mutually address and identify each other. They are also able to interface, interact and communicate with their environment and with other smart systems [4]. Sensors and actuators are part of the Smart Systems Ecosystem that will become, most of the times, its central component. Different approaches have been proposed in the context of E-learning, among those we highlight the development of systems based on context and emotional sensing.

Zarrad and Zaguia [5] propose a smart system with an architectural approach in which modalities are selected dynamically based on the operating context. The system considers context from different features including lighting, noise level, network connectivity, communication costs, communication bandwidth, and the social and medical situations.

Emotions have an important role in learning, and are significantly related with students' motivation, learning strategies, cognitive resources, self-regulation and academic achievements [6]. Several contributions were developed based on the identification of a user's emotional state with the goal of improving the eLearning experience. The Framework for Improving Learning Through Webcams And Microphones (FILTWAM) [7] was developed with the aim of interpreting the emotional state of users in an e-learning environment by using webcams and microphones. This study focused only on facial mood detection and presents a structured framework composed by the following layers: learner, device, application and data with the most complex application layer on top of it and including an affective computing tool. In that same direction, Santos et al. [8] conducted a research in computer assisted language learning, using an Arduino based infrastructure that could sense the learners' affective state and deliver interactive recommendations. For that it was developed the Ambient Intelligence Context-aware Affective Recommender Platform (AICARP), a platform similar to the one that will be developed in the context of this study. A particularity of the platform is the sensor module that receives biosignals from the learner and information from the environment and a recommendation module that takes advantage of the intelligence of the environment using internal and/or external actuators to offer interactive personalized support to the learner. This overview brings the relevant smart systems related to the work developed in the scope of the proposed research question. In particular we look to design a smart system that analyses data from bodily-connected sensors, analyses such data and by reasoning over selected algorithms identifies a person's emotional state. Such systems, while connected with other similar smart systems, can produce markers that give indications of common properties of different people while listening to music and performing a learning task (e.g. attending an E-learning course).

# 3 State of the Art

## 3.1 Background Music

The "Mozart effect", i.e. the scientific finding that listening to Mozart's piano sonata enhanced spatial reasoning [9], was supported by McCuntheon [10], while other studies found no support [11, 12]. Despite that, the study conducted to new research and subsequent hypothesis in the field. A study conducted by Thompson et al. [13] suggested that the "Mozart effect" performs a sort of manipulation of arousal and mood that conducted to enhanced performance. According to Day et al. [14], the effect of background music is rather specific and thus, needs to be tested on a selected target situation [14].

Concerning computer based environments, it was visible that, when reading news from a pocket computer in a distracting environment, the reading rate and efficiency was improved in the presence of fast music than slow music [5]. A study by Thompson et al. [13] revealed that background music disrupted the ability of writing fluently.

In order to study the effect of background music in virtual environments for learning Richards et al. [15], it was created a specific environment for a history lesson. Results revealed that listening to the soundtrack of 'Oblivion' conducted to an improvement in the number of facts remembered by the participants.

It is important to notice that the majority of the reported findings about background music are not representative of computer-based environments. Thus, considering the actual impact of technology in daily life it is crucial to continue progressing on that specific research field.

## 3.2 Emotional State and Attention Detection Using Electrocardiogram

Physiological measurements, along with face recognition, or other measurements (e.g. posture and body movements) are the indirect clues about a person's emotional states since those cannot be assessed with some sort of direct measurements. Because of that, biosignals, such as electrocardiogram (ECG), electromyogram (EMG), electroencephalogram (EEG), galvanic skin response (GSR), blood volume pressure (BVP), heart rate (HR) or heart rate variability (HRV), temperature (T), respiration rate (RR), become the most important vital signals of the human body for an emotional assessment. Although the majority of the studies use different biosignals in conjunction with ECG in order to detect the emotional states [16–18], there are studies focused only in relating ECG with the expression of human emotions [3, 19].

Emotions and Autonomic Nervous System (ANS) are interrelated. ANS activity is viewed as the major component of the emotions in many recent theories about emotional assessment [20]. ANS is composed by the sympathetic and parasympathetic nervous system components [20]. The sympathetic nervous system is stimulated by changes in external environment, reacting mostly to any situation perceived as an emergency [20]. The primary function of parasympathetic nervous system is for example to coordinate activities related to the restoration and conservation of the body [21]. Both systems interact antagonistically causing the variations in interbeat intervals of consecutive heart beats [3]. Parasympathetic nervous system is more active in calm periods decreasing heart rate (HR) [17, 22]. However sympathetic activity is predominant when arousal increases conducting to the acceleration of HR [17, 22].

HRV is a measure of the continuous interplay between sympathetic and parasympathetic influences on heart rate that yields information about autonomic flexibility and thereby represents the capacity for regulated emotional response [3].

Several HRV parameters can be measured resulting from time domain analysis or frequency domain analysis. Among those parameter is the ratio between LF/HF, which is known as the active indicator of the sympathetic nervous system activity. High values of LF/HF correspond to the predominance of sympathetic activity and thus high arousal in opposition to parasympathetic activity for which the ratio LF/HF is lower, corresponding to a low arousal [23].

Progressing further towards the relevance of such parameters for the current study, it is important to value the physiological signals from ECG as a good source to analyze and predict the lack or presence of cognitive attention in individuals during task execution [3] that can achieve similar performance than EEG. Scientific evidence shows that there is a close relation between variability in heart rate and attention in adults [24], infants [25] and children [26]. It also has been shown that during tasks that require attention, HF and LF parameters of HRV have a decrease [27].

## 4 Experiment

The proposed experiment was designed to study emotional arousal and attention in the context of eLearning and was prepared as follows. Ten individuals where enrolled in the study (1 female and 9 males). Their ages were between 32 and 47. The participants have different educational backgrounds, all completed at least high school, and none was employed in the aquaculture business. This fact was not considered relevant since the aim was not the course contents but it was considered of relevance as they would be similar to an aquaculture candidate willing to have training in such matters.

The experiment was conducted in the scope of the AquaSmart project [28], which has the focus on data mining and big data technologies applied to the aquaculture sector. AquaSmart helps companies to transform raw data into knowledge via accurate business-driven analytical models in a seamless and efficient process. The AquaSmart project has a component of training and that leads to the relevance of such experiment in the scope of the project's training framework.

The materials for the execution of the experiment were adopted from AquaSmart training program and adapted for the hereby described experimental setup. The study comprises two e-learning courses: (a) Introduction to AquaSmart (course 1); (b) History of aquaculture (course 2). The second course was delivered with an extract from the beginning of the first part of Bach's Brandenburg Concerto N° 2 as background music, which was rated as expressing happiness [29]. Both courses consisted of a learning component composed by 4 slides followed by a test with 5 questions about the learned materials. Previous studies already evaluated the impact of music in a person's mood, in this case it is aimed a step further in evaluating music's impact on attention [30].

During the sessions ECG measurements were executed using a Olimex SHIELD-EKG-EMG. In the end of the session a questionnaire was performed to quantify: (a) the attention of the participants in each training course; (b) the likability of each course; (c) the difficulty of each course.

The sample was divided into two groups (Fig. 1): (a) a group that performed the training course 1 with no music followed by the training course 2 with music in an automatic form, i.e., with a pre-determined time interval between slides; (b) a group that performed the training course 1 with no music followed by the training course 2 with music in an interactive form, i.e., requiring by the user to press the "Next" button of the interface to move to the next slide.

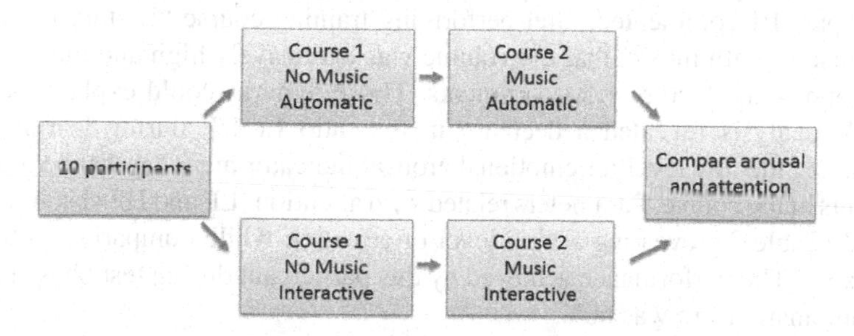


**Fig. 1.** Experimental design.

The participants were informed about the procedures and while agreeing each signed a consent form. The procedural execution started by installing clinical electrodes to his/her left arm, right arm and right leg. Heart rate was measured 30 s before the start of each e-learning training course for baseline purposes and data was collected during the attendance to the e-learning courses. After the session, each subject rated the level of difficulty, attention and motivation towards each training course on a seven-point scale instantiated from "very little" (1) to "very much" (7).

Heart Rate Variability was measured, according to specification, as beat to beat variation in the heart rate (R-R intervals) of ECG. Such measurement quantifies the interplay between sympathetic and parasympathetic system of ANS, and it was acquired using Artifact software [2]. Special emphasis was given to LF and HF of HRV, related with attention, and the ratio LF/HF indicator of arousal (more details in Sect. 3.2) that were measured while resting and along the proposed learning tasks.

## 5 Results

As an example, the results of participant 1 (P1), concerning training course "AquaSmart" (course 1) with no music, revealed that LF/HF ratio increased during learning phase, as expected. Since LF and HF values, calculated in milliseconds squared, decreased; it can support that such participant was more attentive during the learning stage (Table 1). The performance achieved during the test phase for training course 1 in an automatic form was 80%.

**Table 1.** Training course 1 with no music for P1.

| | Resting | Learning | Test |
|---|---|---|---|
| LF ($ms^2$) | 8074.65 | 4899.53 | 1453.03 |
| HF ($ms^2$) | 6336.96 | 2479.01 | 4787.75 |
| LF/HF | 1.27 | 1.98 | 0.95 |

The next tables present LF and HF measurements for Course 1 (no music) and Course 2 (with music) respectively for the same person designed as P1.

Participant P1 commented, after performing training course "History of Aquaculture" (course 2) with music, that the volume was excessively high and that he usually does not appreciate history related contents. Those remarks could explain the reason why HRV analysis revealed a decrease in the ratio LF/HF during learning phase (Table 2), i.e., the low level for emotional arousal indicator meaning that he didn't like the contents of the course. Parameters related with attention (LF and HF absolute values) decreased (Table 2) revealing a high level of attention while comparing to the other course stages. The performance achieved by this participant during test phase in course 2, in an automatic form was 60%.

**Table 2.** Training course 2 with music for P1.

| | Resting | Learning | Test |
|---|---|---|---|
| LF ($ms^2$) | 7377.99 | 2868.48 | 2996.58 |
| HF ($ms^2$) | 4995.74 | 2761.50 | 2341.79 |
| LF/HF | 1.48 | 1.04 | 1.28 |

Additionally, from the collected information of all questionnaires participants it was concluded that some of them like to study with music and others not. From the data collected using the ECG, it was possible to infer the same. This means that music helps in the learning and test phases only of a part of the participants. Such conclusion is in line to the study made by of Gião et al. in [30], which says that a same music can result in different emotional states to each person. This is also in line to the standard deviation of the graph of Fig. 2, where with music (Fig. 2 – right graph) it has a $\sigma = 4.67$ and without music (Fig. 2 – left graph) it has a $\sigma = 2.48$. This means that there is more variability of values in the first case (course with music) representing its tendency for amplified results, which implies also an amplified arousal (emotional intensity). That means that people feel the music reacting accordingly, independently if in a positive or negative way. In conclusion could be said that despite being in a positive or negative way, classic background music enhances intensity of emotion in eLearning experiences.

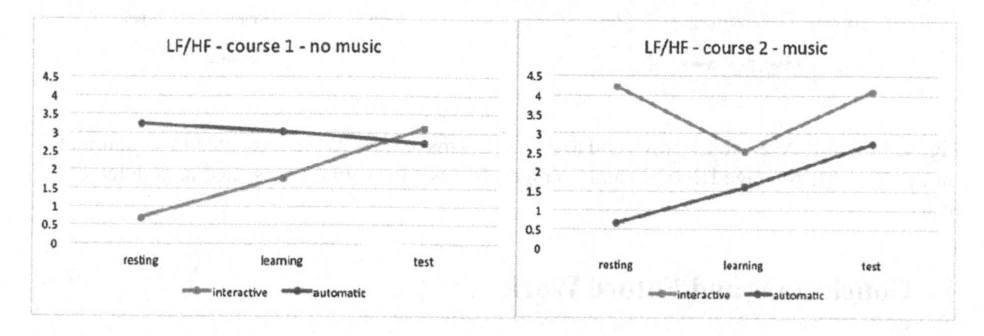


**Fig. 2.** (At Left) Average LF/HF ratio for course 1 with no music presented in both automatic and interactive way; (At Right) Average LF/HF ratio for course 2 with music presented also in both ways.

When analyzed the engagement of students in the courses with music in relation to the one without can be verified that a higher commitment attention it is obtained in courses with music. That can be verified in Fig. 3 by lower average LF and HF values in absolute terms in courses with music.

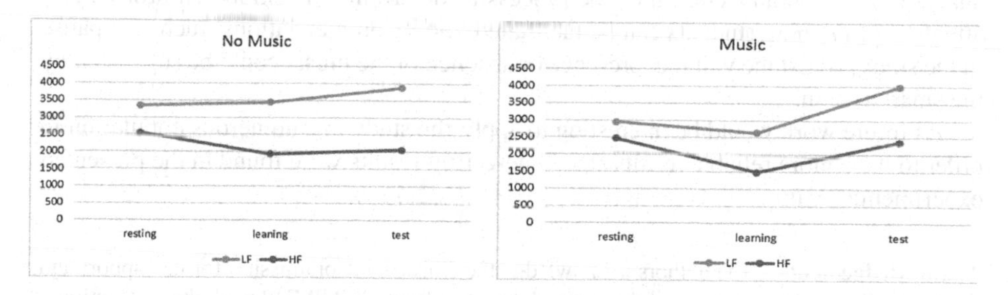


**Fig. 3.** (At Left) Average LF ($ms^2$) and HF ($ms^2$) for course 1 with no music; (At Right) Average LF ($ms^2$) and HF ($ms^2$) for course 2 with music.

Figure 4 presents the average LF and HF values of all the students, from which could be inferred that students were more attentive or engaged in the automatic course. This means, that since they had not the possibility of going back or forward to see the contents again. It indicates that in order do not fail in the test they tried to be more attentive to the contents. This is explained by the reducing in time of the LF and HF absolute values. The graph in the left of Fig. 4 shows that in relation to the "automatic" course. On the other hand, the graph in the right part of Fig. 4 shows the opposite.

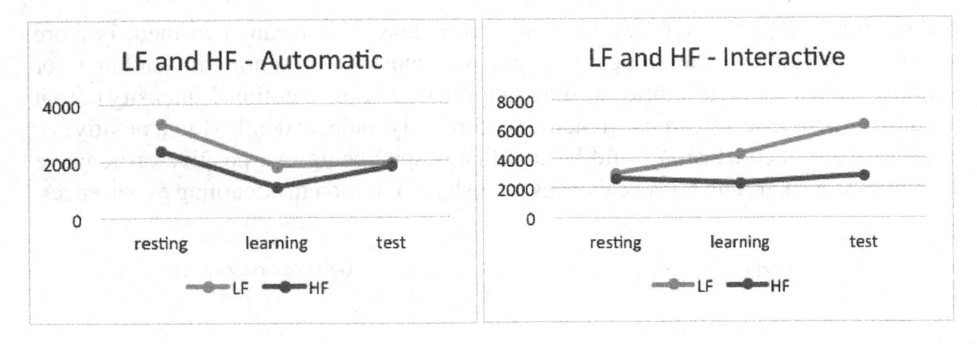


**Fig. 4.** (At left) Average LF ($ms^2$) and average HF ($ms^2$), for courses presented in an automatic form; (At right) Average HF ($ms^2$) and average LF ($ms^2$) for courses presented in an interactive form.

## 6 Conclusions and Future Work

The use of wearables able to obtain LF and HF values from a ECG potentiates the creation of smart systems. This can be applied to the area of education, which will allow the system to sense the students' physiological signals enabling the possibility of inferring particular situations about the level of the attention of the student. Through this kind of systems could be implemented specific recommendation services that will suggest particular contents or topics to recover the students attention.

A smart system prototype able to conduct analysis of arousal and attention based on biosignals using wearables to measure ECG is proposed with incorporation of the analyzed HRV parameters. Thus, the process to determine the emotional arousal and attention of learning students can be facilitated and recommendations, such as a pause in the study, adjust the volume, presence or absence of the music could be suggested by the smart system.

As future work would be interesting to apply the study to a numerous population in order to have more reliable results since interesting results were found in the presented experimental setup.

**Acknowledgements.** The authors acknowledge the European Commission for its support and partial funding and the partners of the research projects from ERASMUS+: Higher Education – International Capacity Building - ACACIA – Project reference number – 561754-EPP-1-2015-1-CO-EPKA2-CBHE-JP, (http://acacia.digital); and Horizon2020 - AquaSmart – Aquaculture Smart and Open Data Analytics as a Service, project number - 644715, (http://www.aquasmartdata.eu/).

## References

1. Khan, B.: Managing E-Learning: Design, Delivery, Implementation and Evaluation. Information Science Publishing, Hershey (2005)

2. Kaufmann, T., Sütterlin, S., Schulz, S.M., Vögele, C.: ARTiiFACT: a tool for heart rate artifact processing and heart rate variability analysis. Behav. Res. Methods **43**(4), 1161–1170 (2011)
3. Appelhans, B.M., Luecken, L.J.: Heart rate variability as an index of regulated emotional responding. Rev. Gen. Psychol. **10**(3), 229–240 (2006)
4. European Commission ICT Work Programme 2007-08
5. Zarrad, A., Zaguia, A.: Building a dynamic context aware approach for smart e-learning system. In: 2015 Second International Conference on Computing Technology and Information Management (ICCTIM), pp. 144–149. IEEE (2015)
6. Pekrun, R., Goetz, T., Titz, W., Perry, R.P.: Academic emotions in students' self-regulated learning and achievement: a program of qualitative and quantitative research. Educ. Psychol. **37**(2), 91–105 (2002)
7. Bahreini, K., Nadolski, R., Westera, W.: Towards multimodal emotion recognition in e-learning environments. Interact. Learn. Environ. **24**(3), 590–605 (2016)
8. Santos, O.C., Saneiro, M., Boticario, J.G., Rodriguez-Sanchez, M.C.: Toward interactive context-aware affective educational recommendations in computer-assisted language learning. New Rev. Hypermedia Multimedia **22**(1–2), 27–57 (2016)
9. Steele, K., Bass, K., Cook, M.: The mystery of the Mozart effect: failure to replicate. Psychol. Sci. **10**(4), 366–369 (1999)
10. McCuntheon, L.: Another failure to generalize the Mozart effect. Pshychol. Rep. **87**(1), 325–373 (2000)
11. Ransdell, S., Gilroy, L.: The effects of background music on word processed writing. Comput. Hum. Behav. **17**, 141–148 (2001)
12. Kallinen, K.: Reading news from a pocket computer in a distracting environment: effects of the tempo of background music. Comput. Hum. Behav. **18**, 537–551 (2002)
13. Thompson, W.F., Schellenberg, E.G., Hussain, G.: Arousal, mood and the Mozart effect. Pshychol. Sci. **12**(3), 248–251 (2001)
14. Wilson, T., Brown, T.: Reexamination of the effect of Mozart's music on spacial-task performance. J. Psychol. **131**(4), 365–370 (1997)
15. Richards, D., Fassbender, E., Bilgin, A., Thompson, W.B.: An investigation of the role of background music in IVWs for learning. ALT-J **16**(3), 231–244 (2008)
16. Kim, J., André, E.: Emotion recognition based on physiological changes in music listening. IEEE Trans. Pattern Anal. Mach. Intell. **30**(12), 2067–2083 (2008)
17. Schaaff, K., Adam, M.T.P.: Measuring emotional arousal for online applications: evaluation of ultra-short term heart rate variability measures. In: Humaine Association Conference on Affective Computing and Intelligent Interaction, pp. 362–368 (2013)
18. Kreibig, S.D.: Autonomic nervous system activity in emotion: a review. Biol. Psychol. **84**, 394–421 (2010)
19. Katsis, C.D., Katertsidis, N., Ganiatsas, G., Fotiadis, D.I.: Towards emotion recognition in car-racing drivers: a biosignal processing approach. IEEE Trans. Syst. Man Cybern. Part A: Syst. Hum. **38**(3), 502–512 (2008)
20. McCorry, L.K.: Physiology of the autonomic nervous system. Am. J. Pharm. Educ. **71**(4), 78 (2007)
21. Freeman, J.V., Dewey, F.E., Hadley, D.M., Myers, J., Froelicher, V.F.: Autonomic nervous system interaction with the cardiovascular system during exercise. Prog. Cardiovasc. Dis. **48**(5), 342–362 (2006)
22. Diffen. Parasympathetic vs. Sympathetic Nervous System. http://www.diffen.com/difference/Parasympathetic_nervous_system_vs_Sympathetic_nervous_system. Accessed 04 Feb 2017

23. Fruhata, T., Miyachi, T., Adachi, T.: Doze driving prevention system by low frequency stimulation and high density oxygen with fragrance of GF (Grape Fruit). In: König, A., Dengel, A., Hinkelmann, K., Kise, K., Howlett, R.J., Jain, L.C. (eds.) KES 2011. LNCS (LNAI), vol. 6883, pp. 11–20. Springer, Heidelberg (2011). doi:10.1007/978-3-642-23854-3_2
24. Porges, S.W., Raskin, D.C.: Respiratory an heart rate components of attention. J. Exp. Psychol. **81**, 497–503 (1969)
25. Richards, J.E., Casey, B.J.: Heart rate variability during attention phases in young infants. Psychophysiological **28**, 43–53 (1991)
26. Porges, S.W.: Heart rate indices of newborn attentional responsivity. Merill-Palmer Quart. **20**, 231–254 (1974)
27. Tripathi, K.K., Mukundan, C., Mathew, T.L.: Attentional modulation of heart rate variability (HRV) during execution of PC based cognitive tasks. Ind. J. Aerospace Med. **47**(1), 1–10 (2003)
28. AquaSmart. http://www.aquasmartdata.eu/. Accessed 04 Feb 2017
29. Kallinen, K.: Emotional ratings of music excerpts in the western art music repertoire and their self-organization in the Kohonen neural network. Psychol. Music **33**(4), 373–393 (2005)
30. Gião, J., Sarraipa, J., Francisco-Xavier, F., Ferreira, F., Jardim-Goncalves, R., Zdravkovic, M.: Profiling based on music and physiological state. In: Mertins, K., Jardim-Gonçalves, R., Popplewell, K., Mendonça, J.P. (eds.) Enterprise Interoperability VII. PIC, vol. 8, pp. 123–135. Springer, Cham (2016). doi:10.1007/978-3-319-30957-6_10

# A System for Driver Analysis Using Smartphone as Smart Sensor

Rui Daniel Vilaça[1], Rui Araújo[2], and Rui Esteves Araújo[3(✉)]

[1] Faculty of Engineering, University of Porto,
Rua Dr. Roberto Frias, 4200-465 Porto, Portugal
rdvilaca@fe.up.pt

[2] MCA and Associados, Lda Rua D. Manuel II, S/44, 4050-345 Porto, Portugal
ruka.araujo@gmail.com

[3] INESC TEC and Faculty of Engineering, University of Porto, Porto, Portugal
raraujo@fe.up.pt

**Abstract.** This work is focused on the development of system able to keep tracking driver's behavior without a black box device mounted inside the car. Firstly, we intend to explore the data from GPS (Global Positioning System), accelerometer, gyroscope and magnetometer for a full characterization of the vehicle dynamics. Secondly, we develop an event detector that determines and classifies distinct kind of maneuvers, like turns, lane change, U-turns, among others. Finally, we developed a simple aggressiveness classifier using fuzzy logic. Experiments have been conducted and the initial results of the system were found to be encouraging on the implementation of a non-intrusive system for driver analysis.

**Keywords:** Intelligent systems · Smartphone sensor · Driver behavior · Vehicle dynamics

## 1 Introduction

According to a study from the World Health Organization [1] the injuries caused by traffic accidents are one of the main causes of death, the majority being among young people between the ages of 15 and 29 years old. This circumstance motivates the scientific community to the continuous development of technologies and technical solutions that contribute to attenuate the numbers related to road accidents. One possible solution is the development of driver support and/or evaluation systems in on-board computers or mobile computing devices, which can provide almost instantaneous support to the driver and contribute to the driver's awareness about driving risks.

In recent years, there has been a highly significant growth in the usage of mobile phones with computer-like capabilities. These units, commonly known as smartphones, stand out by their high capacity of processing and connection to several high-speed networks, as well as for the ability of sensing information from movement and position.


L.M. Camarinha-Matos et al. (Eds.): DoCEIS 2017, IFIP AICT 499, pp. 103–110, 2017.
DOI: 10.1007/978-3-319-56077-9_9

In this context, the use of the provided sensors is an opportunity for the development of an intelligent system capable of monitoring, evaluation and assistance to the driver of road vehicles.

Studying the specialized literature, we found works with similar objectives, such as: driver evaluation [2–6], driving support (in real-time or not) [7, 8], increase vehicle performance [7] or traffic control [9, 10]. These types of works can also help reduce pollution and the number of road accidents.

This paper presents a system with the objective of evaluating the aggressiveness of the driver, focusing mainly on the characterization of vehicle dynamics, using one smartphone as a smart sensor. The basic system requirements are: Plug and Play – easy usability for the driver; No extra costs or additional devices and Non-intrusive.

At this early stage the challenge of characterizing the dynamics of the vehicle with a non-oriented smartphone is solved. As a complement, two applications were developed for validation of results and emulation of a simple aggressive classifier.

## 2 Smart System

In the developed countries, most of the population has, at least, one smartphone. Also, these machines have an increasing processing capacity and more types of sensors with better quality in terms of signal acquisition. Such capabilities make it a platform with high potential for implementation of smart systems.

Generally, this type of equipment allows the acquisition of image, recording of sound with the microphone, smartphone movement characterization with the accelerometer and gyroscope, location and orientation through GPS and magnetometer and detection of use with the presence sensor.

Thus, for the system in question, due to the objective of creating a non-intrusive system and to preserve the users' privacy, the use of the camera and microphone is ruled out, focusing only on GPS, accelerometer, gyroscope and magnetometer.

The GPS operates with 1 Hz sample frequency and allows to obtain the position in terms of the Geographic Coordinate System (longitude, latitude, altitude), and speed. Its use requires significant battery consumption.

In the remaining sensors, the sampling frequency can reach up to hundreds of Hz. The sampling frequency is a trade-off between the precision of information and battery consumption. For this system, a sampling rate of 10 Hz was defined.

The accelerometer and gyroscope allow obtaining, respectively, the acceleration in m/s and the angular velocity in rad/s, represented in terms of the smartphone coordinate system. Due to the no strict orientation requirements of the smartphone in order to ease the user experience, it is necessary to perform a virtual orientation of the signals to characterize the movement of the vehicle. In addition, the signals from the accelerometer require a more efficient filtering due to high sensitivity to small movements. The use of the gyroscope comes with the problem that not all smartphones have one.

The use of the magnetometer allows the measurement of the magnetic field in μT around of the device, which is not necessarily the Earth's magnetic field due to the environment around it, a notorious situation in a motorized vehicle due to the electromagnetic materials. Calibrating the magnetometer to the surrounding environment allows obtaining a more precise orientation of the vehicle.

For data storage and processing, the operating systems available for smartphones allow a real-time processing of the data or store in files for further processing performed locally or on an external machine.

## 3 Related Work

A recent bibliography research allows us to conclude that there are several scientific papers describing systems with characteristics similar to the one presented in the present work. However, there are aspects related to the central objective of the work, the operating conditions or signal processing techniques that deserve a more detailed discussion and that contribute to support the present work.

The authors present, in [2], a work divided in two parts, with the main objective being the development of a system adaptable to mobile devices with the ability to detect aggressive behaviour. It is defined as aggressive behaviour: speeding, abrupt acceleration and braking, nonconformity with the minimum distance for the front vehicle and overtaking in heavy traffic. In the first part, it is validated experimentally a mathematical model based on the initial hypothesis that the aggressiveness operates as a linear filter on the probability distribution function of the input signals. The mainly conclusions were that aggressiveness affects the mean, standard deviation and range of values of lateral and longitudinal acceleration, but maintains the waveform of the acquired signals. Also, it said that the results are independent of the road type and the driver.

In the second part the model is tested through the development of an aggressive classifier, a GPS data logger is used for data acquisition and a success rate of 92.3% is achieved identifying the aggressive lap in a total of 10 laps of the same driver.

From this work, it is important to retain the aggressive behaviours, the statistical indicators affected with the aggressiveness and the independence before the road pavement and the driver. In addition, it serves as validation of the hypothesis of creating a non-intrusive aggressiveness classifier applicable to smartphones.

The document [3] explains the development of MIROAD, a smartphone application for detecting aggressive driving. It differentiates itself from this work using the camera and requiring a specific positioning of the smartphone. For detection of driving patterns, lateral and longitudinal movement is characterized by the fusion of accelerometer, gyroscope and magnetometer data.

The system allows detecting the following events: normal/aggressive right/left shift, normal/aggressive U-turn, aggressive right/left lane change, aggressive acceleration/braking, speeding and device removal. For detection and classification, the authors used triggers and pattern recognition through dynamic time wrapping (DTW).

This work presents a lot of similarities to the one developed previously, providing useful information regarding signal acquisition and processing, and event detection and classification. The main difference is the restriction of the smartphone position so that is oriented with the vehicle. The removal of this restriction is the novelty of the work developed here.

Finally, in [4] it is presented a tool applicable to mobile applications, V-sense, which aims to detect and differentiate different maneuvers. The maneuvers detected are:

lane change, turns and driving in curvy roads. Through the accelerometer, it is calculated the distribution of the gravity force and, from there, the angles of the rotation to orient the smartphone coordinate system with the vehicle's one. With the oriented gyroscope signal the maneuver classification is done considering the bumps duration and the value of maximum and minimum. In some cases, the horizontal displacement calculated through GPS and accelerometer data is used as decisive factor.

The major differentiation of this work is the calibration of the signal to the vehicle coordinate system and the maneuvers classification with the oriented gyroscope signal and horizontal displacement.

## 4 System Architecture

The system developed in this work presents a modular architecture where two parts are clearly differentiated. The first on is responsible for characterizing the movement of the vehicle during a recorded trip and, it is called the "Vehicle Black-Box". The second on is responsible for extracting useful information from processed signals, consisting of two main modules: Event Detector and Aggressiveness Classifier. In Fig. 1 it is represented a layout with the architecture and information flow of the system.

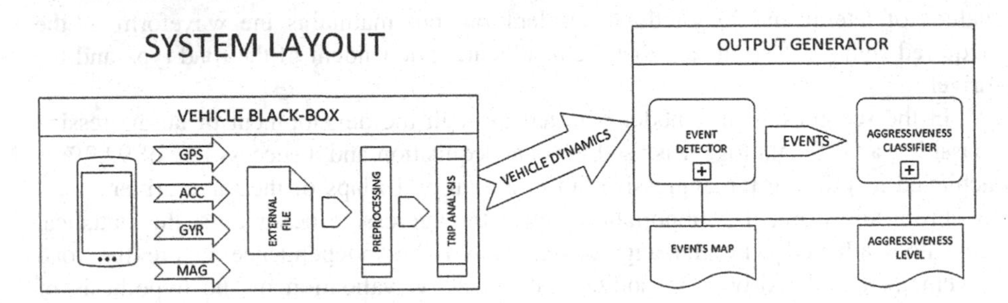


**Fig. 1.** System architecture and information flow.

One of the central aspects for orientation manipulation is the definition of the necessary coordinate systems, which is represented in Fig. 2. The correct definition, followed by a precise virtual orientation of the signals, gives the system freedom relatively to the smartphone's position. Furthermore, it allows one to introduce, in future work, an automatic mechanism that detect phone handling while driving.

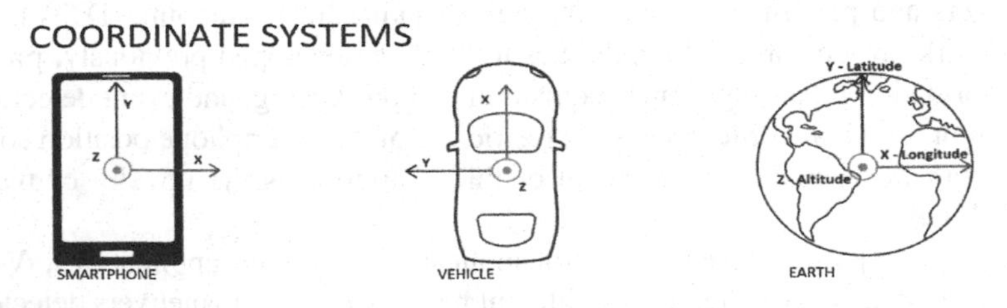


**Fig. 2.** Representation of coordinate systems: smartphone, vehicle and earth.

### 4.1 Vehicle Black Box

This part of the system is responsible for characterizing the dynamics of the vehicle during a trip using the signals acquired from smartphone sensors. For correct operation, it is necessary to place the device in a stable place during the entire recording session. A sampling of the last value read by each sensor is done at 10 Hz. Although GPS has a maximum sampling frequency of 1 Hz, this method is applied to obtain a full data sample at determined times instead of having individual sensor samples in different instants of time.

From the GPS data, in a X-size sample, the velocity signal of 1 Hz is transformed into a signal of 10 Hz using a linear extrapolation:

$$R(i) = \frac{1}{2 * R + 1} \sum_{k=i-R}^{i+R} V(k), \forall i \leq X \tag{1}$$

A reference signal for the acceleration and jerk of the vehicle is also calculated through the first and second derivatives. The orientation of the vehicle is calculated using the pair (longitude, latitude) through the slope of the line drawn using three consecutive instances.

A filter is applied to the signals from accelerometer based on the exponential average,

$$R(i) = \alpha * A(i) + (1 - \alpha) * R(i - 1), \forall 1 < i \leq X, R(1) = k, \tag{2}$$

eliminating most of the noise without removing relevant information. The gyroscope disregards the use of filters due to the low susceptibility to noise.

The next step consists in the virtual orientation of the signals to the vehicle coordinate systems. For this, every time the speed is zero or constant, the distribution of acceleration in the three axes according to the smartphone coordinate system is analysed and a weighted average for such distribution is calculated. It is assumed that the obtained vector corresponds to the gravity force and it is completely represented in the Z axes of the vehicle. With this assumption, the rotation matrix for the vehicle XY plane is calculated. In the new oriented signals, the rotation speed of the vehicle is represented in the Z axe of the vehicle coordinates system and the module of vehicle's acceleration is represented by the module of X and Y values.

In order to differentiate the longitudinal and lateral acceleration, all values of the acceleration derived from the GPS speed with a significant value are selected. Those values represent the times when it can be assumed that all the acceleration is longitudinal. For those times, it is calculated the angle between the acceleration vector in XY plane and the X axe of the vehicle. The longitudinal acceleration is derived to compute the jerk.

Having properly characterized the position (longitude, latitude), speed, longitudinal and lateral acceleration, rotation speed and jerk, its assumed that the vehicle dynamics are completely characterized and some travel statistics are computed, such as average, standard deviation, maximum and minimum values.

### 4.2 Event Detector

Using the previously calculated vehicle dynamics, the different events are detected. It is considered as an event: the U-turn, curve, turn and lane change. The direction of the change, i.e., left or right, is also differentiated.

For the event detection, an algorithm like the one presented in [4] is used. Rotation speed is used as trigger value and when the $\pm\delta$ is exceeded, it is defined that an event started $t_{ini}$ instants ago, and will only end when it has $t_{end}$ consecutive instants below the limits as is shown in Fig. 3.

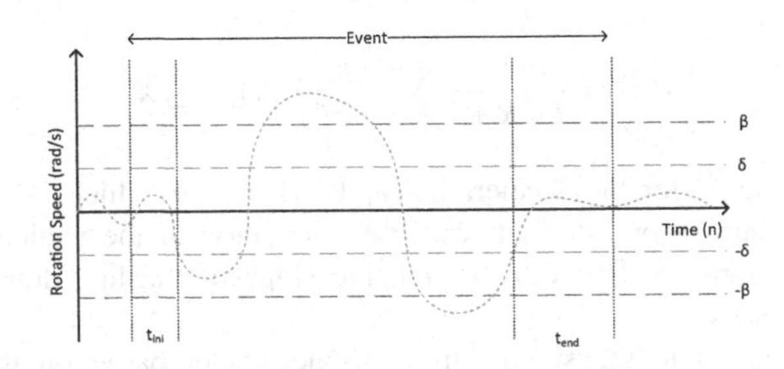


**Fig. 3.** Example for the event detector algorithm.

To classify the event, additional information such as horizontal displacement, vehicle rotation during the entire event, number and type of bumps – time slots with a consecutive minimum duration at which the rotation speed module is greater than β. Taking all the information calculated, the classification is done using a decision tree with number of bumps, total rotation, type of bumps and horizontal displacement as decision factors.

### 4.3 Aggressiveness Classifier

In this last module, we intend to assign an aggressiveness index to each driver according to their behaviour. For this a level of aggressiveness is attributed to all the events detected in the previous module. It also analyses the percentage of time it was in excessive of speed, at this stage only comparing to the maximum limit applied to the fastest roads of the country.

The fuzzy logic system used has two linguistic variables, speed and square of the jerk. The first uses six membership functions and the second four. The output linguistic variable has three membership functions, assigning an aggressiveness index between 0 and 100. With the new event index, the driver's aggressiveness index is updated through an exponential average, giving more importance to the most recent events.

## 5 Results

For the initial tests of the system two different sets of trips were used. One of the sets consisted of three trips recorded on a previously defined circuit. The other corresponded to a set of trips recorded by three different drivers during different periods of time.

The previously defined circuit consists of twenty maneuvers previously identified with the desired output for each one. The driver was asked to perform three laps, two with normal behavior and one with aggressive behavior. For all of them, the system detected all the maneuvers performed, and obtains a seventy-five percent success in identifying the type of event in the first two cases and sixty percent in the most aggressive lap. In the aggressiveness classifier, for the two normal laps, the mean score was thirty while for the aggressive one the mean score was seventy-five points.

The other set of trips allowed to intensively teste the event detector, presenting very favorable results. Two examples are shown in Fig. 4 with their respective rotation speed. As for the level of aggressiveness, the temporal variation of the behavior of the two drivers for different coefficients applied in the exponential average used in the formula are presented in Fig. 5. The red line, the one with higher coefficient, shows two distinct profiles, the left one with an inconsistent driving style, and the right one with a steadier driving style. The other lines, with smaller coefficient, do not show that difference because the "mean behavior" is similar in both drivers. Calibrating this parameter allows to set whether to have an analysis with a higher weight in the most recent history, or to allow older trips to significantly influence the index.

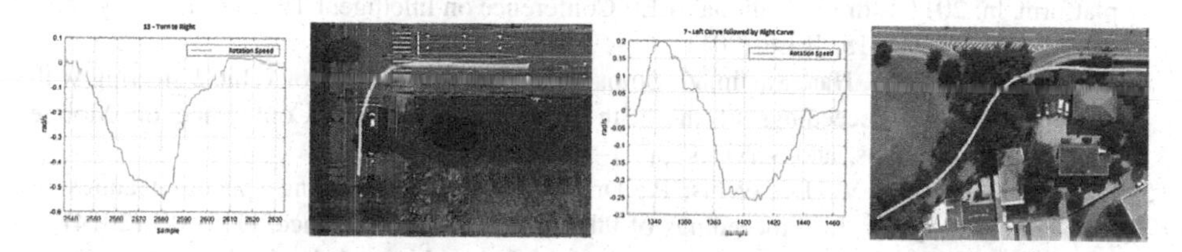

**Fig. 4.** Event detector example – turn to right and left curve followed by right curve.

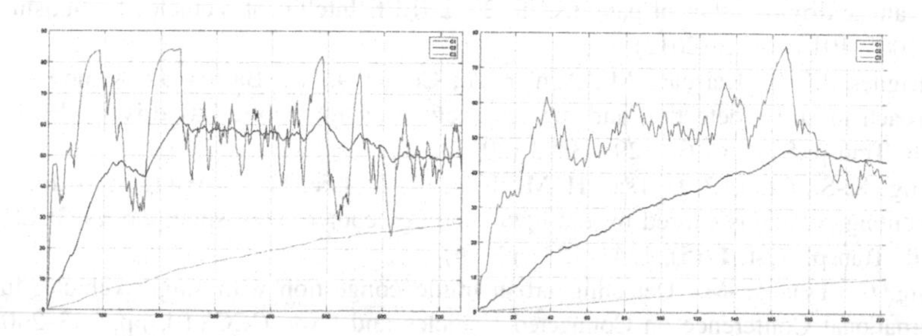

**Fig. 5.** Evolution over time of aggressiveness level for two different drivers with three different coefficients in exponential average. (Color figure online)

## 6 Conclusion

This paper allows concluding that it is possible to characterize the dynamics of a vehicle using the sensors of the smartphone. The accuracy obtained in this process is sufficient to apply it as a basis in the development of applications with the purpose of monitoring and/or assessment of driving style.

Our tests have confirmed this claim by showing that different experiment can indeed capture distinct driving responses from the participants. These responses can be used by the aggressive classifier to determine a score for the driver during a trip. This leads us to believe that our system can indeed help the driver to improve his driving performance.

For future work, it is necessary to develop one algorithm to detect the smartphone movement during the trip and take actions for correcting the orientation of the acquired signals.

## References

1. World Health Organization. Road traffic injuries, May 2016. http://www.who.int/mediacentre/factsheets/fs358/en/. Accessed 30 June 2016
2. González, A.B.R., Wilby, M.R., Díaz, J.J.V., Ávila, C.S.: Modeling and detecting aggressiveness from driving signals. IEEE Trans. Intell. Transp. Syst. **15**(4), 1419–1428 (2014)
3. Johnson, D.A., Trivedi, M.M.: Driving style recognition using a smartphone as a sensor platform. In: 2011 14th International IEEE Conference on Intelligent Transportation Systems (ITSC), pp. 1609–1615. IEEE (2011)
4. Chen, D., Cho, K.-T., Han, S., Jin, Z., Shin, K.G.: Invisible sensing of vehicle steering with smartphones. In: Proceedings of the 13th Annual International Conference on Mobile Systems, Applications, and Services, pp. 1–13. ACM (2015)
5. Zylius, G., Vaitkus, V., Lengvenis, P.: Driving style analysis using spectral features of accelerometer signals. In: Proceedings of 9th International Conference, ITELMS (2014)
6. Eren, H., Makinist, S., Akin, E., Yilmaz, A.: Estimating driving behavior by a smartphone. In: 2012 IEEE Intelligent Vehicles Symposium (IV), pp. 234–239. IEEE (2012)
7. Araújo, R., Igreja, Â., de Castro, R., Araújo, R.E.: Driving coach: a smartphone application to evaluate driving efficient patterns. In: 2012 IEEE Intelligent Vehicles Symposium (IV), pp. 1005–1010. IEEE (2012)
8. Rodrigues, J.G.P., Kaiseler, M., Aguiar, A., Cunha, J.P.S., Barros, J.: A mobile sensing approach to stress detection and memory activation for public bus drivers. IEEE Trans. Intell. Transp. Syst. **16**(6), 3294–3303 (2015)
9. Huang, K.-S., Chiu, P.-J., Tsai, H.-M., Kuo, C.-C., Lee, H.-Y., Wang, Y.-C.F.: Redeye: preventing collisions caused by red-light running scooters with smartphones. IEEE Trans. Intell. Transp. Syst. **17**(5), 1243–1257 (2016)
10. Wang, C., Tsai, H.-M.: Detecting urban traffic congestion with single vehicle. In: 2013 International Conference on Connected Vehicles and Expo (ICCVE), pp. 233–240. IEEE (2013)

# Multi-criteria Analysis and Decision Methodology for the Selection of Internet-of-Things Hardware Platforms

Edgar M. Silva[1,2(✉)] and Ricardo Jardim-Goncalves[1,2]

[1] Faculdade de Ciências e Tecnologia, Universidade Nova de Lisboa, Lisbon, Portugal

[2] Centro de Tecnologia e Sistemas, UNINOVA, Caparica, Portugal

{ems, rg}@uninova.pt

**Abstract.** The Internet-of-Things (IoT) is today a reality, and Smart Systems have taken advantage of this to improve its own sense, act and control capabilities. IoT is a highly heterogeneous environment composed by a vast number of "things" (sensors, smart objects, etc.). These "things" are based on hardware platforms which can differ widely since manufactures are being capable of develop new devices every day to tackle different application domains. Consequently, a problem emerges regarding which will be the suitable, proper hardware solution for an IoT deployment. Make a right decision is probably one of the toughest challenges for science and technology managers. This work proposes a novel methodology to analyze a set of hardware alternatives based on user's multi-criteria requirements, and advice on the more suitable hardware solution for a specific situation. For proof-of-concept it is used different Arduino boards as hardware alternatives, in which user requirements are based on hardware features. This methodology foresees its use during the development of Smart Systems (e.g.: Transportation, Healthcare) to optimize the selection of hardware platforms.

**Keywords:** IoT · Smart Systems · Decision making · Analytic Hierarchy Process · Arduino

## 1 Introduction

The emerging Internet-of-Things (IoT) technology is becoming more and more important in society. Devices are being used for environment awareness giving to people a more truly interaction with the surrounding world. With advances in micro-electronic technologies, manufactures are being capable to develop small and cheap devices, to tackle different application domains and services based on IoT deployments [1]. Hence, these devices have limited resources, for example low power, small processing and storage capabilities, and for this reason, devices must be chosen properly accordingly to the task they will fulfil. Consequently, a research question arises: **"which is the more suitable device (hardware platform) to implement a certain Smart System solution in IoT?".**


L.M. Camarinha-Matos et al. (Eds.): DoCEIS 2017, IFIP AICT 499, pp. 111–121, 2017.
DOI: 10.1007/978-3-319-56077-9_10

Multi-Criteria Decision-Making (MCDM) is one of the most widely used decision methodologies in sciences, business, government and engineering worlds. MCDM methods can help to improve the quality of decisions by making the decision-making process more explicit, rational, and efficient [2, 3]. MCDM is an approach comprehensive of a set of techniques which provides ranking methods to classify criteria of a certain problem and assist on decision phase. MCDM can tear down complex problems into small, manageable problems, in such way that small sets of data can be analyzed from which is obtained reasonable judgements. Reassembling all pieces is then formed a coherent overall system view that will aid decision makers on their decisions [4]. From this emerges the following hypothesis that could respond to the question in place: **"If during a Smart System design, engineers could use a multi-criteria methodology to rank IoT hardware solutions, then more proper solutions could be selected, leading to a more suitable (i.e.: low-energy consumption, low cost, etc.) design of Smart-Systems."** This work presents a methodology capable of analyse a set of hardware alternatives based on user's multi-criteria requisites and advise which is the more suitable hardware solution for a specific situation. Consequently, improve in overall, a Smart System deployment and operation (see Fig. 1).

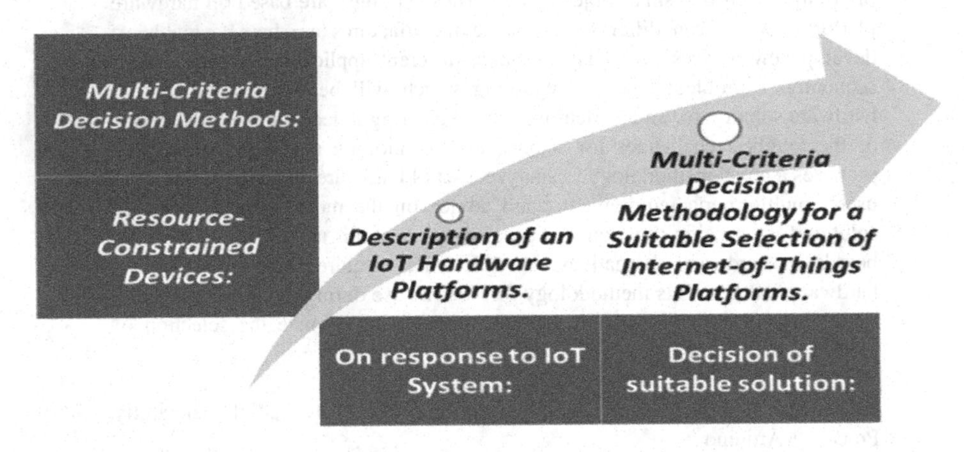


**Fig. 1.** Multi-criteria decision methodology for the selection of IoT hardware platforms.

## 2 Hardware Platforms of Smart Systems

People are living in an environment where Smart Systems are everywhere. These systems can be found in areas such as transportation, manufacturing, healthcare, energy, etc. In a succinct definition, Smart Systems can be seen as a system capable to sense, act and control a certain situation. Smart Systems present unique capabilities (cognitive as physical) to optimize their performance in response to external (sensed) information. With the technological advances in the microelectronic these capabilities have improved widely [5]. Deployment of a Smart Systems can vary in many ways, in scale, functionality, implementations and technologies. Although, developers struggle with the same challenges during the deployments design phase. One of problems arises with the need to choose a suitable hardware platform (device) or platforms between a set of alternatives. Even more when manufacturers are being capable of develop new devices to tackle different application domains. These devices, also known as motes or sensor nodes, have some particular characteristics such as small size, low-power, sensing units, wireless communication and they are cheap [6]. As a result of such features they are categorized as Resource-Constrained Devices (RCD). Multi-Criteria Decision Methods (MCDM) can assist on the development of Smart Systems, in a way that a decision could be provided in terms of what will be a more proper hardware solution for a certain task. For example, sense the environment, perform a system action or apply/compute a control function.

## 3 Literature Review

### 3.1 Hardware Platforms for IoT: Resource-Constrained Devices

With the growing interest of major technology players on Internet-of-Things (IoT), it has been seen a constant evolution of microelectronic technologies, bringing to this domain a series of devices which enables the realization of the IoT concept. These devices, also known as motes or sensor nodes, have some particular features such as small size, low-power, and processing, storage, sensing and wireless communication capabilities. These characteristics, from which comes the categorization as Resource-Constrained Devices (RCD), allow Wireless Sensor Networks (WSN)s to differ from other wireless technologies. Devices are reasonably cheap, and still maintain accurateness/preciseness and reliability. Although, energy efficiency is considered to be the major requirement for these type of devices [6, 7].

### 3.2 Methods for Multi-criteria Decision-Making

Multi-Criteria Decision-Making (MCDM) is a process that takes a decision upon a set of available solutions, assessing which one is more suitable across diverse, contradicting, qualitative and/or quantitative criteria. MCDM methods have been applied to engineering problems, providing useful insights to decision makers, making their decisions more qualified to overcome complex problems [8]. Based in [8–10], as others could be point out, three of the most widely used decision making methods in literature are the: AHP (Analytic Hierarchy Process); PROMETHEE (Preference Ranking Organization Method for Enrichment Evaluation); and ELECTRE (Elimination Et Choix Traduisant la Realite).

Analytic Hierarchy Process (AHP). The Analytic Hierarchy Process (AHP) was proposed by Saaty [11], and it is aimed to solve problems with multiple and conflicting criteria. AHP is a powerful decision making methodology capable of decompose a complex problem into a systematic hierarchy procedure [12], as depicted in Fig. 2.

AHP applies a criteria pairwise comparison based on decision maker judgement, creating the Saaty 1–9 scale to define the criteria priority. In which qualifying a criterion with 1 indicates equal importance, and 9 extremely important. With an $n$ criteria and $p$ solutions it is achieved a decision matrix of $p*n$ size, in which the $\{d_{pn}\}$

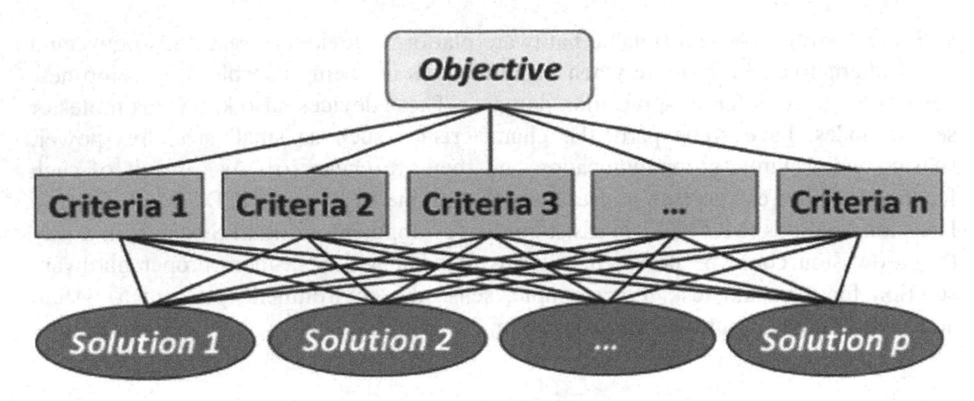


**Fig. 2.** Example of a criteria and solutions hierarchy.

element indicates the value for *p*th solution in respect to the *n*th criteria. The pairwise comparison matrix is a $m*n$ size, in which the $\{v_{mn}\}$ element indicates the value for *m*th criteria in respect to the *n*th criteria. Normalized comparison matrix has its elements calculated as follows: $Nv_{pn} = v_{pn}/w_n$; $w_n = \sum v_{pn}$, $p = 1, 2$ …. The contribution of each criteria $n$ for the problem objective, i.e. the eigenvector, is then computed as: $CW_n = \sum w_n/n$, $n = 1, 2$ … [13]. Identified the eigenvector, is then possible to rank the solutions through a simple matrix multiplication, i.e.: the decision matrix multiplied by the eigenvector.

**PROMETHEE.** Preference Ranking Organization Method for Enrichment Evaluation (PROMETHEE) [14] is a MCDM method, with six different versions based on ranking. The first version, PROMETHEE I, consists on partial ranking; PROMETHEE II on complete ranking; PROMETHEE III performs ranking based on intervals and PROMETHEE IV is a method for the continuous case; PROMETHEE V method has integer linear programming and net flows; finally the PROMETHEE VI method includes a representation of the human brain [13, 15]. The PROMETHEE methodology can be described in five steps, as Fig. 3 shows. The decision maker chooses a preference function defined into two independently actions (e.g. Boolean function – value 0 or 1), and then it is applied a solutions comparison. The comparison results and criteria values are used to form a matrix, on which methods can then be applied. For example, PROMETHEE I to create a partial solutions rank and PROMETHEE II to obtain a final solutions ranking [16]. PROMETHEE methods I and IV use preference functions which are more suitable for qualitative criteria. Methods type III and V are normally chosen for quantitative criteria problems. Type II and VI preference functions are used with less frequency.

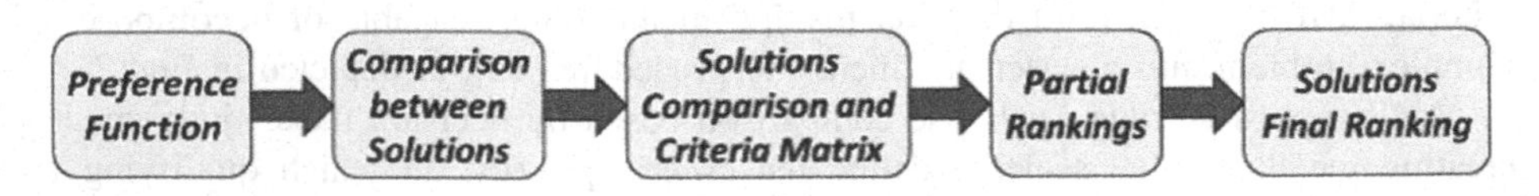


**Fig. 3.** PROMETHEE methodology.

**ELECTRE.** Elimination Et Choix Traduisant la Realite (ELECTRE – Elimination and Choice Expressing the Reality) method was created by Bernard Roy [17]. Similar to the previous decision method, ELECTRE has also six variations: type I, II, III, IV, IS and TRI. All ELECTRE methods are based on the same background, but have different process. In ELECTRE I and IS a single solution or a group of solutions are selected and assigned as a possible solution. ELECTRE II was developed to deal with problems that need to rank solutions from the best to the worst solution. ELECTRE III introduced pseudo-criteria and fuzzy binary outranking relations. ELECTRE IV method was possible to rank solutions without the use of relative criteria coefficients (the only ELECTRE method that did not use it). Finally, ELECTRE TRI assigns categories to solutions [8, 14]. ELECTRE is a preference-based model, performing a pair-wise comparison between solutions. The model assumes three types of relations between solutions: is preferred to; is indifferent to; and incomparable to. ELECTRE brings the notion of thresholds values, used to set concordance and discordance indexes. These indexes are used to classify a solution in relation to another.

**Summary.** The Analytic Hierarchy Process (AHP) follows a prioritization theory, dealing with complex problems that need multi criteria consideration simultaneously. Although, AHP presents some disadvantages. The pairwise comparison based on decision makers' subjective judgment, and the criterion weight has a direct impact on the final score. The PROMETHEE method is consistent, easy to use and does not need great interaction with decision makers. Although, PROMETHEE presents as downside the incapability to react when new alternatives are introduced. The method II is unable to present a preferred solution in a bi-criteria problem, after the method I have been applied. Generally, ELECTRE method does not frequently leads to the case in which one solution stands out from the others. For this reason the method is considered to be more suitable for problems with several solutions and not so many criteria. Also, it is considered to be difficult to understand, and thresholds definition is a problem since there is no "correct" value.

## 4 Proposed Methodology

This section will describe the proposed methodology which enables users, developers and researchers to assess which hardware platform is more suitable for their own application. The methodology provides means to indicate restrictions if any and/or criteria optimization methods, combined with a criteria prioritization based on decision maker judgement.

Possible solutions, i.e. hardware platforms, are represented by set $\boldsymbol{S}$ in (1):

$$S = \{s_1, s_2, \ldots, s_n\}; n \in N. \tag{1}$$

The amount of alternatives is a finite number $\boldsymbol{n} \in \boldsymbol{N}$ and is defined by the user, i.e. users select the solutions that will be analyzed. Platforms have a vast number of features, $m \in N$, although user may or may not consider all of them as important. The set of features/criteria, called $\boldsymbol{F}$, is defined as in (2).

$$F = \{f_1, f_2, \ldots, f_m\}; m \in N. \tag{2}$$

With the sets of possible solutions and criteria defined is then possible to create an Evaluation Table. The Evaluation Table, $\boldsymbol{Ev_{table}}$, given by (3), is an $\boldsymbol{n}$-by-$\boldsymbol{m}$ matrix. Rows represent the contemplated solutions and the columns the assessment features. The element $v_{i,j}$ presents the value for feature $\boldsymbol{j}$ of the alternative $\boldsymbol{i}$.

$$Ev_{table} = \begin{bmatrix} v_{1,1} & v_{1,2} & \cdots & v_{1,m} \\ v_{2,1} & v_{2,2} & \cdots & v_{2,m} \\ \vdots & \vdots & \ddots & \vdots \\ v_{n,1} & v_{n,2} & \cdots & v_{n,m} \end{bmatrix} = \left(v_{i,j}\right) \in \mathbb{R}^{n \times m}. \tag{3}$$

Hardware platforms, in our case the possible alternatives, present a vast number of features, and each feature has its own units. Features values are treated as dimensionless. Hardware platforms, in our case the possible alternatives, present a vast number of features, and each feature has its own units. Features values are treated as dimensionless. To perform an analysis on a set of hardware alternatives, considering user's multi-criteria requisites, to obtain an advise on which is the more suitable hardware solution for a specific situation, users must indicate restrictions if any and/or criteria optimization method (minimize or maximize). Currently the proposed methodology presents seven types of restrictions/optimization, from now on call Assessment Constraints, which can be applied by users to assess the more suitable solution(s). The Assessment Constraints set, $\boldsymbol{AC}$, is defined in (4). Each constraint can be applied to one or more criteria. Two or more constraints can be applied to single feature/criteria.

$$AC = \{min, max, mustHave, cannotHave, lessThan, equal, greaterThan\} \tag{4}$$

Assessment Constraint ***min*** and ***max*** (optimization methods) are used respectively when user's purpose is to minimize or maximize values of a feature. Constraint ***mustHave*** is used for cases in which a feature must be present. Constraint ***cannotHave*** is used in the opposite case, i.e. criteria/feature cannot be available in the final proposed solution(s). Constraints ***lessThan***, ***equal*** and ***greaterThan***, enables users to set threshold values or look for a specific feature value. When users apply the Assessment Constraints, points are assigned. In the case of ***min*** and ***max***, points are assigned from 1 to n points. For instance, in a minimization applied to a feature $f_j$, solution with the smallest value for $f_j$ get n points, and the solution with the higher value in $f_j$ get 1 point. The same principle is used for maximization, 1 point for the lowest $f_j$ value and n points for the higher. For the other Assessment Constraints, it is assigned 0 or n points. For example, the ***mustHave*** case, a solution with $f_j$ value different from zero gets n points, if the $f_j$ value is zero (which indicates that the feature is not present in that solution) it get 0 points. When more than one Assessment Criteria is applied to a single feature, assigned points are multiplied. Applying the selected Assessment Criteria to the Evaluation Table, $\boldsymbol{Ev_{table}}$ is formed the Score Table, $\boldsymbol{ST}$ formalized next in (5):

$$ST = \begin{bmatrix} p_{1,1} & p_{1,2} & \cdots & p_{1,m} \\ p_{2,1} & p_{2,2} & \cdots & p_{2,m} \\ \vdots & \vdots & \ddots & \vdots \\ p_{n,1} & p_{n,2} & \cdots & p_{n,m} \end{bmatrix} = \left(p_{i,j}\right) \in \mathbb{R}^{n \times m}. \quad (5)$$

To this point, users are able to optimize features using minimization or maximization methods. Are able to set binding rules between solutions and its features, as well as define thresholds for feature values. However, it is not yet possible to assign preference between features, i.e. features priority. To achieve this goal, the proposed methodology takes use of Analytic Hierarchy Process (AHP) presented in Sect. 3.2. AHP applies to features a pairwise comparison based on decision maker (users) judgement, using the Saaty 1–9 scale to establish criteria priority. Qualifying a feature with 1 indicates equal importance, and 9 extremely important. The AHP comparison matrix, $\boldsymbol{AHP_{table}}$ is given in (6). The element $\delta_{k,j}$ represents the comparison value for feature $\boldsymbol{j}$ and feature $\boldsymbol{k}$. It is also important to note that the comparison value, $\delta_{k,j}$ has the inverse value of $\delta_{j,k}$ Specifically, the impact of feature $\boldsymbol{j}$ in feature $\boldsymbol{k}$ has the inverse importance of feature $\boldsymbol{k}$ in feature $\boldsymbol{j}$. The AHP comparison matrix, $\boldsymbol{AHP_{table}}$ is then normalized as presented in (7). The contribution of each feature/criteria on the selection of a proper solution is given by the eigenvector and it is computed as described in (8). The final solution ranking, $\boldsymbol{FSR}$ is then given by (9).

$$\boldsymbol{AHP_{table}} = \begin{bmatrix} \delta_{1,1} & \delta_{1,2} & \cdots & \delta_{1,m} \\ \delta_{2,1} & \delta_{2,2} & \cdots & \delta_{2,m} \\ \vdots & \vdots & \ddots & \vdots \\ \delta_{m,1} & \delta_{m,2} & \cdots & \delta_{m,m} \end{bmatrix} = \left(\delta_{k,j}\right) \in \mathbb{R}^{m \times m}. \quad (6)$$

$$\boldsymbol{NAHP_{table}} = \frac{\left(\delta_{k,j}\right)}{\sum_{k=1}^{m}\left(\delta_{k,j}\right)}; k = 1, \ldots, m; j = 1, \ldots, m. \quad (7)$$

$$\boldsymbol{E_{vector}} = \frac{\sum_{k=1}^{m}\left(N\delta_{k,j}\right)}{m}; k = 1, \ldots, m; j = 1, \ldots, m. \quad (8)$$

$$\boldsymbol{FSR} = \left(\boldsymbol{E_{vector}} * \boldsymbol{ST^{T}}\right)^{\boldsymbol{T}}. \quad (9)$$

## 5 Case Study on Hardware Platforms

The selected hardware platforms chosen to validate the proposed methodology are based on Arduino platform. Five different devices were selected, represented by set $\boldsymbol{S}$, as described in (10). Arduino devices present several features and characteristics suitable for the development of sensor networks. Among the vast number of features, were selected eight to serve as assessment criteria, as described in (11).

Table 1 summaries the selected hardware platforms and values for each feature. Possible solutions and features are placed in the same order as given in (10, 11).

$$\boldsymbol{S} = \{Uno\,Rev3, Mega\,2560\,Rev3, DUE, Mini\,05, Micro\}. \tag{10}$$

$$\boldsymbol{F} = \left\{ \begin{array}{c} Energy[mA], Cost[€], Clock[MHz], \\ Memory[KB](Data\,RAM), Memory[KB](Program\,Flash), \\ Used\,Memory[bytes](RAM), \\ Used\,Memory[bytes](Program\,Flash), Weight[g] \end{array} \right\} \tag{11}$$

The values for features *Cost, Clock speed, Data Ram Memory, Program Flash Memory and Weight*, respectively $(f_2, f_3, f_4, f_5\, and\, f_8)$ were retrieved from Arduino page[1]. Values for features *Used Memory Ram and Program Flash* when an *Empty application* is installed on the selected Arduino devices, respectively $(f_6\, and\, f_7)$ were retrieved using Arduino IDE version 1.5.7. Arduino IDE was used to program the *Empty application* into devices. *Empty application* is the Arduino operating system without any kind of operation, mainly the operating system footprint. Values of feature *Energy* $(f_1)$ were obtained by placing a multimeter between the power source and the Arduino and then was read the current consumed (milliamps). The Evaluation Table, $\boldsymbol{Ev_{table}}$ can be obtained directly from Table 1.

**Table 1.** Hardware solutions versus assessment criteria.

| Solutions: | Features/Criteria: | | | | | | | |
|---|---|---|---|---|---|---|---|---|
| | $f_1$ | $f_2$ | $f_3$ | $f_4$ | $f_5$ | $f_6$ | $f_7$ | $f_8$ |
| $s_1$ | 25.5 | 20 | 16 | 2 | 32 | 9 | 450 | 25 |
| $s_2$ | 34.2 | 35 | 16 | 8 | 256 | 9 | 642 | 37 |
| $s_3$ | 113 | 36 | 84 | 96 | 512 | 0 | 10492 | 36 |
| $s_4$ | 31.5 | 14 | 16 | 2 | 32 | 9 | 450 | 6 |
| $s_5$ | 41 | 18 | 16 | 2.5 | 32 | 157 | 4264 | 13 |

To define the Analytic Hierarchy Process (AHP) table, $\boldsymbol{AHP_{table}}$ reflecting user criteria importance/preference, was performed a research using popular search tools (IEEE eXplore, Harzing's Publish and BASE). The research was based on titles that contain some key words. The key words "Internet" and "Things" were stuck and varying worlds such "energy", "cost", "weight", etc. The results are summarized in Table 2. The AHP table, $\boldsymbol{AHP_{table}}$ presented in (12) is formed basically from the preformed research. Notice that this AHP table will be used in the application scenario in Sect. 6. It also reflects a preference example, since criteria importance can differ widely from one user to others, and it also depends on application purposes.

[1] https://www.arduino.cc/.

**Table 2.** Research results summary: a criteria preference/importance.

| | Energy | Cost | Processor | Memory | Weight |
|---|---|---|---|---|---|
| Percentage [%] | 83.9 | 9.6 | 3.5 | 3 | 0 |

$$AHP_{table} = \begin{bmatrix} 1 & 2 & 3 & 4 & 4 & 4 & 4 & 9 \\ 1/2 & 1 & 2 & 2 & 2 & 2 & 2 & 9 \\ 1/3 & 1/2 & 1 & 1 & 1 & 1 & 1 & 9 \\ 1/4 & 1/2 & 1 & 1 & 1 & 1 & 1 & 9 \\ 1/4 & 1/2 & 1 & 1 & 1 & 1 & 1 & 9 \\ 1/4 & 1/2 & 1 & 1 & 1 & 1 & 1 & 9 \\ 1/4 & 1/2 & 1 & 1 & 1 & 1 & 1 & 9 \\ 1/9 & 1/9 & 1/9 & 1/9 & 1/9 & 1/9 & 1/9 & 1 \end{bmatrix}. \tag{12}$$

From $\boldsymbol{AHP_{table}}$ is possible to compute the normalized AHP comparison matrix, $\boldsymbol{NAHP_{table}}$ from which the eigenvector, i.e. importance of each feature on selecting the more suitable solution, is obtain. The eigenvector is presented in (13).

$$\boldsymbol{E_{vector}} = \{0.32 \quad 0.18 \quad 0.10 \quad 0.10 \quad 0.10 \quad 0.10 \quad 0.10 \quad 0.01\} \tag{13}$$

## 6 Application Example: Small Storage Unit

For application example was selected one scenario, a Small Storage Unit, where it is needed a hardware platform which presents a small energy consumption, it is cheap and has a large amount of memory available. Since there is no available data regarding energy consumption for the specific application, let's consider the energy consumption of the operating system footprint (in this case Arduino operating system). Accordingly, is then possible to build the Assessment Constraints set, $\boldsymbol{AC_{scenario}}$ presented in (14), and the Score Table, $\boldsymbol{ST_{scenario}}$ as described in (15):

$$\boldsymbol{AC_{scenario}} = \{min_{f_1}, min_{f_2}, max_{f_4}, max_{f_5}\} \tag{14}$$

The $\boldsymbol{AC_{scenario}}$ presented in (14) minimizes the first criteria (energy consumption), $min_{f_1}$ to classify the available hardware platforms in energy consumption terms. Constraint, $min_{f_2}$ to minimize the cost, and constraints $max_{f_4}$ *and* $max_{f_5}$ to maximize the available memory.

$$\boldsymbol{ST_{scenario}} = \begin{bmatrix} 5 & 3 & 1 & 2 & 3 & 1 & 1 & 1 \\ 3 & 2 & 1 & 4 & 4 & 1 & 1 & 1 \\ 1 & 1 & 1 & 5 & 5 & 1 & 1 & 1 \\ 4 & 5 & 1 & 2 & 3 & 1 & 1 & 1 \\ 2 & 4 & 1 & 3 & 3 & 1 & 1 & 1 \end{bmatrix} \tag{15}$$

**Results Comparison.** Table 3 presents the final classification results for each available hardware solution. Three different evaluations are available to perform a methods comparison. The first column shows the results when AHP method, alone, is applied to features values. As discussed in Sect. 3.2, this method presents a direct influence of feature values on the final score. Also, it is not able to reflect user assessment constraints (e.g. min, max). Considering, the objective to minimize all selected criteria, this method advice ***Uno Rev3***. But the scenario also includes memory maximization, leading to ***DUE*** solution. Normally, using AHP method the solution with the higher score is considered the best one. In this scenario that is wrong, excessive energy consumption and cost. Consequently, the result is inconclusive for the presented case study. Since the application objective is more than that, the midway result, $\boldsymbol{ST_{scenario}}$ which consists only on assessment constraints enforcement, indicates that the proper solution is ***Mini 05***. The complete proposed methodology execution indicates the same solution, ***Mini 05***, but reflects criteria preference and removes all solutions draws. This methodology brings to developers the capability to select the proper solution for a certain task from a set of available solutions. Provides methods to select criteria importance and it is also able to apply constraints functions to features, improving solutions assessment.

**Table 3.** Application example results.

| | AHP | $\boldsymbol{ST_{scenario}}$ | R |
|---|---|---|---|
| ***Uno Rev***3 | 61.38 | 17 | 2.93 |
| ***Mega*** 2560 ***Rev***3 | 107.81 | 17 | 2.40 |
| ***DUE*** | 1124.9 | 16 | 1.77 |
| ***Mini*** 05 | 61.96 | 18 | 2.96 |
| ***Micro*** | 448.97 | 16 | 2.24 |

## 7 Conclusion and Future Work

There are many features/criteria which can influence the overall performance of an application running in a hardware platform. Also, developers have different perspectives of which feature is more important. The proposed methodology enables users, developers and researchers to assess which hardware platform is more suitable for their own application, depending on the application scenario. It is possible to indicate restrictions (e.g.: must have a certain feature, thresholds for feature values), and apply criteria optimization (minimize or maximize). The solutions ranking is attained using a well-known decision making method, Analytic Hierarchy Process (AHP), aimed to solve problems with multiple and conflicting criteria. Methodology results were able to point out the better/proper solution from the available ones. It also presents a more clear solution ranking compared to the AHP method, which presents an ambiguous result. Future work involves the extension of the analysis to different hardware elements/features. A further step will consider the inclusion of hardware platforms formal descriptions, to facilitate the inclusion process of new criteria.

**Acknowledgments.** This research work was partially supported by funds provided by the European Commission in the scope of FoF/H2020-636909 C2NET.

## References


1. Silva, E.M., Maló, P.: Iot testbed business model. Adv. Internet Things **4**, 37–45 (2014)
2. Boucher, T.O., Mcstravic, E.L.: Multi-attribute evaluation within a present framework and its relation to the analytic hierarchy process. Eng. Econ. **37**, 55–71 (1991)
3. Wang, X., Triantaphyllou, E.: Ranking irregularities when evaluating alternatives by using some electre methods. Omega **36**(1), 45–63 (2008)
4. Multi-criteria analysis: a manual (2009). http://eprints.lse.ac.uk/12761/1/Multi-criteria_Analysis.pdf. Accessed 20 Oct 2016
5. Jones, A., Subrahmanian, E., Hamins, A., Grant, C.: Humans' critical role in smart systems: a smart firefighting example. In: IEEE Internet Computing, vol. 19, no. 3, pp. 28–31, May–June 2015. doi:10.1109/MIC.2015.54
6. Yick, J., Mukherjee, B., Ghosal, D.: Wireless sensor network survey. Comput. Netw. **52**(12), 2292–2330 (2008)
7. Moschitta, A., Neri, I.: Power consumption assessment in wireless sensor networks. In: ICT-Energy-Concepts Towards Zero–Power Information and Communication Technology (2014)
8. Kolios, A., Mytilinou, V., Lozano-Minguez, E., Salonitis, K.: A comparative study of multiple-criteria decision-making methods under stochastic inputs. Energies **9**, 566 (2016)
9. Sabaei, D., Erkoyuncu, J., Roy, R.: A review of multi-criteria decision making methods for enhanced maintenance delivery. Procedia CIRP **37**, 30–35 (2015). http://dx.doi.org/10.1016/j.procir.2015.08.086, ISSN 2212-8271
10. Aruldoss, M., Lakshmi, T.M., Venkatesan, V.P.: A survey on multi criteria decision making methods and its applications. Am. J. Inf. Syst. **1**(1), 31–43 (2013)
11. Saaty, T.L.: A scaling method for priorities in hierarchical structures. J. Math. Psychol. **15** (3), 234–281 (1977)
12. Aldlaigan, A., Buttle, F.A.: A new measure of bank service quality. Int. J. Serv. Ind. Manag. **13**, 362–381 (2002)
13. Figueira, J., Greco, S., Ehrogott, M.: Multiple Criteria Decision Analysis: State of the Art Surveys. International Series in Operations Research & Management Science Edition, vol. 78. Springer, New York (2005)
14. Brans, J., Vincke, P.: A preference ranking organization method: the PROMETHEE method for multiple criteria decision-making. Manag. Sci. **31**, 647–656 (1985)
15. Mela, K., Tiainen, T., Heinisuo, M.: Comparative study of multiple criteria decision making methods for building design. Adv. Eng. Inform. **26**(4), 716–726 (2012)
16. Mateo, J.R.S.C.: Multi-criteria Analysis in the Renewable Energy Industry. Springer, London (2012)
17. Roy, B.: The outranking approach and the foundations of electre methods. Theory Decis. **31** (1), 49–73 (1991)

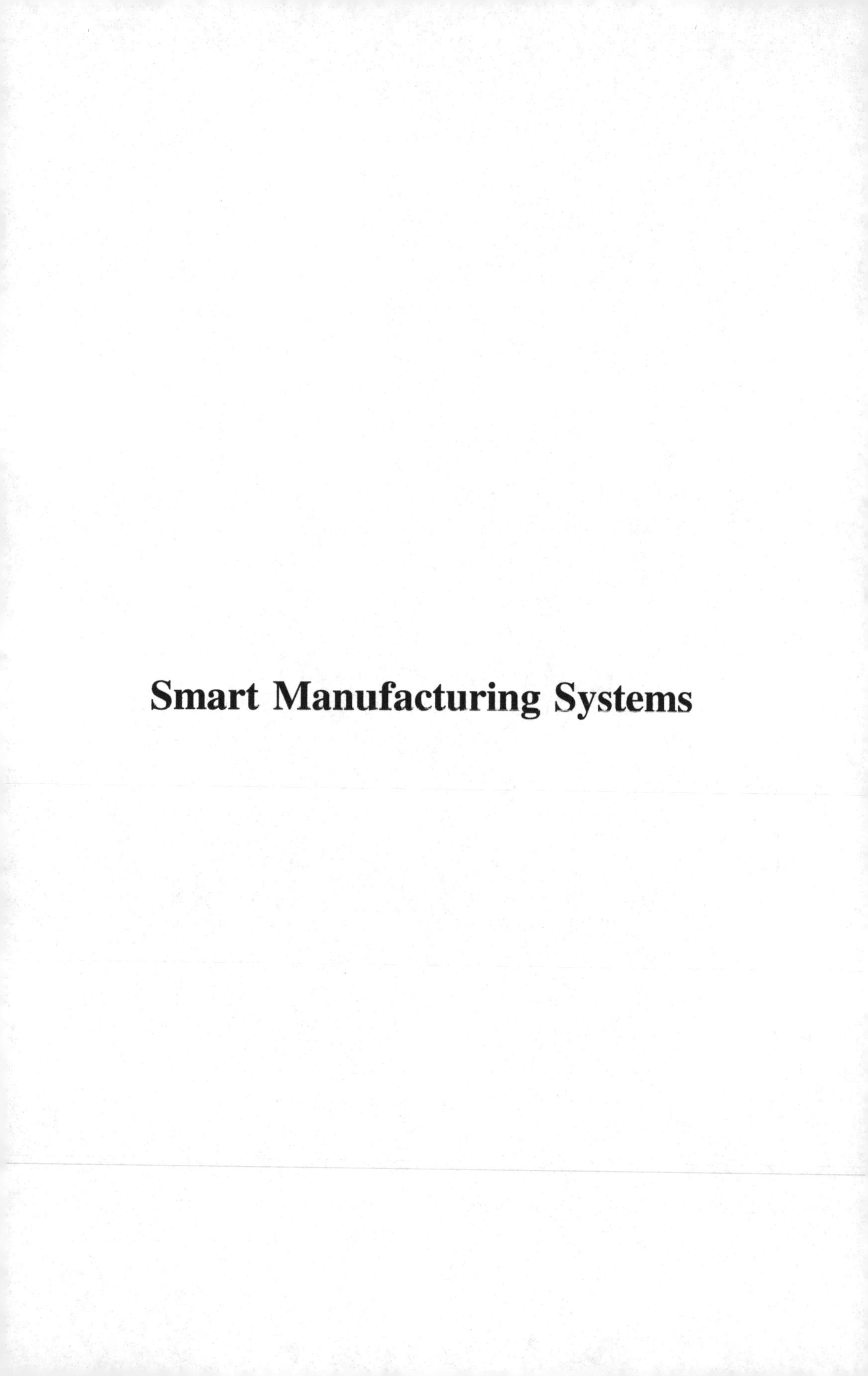

# Smart Manufacturing Systems

# Dynamic Simulation for MAS-Based Data Acquisition and Pre-processing in Manufacturing Using V-REP

Ricardo Silva Peres[1,2(✉)], Andre Dionisio Rocha[1,2], and Jose Barata[1,2]

[1] Centre of Technology and Systems (CTS) - UNINOVA, FCT Campus, Monte de Caparica, 2829-516 Caparica, Portugal
{ricardo.peres, andre.rocha, jab}@uninova.pt
[2] Faculdade de Ciências e Tecnologia – Universidade Nova de Lisboa, Monte de Caparica, 2829-516 Caparica, Portugal

**Abstract.** With the advent of the Industry 4.0 movement, smart multiagent-based cyber-physical systems (CPS) are being more and more often proposed as a possible solution to tackle the requirements of intelligence, pluggability, scalability and connectivity of this paradigm. CPS have been suggested for a wide array of applications, including control, monitoring and optimization of manufacturing systems. However, there are several associated challenges in terms of validating and testing these systems due to their innate characteristics, emergent behavior, as well as the availability and cost of physical resources. Therefore, a dynamic simulation model constructed in V-REP is proposed as a way to test, validate and improve such systems, being applied to a data acquisition and pre-processing scenario as one of the key aspects of the interaction between a CPS and the shop-floor.

**Keywords:** Simulation · V-REP · Manufacturing · MAS · CPS · Data acquisition

## 1 Introduction

In recent years, the manufacturing field has been undergoing profound changes, in great part due to the developments in the information technologies fields, in regards particularly to Cloud Computing and the Internet of Things.

An evidence of that is the emergence of the Industry 4.0 movement [1], advocating the existence of highly interconnected, intelligent devices and the growth in the amount of data shared between them. This led to the emergence of several new solutions based on smart cyber-physical systems (CPS), focused on translating these characteristics into added value for the manufacturers.

The work detailed in the following sections is motivated by such a case, more specifically whether or not a smart CPS-based predictive maintenance system can improve productivity and reduce operation expenditures in Industry 4.0 systems. Playing a pivotal role in such an approach, the simulation of the data acquisition and pre-processing stage of these systems can be seen as a first step towards validating and


L.M. Camarinha-Matos et al. (Eds.): DoCEIS 2017, IFIP AICT 499, pp. 125–134, 2017.
DOI: 10.1007/978-3-319-56077-9_11

measuring the potential impact that these solutions may have in real environments, being thus the focus of the research documented in the subsequent sections.

As such, the work detailed in the remaining sections is organized as follows. Subsections 1.1 and 1.2 contextualize simulation in regards to Smart Systems, and provide an overview of the related work in the current literature. Afterwards, Sect. 2 describes the simulation model, while Sect. 3 provides the details on a specific application case. Finally, Sect. 4 presents the discussion of the results obtained from the simulation environment, followed by some brief conclusions and description of future work in Sect. 5.

### 1.1 Simulation in Smart Systems

In the last few years the Smart Industrial Cyber-Physical Systems emerged as an important trend in the new manufacturing ecosystem. In this sense, there is a demand to support customization and the capability to learn and adapt the production systems in real time. This stems from the need to deal with disturbances and unpredictability in production, which lead into the development and research of very dynamic systems.

Consequently, the traditional simulation environments need to evolve in the same direction, otherwise the demonstration and validation of the added value presented by these new Smart Systems cannot be done. Hence, the simulation of these systems will allow researchers and developers of such solutions to improve and keep improving their own solutions and will also allow the migration of these solutions to real production environments.

### 1.2 Related Work

**Emergent Production Paradigms.** That evolution has brought new production paradigms such as evolvable systems [2], reconfigurable distributed processing, among others. In short, adaptive and reconfigurable systems, using optimization strategies to deal with unexpected events and adversities. The strongest branch of research lies on the multi-agent systems, MAS, with several implementations of solutions following these paradigms [3–6].

**Simulation in Self-adaptable Manufacturing Systems.** In the current market and with the emergent needs faced by the manufacturing companies, such as mass customization, small sized batches, etc. simulation appears as a cost effective and valuable tool. Due to the current trends of distributed, intelligent control systems and/or reconfigurable manufacturing systems [7, 8] and all the associated distributed software or functionalities such as diagnosis [9], monitoring [10], among others, the traditional rigid simulation environments are not up to par to simulate these emergent systems with their inherent requirements.

These new proposed production paradigms [11–13] can be simulated and tested in two different stages, during the design and after the development. During the design of the solution it is often important to test and tune the algorithms and the behavior of the logic system. At this stage a software like Netlogo [14, 15] or AnyLogic [16] is capable to mimic the behaviors and results from the logic point of view, where the developers are more focused on the interactions and behaviors of each distributed entity in order to validate and evaluate the overall behavior of the system.

After the development of the logic system, for instance the software responsible for abstracting a production line, it is important to mimic a real industrial line before deploying the technology in a production ready state. This previous assessment using simulation is not only important to validate the solution but also to demonstrate the expected results regarding the usage of these new solutions due to their partially unpredictable and non-deterministic components, based on self-organized and self-adaptable behaviors. Henceforth a new kind of simulation model with intelligent entities in the loop, capable of meeting these new requirements of adaptability and flexibility in run-time is required. With this goal in mind, V-REP [17] appears as an effective possibility to meet these new simulation needs, being based on a modular, distributed architecture and presenting high versatility and flexibility across a wide array of applications.

## 2 Simulation Model

Typically, one of the main goals when designing a simulation model is to maintain a balance between making it as close as possible to a real system and managing its complexity, so that it can be easily altered and experimented with. Despite not representing any concrete system, the model described in this section still abides by these principles, being however focused on providing useful data so that acquisition and processing systems can be quickly tested. Furthermore, another critical point in this modelling process was making sure that the model could be easily extendible to accommodate additional data sources (e.g. sensors and parameters), stations or functionalities, as well as dynamically changeable to cope with new conditions or parameters in run-time. The model itself consists in a simple production line, in which a conveyor belt transports parts from a source to a working station shared by two robots, as seen in Fig. 1.

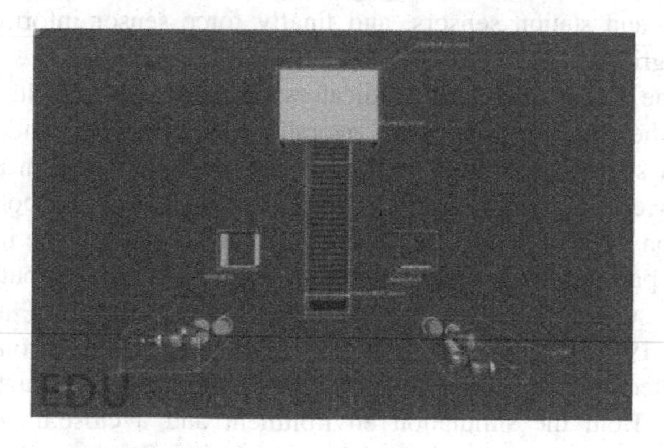


**Fig. 1.** Overview of the simulation model

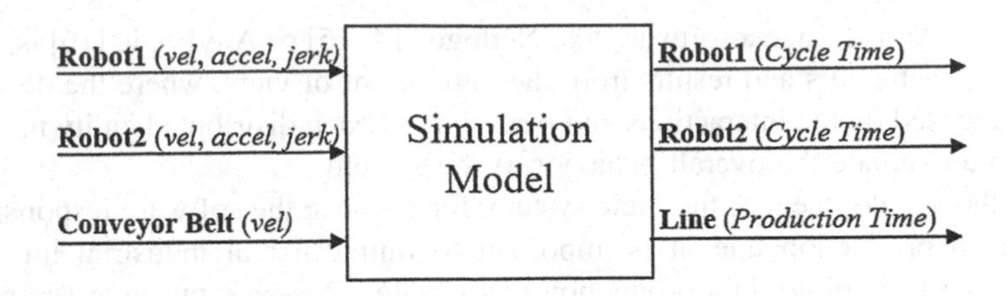


**Fig. 2.** Simulation model - inputs and outputs

Once the part has arrived at the station sensor, *robot1* performs a welding operation on it (see Fig. 3), followed by *robot2* picking it up and moving it to a storage box (*Pick and Place* operation, or *P&P*).

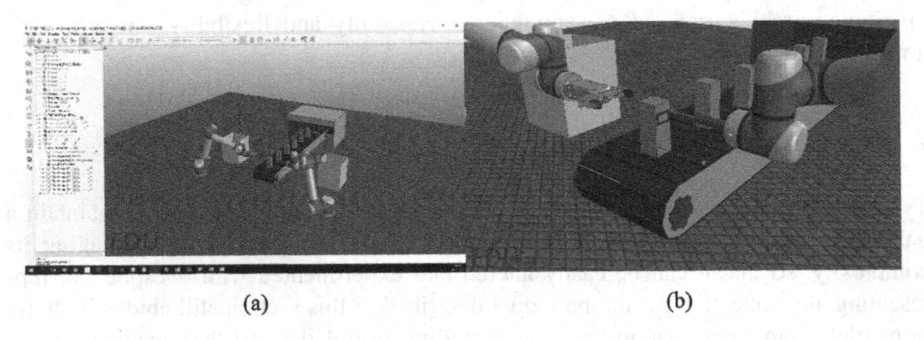

(a) (b)

**Fig. 3.** Simulation environment (a) and process close-up (b)

Model design entails a series of key steps, including the clear definition of the system boundaries, its actors, input variables and quantitative criteria or performance measures on the basis of which the performance of different configurations can be compared and evaluated [18]. For this model, the input parameters define the *velocity* for the conveyor belt, as well as for each of the robots. The meaningful data points to be extracted include each robot's *state* (e.g. picking, welding), *part presence* information from the source and station sensors, and finally force sensor information from the picking robot's gripper.

Therefore, the Key Performance Indicators (KPI) associated with measuring the performance of the system consist in the overall *Production Time*, and the *Cycle Time* of each robot. A summary of these inputs and outputs is provided in Fig. 2.

In this instance the *Cycle Time* indicates the time it takes each robot to carry out its respective task on the part, while the *Production Time* represents the interval between the moment the part arrives at the workstation and the instant it is put into storage.

To construct such a model, the Virtual Robot Experimentation Platform (V-REP) was chosen [17, 19]. V-REP is a versatile 3D robot simulator based on a modular and distributed architecture, capable of concurrently simulating control, actuation, sensing and monitoring. Both the simulation environment and a close-up of the welding operation during simulation run-time can be seen in Fig. 3.

The scene was built using two UR5 flexible robots (one coupled with an RG2 gripper and the other with a welding gun) and a simple conveyor with two part-presence sensors, one at the source and one at the workstation.

The simulation model was programmed using a collection of Lua scripts, each associated to its own scene object. These scripts are responsible for controlling the system execution, with the resources communicating via signals using V-REP's own API. These communications can be observed in Fig. 4.

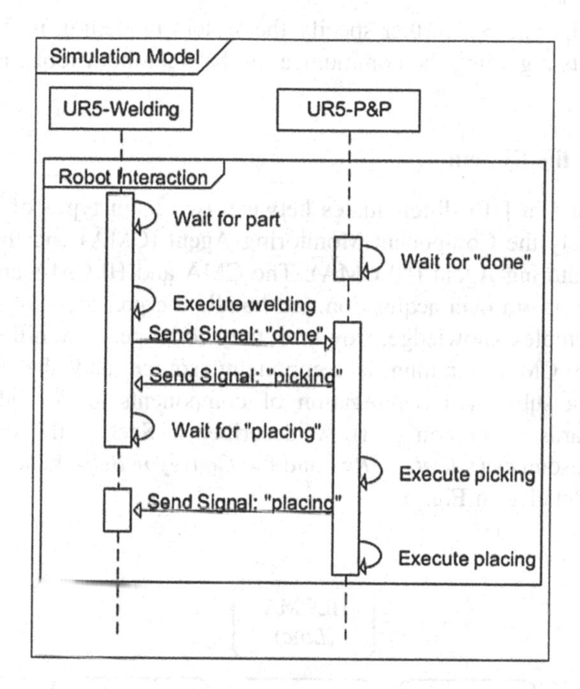


**Fig. 4.** Robot interactions within the simulation model

Furthermore, the simulator allows the connection of external clients with relative ease via V-REP's Remote API, which supports several programming languages including Java, C/C++ and Python.

In this context, Sect. 3 provides a detailed description of an application case in which this API was used in order to connect the simulation environment to an external MAS-based data acquisition and pre-processing CPS implemented in Java. This description is focused on the system modelling and communication aspects of the integration process.

## 3 Application Case

Industry 4.0 manufacturing systems typically comprise highly interconnected, modular and flexible intelligent components producing large amounts of data at often astounding rates. This makes data acquisition an essential stage in several applications, such as control, monitoring, preventive maintenance and energy optimization. Moreover, solutions proposed regarding these topics tend to rely on data mining and analysis techniques, which not only make it crucial for raw data to be extracted, but also preprocessed in order for it to be translated into a potential business advantage.

With this in mind, the MAS-based data acquisition and preprocessing CPS described in [10], [20] was instantiated in order to collect data from the simulation model and compute more complex values from the raw data readily available from the simulated shop floor.

Subsections 3.1 and 3.2 further specify the system modelling in the CPS and the integration aspects regarding the communication between both tools, respectively.

### 3.1 Modelling the System

The CPS proposed in [10] differentiates between two main types of agents for data acquisition, namely the Component Monitoring Agent (CMA) and the Higher-Level Component Monitoring Agent (HLCMA). The CMA and HLCMA are similar in the sense that both perform data acquisition and handle the preprocessing of raw data to generate more complex knowledge. However, they both operate at different abstraction levels, with the CMA handling a component/resource and the HLCMA being responsible for a subsystem (aggregation of components and/or other subsystems cooperating towards a common goal). As described in Sect. 2, the simulation model encompasses 3 resources (*UR5#1*, *UR#2* and the *Conveyor Belt*), hence resulting in the agent topology detailed in Fig. 5.

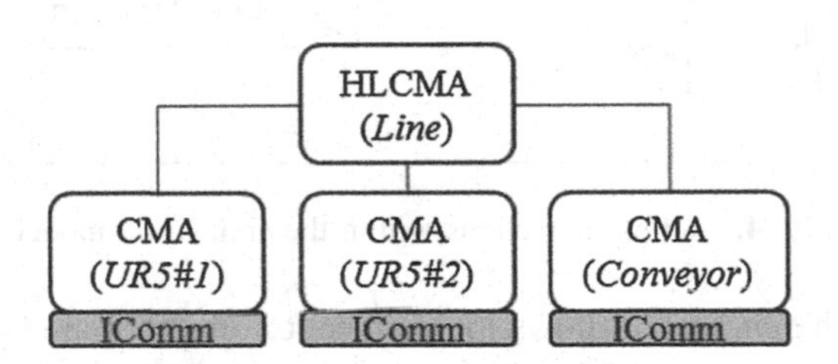


**Fig. 5.** Agent topology

Agents are initially deployed in accordance to an XML blueprint file, which should be modelled in order to provide an accurate description of the system. This description is then interpreted by the MAS to understand not only the system topology in terms of available resources, but also regarding the relevant values to be extracted or calculated.

As an example, the *Cycle Time* relative to the welding operation can be represented in this XML file as:

```
<Timespan id="welding_time" source="agent">
  <TimespanMapping>
     <StartConditions>
     <StateValue id="welding" value="true"/>
     </StartConditions>
    <EndConditions>
    <StateValue id="welding" value="false"/>
    </EndConditions>
  </TimespanMapping>
</Timespan>
```

### 3.2 Communication

As it can be seen in Fig. 5, each agent has a generic hardware communication interface that should be implemented for each application case, thus allowing the actual CPS to be generic and easily integrated into different scenarios independently from the underlying technologies and protocols. For this case the interface was implemented using V-REP's remote Java API, and allows data to be acquired by two different means, more specifically either periodically polling of relevant data, or subscription to data values on value-change events. For this application case, data acquisition follows the latter.

On V-REP's side, an additional script was developed in order to act as a server for the simulation data. On each pass the script polls the simulation data and checks if it has changed compared to the last value. If so, it sets up a signal with the new value so that it can be read by the CPS, acting similarly to what one can find for instance in an OPC-UA client-server subscription model.

## 4 Discussion of Results

In order to validate the repeatability of the simulation results, as well as the capability of the MAS to acquire raw data from the simulator and preprocess it accurately, five different runs were executed. Each simulation run was executed until twenty samples of the *cycle time* for each robot (*Welding* and *P&P*) as well as the *production time* for the line were extracted, thus resulting in a total of one hundred samples across all runs for each KPI. The results can be observed in Fig. 6.

For each of the KPIs represented in Fig. 6, the standard deviation $\sigma$ was calculated through:

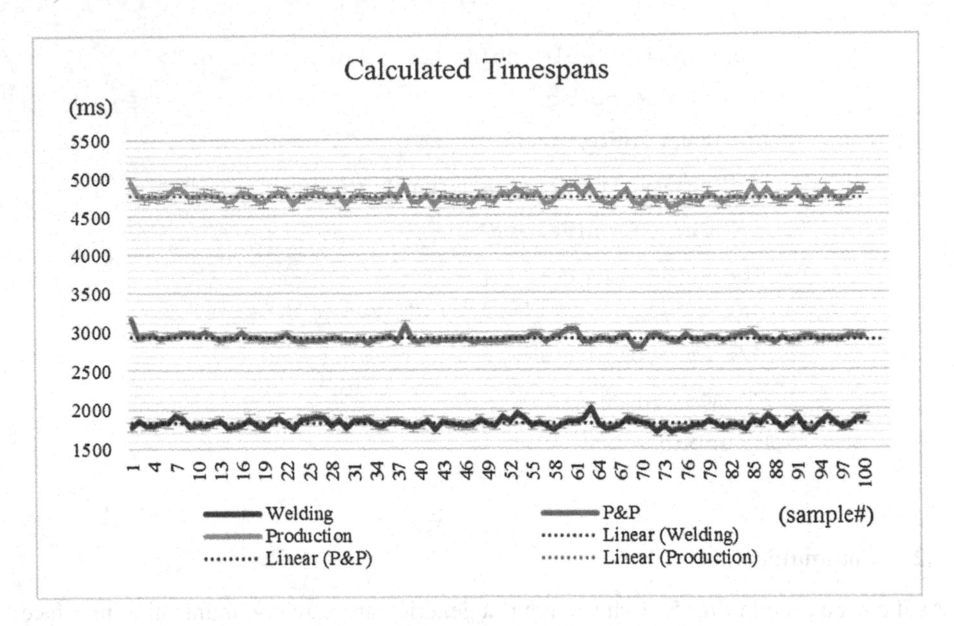


**Fig. 6.** Timespans calculated by the MAS

$$\sigma = \sqrt{\frac{\sum (x - \mu)^2}{N}}. \quad (1)$$

The application of Eq. 1 resulted in standard deviation values of 56.099 ms, 49.639 ms and 68.977 ms for the *Welding*, *P&P* and *Cycle Time* variables, respectively. In order to measure the repeatability of the results, the coefficient of variation $C_v$ of each KPI was calculated using:

$$C_v = \frac{\sigma}{\mu}. \quad (2)$$

Through Eq. 2 each coefficient of variation was obtained, resulting in 3.07% for *Welding*, 1.45% for the *Production Time* and 1.70% for *P&P*. The coefficient of variation has been shown to be a good indicator of repeatability in non-negative datasets [21], as such these percentages suggest that the repeatability of the simulation results is reliable, with the extracted data presenting relatively low levels of variability.

## 5 Conclusion and Future Work

A dynamic and extendable simulation model was proposed for the validation of MAS-based CPS data acquisition and pre-processing solutions to be applied in the context of Industry 4.0.

The model was implemented in V-REP, a versatile and modular 3D robot simulator capable of concurrently handling the simulation of control, actuation, sensing and monitoring. This implementation consisted in a conveyor belt continuously transporting parts from a source to a workstation, where two UR5 robots would weld the part and then move it to a storage box.

An example of an application case was also shown, consisting in the integration of a MAS-based data acquisition and pre-processing CPS with the simulation environment, from which it extracted raw data such as robot states and calculated relevant KPIs, namely cycle times and overall production time. Through the analysis of the results it can be concluded that the simulation model implemented in V-REP presents reliable levels of repeatability with low degrees of variance in the collected data across different runs, hence possibly being a viable solution to validate these kinds of systems in the context of Industry 4.0.

Looking at the application of dynamic simulation for data acquisition and pre-processing as a first step, future work will include the extension of the simulation model to accommodate the testing of plug and produce and reconfiguration functionalities, aiming at providing a flexible environment to validate not only systems tailored for data extraction, but also other contexts such as shop floor control and optimization.

## References


1. Lee, J., Bagheri, B., Kao, H.A.: A cyber-physical systems architecture for Industry 4.0-based manufacturing systems. Manuf. Lett. **3**(December), 18–23 (2015)
2. Frei, R., Barata, J., Onori, M.: Evolvable production systems context and implications. In: IEEE International Symposium on Industrial Electronics, ISIE 2007, pp. 3233–3238 (2007)
3. Rocha, A.D., Barata, D., Di Orio, G., Santos, T., Barata, J.: PRIME as a generic agent based framework to support pluggability and reconfigurability using different technologies. In: Technological Innovation for Cloud-Based Engineering Systems, pp. 101–110 (2015)
4. Barbosa, J., Leitão, P., Adam, E., Trentesaux, D.: Structural self-organized holonic multi-agent manufacturing systems. In: Mařík, V., Lastra, J.L.M., Skobelev, P. (eds.) HoloMAS 2013. LNCS (LNAI), vol. 8062, pp. 59–70. Springer, Heidelberg (2013). doi:10.1007/978-3-642-40090-2_6
5. Colombo, A.W., Schoop, R., Neubert, R.: An agent-based intelligent control platform for industrial holonic manufacturing systems. IEEE Trans. Ind. Electron. **53**(1), 322–337 (2006)
6. Lepuschitz, W., Zoitl, A., Vallée, M., Merdan, M.: Toward self-reconfiguration of manufacturing systems using automation agents. IEEE Trans. Syst. Man Cybern. Part C Appl. Rev. **41**(1), 52–69 (2011)
7. Rocha, A., et al.: An agent based framework to support plug and produce. In: 2014 12th IEEE International Conference on Industrial Informatics (INDIN), pp. 504–510 (2014)
8. Ribeiro, L., Barata, J., Onori, M., Hanisch, C., Hoos, J., Rosa, R.: Self-organization in automation - the IDEAS pre-demonstrator. In: IECON 2011 - 37th Annual Conference on IEEE Industrial Electronics Society, pp. 2752–2757 (2011)
9. Rocha, A.D., Monteiro, P.L., Barata, J.: An artificial immune systems based architecture to support diagnoses in evolvable production systems using genetic algorithms as an evolution enabler

10. Rocha, A.D., Peres, R., Barata, J.: An agent based monitoring architecture for plug and produce based manufacturing systems. In: 2015 IEEE 13th International Conference on Industrial Informatics (INDIN), pp. 1318–1323 (2015)
11. Leitão, P., Restivo, F.: ADACOR: a holonic architecture for agile and adaptive manufacturing control. Comput. Ind. **57**(2), 121–130 (2006)
12. Neves, P., Barata, J.: Evolvable production systems. In: IEEE International Symposium on Assembly and Manufacturing, ISAM 2009, pp. 189–195 (2009)
13. Koren, A.Y., Heisel, U., Jovane, F., Moriwaki, T., Pritschow, G., Ulsoy, G., Van Brussel, H.: Reconfigurable Manufacturing Systems. http://www-personal.umich.edu/~ykoren/uploads/Reconfigurable_manufacturing_systems_1999_keynote_paper.pdf. Accessed 08 June 2014
14. Barbosa, J., Leitão, P.: Simulation of multi-agent manufacturing systems using agent-based modelling platforms. In: 2011 9th IEEE International Conference on Industrial Informatics, pp. 477–482 (2011)
15. Barbosa, J., Leitão, P.: Modelling and simulating self-organizing agent-based manufacturing systems. In: IECON 2010 - 36th Annual Conference on IEEE Industrial Electronics Society, pp. 2702–2707 (2010)
16. Zheng, N., Lu, X.: Comparative Study on Push and Pull Production System Based on AnyLogic
17. Rohmer, E., Singh, S.P.N., Freese, M.: V-REP: a versatile and scalable robot simulation framework. In: IEEE International Conference on Intelligent Robots and Systems (IROS), pp. 1321–1326 (2013)
18. Maria, A.: Introduction to modelling and simulation. In: Winter Simulation Conference, pp. 7–13 (1997)
19. Freese, M., Singh, S., Ozaki, F., Matsuhira, N.: Virtual robot experimentation platform V-REP: a versatile 3D robot simulator. In: Ando, N., Balakirsky, S., Hemker, T., Reggiani, M., Stryk, O. (eds.) SIMPAR 2010. LNCS (LNAI), vol. 6472, pp. 51–62. Springer, Heidelberg (2010). doi:10.1007/978-3-642-17319-6_8
20. Rocha, A.D., Peres, R.S., Flores, L., Barata, J.: A multiagent based knowledge extraction framework to support plug and produce capabilities in manufacturing monitoring systems. In: ISMA 2015 - 10th International Symposium Mechatronics its Applications (2016)
21. Pryseley, A., Mintiens, K., Knapen, K., der Stede, Y., Molenberghs, G.: Estimating precision, repeatability, and reproducibility from Gaussian and non-Gaussian data: a mixed models approach. J. Appl. Stat. **37**(10), 1729–1747 (2010)

# Enhancing Dependability and Security of Cyber-Physical Production Systems

Hessamedin Bayanifar(✉) and Hermann Kühnle

Otto Von Guericke Universität Magdeburg, Magdeburg, Germany
{hessamedin.bayanifar,hermann.kuehnle}@ovgu.de

**Abstract.** Despite all its potentials, new industrial revolution enabled by cyber-physical systems (CPS), still has major concerns and obstacles to overcome with regards to dependability and security on its way to be fully appreciated. This study targets these concerns by proposing a generic model for intelligent distributed dependability and security supervision and control mechanisms to enable components to autonomously meet their own security and dependability objectives through real-time distributed improvement cycles, using multi-agent systems approach to enable full exploitation of the model's evolution capabilities.

**Keywords:** Cyber-physical production systems · Smart manufacturing unit · Dependability and security · Multi-agent systems

## 1 Introduction

Smart distributed manufacturing systems consist of a large number of widely dispersed loosely-coupled yet collaborating heterogeneous components that are vastly connected to and communicate with cyber space. To enhance their capabilities, these systems try to exploit smart properties through enhancing their own intelligence and processing power, or via accessing the internet and its vast options to enhance these properties. On the one hand, using these properties and enhancing capabilities can offer manufacturing enterprises a plethora of opportunities and strategic advantages, on the other hand, however, such vast diversity and total exposure to cyber space, as well as versatility of processes and structures, raise major vulnerabilities as dependability and data security issues that may lower the motivation to rely on these enormous capabilities. In order to increase acceptance, three research hypotheses are posited:

*H1: Additional properties and enhanced capabilities increase risks in playing manufacturing on the base of smart manufacturing units. Multi-agent software module patterns are most appropriate for supporting and optimising smart manufacturing units' applications and risks may be lowered by continuously monitoring and analysing the readiness (maturity) of properties and the cleanness of data flows (failure probabilities multiply).*


L.M. Camarinha-Matos et al. (Eds.): DoCEIS 2017, IFIP AICT 499, pp. 135–143, 2017.
DOI: 10.1007/978-3-319-56077-9_12

*H2:* *A powerful tool for monitoring and analysing dependability as well as maturities of properties and units can be established on the base of dynamic modelling. The model architecture may be specified by expected contributions and resulting requirements. The desired features for tool implementations may be provided by Multi-agent design pattern.*

*H3:* *A decision table for failure risks, disturbance cases, decisions for actions and inputs for learning may be established. Using agents, adequate architecture may be implemented into a learning experimentation environment for simulation or manufacturing scenarios and evaluations of risks.*

Following the proposed three hypotheses the research question is formed as weather an agent-based distributed dependability and security supervision and control that has inherited smart system properties, can enhance overall dependability and security of the system. The corresponding factor to analyse the performance of the adopted approach is to see to what extent it promotes or enable flexibility, responsiveness, learning capability, scalability, level of autonomy, reconfigurability, and reusability of the model's modules. To assure maximum dependability throughout an enterprise, the adopted approach must be able to deal with all incorporated components, information flows among them as well as the cyber areas, networks, databases and servers. To this goal, a distributed Dependability and Security Model (Fig. 1) is introduced for covering the entire system, every units and components down to all levels of detail (LoD). The model includes a core model, a control loop, and a connection to the virtual world.

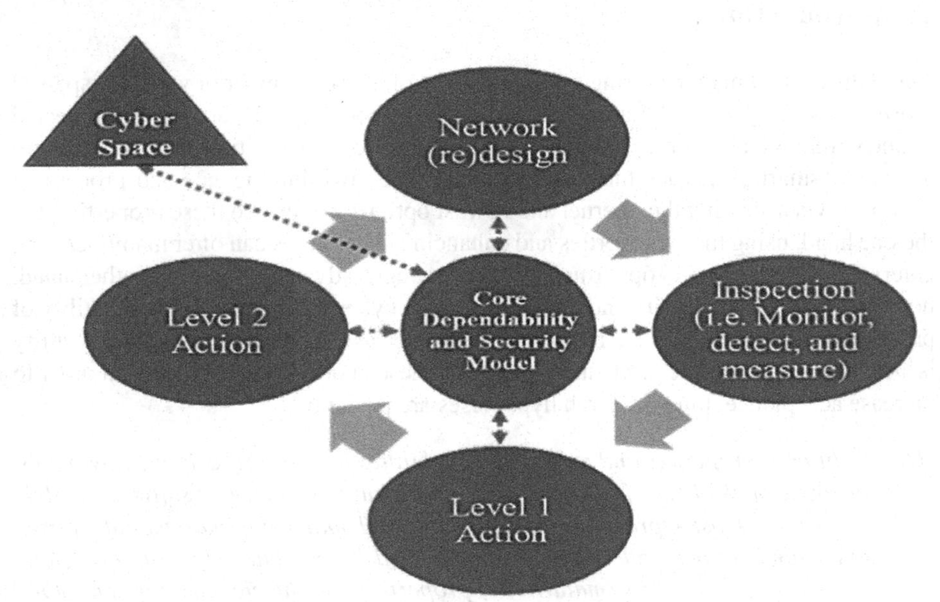


**Fig. 1.** Smart dependability and security architecture

The **core model** consists of two main parts: *object description*, and *risk model*, where the former focuses more on objects' context and self-awareness, and imports data about object's environment, collaborations, functions and modules, objectives, application and task description, etc. that are needed to develop risk model, and the latter covers accordingly all Dependability and Security parameters, vulnerabilities and risks, and the ways of measuring and dealing with them. The *core model*, in other words, feeds the improvement process to be done by the *control loop*. The relevant data for object description section are imported from the cloud or sensed as a part of the object's self-/context-awareness. The risk model contains a scalable feature for considering possible risks, assessing their criticality and their possible effects on the object or the system in total (e.g. Failure Mode, Effect and Criticality Analysis (FMECA)/Fault Tree Analysis (FTA)). Self-optimizing occur via sharing knowledge with all other smart units, and updating its own structure and database through continuous feedbacks (control loops).

**The Control Loop** invokes the process of Inspection (i.e. *Monitoring*, *Detecting*, and Identifying and *Measuring*), and Reaction (i.e. giving *Alarms*, taking *Action*, and doing the *Reconfiguration* afterwards) in real-time. All steps can be carried out fully- or semi-autonomously by smart objects through this attached core model, which is located in the cyber space and is in collaboration with all other models. This gives the components all abilities to collaborate with the common objective of raising and maintaining the dependability and security of the total system.

## 2 Relationship to Smart Systems

The suggested solution aims at improving the dependability and security of smart systems with its focus on (but not limited to) cyber physical production systems (CPPS). To this aim, we introduce a generic dependability model and architecture that tries to harness the capabilities and properties of smart systems, that are elaborated in [1] (i.e. interoperability, autonomy, scalability, modularity, heterogeneity, reconfigurability, and context-awareness), for maximising its performance and versatility. In other word, it aims to be mechanism equipped with smart systems' properties, to ensure the dependability and security of smart systems. In order to enable these properties, Intelligent Agents are to be summoned as the implementation toolset for our model.

## 3 Review of Literature

Smart Distributed Manufacturing systems, enabled by Cyber-Physical Systems, have major structural similarities, as they both address three main layers: physical layer, cyber layer, and data communication and integration layer. Each of these layers has its own concerns with regards to dependability and security. Accordingly, many studies tried to point out these issues or suggest countermeasures [2–6], or to point out and to evaluate the cyber-physical vulnerabilities and their impacts on manufacturing systems [7]. In [8], the approach tries to design and to implement a robust cyber physical system, and [9] attempts to model ontology-based dependability in CPSs using FMEA techniques are outlined. Mathematical approaches were used in [10] to model security risks of CPSs and to quantitatively evaluate their risks. Dependability of self-optimizing systems was studied and analysed in [11], where various methods covering conceptual design, and development phases were introduced. Authors in [12] used systems' context awareness to increase security in information access, by asking questions like: *who* wants *what* information, *how*, from *where*, and *when*? This study, on the other hand, tries to propose a generic model, and an agent-based structure for developing an intelligent and autonomous distributed dependability and security supervision, and control in CPPS. The model, given its structure, and enabled by intelligent agents, is expected to bring more flexibility and responsiveness. Also, higher autonomy and scalability is predicted through context-awareness and summoning the capabilities of intelligent agents. Moreover, due its decentralized real-time monitoring and control, higher coverage and improved stability is likely to be achieved.

# 4 Research Contribution and Innovation

This section describes the multi-agent based architecture of the dependability and security mechanism explained in the introduction, and elaborates on the way smart systems properties are manifested and practised in this model. In the end an example shows how the model can function in a given condition using its capabilities enabled by the properties it possesses.

## 4.1 Multi-agent Systems (MAS) for Model Architecture

The dependability and security model as described, with the potential properties mentioned, requires a toolset to enable such properties. Accordingly, Multi-Agents Systems (MAS) can be a decent candidate, since intrinsically, intelligent agents (IA) demonstrate responsiveness, proactiveness, goal-orientation, social-ability, scalability, flexibility, robustness, self-configuration, adaptability/re-configurability, along with their decentralized architecture and learning capabilities [13]. After determining the tool, the model is to be translated into an architecture composed of interacting agents. The first step would split the model task-wise onto single agents. Doing so, the following Table 1 and Fig. 2 demonstrate how to introduce agents and their task descriptions, as well as their overall structure and collaborations' relations.

Data captured by sensors are sent to the VO (virtual object), which is the cyber representative or digital twin of the component (e.g. industry 4.0 component). Then, these data are filtered (to omit unnecessary data), monitored, and at the same time this stage is being fed into a status/topology manager to gain the approved context for comparing the filtered sensed data to detect anomalies. When detected, info is sent to the level one analyser, where in collaboration with assessment agent and database, the risk will be identified, and its severity will be measured as the sum of possible losses it can cause to other components or parts of the system. Risk data will be sent to the decision-making agent (DM level one) for making decisions on the appropriate actions, e.g. alarms, and send the command to actuators. However, if the problem requires more advanced analysis, it will be sent to the analyser level two, where harder problems can be analysed, and negotiations with other agents might be necessary to make the right measurements and analyses to provide accurate data for DM level 2. In decision making

**Table 1.** Applied agents and their task description

| | Agents | Task description |
|---|---|---|
| Core model | Status manager | Updates the status of components. The main part of context-awareness. Knows the approved context, authorities, topography of the system, etc. |
| | Database | Stores components models and risk data. Data are stored here in modules for each type of risk to be accessed by analysers and assessment agent. Other components when authorized, can have access to some of the data during negotiation or when they are new to the system, to get updated with vulnerabilities and measures, etc. |
| | Assessment agent | Receives data from analyser and assesses the risk through negotiating with other components' assessors, receiving context data from TS manager, and having access to the database and the object model |
| | Interface | For negotiations between agents of other components. Updating the topography and context information, more accurate and global risk assessment, providing access to databases of other components. |
| Monitoring/ detection | Data filter | Filtering out redundancies. Looking for useful data among loads of data |
| | Monitoring | Looking for anomalies and risks, by comparing the current-state sensed data with current-state approved context. Then sends the detected cases to level one analyser |
| Measurement | Data analyser lvl 1 | For simpler problems/quicker responses. Data analyser does the identification and measurement of risks. Lvl 1 analyser does the simple analysis and communicates with assessment agent to provide proper input for DM level one. It then provides feedback to the core model |
| | Data analyser lvl 2 | For more complicated problems, analyser level 1 sends the case to analyser level two with more abilities. If needed this analyser practises negotiations with other components agents to provide best global data of the risk to feed the DM level 2 |
| Alarm/action/ reconfiguration | Decision-making lvl 1 | For simpler reactions/quicker responses. After supplied with proper risk information, releases alarm to right entities and send proper commands to actuators to apply right corrective actions, and reconfiguration when needed. Feedback is then being provided to the core model |
| | Decision-making lvl 2 | Provides higher lever reactions, and reconfigurations, and more advanced alarms for more complicated issues. If needed asks for collaboration of other agents and components resources to solve a problem. Feedback is then being provided to the core model |

level two, more complex actions, and if required, negotiations with agents of other components (e.g. for sharing resources in fixing an issue), take place. After the actions are carried out, and issues are confirmed to be solved, the result is fed into the core model to update the risk assessor.

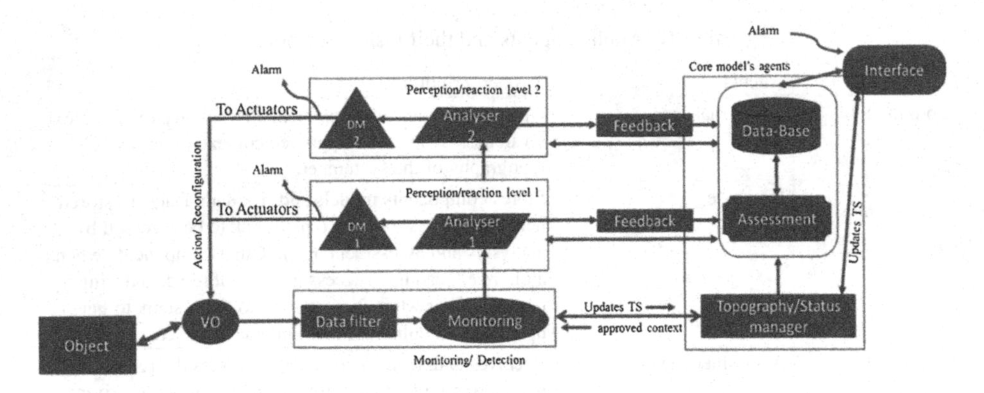


Fig. 2. Proposed agent architecture for components' intelligent dependability and security

## 4.2 Properties, as Seen in the Model, and an Agent-Based Example

Table 2 shows the expected contribution of the smart systems properties and accordingly the resulting requirements.

An example can consider a job-shop unit with several automated machines and conveyors, using multi-agent systems to control their production system. Simultaneously, along with data form sensors, machines and controllers' interactions will be checked via "monitoring agents", receiving all information being send and captured by controller units and machines. Two of the possible risks can be either one of the controller agents itself be compromised by an adversary, or something unintentionally occurs to one of the machines. Some cyber security risks associated with the former might be cloning, repudiation, MITM attack, DDoS, eavesdropping, and some risks concerning the machines (can be partial or full breakdowns, connectivity loss, etc.). Taking the breakdown of one of the machines as an example, the monitoring agent will notice the change in the system in real-time (e.g. the number of parts passing across a specific sensor), the analyser will identify the issue (i.e. in this case breakdown of a machine (machine B, in Fig. 3)) and will measure the impact on the system and its components it collaborates with, and will send the data to decision the maker agent: Alarms will be published to the right entities (e.g. controllers, maintenance centre, spare part inventory, etc.). The machine will be stopped and called unavailable. The request will be sent to other agents for availability of another machines to do the task instead of the broken-down machine and after locating the alternative machine, ways (e.g. conveyors, AVGs, etc.), will be found to send the parts to the new alternative machine. And finally, a feedback will be sent to the core model for updating the data base and the assessment model, and for generating reports.

**Table 2.** Control loop steps and the relationship with smart manufacturing systems' properties

| | Contribution | Requirements/method |
|---|---|---|
| **Monitoring**: Constantly checking the related parameters defined in the core model to find risks | | |
| Context-awareness | Parameters are derived from the context | Via sensors/cloud |
| Interoperability | Hierarchical/hierarchical collaboration | Shared semantic/ontology |
| Autonomy | Autonomously done by Intelligent Agents | Core model/sensors |
| Modularity | Resources to be used in various setups | Modular resources |
| Scalability | Resources to be added or removed | Registering mechanism |
| Heterogeneity | Mechanism differs based on object type | Via core model |
| **Detecting**: Finding anomalies/risks by comparing real-time status with approved parameters | | |
| Context-awareness | Current vs Approved context comparison | Via sensors/core model |
| Interoperability | Hierarchical/hierarchical collaboration | Shared semantic/ontology |
| Autonomy | Done by agents, and aided by core model | Agents/data base |
| Modularity | Data-base and model to be re-useable | Modular risk Data-base |
| Scalability | To be seen data-base/detection methods | Updateable core model |
| Heterogeneity | Methods differs based on object type | Set in core model |
| **Measuring**: When detected, identifying the risk type, its severity, impacts, occurrence frequency, using assessments methods such as FMEA/FTA | | |
| Context-awareness | Finding global measure based on context | Updating context data |
| Interoperability | Objects negotiating to find global measure | Via agents/model |
| Autonomy | Done by agents and by using risk model | Risk model in the core |
| Modularity | Measurement resources to be shared reused in various setups | Risk categories to be modularly saved |
| Scalability | Criteria/data-base to be changed/updated | Scalable risk model |
| Heterogeneity | Criteria and risk model differ object-wise | Risk model definition |
| **Alarm**: Alarming right entities, i.e. people, or components that may be affected by the risk | | |
| Context-awareness | Right entities are known through context | Context update in model |
| Interoperability | In carrying alarm to various entities | Semantic definition |
| Autonomy | Done by agents after measuring risks | Agent collaboration |
| Modularity | Agents/functions to be used in new setups | Modular alarm resource |
| Scalability | methods/agents to be added or removed | Scalable alarm resource |
| Heterogeneity | Mechanism tries to stay the same for all | Semantic definition |
| **Action**: Making globally optimum decisions and defending against/fixing measured risks | | |
| Context-awareness | Optimum decision/reaction context-wise | Via sensors/models |
| Interoperability | Sharing resources/information for taking optimum decision and action | Via semantic and ontology definition |
| Autonomy | To be done autonomously by agents | Agent collaboration |
| Modularity | Resources to be mixed in various setups | Modular agents/actuators |
| Scalability | Agents/actuators/models to be scalable | Registering mechanism |
| Heterogeneity | Actions/resources differ by object type | Via model/resources |
| **Reconfiguration**: Providing feedback to the model and preparing the component to be reused | | |
| Context-awareness | To be done based on context requirements | Via the model |
| Interoperability | Providing understandable feedbacks to others/receiving required data | Via model and semantic definition |
| Autonomy | Semi/fully autonomous and done by agents | Model/agents |
| Modularity | Components may be reused in new setups | Modular components |
| Scalability | Extended/reduced structure in new setup | Scalable model/object |
| Heterogeneity | Done based on object type<br>Same feedback mechanism | Via model<br>Same agent functionality |

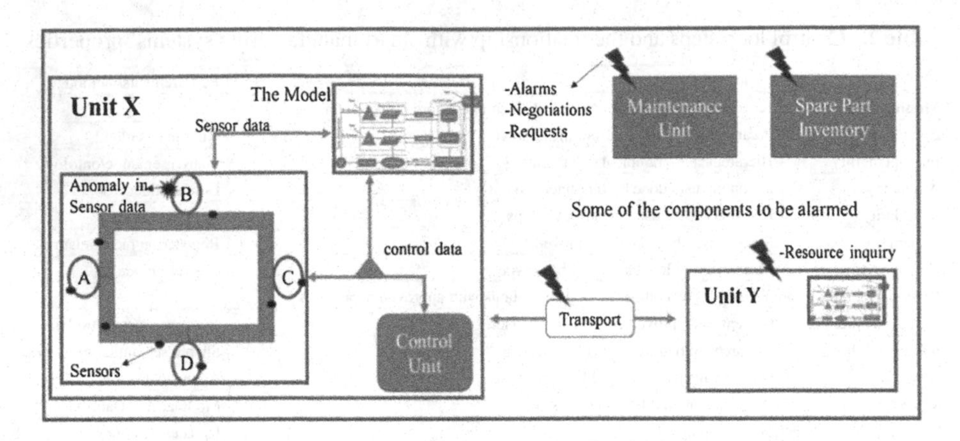


**Fig. 3.** Experiment setup and scheme in a smart manufacturing unit breaks down

## 5 Conclusion

Based on available technologies, a structure based on multi-agent systems was suggested as the dependability and security model. Moreover, intelligent agents are shown to be capable of equipping the process of dependability and security supervision and control with the smart systems' properties to improve it and make it more efficient. The next step of the study would be testing the model in various cases and extending its performance and capabilities (H3). One case is simulating the example described above, and the other will focus on the data security risks (e.g. intrusion attack, DDoS attack) on one component to test the models performance in detecting and blocking it, and after disinfection, reconfiguring the component to be used again by the system. The experiment is to be done by simulating DDoS attack i.e. by overloading and increasing data traffic, assessing the models' reactions in handling the risk and its learning progresses for feedback.

## References

1. Kühnle, H., Bitsch, G.: Smart Manufacturing Units. pp. 55–70 (2015)
2. Huang, S., et al.: Cyber-physical system security for networked industrial processes. Int. J. Autom. Comput. **12**(6), 567–578 (2015)
3. Wells, L.J., et al.: Cyber-physical security challenges in manufacturing systems. Manuf. Lett. **2**(2), 74–77 (2014)
4. Wu, G., Sun, J., Chen, J.: A survey on the security of cyber-physical systems. Control Theory Technol. **14**(1), 2–10 (2016)
5. Zhang, L., Wang, Q., Tian, B.: Security threats and measures for the cyber-physical systems. J. China Univ. Posts Telecommun. **20**(Suppl. 1), 25–29 (2013)
6. Karnouskos, S.: Chapter 6 - Industrial agents cybersecurity. In: Industrial Agents, pp. 109–120. Morgan Kaufmann, Boston (2015)
7. DeSmit, Z., et al.: Cyber-physical vulnerability assessment in manufacturing systems. Proc. Manuf. **5**, 1060–1074 (2016)

8. Hu, F., et al.: Robust cyber-physical systems: concept, models, and implementation. Future Gener. Comput. Syst. **56**, 449–475 (2016)
9. Sanislav, T., Mois, G., Miclea, L.: An approach to model dependability of cyber-physical systems. Microprocess. Microsyst. **41**, 67–76 (2016)
10. Orojloo, H., Azgomi, M.A.: A method for modeling and evaluation of the security of cyber-physical systems. In: 2014 11th International ISC Conference on Information Security and Cryptology (2014)
11. Dorociak, R., Gausemeier, J.: Methods of improving the dependability of self-optimizing systems. In: Gausemeier, J., Rammig, F.J., Schäfer, W., Sextro, W. (eds.) Dependability of Self-optimizing Mechatronic Systems. LNME, pp. 37–171. Springer, Heidelberg (2014). doi: 10.1007/978-3-642-53742-4_3
12. Wan, K., Alagar, V.: Context-aware security solutions for cyber-physical systems. Mobile Netw. Appl. **19**(2), 212–226 (2014)
13. Unland, R.: Software agent systems A2 (Chap. 1). In: Leitão, P., Karnouskos, S. (eds.) Industrial Agents, pp. 3–22. Morgan Kaufmann, Boston (2015)

# Features Extraction from CAD as a Basis for Assembly Process Planning

Baha Hasan(✉) and Jan Wikander

Department of Machine Design, The Royal Institute of Technology (KTH), Stockholm, Sweden
bahasan@kth.se

**Abstract.** This paper describes a novel approach to recognize product features, which are significant for Assembly Process Planning (APP). The work presented in this paper is a part of a larger effort to develop methods and tools for a more automated and bidirectional link between product CAD and the different processes and resources applied in APP. APP is the phase, in which the required assembly processes and resources are determined in order to convert a product to fully assembled or semi-assembled product. Product features will be extracted from the SolidWorks (SW) CAD file using SW - Application Programming Interface (API). SW-API is an interface that allows the exchange of data between CAD design and different software applications. The work includes automatic recognition for assembly knowledge, geometry and non-geometry knowledge (dimensions, geometrical tolerances, and kinematic constraints) in assembly design, which are relevant for assembly process and resources. Recognition algorithms have been developed by using visual basic. Net (VB.net). A case-study example is included for illustration of the proposed approach.

**Keywords:** Features · Form features · Assembly Process Planning · CAD · SolidWorks · API

## 1 Introduction

Assembly is one of the most complex tasks in manufacturing environment; it consumes a considerable percentage of total manufacturing and labor costs [1]. Assembly Process Planning (APP) is an important and decisive link phase between assembly design and manufacturing. The ultimate final aim of APP is to determine the required assembly processes and resources to convert an assembly design into an assembled or semi-assembled product [2]. The integration of product design and assembly process planning (APP) thus has a strong impact on product realization, as also discussed by Du and Zha [3].

Product features have been used by several researchers to improve the efficiency of process planning both in manufacturing and assembly [2, 4–6]. Features are used when designing a product (Feature-based design), while in process planning features are to be extracted from the product model. The extracted features represent a natural link


L.M. Camarinha-Matos et al. (Eds.): DoCEIS 2017, IFIP AICT 499, pp. 144–153, 2017.
DOI: 10.1007/978-3-319-56077-9_13

between product design and process planning domains and provide a valuable mechanism for information exchange [7]. The work presented in this paper is a part of a larger effort to develop methods and tools for a more automated and bidirectional link between product CAD and the different processes and resources applied in APP. This work is based, in its first stage, on extracting assembly knowledge from the CAD file by using feature recognition techniques. The second stage is to model the extracted assembly knowledge to support integration with APP. The third and last stage is to share assembly knowledge via a layered ontology structure, which will serve as a communication mean between assembly design and APP. Figure 1 illustrates the overall integration framework between product design assembly and APP.

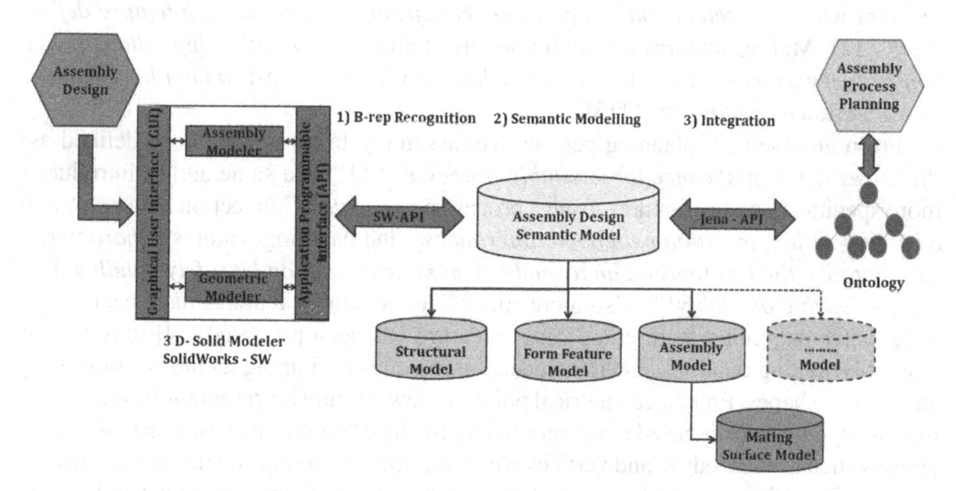


**Fig. 1.** Proposed framework for integrating product features into APP.

In Fig. 1, the three stages of recognition, modelling, and integration are illustrated. In the first stage, geometrical and assembly knowledge are recognized from SolidWorks (SW) - CAD models. An intermediate modelling stage follows where the recognized knowledge is modelled in assembly design semantic model. Assembly design semantic model is composed from several sub-models to separate the recognized geometry knowledge from assembly knowledge.

Assembly knowledge and geometrical knowledge are modelled in a form feature model and a mating surface model, respectively, while part-assembly relations are modelled in structural model. The last stage will be to structure and store the recognized assembly and geometric semantic knowledge according to a well-defined ontology in order to facilitate sharing and integrating this knowledge with the APP. More details about the proposed integration framework are presented in our previous publications [8–10]. This paper is mainly concerned about the first stage of the integration framework; the recognition of product features from SW - CAD software by using Application Programmable Interface (API).

Product features are classified into low-level features (form features), which are basic geometrical and topological entities such as holes, slots, notches etc. and high-level features, which are characterized by both a form and a specific application (machining, assembly, tolerance etc.). For example "machining features" can be defined as geometrical and topological entities significant for manufacturing function. The conversion of form features (low-level) into application-specific (high-level) features in terms of functionality, manufacturing and assembly is the overall aim of feature-based modelling [11].

In this context, an assembly feature could be realized as an association between two form features from different parts [12]. Assembly features have been used either in assembly design or assembly process planning. Many definitions are reported in published literature for assembly features either from assembly design or APP perspectives. From assembly design perspective assembly features are *"Mating pairs of form features with parameters and compatibility constraints as part of each feature definition"* [12]. Mating features or conditions are defined as *"relationships that involve contact between parts, as well as relationships in which two parts do not have contact (e.g. clearance conditions!)"* [13].

From an assembly planning perspective, assembly features have been defined as: *"features with significance for assembly processes"* [4]. The same author introduced more specific assembly features, from a process perspective: Connection features *"such as final position, insertion path/point, tolerances"* and handling features *"characteristics that give the locations on an assembly component that can be safely handled by a gripper during assembly!"*. Also more specialized assembly features have been introduced such as joining features. A joining feature has been proposed [14] to represent assembly/joining relations, and it includes joining entities, joining methods, constraints and groove shapes. From a geometrical point of view, feature representation is classified into two types: surface-based or volume-based. Surface features are based on topological entities such as face, edges and vertices with functional meanings on the part boundary; this representation is known as boundary representation (B-rep). Volumetric features are based on three-dimensional geometrical primitives such as sphere and cylinder. This representation is known as constructive solid geometry (CSG). Based on this classification, feature recognition (FR) approaches can be classified as well into B-rep based approaches and CSG. From an engineering point of view FR systems are divided into two methods; external and internal methods [15].

In internal methods the API of the CAD software is used in order to extract topological, geometrical and assembly information relating to a part or an assembly. While in external methods, a CAD model file is exported in a neutral data format (e.g. STEP, IGES, ACIS). The exported file is then translated using compilers (interface programs) to be compatible for a specific application (e.g. commercial CAM system). Both methods have been used by researchers for FR. In our approach to extract product features in mechanical assembly an internal B-rep CAD recognition approach is proposed. The overall approach is illustrated in Fig. 2. In the Fig. 2, four-stage approach of recognition is proposed. In the first stage feature-based modelling; form features from each part in the assembly are recognized, where each part is composed at least from single form feature. The B-rep modelling is the next stage, in which each form features is decomposed into its basic geometrical and topological entities (faces, surfaces, edges, etc.).

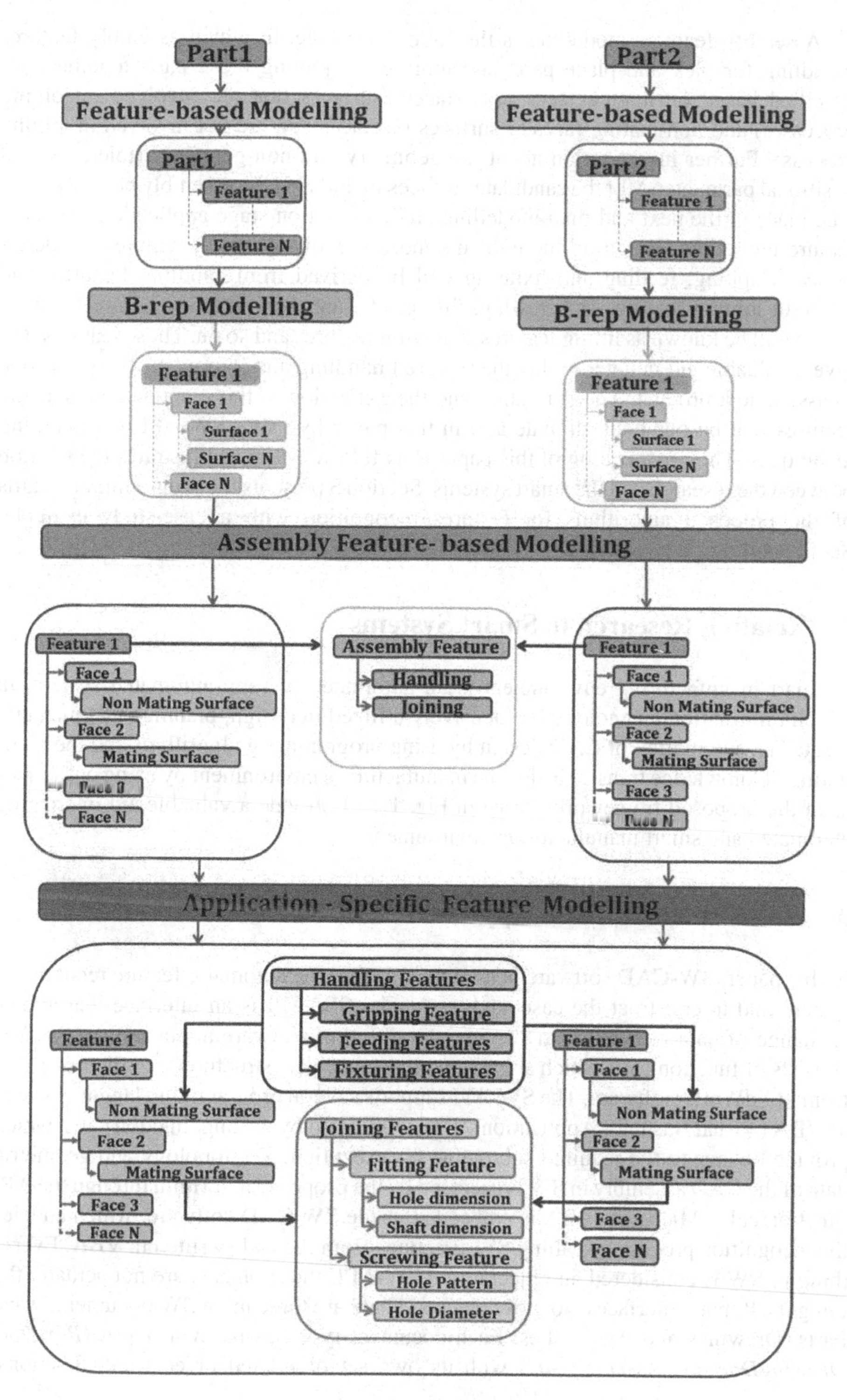


**Fig. 2.** Proposed methodology for extracting product features in an assembly.

Assembly features modelling is the following stage, in which assembly features (handling for pick and place parts and joining for joining those parts together) are specified based on mating faces or surfaces (surfaces that are involved in joining processes) and non-mating faces or surfaces (surfaces that are not involved in joining process). Further investigation about the geometry and non-geometry (tolerance and positional parameters) for the candidate surfaces included in the assembly processes will take place in the next and final modelling and recognition stage application - specific feature modelling. Based on these details more specific assembly features are determined: gripping, feeding and fixturing will be derived from handling features, and different joining features as screwing, fitting, etc. are derived from joining features, which will be known as fitting features, screwing features and so on. Those features will give a valuable aid in determining the required handling and joining processes needed to assemble a product. The derivation and the extraction of those application-specific features will be our final ultimate aim in this paper by using SW-API programming techniques. The organization of this paper is as follows: Sect. 2 illustrates the relation between the research and the smart systems. Section 3 presents the programming details of the proposed algorithms for features recognition with a case-study example. Section 4 draws a conclusion.

## 2 Relating Research to Smart Systems

A smart manufacturing environment is an automated or semi-automated system, in which information technology is extensively utilized in design, planning and management. The automation of CAD design by using programming algorithms and the facilitating of knowledge transfer inside the manufacturing environment by using ontologies as in the proposed framework shown in Fig. 1 will provide a valuable aid in creating automated and smart manufacturing environment.

## 3 SW-API Recognition Approach

In this paper, SW-CAD software is used to construct the automatic feature recognition system and to construct the case-study example. SW-API is an interface that allows exchange of data between CAD design and different software applications. SW-API consists of function calls which are used to access the data structures and their contents from SolidWorks software. The SW-API supports several programming languages such as VBA (Visual Basic for Application). The API is used by writing function calls, which provide linkage to the required subroutine for execution. The topology and geometric data of the CAD assembly in SW is accessed by the proposed algorithm through the API function calls. Microsoft VBA is embedded inside SW CAD software, which enables the recognition process by calling SW-functions from the code written in VBA. Everything in SW is considered an object to the SW-API, those objects are not actually the thing itself, but "interfaces" to those objects. There are three main SW document types: Parts, Drawings and Assemblies. Each document type has its own object (*PartDoc*, *DrawingDoc* and *AssemblyDoc*) with its own set of related objects and functions.

For example the *AssemblyDoc::Mate2* object exists in *AssemblyDoc* document because adding and extracting mate relation is specific to *AssemblyDoc*. Beside the specific objects, which are belonging to a specific document, general objects are available on *ModelDoc2* such as *Feature*, *FeatureManager*, *Configuration*, *SelectManager* and *Sketch*. Those general objects could be accessed by different documents in the SW-API. More details about fundamentals of SW-API are presented in [16]. The case-study model is a three-button panel assembly; the exploded view of the case study assembly is shown in Fig. 3 with its seven parts (the base part, the three buttons, the cover and the screws) and the candidate surfaces, which are significant for the assembly processes. In Fig. 3 for each part some of its surfaces are indicated by labeling the surface with its type (Pla is referred to planner surface and Cyl is referred to circular surface), its part and its number. To make the case-study easier to follow, only the surfaces correspond to one button (red button) and one screw are indicated in the Figure below. For each part in the case-study assembly, the candidate surfaces will be classified into mating surfaces (surfaces involved in mating relations) and handling surfaces (surfaces that will be used in gripping process. Figure 4 illustrates a surface model for the case-study assembly with all the required knowledge needs to be extracted from CAD model in order to facilitate APP.

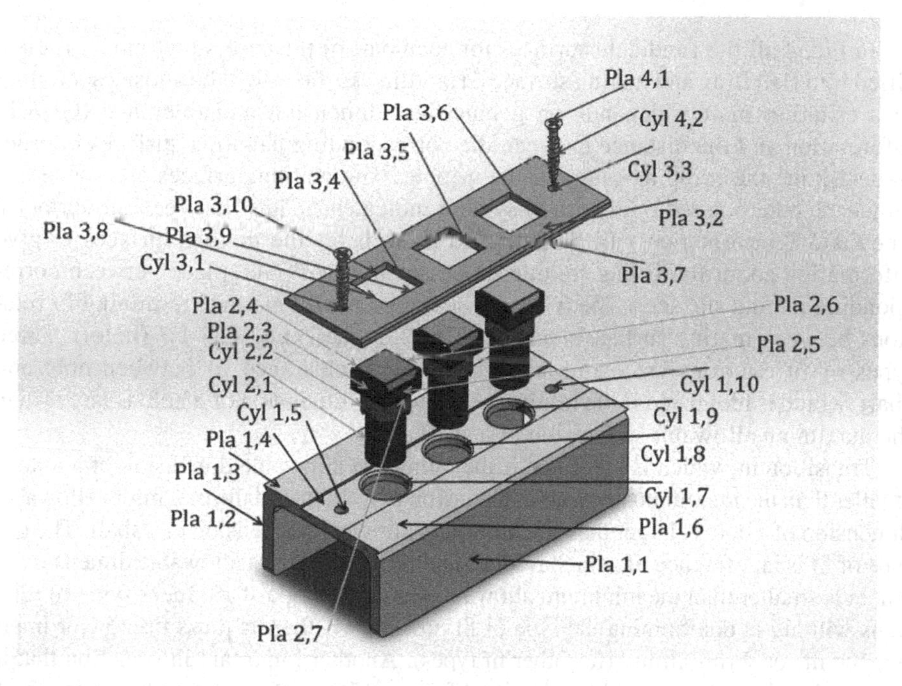


**Fig. 3.** Exploded view of the three-button assembly with the indicated surfaces

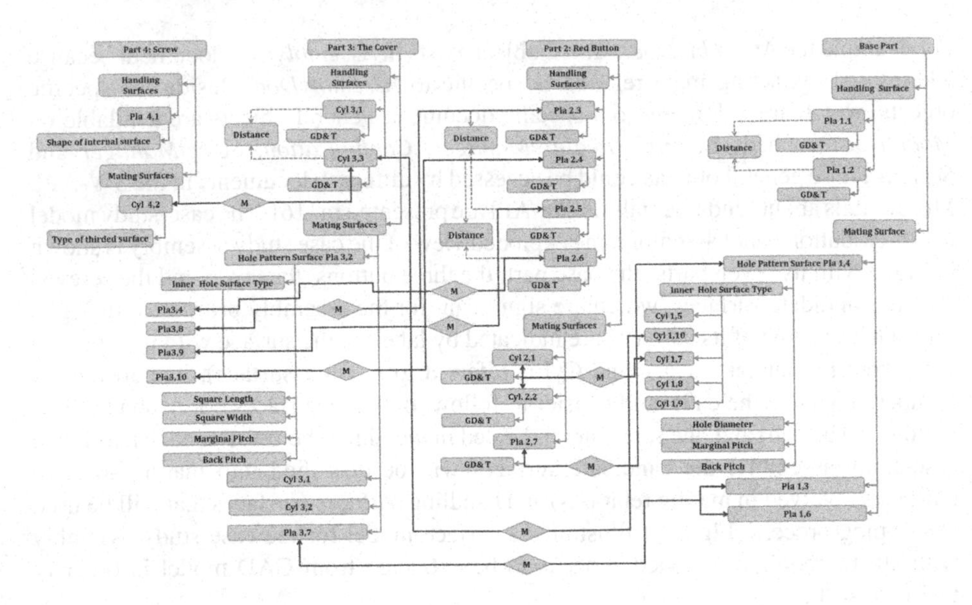


**Fig. 4.** Surface model for the case-study assembly.

In Fig. 4 all the candidate surfaces for each part of the case-study model is classified into Handling and mating surfaces. Handling surfaces will be chosen according to a criterion mainly depends on geometrical dimension and tolerance (GD&T) information and the distance between the corresponding handling surfaces in order to determine the gripping range of the gripper. The mating surfaces are connected with each other via red lines with M symbol indicates "mates" between the surfaces. The GD&T information will be extracted as well for the mating surfaces to give information about the fitting joining processes that will take place between corresponding mating surfaces. The (GD&T) information will aid in determining fit relations between mating surfaces (surfaces Cyl 2,2 (shaft) and Cyl 1,7 (hole)). Three types of fit relations are mentioned in literature, clearance fit between hole and shaft, which is identified if the minimum allowable dimension of a hole is larger than the maximum allowable dimension of a shaft.

Transition fit, which is identified if the minimum allowable dimension of a hole is smaller than the maximum allowable dimension of a shaft, and the maximum allowable dimension of a hole is larger than the minimum allowable dimension of a shaft. The last type of fit is interference fit, which is identified if the maximum allowable dimension of a hole is smaller than the minimum allowable dimension of a shaft. Identifying fit relations will aid in determining the type of fit process whether its press fitting (for interference fit) or shrink fitting (for other fit types). Another important information that is indicated in Fig. 3 and could be identified from mating surfaces is the hole pattern. A hole pattern has several parameters such as hole diameter, hole type, back pitch (the distance between the centers of successive holes) and marginal pitch (the distance between hole-center and surface's edge) that could help to determine a specific joining process. One of these attributes is the hole type. Identification of a hole as threated determines screwing as a joining process to be selected in APP.

A high-level flowchart of the recognition algorithm is shown in Fig. 5. This high level description consists of nine steps. The first step is to read the CAD assembly file from SolidWorks software. The second step is to extract all the surfaces from the assembly by traversing the feature manager tree using the SW-API functions and methods shown beside the corresponding block in the Fig. 5. The third step is to extract the corresponding dimensions and tolerances for all the surfaces in the assembly. Each of the methods and objects indicated beside the related block has access to a specific type of dimension or geometrical tolerance. Dimensions and tolerances have to be assigned to the SW-CAD part document by using DimXpert manger in order to be extracted by this method. The fourth step is to get Persistent IDs for all the surfaces extracted in the second step.

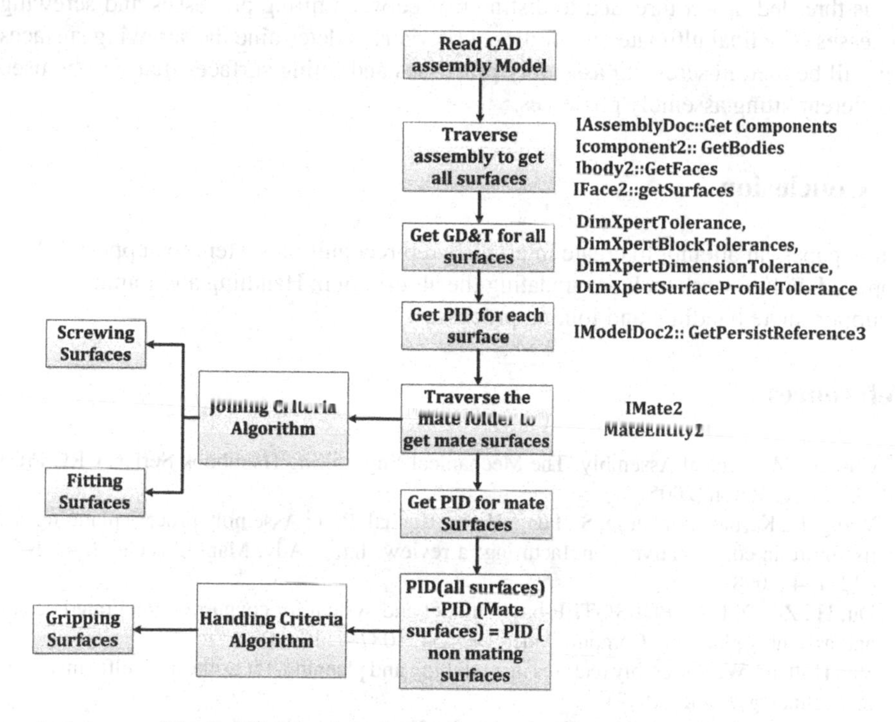


**Fig. 5.** High level description of the recognition algorithm.

Persistent IDs (PID) are a unique identifier for any object and they will be used to identify surfaces in the assembly. The fifth step is to extract the mating surfaces by locating the mate folder in the feature manager tree using feature traversal. This can be achieved by using *IFeature::GetSpecificFeature2* to get *IMate2*, then getting information about the mates, like the type (*IMate2::Type*) and the corresponding reference mating entities. This can be done by using *IMate2::MateEntity*, which will give a pointer to *IMateEntity2*, then use *IMateEntity2::Reference* to get the pointer to the actual reference entity. The sixth step is to get PID for each mate surface extracted in the previous step.

The seventh step is to find the non-mating surfaces by subtract the PID of all surfaces from the PID of the mate surfaces, the result will be the PID for the non-mating surfaces. The eighth step will be to investigate the non-mating surfaces to determine the handling surfaces; this will be achieved by determining the handling criteria. The handling criteria will be to find the non-adjacent parallel surfaces, which are suitable for gripping. Additional parameters in these criteria will be a threshold for the surface profile tolerance for those surfaces. Surfaces with surface profile within a specified range will be chosen for gripping and will be indicated as gripping surfaces.

The ninth and the last step will be the joining criteria for the mating surfaces. The joining criteria will be based mainly on the dimensions of the cylindrical mating surfaces to determine the type of fitting joining processes and on the type of the mating surface if it is threaded or not threaded to distinguish between fitting processes and screwing processes. The final ultimate aim of these criteria is to determine the screwing surfaces that will be used in screwing assembly processes and fitting surfaces that will be used in different fitting assembly processes.

## 4 Conclusion

In this paper, an attempt to create internal-B-rep recognition system to support APP is proposed. Future work includes updating the algorithm in Handling and joining criteria to support more handling and joining processes.

## References

1. Cho, S.: Mechanical Assembly. The Mechanical Engineering Handbook Series, CRC Press LLC, Boca Raton (2005)
2. Wang, L., Keshavarzmanesh, S., Feng, H.-Y., Buchal, R.O.: Assembly process planning and its future in collaborative manufacturing: a review. Int. J. Adv. Manuf. Technol. **41**(1–2), 132–144 (2008)
3. Du, H., Zha, X.F.: A PDES/STEP-based model and system for concurrent integrated design and assembly planning. Comput. Aided Des. **34**, 1087–1110 (2002)
4. Van Holland, W.: Assembly features in modelling and planning. Ph.D. thesis, Delft University of Technology, Delft (1997)
5. Bronsvoort, W., Van Holland, W., Jansen, F.: Feature modelling for assembly. In: Straßer, W., Wahl, F. (eds.) Graphics and Robotics, pp. 131–148. Springer, Heidelberg (1995)
6. Michael, L.: Planning and Scheduling in Manufacturing and Services. Springer, Heidelberg (2009)
7. Dartigues, C., Ghodous, P., Gruninger, M., Pallez, D., Sriram, R.: CAD/CAPP integration using feature ontology. Concurr. Eng. **15**(2), 237–249 (2007)
8. Hasan, B., Wikander, J.: Product feature modelling for integrating product design and assembly process planning. World Acad. Sci. Eng. Technol. Int. Sci. Index 118, Int. J. Mech. Aerosp. Ind. Mechatron. Manuf. Eng. **10**(10), 1725–1735 (2016)
9. Hasan, B., Wikander, J., Onori, M.: Assembly design semantic recognition using SolidWorks-API. Int. J. Mech. Eng. Robot. Res. **5**(4), 280–287 (2016)

10. Hasan, B., Wikander, J., Onori, M.: Ontological approach to share product design semantics for an assembly. In: Proceedings of 8th International Joint Conference on Knowledge Discovery, Knowledge Engineering and Knowledge Management, vol. 2, pp. 104–111 (2016)
11. Stefano, P., Bianconi, F., Angelo, L.: An approach for feature semantics recognition in geometric models. Comput.-Aided Des. **36**, 993–1009 (2004)
12. Shah, J., Rogers, M.: Assembly modeling as an extension of feature-based design. Res. Eng. Des. **5**, 218–237 (1993)
13. Mullins, S., Anderson, D.: Automatic identification of geometric constraints in mechanical assemblies. Comput.-Aided Des. **30**(9), 715–726 (1998)
14. Kim, K.: Assembly operation tools for e product design and realization. Ph.D. dissertation, University of Pittsburgh, Pittsburgh (2003)
15. Miao, H.K., Sridharan, N., Shah, J.J.: CAD-CAM integration using machining features. Int. J. Comput. Integr. Manuf. **15**, 296–318 (2002)
16. SolidWorks Help (2017). http://help.solidworks.com

# Safety Active Barriers Considering Different Scenarios of Faults in Modern Production Systems

Jeferson A.L. de Souza(✉), Diolino J. Santos Fo,
Reinaldo Squillante Jr., Fabricio Junqueira, Paulo E. Miyagi,
and Jose Reinaldo Silva

University of São Paulo, São Paulo, Brazil
{jeferson.souza,diolinos,reinaldo.squillante,fabri,
pemiyagi,reinaldo}@usp.br

**Abstract.** Modern production systems, inserted in a context of high competitiveness, in accordance with policies of sustainability and people protection, as well as being integrated with other (smart) systems, makes complexity an inherent factor in any modern production system. Complexity is reflected in hardware, software and labour qualification for both the design and operation of such systems, resulting in the impossibility of (i) the prediction of all achievable states; (ii) the design of all integrated systems, (iii) non-existence of hardware faults and (iv) absence of human operating errors. Depending on the productive process under analysis, different scenarios, considering the combination of operational errors, faults in field components or even faults in system integration can lead to situations of serious risks for the environment, man and facilities. The bow-tie technique can elicit different scenarios of occurrence of faults and their dynamic evolution, by the results of other risk analysis techniques, such as FMEA, FTA and ETA. The concept of Safety Instrumented Systems, along with the concept of Safety Barriers could be a solution for these problems. This paper proposes the use of Petri nets for formal modeling and the generation of control algorithms, by the simplification of several scenarios of faults fault scenarios listed by a team in the process.

**Keywords:** Modern production systems · Safety Instrumented Systems · Risk analysis · Faults · Scenarios of faults · Safety barriers · Petri nets

## 1 Introduction

In this first decade of the century XXI many studies have indicated that automation processes are undergoing transformations that have been strongly influenced by the advance of technology and computing resources, becoming increasingly complex due to their dynamic and needed to address issues such as global market competitive production and technology used, among other factors (Chen and Dai 2004; Santos Filho et al. 2000; Wu et al. 2008). Given this new scenario, industrial processes and their control are becoming more and more complex. Additionally, organizations have


L.M. Camarinha-Matos et al. (Eds.): DoCEIS 2017, IFIP AICT 499, pp. 154–164, 2017.
DOI: 10.1007/978-3-319-56077-9_14

focused on policies to achieve and to demonstrate people's safety and health, environmental management system, and the capability in risk management.

In a globalized and competitive environment in which organizations are inserted, it is essential to adopt strategic plans and operational practices that ensure the ability to adapt rapidly and consequent change of the systems-productive but hitherto conceived. The expectation is that in addition to result in a process with effective cost reduction, high product quality and flexibility of production lines, and reduction of new products and delivery development times (Santos Filho et al. 2000; Chen and Dai 2004; Wu et al. 2008), also causes the reduction of environmental impacts of the process. The results of this new scenario are Productive Systems (SPs) that perform highly complex processes (Sampaio 2011; Ferreira et al. 2014) that might not be achievable by conventional production methods (Mazzolini et al. 2011). Because of this complexity inherent in any modern production system, some states, though undesirable, can be achieved, it could be mentioned: the fault states of components, design flaws, or operational errors, including intentional, and environmental events that involve the system. Such occurrences could result, depending on the complexity of the SPs, serious risks to the physical integrity of people, the environment and economic losses resulting from damage to the equipment itself (Sallak et al. 2008). Although many studies have been presented both for diagnosis and for the treatment of faults (Morales et al. 2007; Ru and Hadjicostis 2008; Wang et al. 2008; Zhang and Jiang 2008; Summers and Raney 1999; Sallak et al. 2008; Squillante Jr. et al. 2013; Souza et al. 2014; Souza et al. 2016; Peters et al. 2016) accidents continue to occur. In this context, according to specialists, the use of Safety Instrumented Systems - SIS is a solution to this problem in that aims to reduce the risks associated with SPs by successive risk reduction layers, which can be implemented by safety control systems that operate independently of the Basic System Process Control - BPCS. In general, the role of a SIS is to monitor through security sensors, critical events in the industrial process and indicate alarms or perform preset actions through security actuators, for the prevention of accidents or mitigation of the consequences generated by these events (Goble 1998). SIS are referenced in standards such as IEC 61508 and IEC 61511, that lists the performance requirements and the life cycle for a design of a SIS. However, the standards make no mention of methods for SIS implementation. Another concept that promotes de risk reduction of a SP is that of safety barriers. Safety barriers are defined as any physical or non-physical means used in order to prevent, control or mitigate undesirable events or accidents (Sklet 2006). It may be from human action to a complex logical system. This paper proposes a method for the implementation of SIS, in reference to IEC 61508 and IEC 61511, considering different scenarios of faults, making use of safety barriers both to prevent and to mitigate faults. By associating the occurrence of a fault to an event, and by the fact a fault promotes the change of the process state, scenarios can be modeled by Petri nets (PN). Several isolated models are obtained, representing a great difficulty to implement the control algorithm. The hypothesis lies in the interrelationship of the models that represent the different scenarios of faults, which can result in a considerable simplification both for the good properties verification and formal validation of the algorithm, as well as for control codes transcription. This paper is organized as follows: Sect. 2 presents the Relationship to Smart Systems, Sect. 3 the Fundamental Concepts of some Risk Analysis techniques, Petri nets and Fuzzy Logic.

Section 4 presents the Proposal with partial results, obtained by the application of the method in a gas compression station. Finally, Sect. 5 presents the Conclusion and References.

## 2 Relationship to Smart Systems

This work has close relationship with smart systems as it proposes a safety control system considering different scenarios of faults, by the use of safety barriers, in reference to IEC 61508/IEC 61511. The work proposes that such barriers be fully implemented by programmable control systems, rather than passive physical barriers or exercised by human activity. The barriers interrelationship characterizes such a system as a "smart system", in line with the event proposal, contributing to the reduction of the inherent risk, presented in any modern production system, with the objective of protecting man, the environment and facilities.

## 3 Fundamental Concepts

### 3.1 Risk Analysis Techniques

This section introduces the fundamental concepts of some risk analysis techniques, such as FMEA, FTA, ETA HAZOP and the Bow Tie Diagram. The FMEA technique, described in IEC 60812 (IEC 2006), consists of a detailed and systematic study of possible failures of the components of a system. The failure modes of each component are identified, and a severity level is associated with its effect, and assess the likelihood of their occurrence. The FMEA also discusses actions to prevent, eliminate, mitigate and control the causes and consequences of failures. (Lewis 1995). Another technique used is the Fault Tree - FTA, deductive reasoning methodology described in IEC 61025 (2008) that part of a top event, which is the occurrence of a specific fault in a system, which aims to determine the relationship fault logic components and/or operational human errors that may be associated with the occurrence of the top event. The analysis is done from the construction of a logical tree. In this way, one obtains a graph which can be used to identify all possible causes for the occurrence of a fault (Modarres et al. 2010). The graph enables an analysis of the "top-down", which results in understanding how the event occurred. In the analysis "bottom-up" it has been "why" of the event. The advantage of FTA on the FMEA is that one can have a combination of several elements or multiple failure modes, the graph connected by logic elements such as "and" and "or". The study of operability and risks, or HAZOP (HAZard and OPerability studies) defined in IEC 61882 (IEC 2001) was developed for efficient and detailed examination of the variables of a process having a strong resemblance to the FMEA. Hazop identifies the ways in which the process equipment may fail or be improperly operated. It is developed by a multidisciplinary team, being guided by the application of specific words - words guide - each process variable. Thus, to generate the deviation of operational standards, which are analyzed in relation to their causes and consequences. Event trees (DIN 25419 1985-11) are frequently used in the analysis of sequences of events, including human activities, that can lead to disasters or undesirable events. The activity is sometimes called cause consequence analysis, and is used more frequently in safety studies. In this analysis, a basic event, resulting from a specific failure of a human equipment or error, called the initiating event, is used to determine one or more subsequent states of possible faults (Rausand 2011; Villemeur 1992). In this way, the ETA considers the action to be taken by the operator or the process response to the initial event. As in FTA, the study is performed through a tree, starting from the initiating event, in order to quantify the probabilities of failure in the system. Explore responses through a single initiating event and lays a path for assessing probabilities of the outcomes and overall system analysis. It may be applied to a system early in the design process to identify potential issues that may arise rather than correcting the issues after they occur. A widely technique for modeling complete accident scenarios, based on the identification of the main "causes" and "consequences", given a top event (TE) or risk. Basically, the principle of this technique is to construct a tree type, called a bow-tie diagram, due to its special shape (Badreddine and Ben Amor 2010). The bow-tie diagram is based on two parts, as shown in Fig. 1. The left side of the diagram represents the fault tree (FT) that defines all the possible combinational logical relationships between the causes of the TE. In addition, the right side of the diagram represents the event tree (ET) that defines all the possible undesired consequences of TE.

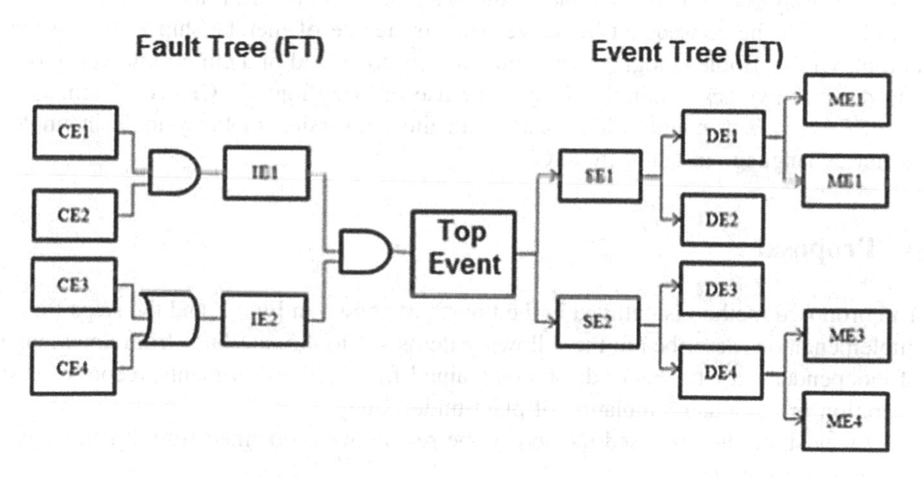


**Fig. 1.** Example of a bow-tie diagram

### 3.2 Petri Nets

Petri net (PN) as a graphical tool and mathematics provides a uniform way for model, analysis and design of Discrete Event Systems - SEDs (Adam et al. 1998; Nassar et al. 2008; Zurawski and Zhou 1994), being effective as a description of technical and specification processes (Hamadi and Benatallah 2003; Morales et al. 2007). It provides a representation that can be used both as a conceptual model and functional model of a system that can analyze and validate the operation of the system at each stage of its development cycle. The PN can also be used as a design tool, allowing for easy interpretation and identification of processes and their dynamic behavior and/or systems being modeled (Nassar et al. 2008). The models based on NP can be used for qualitative and quantitative assessment involving the analysis of the behavioral properties and performance measure, respectively. Moreover, with the development of software simulators (Zurawski and Zhou 1994), has provided tools for editing and analysis of these models. Enables the representation of the system dynamics and structure at various levels of abstraction, according to the with-complexity system (Nassar et al. 2008). It is able to model synchronization process, the occurrence of asynchronous events, competitors and conflict operations, or resource sharing (Adam et al. 1998; Nassar et al. 2008).

### 3.3 Fuzzy Logic

Fuzzy logic is becoming useful in modeling of nonlinear systems, or when the use of differential equations becomes too complex, or even in processes whose knowledge of the dynamic behaviour is not yet fully understood. Fuzzy systems are based on the human knowledge or a set of rules that are designed to mimic human reasoning in control decisions. Questions like "If … (conditional) So … (consequent)" are formulated process experts in analysis, and control actions are defined from the responses, and in its May-ria, systems multiple inputs to a single output. All rules are processed in parallel, with the consequent be active with its degree of membership in the system output. Unlike Boolean logic, fuzzy numbers are contained in a closed interval 0 to 1, and may take values within this range. The use of fuzzy logic in CPs is referenced in the IEC61131-7 standard, which deals with the conversion of fuzzy logic in implementable language in commercial CPs.

## 4 Proposal

The proposed method is outlined in the flowchart shown in Fig. 2, and the steps for its implementation, described in the following items 4.1 to 4.5, are built from knowledge of independent experts and/or database obtained from field experiments, record of past operation or computer simulation of plant under study.

To illustrate the proposed method, some results were obtained from a natural gas compression station.

### 4.1 FMEA – Determination of the Critical Elements

To determine the critical elements of the process under study FMEA is applied as a first evaluation insofar as it associates a level of severity to the occurrence of fault in an isolated and component-centered way, and proposes actions to prevent and mitigate the effects. Faulted components that pose risks to operators, the environment and equipment, besides violating the legislation, receives maximum severity. In the proposed

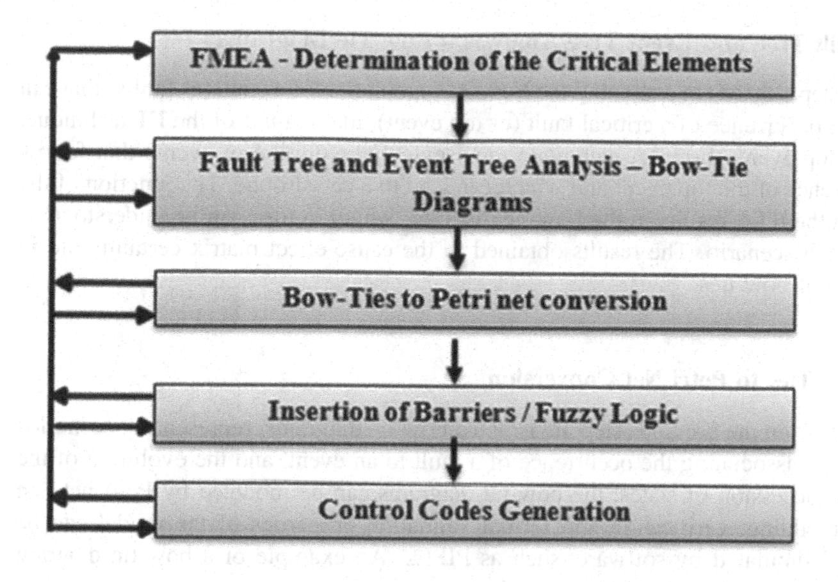


**Fig. 2.** Flowchart of steps of the proposed method.

method, it is part of the FMEA to associate sensors to detect the occurrence of fault of the critical element, and criteria to confirm the occurrence of failure, such as 2003 voting criteria. Based on the results, a cause/effect matrix is elaborated, in which the lines represent the initializing events of the sensors associated to the occurrence of faults and columns the respective actions proposed for prevention and/or mitigation. Such information is obtained with the assistance of a team of experts in the process under study, according to IEC 61508. Based on the concept of SIS, the Safety Instrumented Functions (SIFs) are listed. An example of cause-effect matrix is presented in Table 1.

**Table 1.** Example of a cause-effect matrix of a SIF

| SIF1 / CAUSES | EFFECTS | STOP TCA | STOP TCB | STOP TCC | STOP TCD | CLOSE XV -001 | CLOSE XV -017 | CLOSE XV -019 | CLOSE XV -020 | CLOSE XV -003 | CLOSE XV -018 | CLOSE XV -005 | CLOSE XV -022 |
|---|---|---|---|---|---|---|---|---|---|---|---|---|---|
| PSHH A | | X | X | | | X | X | X | | X | | X | |
| PSHH B | | | X | X | | X | | | X | X | X | X | |
| PSHH C | | | | X | X | X | | | | X | | X | X |

### 4.2 Fault Tree and Event Tree Analysis – Bow Tie Diagrams

The next step consists in analyzing the logical combination of events (or faults) that can lead to the occurrence of a critical fault (or top event), making use of the FT technique. For each top event, the ETA technique represents the evolution of events that follow the occurrence of the top event and which can lead to a catastrophe. The junction of the FTA with the ETA results in the bow-tie diagram, which in turn can be understood as being a fault scenario. The results obtained in the cause-effect matrix certainly aid in the design of bow-ties.

### 4.3 Bow Ties to Petri Net Conversion

The results from the previous step are isolated bow-tie diagrams, representing scenarios of faults. By associating the occurrence of a fault to an event, and the evolution of the fault to a succession of states, the bow-tie diagrams can be modeled by Petri nets, in which the editing, verification and formal validation properties of the model can be edited and simulated by software, such as PIPE2. An example of a bow-tie diagram obtained in the application example is shown in Fig. 3.

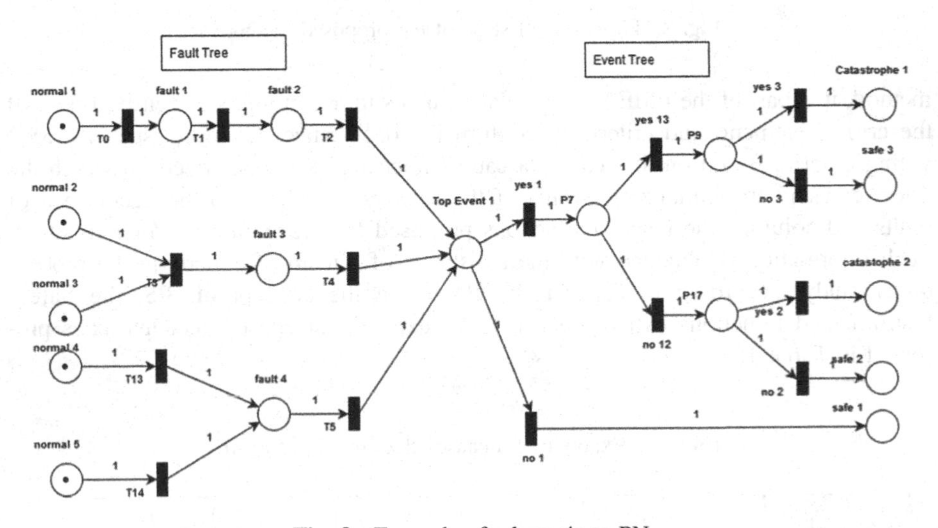


**Fig. 3.** Example of a bow tie to PN.

It can be observed from the obtained models that places and transitions are common to different bow-ties, so that the graphical representation of PN can contribute to the simplification of the final model, integrating all obtained bow-ties. This step is fundamental to the implementation of the proposed method. After the simplification process, the properties for the formal validation of the obtained model must be verified again.

### 4.4 Insertion of Barriers/Fuzzy Logic

After the formal validation of the obtained PN model, the next step consists of associating external actions with the PN states, representing barriers to the evolution of fault states. In the proposed method, only the active barriers, associated with a control system, are considered, in which prevention and mitigation actions are associated with actuators. Such actuators can be defined in the initial stages of FMEA and FTA/ETA. N example of a PN model with barriers is show in Fig. 4.

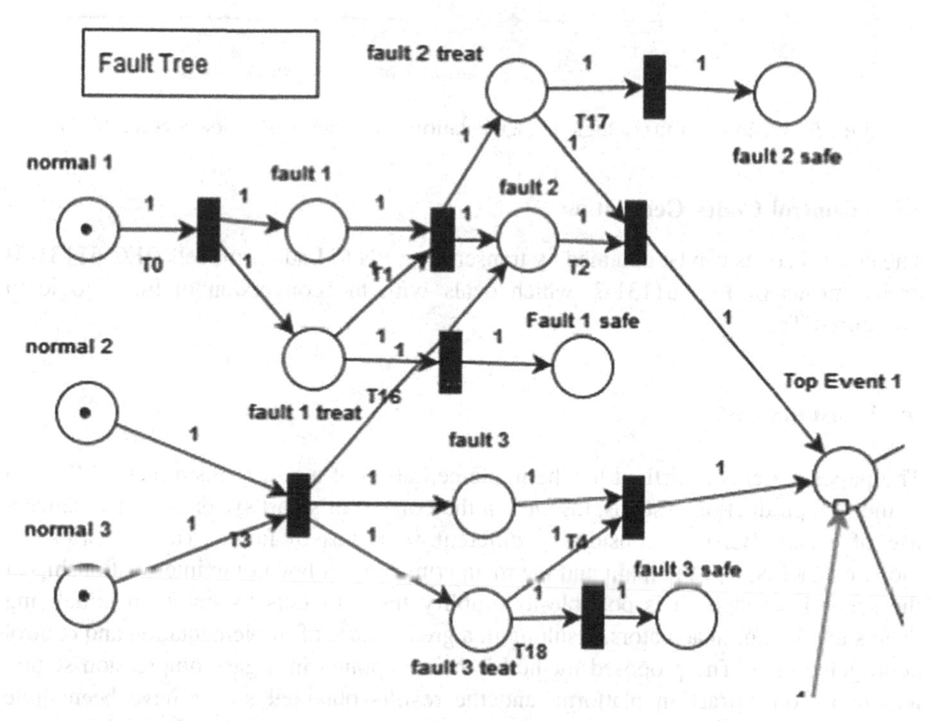


**Fig. 4.** Example of safety barrier in the PN model.

The advantage of simplifying the models in the previous step is that actuators can be common to different bow-tie actions, which certainly minimizes implementation costs. Another advantage is the property of the encapsulation of functionalities, which represents a great advantage in the design process. In terms of the PN structure, the barriers have the function of inserting deadlocks. Fuzzy logic can help in this process since can define the parameters of the variables associated with the occurrence of a fault, as well as represent an anticipatory action of prevention/mitigation action insofar as the analysis of the temporal increment of the variable can be part of the control action. (Souza et al. 2014) presents the contribution of fuzzy logic to the implementation of a SIS. An example of a fuzzy logic implementation is shown in Fig. 5.

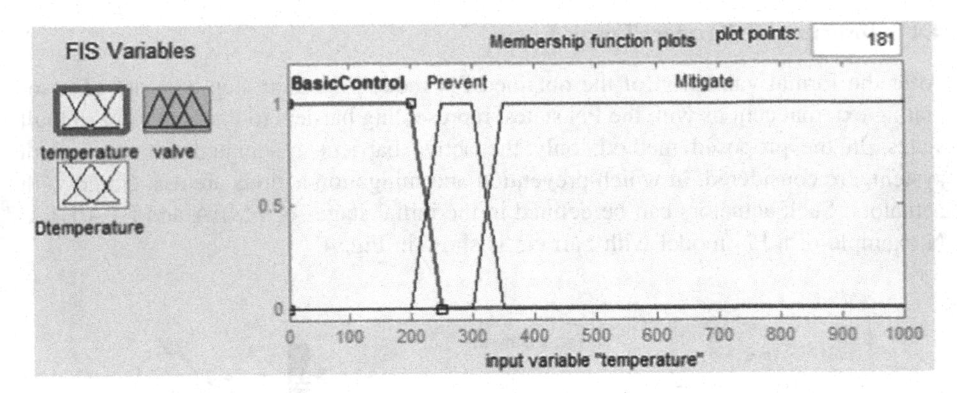


**Fig. 5.** Example of fuzzy logic implementation. Extracted from (Souza et al. 2014)

### 4.5 Control Codes Generation

The control codes can be obtained by transcribing PN to Ladder models (IEC-61131-3) or by means of IEC 61131-7, which deals with the conversion of fuzzy logic to Structured Text.

## 5 Conclusions

The paper presents a method for the implementation of a Safety Instrumented System in modern production systems, inserted in the context of smart systems, by the concept use of Safety Barriers, considering different scenarios of faults. The scenarios are modeled by PN, and the main and the main contribution lies in the interrelationship of the scenario models. It is possible to simplify those models by common initializing events and common actuators, resulting in a greater ease of implementation and control code generation. The proposed method is being applied in a gas compression station and in an oil extraction platform, and the results obtained so far have been quite satisfactory. A limitation of the method are the extensive graphs obtained from the interrelationship of the PN models of the Bow Tie diagrams. A possible solution would be the generation of the models in Colored Petri nets (Jensen 1997) (ISO, IEC 15.909), which would result in more synthetic models, without losing information or complexity of the models, and having as additional benefit the insertion of the fuzzy levels of the associated process variables.

**Acknowledgments.** The authors would like to thank the Brazilian governmental agencies CNPq, FAPESP, and CAPES for their financial support to this work.

## References

Adam, N.R., Atluri, V., Huang, W.: Modeling and analysis of workflows using Petri nets. J. Intell. Inf. Syst. **10**, 131–158 (1998)

Badreddine, A., Ben Amor, N.: A new approach to construct optimal bow tie diagrams for risk analysis. In: García-Pedrajas, N., Herrera, F., Fyfe, C., Benítez, J.M., Ali, M. (eds.) IEA/AIE 2010. LNCS (LNAI), vol. 6097, pp. 595–604. Springer, Heidelberg (2010). doi:10.1007/978-3-642-13025-0_61

Chen, C., Dai, J.: Design and high-level synthesis of hybrid controller. In: Proceedings of IEEE Conference (2004)

DIN 25419: DIN, Event Tree Analysis; Method, Graphical Symbols and Evaluation. Germany (1985)

Ferreira, J.D., Ribeiro, L., Onori, M., Barata, J.: Challenges and properties for bio-inspiration in manufacturing. In: Camarinha-Matos, L.M., Barrento, N.S., Mendonça, R. (eds.) DoCEIS 2014. IAICT, vol. 423, pp. 139–148. Springer, Heidelberg (2014). doi:10.1007/978-3-642-54734-8_16

Goble, W.M.: Control Systems Safety Evaluation & Reliability. ISA - The Instrumentation, Systems and Automation Society, no. 2 (1998)

Hamadi, R., Benatallah, B.A.: A Petri net-based model for web service composition. In: Proceedings of 14th Australasian Database Conference (ADC 2003), Adelaide, pp. 191–200 (2003)

IEC: IEC 60812 - Analysis Techniques for System Reliability - Procedures for Failure Modes and Effects Analysis (FMEA). IEC - International Electrotechnical Commission, Geneva, Switzerland (2006)

IEC: IEC 61025 - Fault Tree Analysis (FTA). IEC - International Electrotechnical Commission, Geneva, Switzerland (2008)

IEC: IEC 61882 - Hazard and Operability Studies (Hazop) - Application Guide. IEC - International Electrotechnical Commission. London, UK, p. 58 (2001). (ISBN 0 580 37625 7)

Jensen, K.: Coloured Petri Nets. Basic Concepts, Analysis Methods and Practical Use. Springer, Heidelberg (1997)

Lewis, E.E.: Introduction to Reliability Engineering, 2nd edn. Wiley, Hoboken (1995)

Mazzolini, M., Brusaferri, M., Carpanzano, E.: An integrated framework for model-based design and verification of discrete automation solutions. In: Proceedings 2011 9th IEEE International Conference on Industrial Informatics, Milan, pp. 545–550 (2011)

Modarres, M., Kaminskiy, M., Krivstov, V.: Reliability Engineering and Risk Analysis: A Practical Guide. CRC Press, Boca Raton (2010)

Morales, R.A.G., Garcia Melo, J.I., Miyagi, P.E.: Diagnosis and treatment of faults in productive systems based on Bayesian networks and Petri net. In: IEEE International Conference on Automation Science and Engineering (Case 7), pp. 357–362 (2007)

Nassar, M.G.V., Melo, J.I.G., Miyagi, P.E., Santos Filho, D.: Modeling and analysis of the material entry flow system in a pickling line process using Petri net. In: ABCM Symposium Series in Mechatronics, vol. 3, pp. 444–453 (2008)

Peters, B., Perez, P., Soares, J., Thomaz, D.: Risk analysis and management based on the performance of safety barriers: case study: fixed platform Polvo A. In: IBP Rio Oil & Gas Expo and Conference, Rio de Janeiro, RJ (2016)

Rausand, M.: Risk Assessment: Theory, Methods, and Applications, 1st edn. Wiley, Hoboken (2011)

Ru, Y., Hadjicostis, C.: Fault diagnosis in discrete event systems modeled by Petri nets with outputs. In: Proceedings of 9th International Workshop on Discrete Event Systems, Goteborg, Sweden (2008)

Sampaio, L.R.: Validação Visual de Programas Ladder Baseada em Modelos. Dissertação de Mestrado - Universidade Federal de Campina Grande, Campina Grande (2011)

Santos Filho, D.J.: Aspectos do Projeto de Sistemas Produtivos. Tese de Livre Docência - Escola Politécnica da USP, São Paulo (2000)

Sallak, M., Simon, C., Aubry, J.: A fuzzy probabilistic approach for determining safety integrity level. IEEE Trans. Fuzzy Syst. **16**(1), 239–248 (2008)
Sklet, S.: Safety barriers: definition, classification, and performance. J. Loss Prev. Process Ind. **19**, 494–506 (2006)
Souza, J.A.L., Santos Fo, D.J., Squillante Jr., R., Junqueira, F., Miyagi, P.E.: Mitigation control of critical faults in production systems. In: Camarinha-Matos, L.M., Barrento, N.S., Mendonça, R. (eds.) DoCEIS 2014. IAICT, vol. 423, pp. 119–128. Springer, Heidelberg (2014). doi:10.1007/978-3-642-54734-8_14
Souza, J.A.L., Santos Filho, D.J., Miyagi, P.E., Silva, J.R., Moscato, L.A., Squillante Jr. R., Sicchar, J.R.: Coloured Petri nets for implementation of safety instrumented systems in critical production systems. In: CLCA 2016 17th Latin American Conference on Automatic Control, 2016, Medellin. Proceedings Book, pp. 27–33. Universidad EAFIT, Medellin (2016)
Squillante Jr., R., Fo, D.J.S., Souza, J.A.L., Junqueira, F., Miyagi, P.E.: Safety in supervisory control for critical systems. In: Camarinha-Matos, L.M., Tomic, S., Graça, P. (eds.) DoCEIS 2013. IAICT, vol. 394, pp. 261–270. Springer, Heidelberg (2013). doi:10.1007/978-3-642-37291-9_28
Summers, A., Raney, G.: Common cause and common sense, designing failure out of your safety instrumented systems (SIS). ISA Trans. **38**, 291–299 (1999)
Villemeur, A.: Reliability, Availability, Maintainability and Safety Assessment, vol. 1 and 2, 1st edn. Wiley, Hoboken (1992)
Wang, X., Chen, G., Xie, Y., Guo, Z.: Fault detection and diagnosis based on time Petri net. In: Proceedings of 8th International Conference on Electronic Measurement and Instruments, Beijing, China (2008)
Wu, B., Xi, L.-F., Zhuo, B.H.: Service-oriented communication architecture for automated manufacturing system integration. Int. J. Comput. Integr. Manuf. **21**(5), 599–615 (2008)
Zhang, Y., Jiang, J.: Bibliographical review on reconfigurable fault-tolerant control systems. Ann. Rev. Control **32**, 229–252 (2008)
Zurawski, R., Zhou, M.: Petri nets and industrial applications: a tutorial. IEEE Trans. Ind. Electr. **41**, 567–583 (1994)

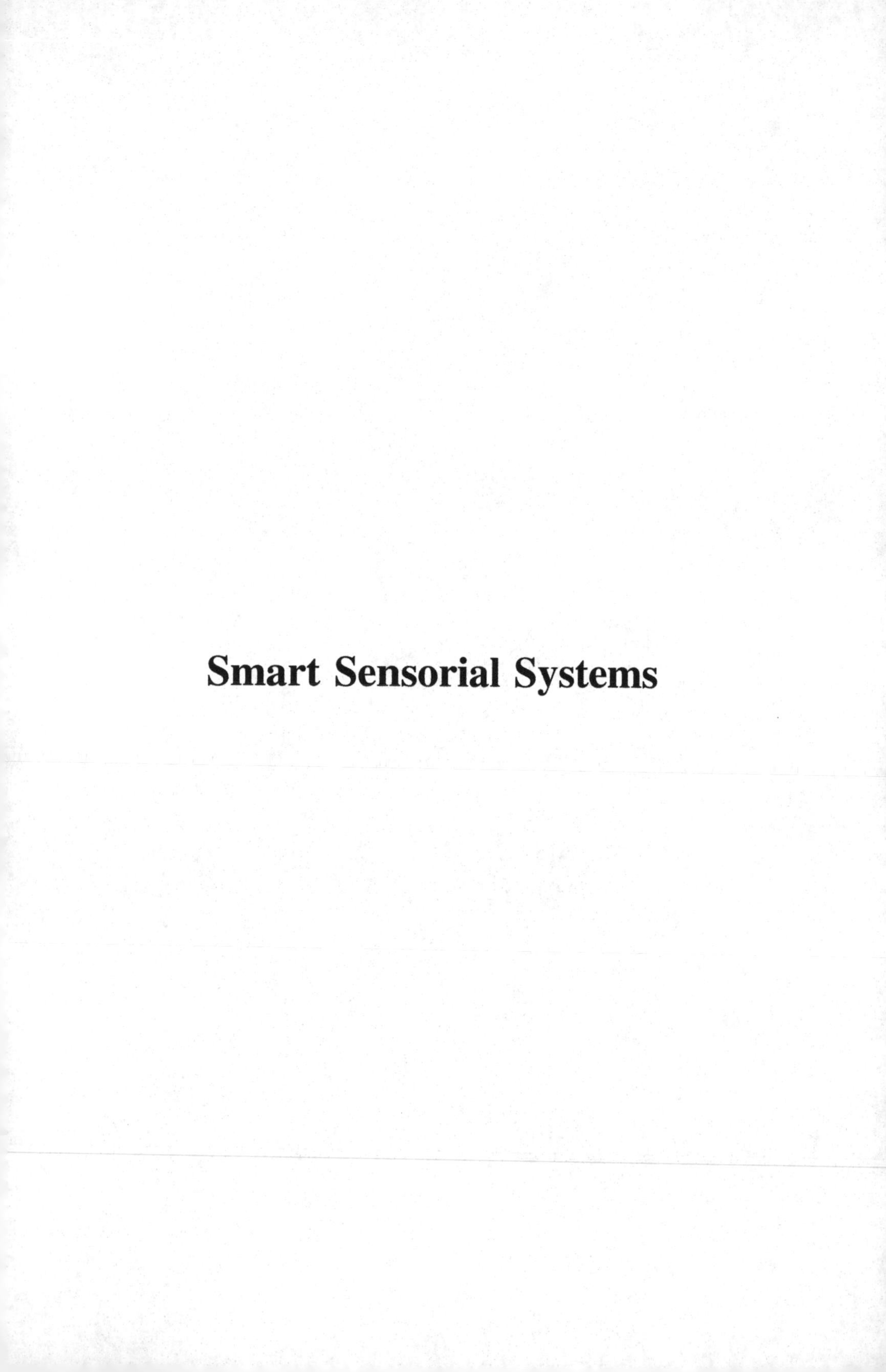

# Smart Sensorial Systems

# Image Analysis as a Tool to Age Estimations in Fishes: An Approach Using Blue Whiting on ImageJ

Patrícia Gonçalves[1,2(✉)], Vitor Vaz da Silva[3,4], Alberto G. Murta[1], António Ávila de Melo[1], and Henrique N. Cabral[2]

[1] Departamento do Mar e Recursos Marinhos, IPMA, Lisbon, Portugal
patricia@ipma.pt
[2] MARE, FCUL, Lisbon, Portugal
[3] Instituto Superior de Engenharia de Lisboa, ISEL-IPL, Lisbon, Portugal
vsilva@deetc.isel.ipl.pt
[4] CTS – UNINOVA, Caparica, Portugal

**Abstract.** Otoliths are the fish bones that allow it to hear sounds and achieve balance. The otolith grows in size as fish grows; ring bands are formed in the otoliths' surface registering periods of rapid and slow growth, opaque bands appear alternating with translucent bands. Age classification was made considering the number of translucent rings in the otolith; one translucent ring was equivalent to one year. The modeling of fish species abundance on the majority of fisheries assessment use age based models. The task of ring counting and ageing is time consuming and may introduce errors that can have a strong impact in stock assessment results. Thus, accurate and precise age estimates are crucial for the effective management and understanding of fisheries resources because recruitment dynamics, growth and mortality estimates relies on these data. The main goal of this study is to produce automatic reading procedures to help researchers, ageing blue whiting fish, minimize ring error count and improve accuracy and precision on age estimation.

**Keywords:** Otoliths · Fish ageing · Image analysis · Image processing · Blue whiting and fisheries science

## 1 Introduction

Otoliths are constituted by three pairs of calcareous structures, the sagittae, lapilli and asterisci, which are found in the inner ear. At the end of the 1960s and the beginning of the 1970s, sagittae otoliths started to be used for age determination and to study the feeding relationships between fish predators and their prey [1]. Since then the modeling of fish species abundance on the majority of fisheries assessment relies on age based models. Accurate and precise age estimates are critical for the effective management and understanding of fisheries resources because dynamic rates of recruitment, growth and mortality depend on these data [2].

Age classification has been based in age readings under stereo-microscopic observation. The interpretation of age reading is a very difficult task and may change among


L.M. Camarinha-Matos et al. (Eds.): DoCEIS 2017, IFIP AICT 499, pp. 167–174, 2017.
DOI: 10.1007/978-3-319-56077-9_15

different readers or as a reader becomes more experienced [3]. Regularly, calibration age reading workshops are conducted aiming the improvement of accuracy and precision of age estimations between the readers of the same species. Quality assurance in the ageing process guarantees the consistency which is uniquely important to this type of research work [4]. The criteria used to count the rings and determine age are made to be objective, but the final classifications are always dependent on the reader's experience, which introduces an interpretation error in the process [5]. This interpretation error can be either biased or random. In combination, process and interpretation error can result in age estimates that may differ by as much as a factor of three among readers [6].

There is an urgent need to validate and calibrate the currently age reading procedure within more effective automatic classification, in order to avoid subjective, and misinterpretation among readers. The image analysis constitutes a valuable tool applied to otoliths ageing and the recorded images allow performing simultaneously other studies, such as age validation. Several image analysis programs had been adapted in order to allow automatic count of daily increments such as RATOC [7], Image Pro Plus [8] and TNPC [9] and also annually increments like Image Pro Plus, TNPC, CAF [4] and ImageJ [10]. Although, the majority of the programs requires a license and is expensive, ImageJ has the advantage of being free software which alows to record macros and plug-ins, making simple to add algorithms to improve species ageing classifications.

The aim of this study is to improve ageing classifications through image analysis making the process based in more objective criteria. In order to achieve that goal a plug-in has been written and added to ImageJ. The otoliths used to test this new approach are from the blue whiting (*Micromesistius poutassou*) and the preliminary results are presented here. This study has been developed in order to know if: ImageJ could be used as an approach to improve the accuracy and precision of blue whiting ageing estimation?

## 2 Contribution to Smart Systems

In fisheries, for stock assessment annually a huge number of otoliths are read and the data for different countries combined. As an example, concerning the blue whiting stock, in 2015, around 102,000 fishes were measured and from these around 30,000 otoliths were processed for age reading, comprising data from 16 countries [11]. The consistent application of the ageing method over years, among readers, and even among laboratories (countries), is a particular requirement of stock assessment studies [2]. The inter-calibration workshops conducted regularly aim to guarantee that these age data can be combined and all the readers, from the different countries, follow the same criteria. The criteria for age estimation, based on ring counts, is defined to be easy to follow and objective. In practice, the accuracy and precision on age classifications is always dependent on the reader experience which reveal some subjectivity on the criteria. The age reading protocols based on images observations, is time consuming and based on decisions which can differ according to the reader experience.

The increasing demand on fisheries science to produce more data, the time involved and the ageing criteria mostly based on human decision, created the requirements to searching new technologies applied to age reading estimation. These new technologies should be easy to apply and available to a wide range of scientists.

The otolith grows in size as fish grows; ring bands are formed in the otoliths surface registering periods of rapid and slow growth, opaque bands appear alternating with translucent bands. Therefore, the otoliths are constituted by different optically zones according to their structural appearance or light properties, e.g., translucent or opaque and the distinction among them allows to identify and count the rings through stereo-microscope observation [12].

ImageJ is an open source software for image analysis and processing, and allows java written *plug-ins* to be added [10]. Thus, new *plug-ins* are easily spread and become available to a huge number of scientists with the purpose of improving fish age classification. A smart system for otolith age classification could provide input from different expert readers for the same otolith, and then adaptively learn to classify correctly the ring identification providing further information within each ring that may be useful for the expert readers' further research. The best automatic system should use the output of a camera with optical magnification, identify the otolith and determine its age by providing a value, its confidence intervals and precision to help the researchers evaluate and decide in a more objective way the age classification. In this study, the automatic system is still in development aiming to improve the age classification on blue whiting.

## 3 Materials and Methods

### 3.1 Sample Collection

Fish samples of blue whiting were collected along the Portuguese coast during 2013, from January to December. Total length (cm) and sex of all blue whiting sampled were recorded and the otoliths were removed from each fish. After, the otoliths were washed and stored dry. Sixty seven blue whiting otoliths were considered for this study. Otoliths were submerged in a 0.1% thymol solution (1 g of thymol for 1 l of filtered and distilled water) for approximately 24 h. The whole otolith was immersed in oil and observed under reflected light against a black background in a stereo-microscope. The otoliths images acquisition was performed using the TNPC software (version 7.0) [9]. For age classification only one otolith of the *sagittae* pair was used. On the current study the left otolith was used. Age classification was made considering the number of translucent rings in the otolith; one translucent ring was equivalent to one year.

### 3.2 Image Analysis

A *plug-in* was developed in Java for the ImageJ software [10] which allowed counting the age rings in the otolith image files. On the left otolith of each pair a transect line crossing the otolith longer axis is drawn (Fig. 1).

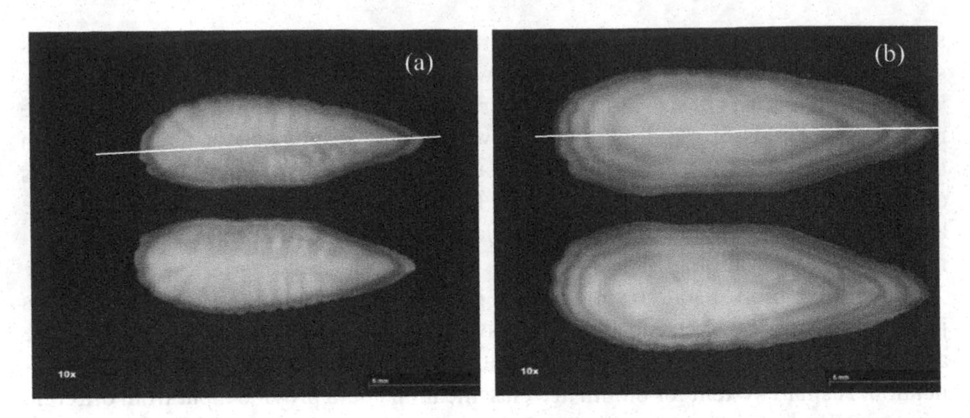


**Fig. 1.** Blue whiting left (top) (with the transect line) and right (bottom) otoliths from fishes with: (a) 1 year old and (b) 4 year old.

The automatic reading evaluation is based on the results from the otolith profile plotted along the longer axis (Fig. 2).

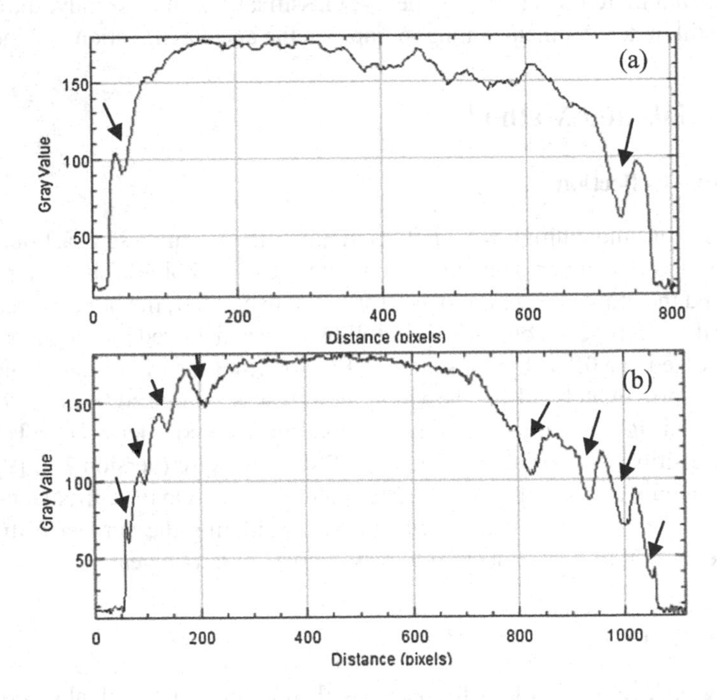


**Fig. 2.** Blue whiting left otolith profile along the long axis from fishes with: (a) 1 year old (Fig. 1a) and (b) 4 year old (Fig. 1b). The arrows indicate the rings position in the profile density plot.

The results from the automatic reading application were recorded and compared to the reader's age estimation.

### 3.3 Statistical Analysis

A test of symmetry [13] for determining if significant differences existed between the ages by readers and the ages by ImageJ was applied. The test of symmetry uses a chi-square-type statistical test to determine if the age-agreement table is symmetric or not. If the age-agreement table is determined to be asymmetric then it can be concluded that there is a systematic difference in ages observed between readers and automatic ageing (ImageJ).

Two statistical indicators were used to measure the precision: (i) the average percent error (APE) [14] and (ii) the coefficient of variation (CV). The APE assumes that the standard deviation of the age estimates are proportional to the mean of the age estimates.

All statistical analyses and plots were performed using the packages FSA and ggplot2 from the statistical environment R [15].

## 4 Results

The length at age by sex was represented taking into account age classifications by readers (Fig. 3a) and by ImageJ (Fig. 3b). There are differences in the growth curves, namely the range of ages is higher in the Image J between 0 and 7, while the range by readers varied from 1 to 6.

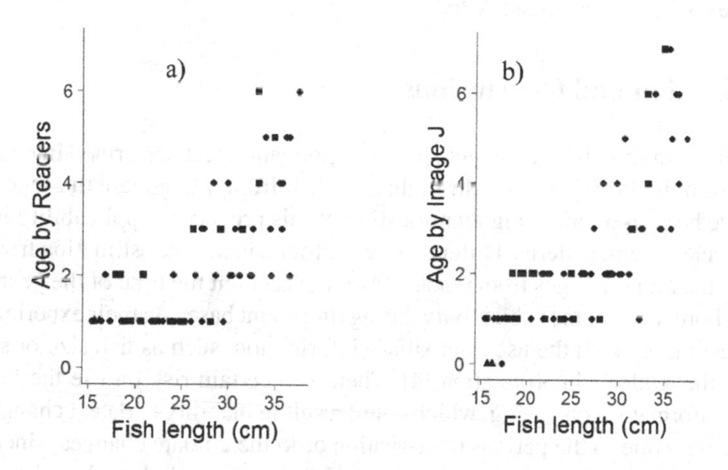


**Fig. 3.** Fish length (cm) by age according to classifications by: (a) readers and (b) ImageJ application. Symbols correspond to fish's sex: ▲ indeterminate (Denomination used when fish gonads show an earlier development stage and is not possible to assign a sex, *i.e.*, to distinguish between female or male.); ● females; and ■ males.

The Hoenig test of symmetry ($p = 0.094$) indicates that there are systematic differences in the assigned ages between readers and ImageJ. The ages estimation based on the automatic procedure are overestimated compared with the ages attributed from the readers (Fig. 4). The same age was attributed by the two procedures in 36 otoliths, in a total of 67.

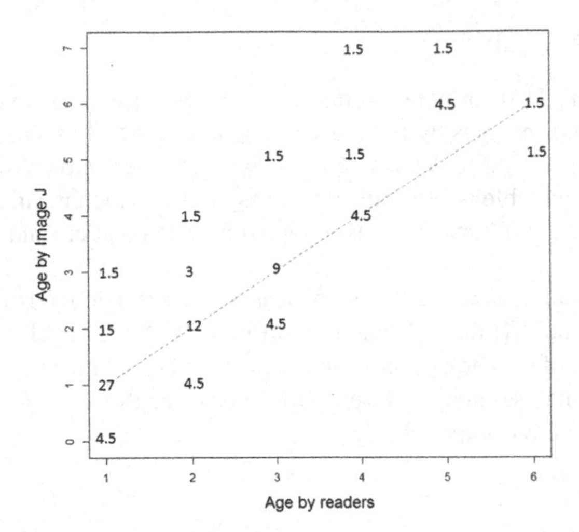


**Fig. 4.** Age-bias plot of estimations by readers and by ImageJ application. The dashed line represents the age-agreement between procedures. The values represent number of otoliths in percentage.

The precision methods reveal a low percentage of agreement (54%) between the two procedures, with CV = 21 and APE = 16.

# 5 Discussion and Conclusions

The results from this study constitute the first application test performed in blue whiting otoliths from the Portuguese coast. In this test fish from a large length range and both sexes have been used, allowing identification of this procedure applicability in otoliths with different growth patterns. Differences were found in the age estimation from readers and using the ImageJ. Ages from readers take into account the time of the year, the type of otolith border and some subjectivity during ring count based on their experience. Ages are also estimated with the use of auxiliary information, such as fish size or sex which may bias the reader's interpretation [4]. There is a certain risk on use the fish size as auxiliary information on ageing, which could result in inability to detect changes on fish growth pattern due to the periods of starvation or to the climate changes, since temperature affects growth. The age estimation from ImageJ was only based on the ring count on the density profile plot which is an advantage. As an example, of this study results, a fish with length higher than 30 cm, was classified as 1 by ImageJ and 2 years by the readers, this underestimation by the program could be due to readers based their classifications on the fish length and this fish could present a different growth pattern which was not detected in readers' classification. Notwithstanding this particular case, ImageJ is overestimating the ages in the majority of this study classifications, which could be due to a higher sensitivity of the software into recognizing patterns in the otolith density across the whole otolith section which could not be so evident through direct image observation or the irregular surface of some otoliths or the software counts is considering the denominated false rings[1] as valid rings. False rings are a common issue in blue whiting and a source of bias amongst the readers [17]. Although, since the age's estimations comparison were based on reader's classifications and not in validated ages those hypothesis need to be further evaluated. In order to validate the *plug-in* ImageJ estimations, it is important to repeat this test using otoliths from fishes in which the absolute age is known. There are also some basic age validations techniques, such as the relations between fish length and otolith length, fish length and otolith weight [2], which could be applied to the samples used on this study as a way to obtain more precision in estimations. This study constitutes a first approach to the blue whiting ageing validation and classification, the facility of incorporate macros on ImageJ will allow improving and calibrating the results until obtaining a good compromise between the automatic procedures and the fish growth.

The future steps, are applying this *plug-in* and provide the age estimations using the information of the otolith type of border and the time of the year when the fish was collected. Also, applying the ImageJ *plug-in* to otoliths with ages validated or in which the age estimation does not present any doubts. These tools will also be tested by experiment readers on blue whiting, as an exercise where the auxiliary information (fish length and sex) will not be provided, during the next international calibration workshop which will be held at Lisbon this June. The authors of this study truly believe that this new approach is a valid and useful application to help, the readers from the different countries, producing more precise and accurate fish age estimations on blue whiting.

**Acknowledgments.** We are thankful to the staff from Instituto Português do Mar e da Atmosfera (IPMA) who collected the blue whiting samples from their survey and from their commercial fishery. The samples used on this study were collected under the Portuguese Biological Sampling Programme (PNAB-EC Data Collection Framework) from IPMA. P. Gonçalves was funded by FCT through the fellowship SFRH/BD/88092/2012. This study had the support of: Fundação para a Ciência e Tecnologia (FCT), through the strategic project UID/MAR/04292/2013 granted to MARE; and IPMA through the Portuguese Biological Sampling Programme (PNAB-EC Data Collection Framework).

## References

1. Tuset, V.M., Lombarte, A., Assis, C.A.: Otolitos de peces del mediterráneo occidental y del atlántico central y nororiental. Sci. Mar. **72**(S1), 7–198 (2008)
2. Campana, S.E.: Accuracy, precision and quality control in age determination, including a review of the use and abuse of age validation methods. J. Fish Biol. **59**(2), 197–242 (2001)
3. Doering-Arjes, P., Cardinale, M., Mosegaard, H.: Estimating population age structure using otolith morphometrics: a test with known-age Atlantic cod (Gadus morhua) individuals. Can. J. Fish. Aquat. Sci. **65**(11), 2342–2350 (2008)
4. Morison, A.K., Robertson, S.G., Smith, D.C.: An integrated system for production fish aging: image analysis and quality assurance. North Am. J. Fish. Manag. **18**(3), 587–598 (1998)

[1] Any opaque area similar to an annual band but interrupted or without a "discontinuity" (if in the central area of the section) [16].

5. Campana, S.E., Moksness, E.: Accuracy and precision of age and hatch date estimates from otolith microstructure examination. ICES J. Mar. Sci.: Journal du Conseil. 1991 Nov 1 **48**(3), 303–316 (1991)
6. Donald, D.B., Babaluk, J.A., Craig, J.F., Musker, W.A.: Evaluation of the scale and operculum methods to determine age of adult goldeyes with special reference to a dominant year-class. Trans. Am. Fish. Soc. **121**(6), 792–796 (1992)
7. RATOC System Engineering Inc. http://www.ratoc.co.jp/ENG/jiseki.html
8. Image Pro Plus. www.mediacy.com/imageproplus
9. TNPC. www.tnpc.fr
10. National Institutes of Health: ImageJ. http://imagej.nih.gov/ij. Accessed 16 Sept 2016
11. ICES: Report of the Working Group on Widely Distributed Stocks (WGWIDE), 31 August–6 September 2016, ICES HQ, Copenhagen, Denmark. ICES CM 2016/ACOM: 16. 500 p. (2016)
12. Casselman, J.M.: Age and growth assessment of fish from their calcified structures–techniques and tools. US Department of Commerce, National Oceanic Atmospheric Administration. National Marine Fisheries Service. Technical report 8: 1–17 (1983)
13. Hoenig, J.M., Morgan, M.J., Brown, C.A.: Analysing differences between two age determination methods by tests of symmetry. Can. J. Fish. Aquat. Syst. **52**, 364–368 (1995)
14. Beamish, R.J., Fournier, D.A.: A method for comparing the precision of a set of age determinations. Can. J. Fish. Aquat. Sci. **38**, 982–983 (1981)
15. R.D.C. Team: R. www.r-project.org. Accessed 16 Sept 2016
16. Peres, M.B., Haimovici, M.: Age and growth of southwestern Atlantic wreckfish Polyprion americanus. Fish. Res. **66**(2), 157–169 (2004)
17. ICES: Report of the Workshop on the Age Reading of Blue Whiting, 10–14 June 2013, Bergen, Norway. ICES CM 2013/ACOM, vol. 53, 52 p. (2013)

# Signal Processing Techniques for Accurate Screening of Wrist Fractures

Ridita Ali[1(✉)], Lyuba Alboul[1], and Amaka Offiah[2]

[1] Centre of Automation and Robotics Research, Sheffield Hallam University, Sheffield, UK
{ridita.ali,l.alboul}@shu.ac.uk

[2] Sheffield Children's NHS Foundation Trust, Sheffield, UK
a.offiah@sheffield.ac.uk

**Abstract.** The common way for doctors to differentiate wrist injury in to a sprain or a fracture, is to take radiographs (X-ray), which expose patients to radiation. The purpose of this study is to explore a non-invasive method to screen for potential fractures. A small, computer run, hand-held system has been developed which consists of a vibration induction mechanism and a piezoelectric sensor for capturing the vibration signals. Two analyzing techniques were considered. The first involves extraction of wavelet coefficients from decomposition of data and the second applies Fast Fourier Transform to the data. Results of both techniques were then cluster analyzed to partition between fracture and sprain. The data were acquired from both the injured and uninjured wrists of six adult patients. This study is currently being evaluated on children's wrists.

**Keywords:** Vibration analysis · Fast fourier transform · Wavelet decomposition · Fracture screening

## 1 Introduction

Use of ionizing radiation (x-rays) to produce radiographs is the most common procedure for doctors to differentiate a wrist injury into a sprain or a fracture. Each wrist radiograph costs around £60 in the United Kingdom and almost 60 percent of those radiographs show no fracture. Thus, a non-invasive technique is very desirable to minimize both radiation exposure and cost.

We have used a small, computer run, hand-held system that generates vibration in the bone under examination, and then captures generated vibration signals by a piezoelectric sensor, which then are analyzed by applying signal processing methods. Various technological systems in both industry and medicine apply either Fast Fourier Transform (FFT) or Wavelet Coefficient extraction to process acquired signals. In this study, we have applied both methods on signals obtained from human wrist bones and then compared the results to find out which method is more accurate to screen for fracture.

Vibration analysis is an established method used in industry for monitoring or diagnosing faults in electric machines by means of signal analysis and signal processing. Mechanical vibrations cause one or more inputs and outputs of a system to change over


L.M. Camarinha-Matos et al. (Eds.): DoCEIS 2017, IFIP AICT 499, pp. 175–182, 2017.
DOI: 10.1007/978-3-319-56077-9_16

time and the signals produced are then evaluated for type and severity of defect [1]. Studying the vibration effects, fault detection and capability of this method of machine maintenance, is an on-going process [2, 3].

By comparison, vibration analysis in the medical field is still in its infancy. In 1932, Lippmann percussed his finger to cause vibration, used his stethoscope to sense it and explained the benefits for using the method in clinical diagnostics. He suggested that analysis of percussion appeared more accurate than x-ray for the diagnosis of fractures [4]. Some researchers developed a finite model of a dry tibia for predicting its natural frequencies of vibration [5]. Nokes published a major study where he proposed the idea to monitor various pathological and trauma-induced conditions by applying a stimulus to bone [6]. Experiments demonstrated that vibration induced by mechanical means can indicate mechanical properties of bones, such as density, biological state, etc. [7]. The influence and results of low-frequency vibrations in the lower limbs have been accurately determined for the analyses based studies [8]. The effects of skin and muscle on the bone's low frequency vibration signals have been demonstrated [9]. We have previously used samples of frequency spectra from adults with and without wrist injuries, to cluster analyze into fractured and sprained bones [10].

Application of vibration analysis is still developing. Additionally, studies are being conducted to find out if its advantages outweigh the disadvantages. However, in the context of health care, the practical applications of vibration analysis are in the near rather that distant future.

## 2 Smart Systems

Smart systems are defined as devices that incorporate the functions of sensing, actuation, control and adaptation. They can describe and analyze situations, and take decisions based on the available data in a predictive or adaptive manner, thereby performing smart (intelligent) actions [11].

Smart systems are found in many applications such as automotive, manufacturing, environment, energy and health care. These systems have the capability to solve very complex problems, including taking over human cognitive functions [12].

In the case of health care and medical applications, smart systems are mainly used to improve the quality of diagnosis and treatment. The system described in this paper shows the utilization of a miniature smart hardware device to produce and procure energy through a medium. The first prototype of the device has been tested to find correlation between vibration frequency and bone density, and has been developed by the Medical Electronics Engineering Group at Sheffield Hallam University's Centre for Automation and Robotics Research facility [13]. The device has been then modified and our group now uses it to differentiate between wrist fracture and strain. The data acquired from this method are stored and analyzed using complex programmed routines through specific software.

## 2.1 System

*Hardware:* The device comprises of the following parts.

- Electrical Vibration Inducer – A sample rate of 50000 samples per second makes the inducer move that produces the impact. Squared signals with a duty cycle of 50 percent at maximum, a frequency of 1 Hz, and an offset of 2.5 were selected for generation of the signals during the impacts.
- Piezoelectric Sensor – The sensor, used for vibration/impact sensing has broad bandwidth, high sensitivity, light weight and low cost features.
- NI myDAQ Assistant – This was used for frequency generation which causes the inducer to initiate and consecutively record the vibration response via the sensor.
- 12 V Battery – This battery was added to remove the portability and electric shocks issues.
- Control Circuit – A small Printed Circuit Board (PCB) was built which consists of amplifier, filter, and power divider to control the circuit for minimal battery power consumption.

*Software:* LabVIEW and MATLAB.

The fundamental principle behind using LabVIEW is for its Integrated Development Environment (IDE) and the Graphical User Interface (GUI), which were used to develop the data acquisition feature for this device. MATLAB was used to process and analyze the acquired data. LabVIEW was designed in such a way that a NI myDAQ Assistant can generate signals to run the inducer and also acquire the corresponding data; in other words, both the generation and acquisition modes of the DAQ Assistant were utilized. Implementation loops were added to create mechanized applications to reduce manual tasks and enhance repeatability. MATLAB is an efficient tool for signal processing and that had been used to perform both FFT and WD.

Thus the system developed for this research, is undeniably a Smart System (Fig. 1).

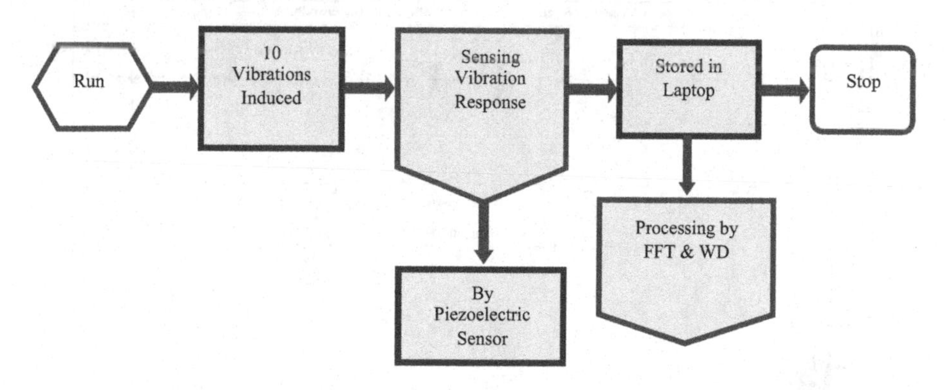


**Fig. 1.** Process flow chart

## 3 Research Methodology and Results

Adult patients, aged between 20 to 70 years old, attending the Emergency Department of a local hospital, following a wrist injury, volunteered to take part in this study. Vibration data were collected from both injured and uninjured wrists of six patients and processed blinded to radiographic findings. This means that at, the time the data were produced and analyzed, it was not known if the injured wrist had a sprain or a fracture. They all sat on a chair during recording. Vibration caused by mechanical excitation, was induced by impacting the most prominent bone in the elbow. The vibration responses were recorded by placing a piezoelectric sensor which has a maximum operating frequency, on a bony prominence at their wrists. The device was intended to induce light vibrations onto the bones for sensing the signals. The vibration inducer had been designed in such a way that its strength can be controlled, if required. All of the volunteers stated after participating in the study that they did not feel any pain due to the vibration induction.

10 vibrations were induced over 10 s in total and each response was recorded at a rate of 50000 samples per second. These data were stored in the computer and then processed by Wavelet Decomposition (WD) and Fast Fourier Transform (FFT).

I. The coefficients were extracted, after decomposition of the acquired data by Level 5 filtration. These were derived by a scaling function "Daubechies 5 (db5)" wavelet family. Two degree polynomials were plotted to find the best fitted curve among the number of coefficients derived from the WD. The equation for each of the polynomials was also calculated.
II. The bone vibration recorded from six patients was converted to a magnitude frequency spectrum by application of FFT processing.

Figures 2 and 3 display the results produced by FFT and WD.

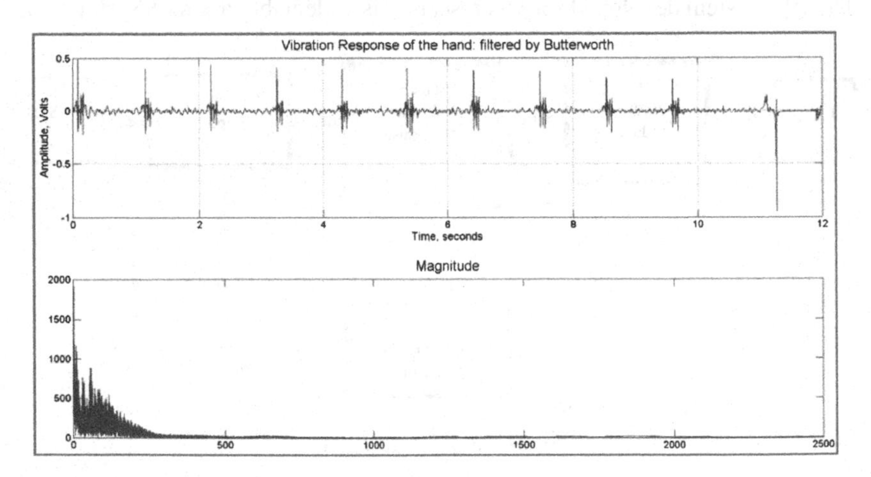


**Fig. 2.** Filtered vibration response and magnitude spectrum by FFT.

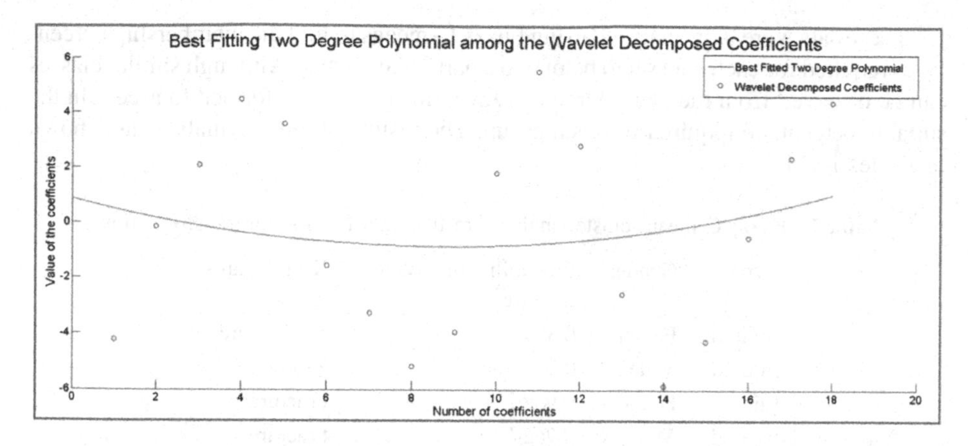


**Fig. 3.** Best fitting two-degree polynomial among 18 wavelet decomposed coefficients.

The frequencies produced by the FFT and the constants derived from the equation produced by the best fitted polynomial curve, were classified using the "dendrogram" (a hierarchical cluster tree diagram) to show which groups are similar. The "dendrogram" was plotted using a linkage of Euclidean distance between each data set. The diagram produced by both FFT and Wavelet Decomposition was similar and is illustrated in Fig. 4.

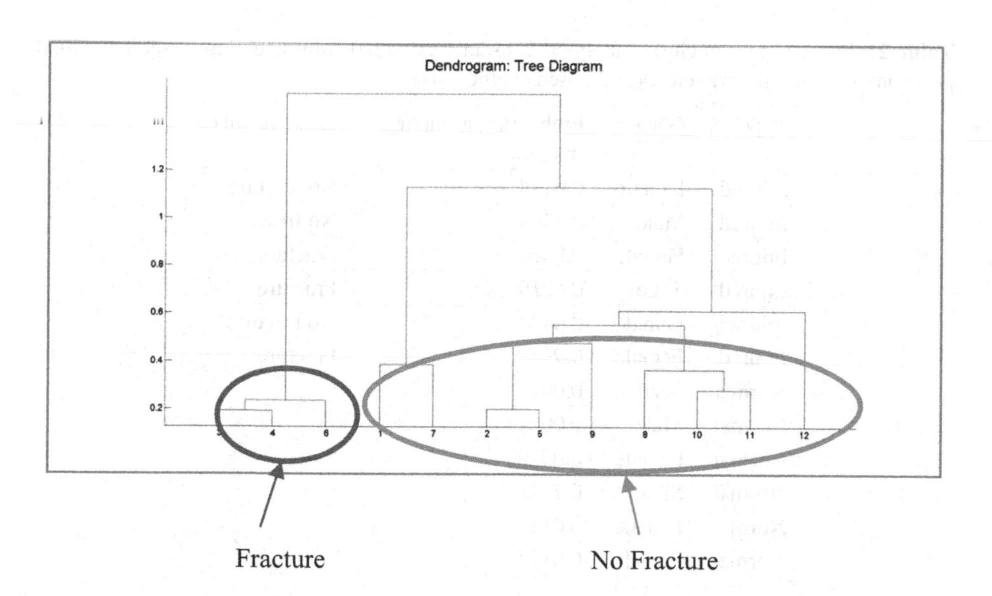


**Fig. 4.** Both FFT and wavelet decomposed coefficients produced the same results by the dendrogram.

These data were then analyzed using Fuzzy C-means to find the membership percentages required for each data set to belong to a particular cluster. Although similar classes can be observed from the dendrogram, Fuzzy C-means was performed to ascertain the suitable percentage required for each group. The results of cluster analysis are shown in Tables 1 and 2.

**Table 1.** Fuzzy C-means cluster analysed results after fast fourier transformations

| Type | Gender | Probability of having fracture | X-ray results |
|---|---|---|---|
| Injured | Female | 0.3397 | No fracture |
| Injured | Male | 0.3227 | No fracture |
| Injured | Female | *0.9836* | Fracture |
| Injured | Male | *0.9624* | Fracture |
| Injured | Female | 0.2429 | No fracture |
| Injured | Female | *0.9275* | Fracture |
| Normal | Female | 0.2758 | |
| Normal | Male | 0.0733 | |
| Normal | Female | 0.3699 | |
| Normal | Male | 0.0520 | |
| Normal | Female | 0.1240 | |
| Normal | Female | 0.1854 | |

**Table 2.** Fuzzy C-means cluster analysed constants extracted from equation of the best-fitted polynomial among the wavelet decomposed coefficients.

| Type | Gender | Probability of having fracture | X-ray results |
|---|---|---|---|
| Injured | Female | 0.0488 | No fracture |
| Injured | Male | 0.0068 | No fracture |
| Injured | Female | *0.9820* | Fracture |
| Injured | Male | *0.6210* | Fracture |
| Injured | Female | 0.0424 | No fracture |
| Injured | Female | *0.9970* | Fracture |
| Normal | Female | 0.0077 | |
| Normal | Male | 0.0011 | |
| Normal | Female | 0.0086 | |
| Normal | Male | 0.2311 | |
| Normal | Female | 0.0189 | |
| Normal | Female | 0.0177 | |

## 4 Discussions

Tables 1 and 2 clearly reflect the techniques used in this study.

WD produced comparatively similar results to FFT, while using both the dendrogram and Fuzzy C-means. However, there are some disadvantages and unreliable results of using the WD technique:

- WD provides both Frequency and Time related information which makes it difficult to use; as the design which induced 10 vibrations must produce vibration responses very similar to each other over the 10 s in order to ensure that results can be processed further.
- In order to make sure the above reason does not affect the accuracy of the data collected, the first and last few sets of data are left out when doing the analysis.
- In addition, there are a sizable number of steps before reaching the concluding stage of Fuzzy C-means cluster.

On the other hand, the frequencies produced by the FFT can easily be fed into the Fuzzy C-means code already built, to produce membership values. Fewer steps were involved in building the FFT codes.

## 5 Conclusion and Further Works

Both FFT and WD techniques produced almost all correct results. However, the FFT is a hassle free and accurate process with less number of steps for future use. The efficiencies are to be checked by a larger volume of data. The discomfort level experienced by each patient for both wrists was recorded and every participant felt that the vibrations would not be uncomfortable to children aged between 0 and 15 years old. It takes less than 5 min to carry out the procedure so it is not a time consuming method. The study is being continued at a local Children's Hospital and aims to recruit a larger number of participants to test the accuracy of the techniques. More signal processing techniques are likely to be assessed once a larger volume of data is available.

## References

1. Brandt, A.: Noise and Vibration Analysis: Signal Analysis and Experimental Procedures, 1st edn. Wiley, Hoboken (2011)
2. Plante, T., Nejadpak, A., Yang, C.X.: Faults detection and failures prediction using vibration analysis. In: IEEE AUTOTESTCON, pp. 227–231 (2015)
3. Plante, T., Stanley, L., Nejadpak, A., Yang, C.X.: Rotating machine fault detection using principal component analysis of vibration signal. In: 2016 IEEE AUTOTESTCON, pp. 1–7 (2016)
4. Lippmann, R.K.: The use of auscultatory percussion for the examination of fractures. J. Bone Joint Surg. **14**(1), 118–126 (1932)

5. Hobatho, M.C., Darmana, R., Barrau, J.J., Laroze, S., Morucci, J.P.: Relationship between mechanical properties of the tibia and its natural frequencies. In: 1988 Proceedings of the Annual International Conference of the IEEE Engineering in Medicine and Biology Society, vol. 2, pp. 681–682 (1988)
6. Nokes, L.D.M.: The use of low-frequency vibration measurement in orthopaedics. Proc. Inst. Mech. Eng. Part H J. Eng. Med. **213**(H3), 271–290 (1999)
7. Alizad, A., Walch, M., Greenleaf, J.F., Fatemi, M.: The effects of fracture and fracture repair on the vibrational characteristics of an excised rat femur. In: 2004 IEEE Ultrasonics Symposium, vol. 2, pp. 1513–1516 (2004)
8. Schneider, A., Coulon, J.M., Boyer, F.C., Machado, C.B., Gardan, N., Taiar, R., Dey, N.: Modelization and simulation of the mechanical vibration effect on bones. In: 2nd International Conference on Computing for Sustainable Global Development (INDIACom), pp. 1968–1972 (2015)
9. Ali, R., Offiah, A., Ramlakhan, S.: Fuzzy C-means clustering to analyze bone vibration as a method of screening fracture. In: 2016 10th International Symposium on Communication Systems, Networks and Digital Signal Processing (CSNDSP), pp. 1–5 (2016)
10. Razaghi, H., Saatchi, R., Offiah, A., Bishop, N., Burke, D.: Spectral analysis of bone low frequency vibration signals. In: 2012 8th International Symposium on Communication Systems, Networks & Digital Signal Processing (CSNDSP), pp. 1–5 (2012)
11. Wan, K., Alagar, V.: Synthesizing data-to-wisdom hierarchy for developing smart systems, p. 473 (2014)
12. Glesner, M., Philipp, F.: Embedded systems design for smart system integration, p. 32 (2013)
13. Razaghi, H., Saatchi, R., Burke, D., Offiah, A.C.: An investigation of relationship between bone vibration frequency and its mass-volume ratio. In: 2014 IEEE International Conference on Acoustics, Speech and Signal Processing (ICASSP), Florence, pp. 3616–3620 (2014)

# TRACEO3D Ray Tracing Model for Underwater Noise Predictions

Rogério M. Calazan(✉) and Orlando C. Rodríguez(✉)

LARSyS, University of Algarve, Campus de Gambelas, 8005-139 Faro, Portugal
{a53956,orodrig}@ualg.pt, http://www.siplab.fct.ualg.pt

**Abstract.** Shipping noise is the main source of underwater noise raising concern among environmental protection organizations and the scientific community. Monitoring of noise generated by shipping traffic is a difficult challenge within the context of smart systems and solutions based on acoustic modeling are being progressively adopted to overcome it. A module of sound propagation stands as a key point for the development of a smart monitoring system since it can be used for the calculation of acoustic pressure, which can be combined with estimates of the source pressure level to produce noise predictions. This paper addresses the usage of the TRACEO3D model for application in such systems; the model validity is addressed through comparisons with results from an analytical solution and from a scale tank experiment. The comparisons show that the model is able to predict accurately the reference data, while a full-field model (normal mode-based, but adiabatic) is only accurate till a certain degree. The results show that TRACEO3D is robust enough to be used efficiently for predictions of sound propagation, to be included as a part of a smart system for underwater noise predictions.

**Keywords:** Shipping noise · Monitoring · Underwater acoustic · Ray tracing · Smart systems

## 1 Introduction

Shipping noise is the main source of underwater noise raising concern among environmental protection organizations and the scientific community. Shipping noise can propagate at long distances (like tens or hundreds of kilometers), thus having the potential to mask and/or disturb biological relevant sounds, such as those vocalized by marine animals for mating, orientation or detection of prays and predators. Monitoring of shipping noise is a difficult challenge due to many factors like, for instance, lack of equipment standardization and the large extensions to be covered (as a reference, the Portuguese EEZ itself has an area of about 1,7 million km2). Recently, the Marine Strategy Framework Directive [1] proposed the adoption of shipping noise modeling to overcome such difficulties. Generally speaking, realistic estimates of noise require transmissions loss (TL) predictions in coastal zones with a complex bathymetry and/or a complex sound speed distribution. To ease the computation load of the predictions they can be based


L.M. Camarinha-Matos et al. (Eds.): DoCEIS 2017, IFIP AICT 499, pp. 183–190, 2017.
DOI: 10.1007/978-3-319-56077-9_17

on the adiabatic coupling of modes and/or the so-called N × 2D modeling, in which the three-dimensional field is constructed using N slices of predictions on a vertical plane, produced with a two-dimensional model. A shipping noise prediction tool based on this approximation, combined with data from an Automatic Identification System (AIS), is discussed in [2] and shows that the system is able to produce relevant estimates of shipping coastal traffic. However, it is well known that even in the simplest case a three-dimensional bathymetry, either by itself or combined with a sound speed field, can induce propagation not confined to a given slice, an effect known as out-of-plane propagation. Modeling of acoustic fields in three-dimensional waveguides, accounting for out-of-plane propagation, had been an active field of research for many years [3–5]. In this context the wedge problem has been an important reference given the availability of an analytical solution [6–8]. Despite the apparent simplicity of the wedge problem the corresponding analytical solution has revealed many interesting features of three-dimensional propagation, such as horizontal refraction, mode coupling and rays propagating up-slope before connecting a source to a receiver. Recent evidence from tank scale experiments fully supports the analytical predictions [9].

Of interest for the topics discussed here is the TRACEO3D ray tracing model, which is able to predict fields of acoustic pressure and particle velocity in environments with elaborate boundaries; the model is under current development at the Signal Processing Laboratory (SiPLAB) of the University of Algarve. This paper looks forward for a robust module of sound propagation for noise predictions, through the comparisons of TRACEO3D predictions with results from an analytical solution of the wedge problem, and with results from a scale tank experiment [5]; both cases are considered to provide a robust reference to test the model's accuracy. The comparisons show that the model is able to predict accurately the reference data, while adiabatic coupling is only valid for a small wedge.

The remainder of this paper is organized as follows: Sect. 2 identifies the relationship of this work to the issue of Technological Innovation for Smart Systems; Sect. 3 compactly describes the TRACEO3D model; Sect. 4 provides a brief description of the analytical solution for the wedge problem, and of the scale tank experiment; Sect. 5 discusses the comparisons; Sect. 6 presents the conclusions and future work.

## 2 Relationship to Smart Systems

Generally speaking, Smart Systems are technologies able to combine data processing with sensing, data exploration and communication, and capable to analyze complex situations in order to take autonomous decisions. Although the availability of sensors is increasing in terms of accuracy and specificity (thus providing a better adaptation to the system objectives) the capability to develop predictions simultaneously with the acquisition of data improves the ability of a given Smart System to deal with real word (mostly unpredictable) situations. Of particular importance for the system is to rely on environmental knowledge to compensate for a given lack of information, or to proceed to a given task given a certain type of previous information. For the specific conditions of underwater acoustics a fundamental component of the Smart System should be a module for predictions of sound propagation, which can be further processed in order for the Smart System to proceed accordingly. Particular examples of such modules can be found in the literature for the case of underwater noise monitoring [2], underwater communications [10] and source tracking [11].

## 3 The TRACEO3D Ray Tracing Model

The TRACEO3D model is a recent three-dimensional extension of the TRACEO ray model [12, 13]. Generally speaking, TRACEO3D produces a prediction of the acoustic field in two steps: first, the Eikonal equation is solved in order to provide ray trajectories; second, ray trajectories are considered as the central axes of Gaussian beams, and the acoustic field is calculated as the coherent superposition of beam influences. The model can take into account the environmental variability in range, depth and azimuth.

## 4 The Wedge Problem

The general geometry of the wedge problem is shown in Fig. 1, with the wedge apex aligned along the Y axis; in the given geometry α stands for the wedge angle, and the source is located at the position (0, 0, zS). Propagation along the positive/negative X axis is known as downslope/upslope propagation, respectively, while propagation along the Y axis is known as cross-slope propagation. Two-dimensional acoustic models can

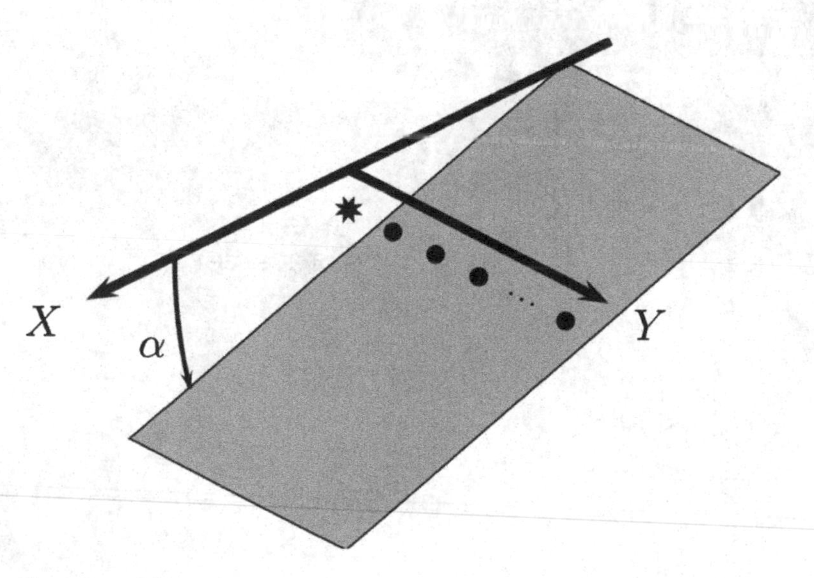


**Fig. 1.** Geometry of the wedge problem; the star indicates the source position; the dots indicate an array in the cross-slope direction.

be used to predict accurately upslope and downslope propagation and the models accuracy had been properly confirmed through comparisons with experimental data. Cross-slope propagation on the other side leads to out-of-plane effects and requires a three-dimensional model.

### 4.1 The Analytical Solution

A detailed description of the analytical solution can be found in [6–8]. Overall, the solution is based on the method of images, where the contribution of each image can be represented in terms of a Bessel function expansion inside an improper integral; numerical implementation of the solution is generally intensive because the convergence of the series is slow, and worsens when small $\alpha$ are considered; in this case the image solution can be replaced with the much faster adiabatic-mode solution. The limits of validity of the adiabatic solution still remain a topic of intense discussion.

### 4.2 The Scale Tank Experiment

The experimental data was obtained from an indoor shallow-water tank of the LMA-CNRS laboratory in Marseille. The tank experiment is described in detail in [5, 14], therefore a compact description is presented in this section. The inner tank dimensions were 10 m long, 3 m wide and 1 m depth. The source and the receiver were both aligned along the across-slope direction, as shown in Fig. 2. The bottom was filled with sand

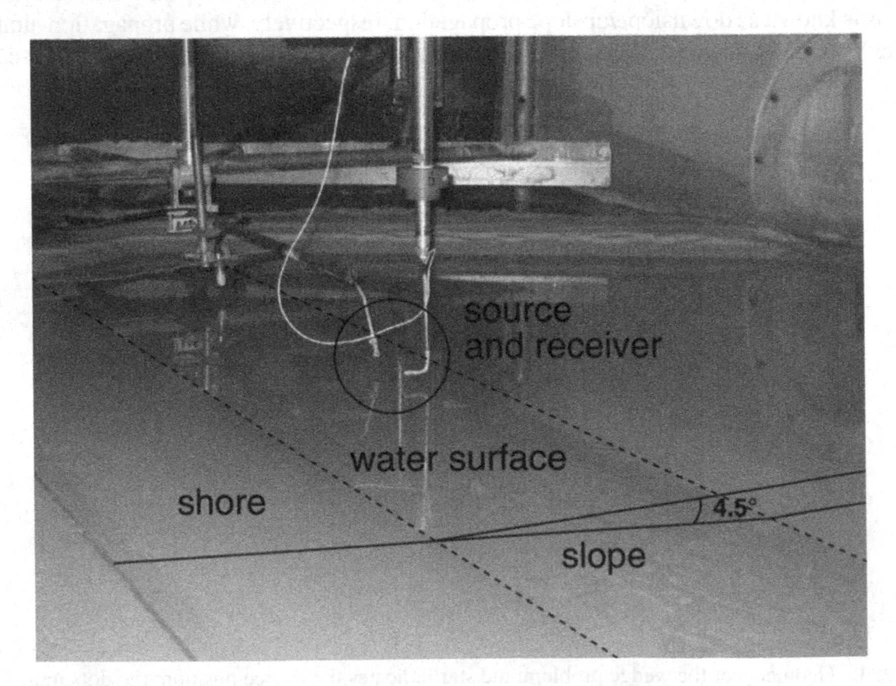


**Fig. 2.** Indoor shallow-water tank of the LMA-CNRS laboratory of Marseille (from [5]).

and a rake was used to produce a slope angle $\alpha \approx 4.5°$; sound speed in the water was considered constant and corresponded to 1488.2 m/s. The bottom parameters corresponded to cp = 1655 m/s, $\rho$ = 1.99 g/cm$^3$ and $\alpha_p$ = 0.5 dB/$\lambda$. The source was located at 8.3 mm depth and bottom depth at the source position corresponded to 44.4 mm. The ASP-H (for horizontal measurements of across-slope propagation) data set was composed of time signals recorded at a fixed receiver depth and at several source/receiver distances starting from r = 0.1 m until r = 5 m in increments of 0.005 m, providing a sufficiently fine representation of the acoustic field in range. Three different receiver depths were considered, namely 10 mm, 19 mm and 26.9 mm, corresponding to data subsets referenced as ASP-H1, ASP-H2 and ASP-H3, respectively. Acoustic transmissions were performed for a wide range of frequencies; however, comparisons are presented only for data from the ASP-H1 subset with the highest frequency (180.05 kHz); this is due to the fact that the higher the frequency the better the ray prediction. It is important to remark that a scale factor of 1000:1 is required to properly modify the frequencies and lengths of the experimental configuration; that implies that the following conversion of units is adopted: experimental frequencies in kHz become model frequencies in Hz, and experimental lengths in mm become model lengths in m; for instance, an experimental frequency of 180.05 kHz becomes a model frequency of 180.05 Hz, and an experimental distance of 10 mm becomes a model distance of 10 m. Sound speed remains unchanged, as well as compressional and shear attenuations.

## 5 Comparisons

The TRACEO3D and KRAKEN models were used to perform TL predictions, which were compared with results from the analytical wedge solution and with measurements from the scale tank experiment. The KRAKEN model is based on normal mode theory [15]; as in the case of TRACEO3D, KRAKEN 3D calculations can be done in two steps: first, modes can be calculated on a two-dimensional grid; second, modes can be coupled along different directions over the grid to produce a 3D prediction. For smooth bathymetries one-to-one exchange of modal energy (i.e. adiabatic coupling) can provide accurate and computationally efficient 3D predictions.

**Table 1.** Parameters for the wedge problem.

| Parameters | Units | Analytical solution | Scale tank experiment |
|---|---|---|---|
| $\alpha$ | ° | 0.5 | 4.5 |
| Frequency | Hz | 50 | 180.05 |
| Sound speed | m/s | 1500 | 1488.7 |
| Source depth | m | 10 | 8.3 |
| Depth at source position | m | 90 | 44.4 |
| Bottom compressional speed | m/s | 2000 | 1700 |
| Bottom compressional density | kg/m$^3$ | 2 | 1.99 |
| Bottom compressional attenuation | dB/$\lambda$ | 0.5 | 0.5 |

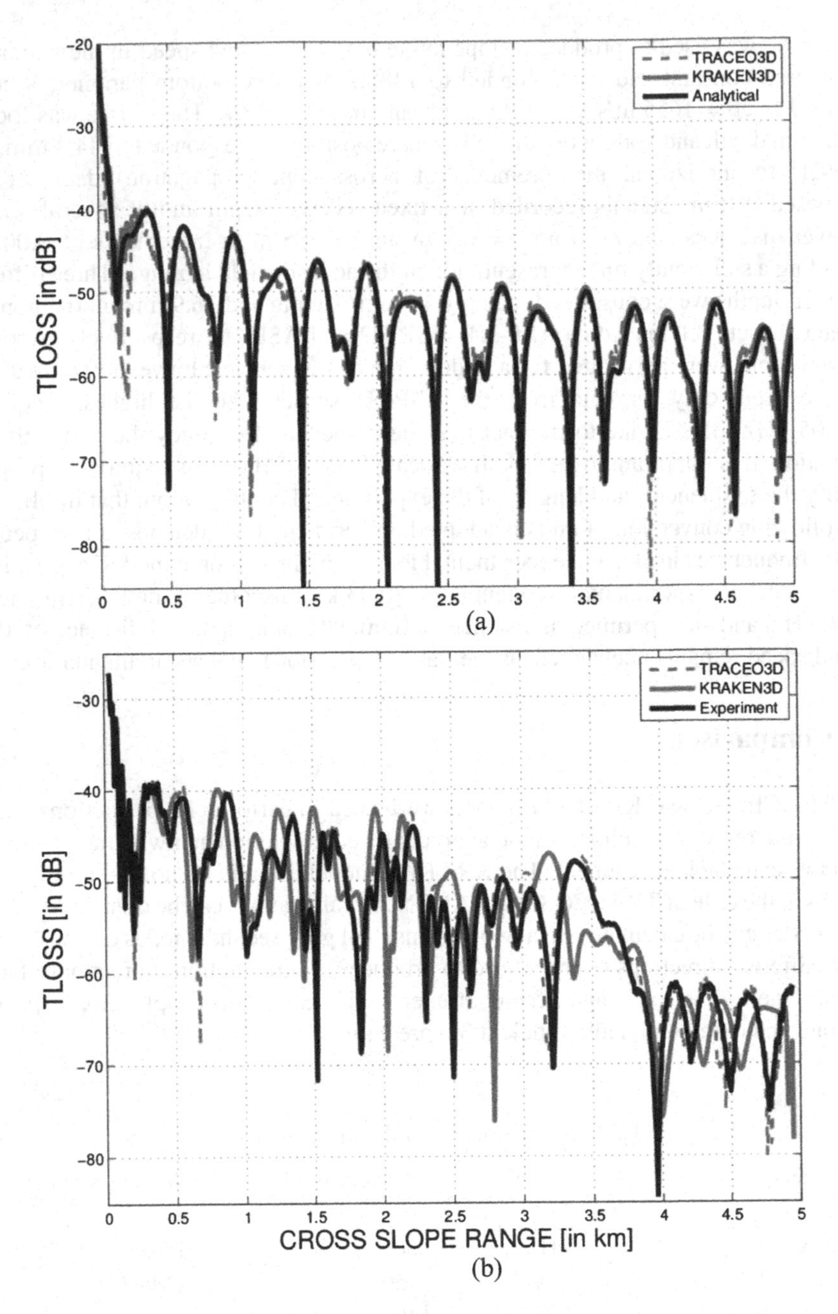


**Fig. 3.** Comparisons for the (a) analytical solution and (b) experimental results.

Waveguide parameters for the analytical solution and for the experimental data are shown in Table 1; it is important to remark the difference in frequencies: for the analytical case the frequency is much lower than for the experimental data; such low value of frequency is important to test whether the ray approximation can be still valid for the parameters of the analytical solution. Comparisons for the analytical solution and for the experimental data are shown in Fig. 3. In Fig. 3(a) one case see clearly that the two models produce accurate predictions, although KRAKEN's prediction is smoother, a fact that can be attributed to the low value of frequency. On the other hand, in Fig. 3(b) KRAKEN's prediction is only accurate at the initial ranges, and quickly starts to diverge due to the failure of the adiabatic prediction to account for the exchange of energy between modes of different orders. TRACEO3D on the other side is able to produce an accurate prediction in both amplitude and phase along the entire cross-slope range, thus showing the capability of the model to deal with an arbitrary wedge slope.

## 6 Conclusions

The discussion presented in the previous sections demonstrated the feasibility of using TRACEO3D as a module of sound propagation for noise predictions, through the comparisons with results from an analytical solution of the wedge problem and measurements from a scale tank experiment. The comparisons show that the model is able to predict the reference data, while adiabatic coupling is only valid for a small wedge. Despite the low frequency limitation, typical of ray theory, TRACEO3D was able to provide an accurate prediction for the analytical (low-frequency) case. The results also indicate that TRACEO3D can be able to deal with arbitrary bathymetries, a feature of fundamental importance for the development of a Smart System for the monitoring of shipping noise. Future work will be dedicated to optimize the current version through parallel computing, allowing decreasing the computational time, and enabling the model to provide fast predictions in an environment with a fine grid. Further improvements will also look for efficient solutions of 3D eigenrays search and fast 3D calculations of particle velocity.

**Acknowledgments.** This work received support from the Foreign Courses Program of CNPq and the Brazilian Navy. Thanks are due to the SiPLAB research team, FCT, University of Algarve and MarSensing. The authors are also deeply thankful to LMA-CNRS for allowing the use of the experimental tank data discussed in this work.

## References

1. Van der Graaf, A.J., Ainslie, M.A., André, M., Brensing, K., Dalen, J., Dekeling, R.P.A., Robinson, S., Tasker, M.L., Thomsen, F., Werner, S.: European marine strategy framework directive - good environmental status (MSFD GES): report of the technical subgroup on underwater noise and other forms of energy, Brussels (2012)
2. Soares, C., Zabel, F., Jesus, S.M.: A shipping noise prediction tool. In: OCEANS 2015, Genova (2015)
3. Tolstoy, A.: 3-D propagation issues and models. J. Comput. Acoust. **4**(03), 243–271 (1996)
4. Jensen, F.B., Kuperman, W.A., Porter, M.B., Schmidt, H.: Computational Ocean Acoustics, 2nd edn. Springer, New York (2011)

5. Sturm, F., Korakas, A.: Comparisons of laboratory scale measurements of three-dimensional acoustic propagation with solutions by a parabolic equation model. J. Acoust. Soc. Am. **133**, 108–118 (2013)
6. Buckingham, M.J., Tolstoy, A.: An analytical solution for benchmark problem 1: the 'ideal' wedge. J. Acoust. Soc. Am. **87**, 1511–1513 (1990)
7. Deane, G.B., Buckingham, M.J.: An analysis of the three-dimensional sound field in a penetrable wedge with a stratified fluid or elastic basement. J. Acoust. Soc. Am. **93**, 1319–1328 (1993)
8. Westwood, E.K.: Ray model solutions to the benchmark wedge problems. J. Acoust. Soc. Am. **87**, 1539–1545 (1990)
9. Sturm, F., Ivansson, S., Jiang, Y.M., Chapman, N.R.: Numerical investigation of out-of-plane sound propagation in a shallow water experiment. J. Acoust. Soc. Am. **124**, 341–346 (2008)
10. Rodríguez, O.C., Silva, A., Zabel, F., Jesus, S.M.: The TV-APM interface: a web service for collaborative modeling. In: 10th European Conference on Underwater Acoustics, Istanbul (2010)
11. Felisberto, P., Rodríguez, O.C., Santos, P., Ey, E., Jesus, S.M.: Experimental results of underwater cooperative source localization using a single acoustic vector sensor. Sensors **13**, 8856–8878 (2013)
12. Rodríguez, O.C.: The TRACEO ray tracing program. Physics Department, SiPLAB, FCT, University of Algarve (2011)
13. Rodríguez, O.C., Collis, J.M., Simpson, H.J., Ey, E., Schneiderwind, J., Felisberto, P.: Seismo-acoustic ray model benchmarking against experimental tank data. J. Acoust. Soc. Am. **132**, 709–717 (2012)
14. Korakas, A., Sturm, F., Sessarego, J.P., Ferrand, D.: Scaled model experiment of long-range across-slope pulse propagation in a penetrable wedge. J. Acoust. Soc. Am. **126**, 22–27 (2009)
15. Kuperman, W.A., Porter, M.B., Perkins, J.S., Evans, R.B.: Rapid computation of acoustic fields in three-dimensional ocean environments. J. Acoust. Soc. Am. **89**, 125–133 (1991)

# Feature Transformation Based on Stacked Sparse Autoencoders for Sleep Stage Classification

Shirin Najdi[1,2(✉)], Ali Abdollahi Gharbali[1,2], and José Manuel Fonseca[1,2]

[1] CTS, Uninova, 2829-516 Caparica, Portugal
[2] Faculdade de Ciências e Tecnologia, Universidade Nova de Lisboa Campus da Caparica, Quinta da Torre, 2829-516 Monte de Caparica, Portugal
{s.najdi,a.gharbali}@campus.fct.unl.pt, jmrf@fct.unl.pt

**Abstract.** In this paper a deep learning based dimension reduction, feature transformation and classification method is proposed for automatic sleep stage classification. In order to enhance the feature vector, before feeding it to the deep network, a discriminative feature selection method is applied for removing the features with minimum information. Two-layer Stacked Sparse Autoencoder together with Softmax classifier is selected as the deep network model. The performance of the proposed method is compared with Softmax and k-nearest neighbour classifiers. Simulation results show that proposed deep learning structure outperformed others in terms of classification accuracy.

**Keywords:** Sleep stage classification · Deep learning · Stacked Sparse Autoencoders · Feature transformation

## 1 Introduction

Sleep occupies significant part of human life. Diagnosis and treatment of sleep related disorders and sleep deficiency is of great importance in the sleep research community. Normal human sleep generally consists of two distinct stages with independent functions known as Non-Rapid Eye Movement (NREM) and Rapid Eye Movement (REM) sleep. In their ideal situation, NREM and REM states alternate regularly, each cycle lasting 90 min on average. NREM sleep accounts for 75–80% of sleep duration. According to the American Academy of Sleep Medicine (AASM) [1], NREM is subdivided into three stages: stage 1 or light sleep, stage 2 and stage 3 or Slow Wave Sleep (SWS). On the other hand, REM sleep accounts for 20–25% of sleep duration. The first REM state usually occurs 60–90 min after the onset of the NREM and lasts a few minutes.

A multiple-parameter test, called polysomnography (PSG), is normally used for analysis and interpretation of multiple, simultaneous physiologic events during sleep. PSG includes several body functions such as electroencephalogram (EEG), electro-oculogram (EOG), chin and leg electromyogram (EMG), airflow signals, respiratory effort signals, oxygen saturation and electrocardiogram (ECG). The data acquired in this way is analyzed by an expert in a clinic or hospital environment. The proportion and distribution of sleep stages can then be used for diagnosis of sleep related problems.


L.M. Camarinha-Matos et al. (Eds.): DoCEIS 2017, IFIP AICT 499, pp. 191–200, 2017.
DOI: 10.1007/978-3-319-56077-9_18

However, visual sleep stage classification is a subjective task and requires much time and effort. On the other hand, there is an increasing interest for at home sleep monitoring technologies that try to modernize sleep analysis and reduce the workload of healthcare centers. Therefore, there is a need for reliable automatic sleep stage classification systems that can efficiently perform sleep scoring.

One of the main challenges of automatic sleep stage classification is to compactly represent the subject's data in the form of a feature vector. This feature vector should be informative and non-redundant enough in order to decrease the computational complexity and facilitate the subsequent classification step. Proper selection of the classifier in order to achieve the highest possible classification accuracy is another challenge in these systems. Looking at the existing literature, It can be noted that a wide range of features including temporal [2–4], spectral [3, 5, 6], linear/non-linear [4, 7, 8] and statistical features [3, 5] were extracted from different subsets of PSG recording and used for sleep scoring. Some of conventional feature transformation methods such as principal component analysis (PCA) [9] and kernel dimensionality reduction (KDR) [10] are used for reducing the dimensionality and enhancing the descriptive power of feature vector. Considering the fact that deep learning methods have found their way into many artificial intelligence applications with successful results reported from academia and industry, the main motivation for the current work was to explore the potential of deep learning for feature transformation and classification in the automatic sleep stage classification area. In view of this, the following research question emerges:

*How should a deep learning technique be used for dimensionality reduction and classification of sleep stages, so that the computational complexity of classification step is reduced and the scoring accuracy is improved?*

In this work, we propose an algorithm that answers this question. This paper is organized as follows: Sect. 2 explains how the proposed algorithm contributes to the smart systems. In Sect. 3, a review of the application of deep learning techniques in sleep stage classification is presented. Section 4, provides detailed description of the proposed algorithm. Simulation results and discussion are presented in Sect. 5. Finally, conclusion and future work are given in Sect. 6.

## 2 Relationship to Smart Systems

Smart systems refer to diverse range of technological systems that can perform autonomously or in collaboration with other systems. These systems have an ability to combine functionalities including sensing, actuating and controlling a particular situation. Based on the information they acquire, they have the ability to perform smart actions such as prediction or deciding and communicate with the user through highly sophisticated user interfaces. Generally, smart systems are made of several components, each having a special purpose. These components include sensors for data acquisition, information transmitting elements, command-and-control units for making decisions, components for transmitting decisions, taken, and finally actuators for triggering required action. These systems have been used to address some problems in diverse areas, such as energy, transportation, security, safety, healthcare, etc. [11].

In healthcare area, the main goal of utilizing smart systems is to improve patient management workflow. This improvement leads to the reduction in the burden of medical staff, consultation time, waiting lists and medical costs. Smart healthcare systems, as advanced technology, are classified into three main categories: remote health monitoring systems (RHMS), mobile health monitoring systems (MHMS) and wireless health monitoring systems (WHMS). Among these three categories, WHMS refer to biosensors that can be worn by the subject and may also include one or both of RHMS and MHMS [12]. Wearable sleep monitoring systems are perfect examples of WHMS. The idea behind them is to use efficient, affordable and medically reliable systems for unsupervised at-home monitoring of patients' sleep data. Typically, wearable sleep monitoring systems have an algorithm that performs automatic sleep stage classification for evaluating sleep quality [13]. This work is developed in such a context to improve the performance of the classification accuracy using deep learning techniques.

## 3 State of the Art

Unlike some of the machine learning areas such as natural language processing and object classification, the potential of deep learning techniques is not fully explored in automatic sleep stage classification. This fact is also noticeable when it comes to the feature transformation for sleep scoring. To the best of our knowledge, there are few research works in this area and in the following we will try to briefly review them.

In [14], the main idea is using hybrid deep learning models to increase the performance of sleep stage classification. Deep Belief Networks (DBNs) are applied on 28 hand-crafted feature set for unsupervised generation of higher level features. For classification another deep structure, namely, Long Short Term Memory (LSTM) is used. In this work, sleep stage classification is regarded as a time series and sequence classification problem, therefore the ability of LSTM models for recognizing the patterns from a sequence of events is mentioned as the reason for using this classifier. The proposed algorithm is tested on two datasets of sleep recording and the features are extracted from EEG, EOG and EMG. The performance of the proposed algorithm (DBN +LSTM) is compared to three other sleep stage classification algorithms, namely DBN only, LSTM only and DBN with Hidden Markov Model (HMM). Simulation results show that two hybrid methods DBN + LSTM and DBN + HMM have significantly better performance than single DBN and single LSTM, while DBN + LSTM performs better than DBN + HMM for both datasets. It has been concluded that LSTM boosted the performance of DBN much better than HMM.

Tsinalis et al. [15] proposed an algorithm for sleep stage classification using on time-frequency analysis based features and Stacked Sparse Autoencoders (SSAE). For each epoch, in total 557 features are extracted from the time-frequency representation of the single channel EEG signal. Complex Morlet wavelet is used for creating time-frequency representation. In this paper, it has been showed that the classification accuracy can be improved by including the features from neighboring epochs. According to the simulation results, the proposed method leads to 1–2% scoring improvement compared to four other sleep scoring algorithms, that don't use deep learning. They succeeded to reduce the gap between the mean performance between S1 (as the most misclassified stage) and all other stages. Also, the adverse effect of inherent class imbalance in sleep data on the classification accuracy is highlighted. Authors tried to alleviate this effect by creating a balanced dataset in which all stages are equally represented.

In [16], the main focus is on generating meaningful data representations from unlabeled data. For this purpose, the performance of a two-layer DBN with 200 hidden units was compared with another feature transformation algorithm consisting of conventional methods (PCA + Sequential Backward Selection (SBR) + Gaussian Mixture Model (GMM)). These feature selection methods are applied on a vector of 28 hand-crafted features. Newly generated feature sets are classified by using Softmax classifier and Hidden Markov Model (HMM) respectively. Experimental results showed that DBN-based feature transformation performs much better the other method for sleep stage classification.

Dong et al. in [17], proposed a practical approach was proposed for mitigating the limitations of single-channel automatic sleep stage classification using Mixed Neural Network (MNN). MNN is a deep learning-based feature transformation and classification technique and is composed of a Rectifier Neural Network (RNN), a Long Short-Term Memory (LSTM) and a Softmax regression. The input to this system is a feature vector with time-frequency domain, statistical and time domain features. Considering temporal dependency of sleep stages to each other, in addition to the features of the current epoch, the features from previous EEG epochs are also fed to the system. In this paper, several alternative electrode placements are explored and finally a convenient single forehead EEG channel together with an EOG channel configuration is proposed for the low-cost, at-home sleep monitoring applications.

## 4 Materials and Methods

Figure 1 shows an overview of proposed sleep stage classification algorithm with proposed feature transformation scheme.

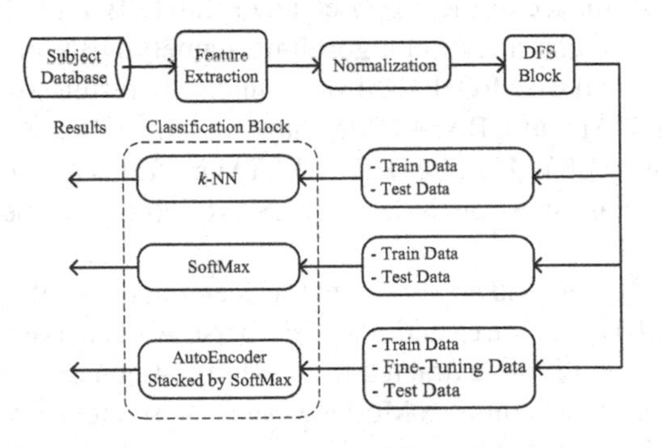


**Fig. 1.** Flowchart of the proposed algorithm.

### 4.1 Data

In this paper, we used a publically available dataset, called ISRUC-Sleep [18]. The data were acquired from 10 healthy adults, including 9 male and 1 female subjects aged between 30 and 58. The recordings were done in Sleep Medicine Centre of the Hospital of Coimbra University (CHUC) during an all-night session (eight hours). The PSG recordings of each subject were scored by two experts according to the AASM manual [1]. All EEG, EOG and chin EMG recordings were performed with a sampling rate of 200 Hz. The quality of the PSG recordings of this dataset have improved through a pre-processing step. In this pre-processing step, (1) a notch filter is applied to eliminate 50 Hz electrical noise, (2) EEG and EOG recordings are filtered using a bandpass Butterworth filter with a lower cut-off frequency of 0.3 Hz and higher cut-off frequency of 35 Hz, and (3) EMG channels are filtered using a bandpass Butterworth filter with a lower cut-off frequency of 10 Hz and higher cut-off frequency of 70 Hz. For the evaluation of our proposed method, we used C3-A2 EEG channel, right EOG and chin EMG channels. In this paper, we used all the data from 10 healthy subjects from ISRUC-Sleep dataset. The number of epochs for these 10 subjects is 954, 941, 824, 794, 944, 853, 814, 1000, 969, and 796. Totally we have 8889 epochs from this database and to avoid overfitting we used all of them.

### 4.2 Feature Extraction

All signals used in this study, were divided into 30-second epochs. A set of features were extracted from each epoch of EEG, EOG and EMG recordings of each subject.

**Table 1.** Summary of the features extracted from PSG recordings.

| Signal | Category | Feature name |
|---|---|---|
| EEG | Time domain (F1 to F12) | Statistical features (minimum value, maximum value, arithmetic mean, standard deviation, variance, skewness, kurtosis, median), zero-crossing rate, Hjorth parameters (activity, mobility and complexity) [20] |
| | Time-frequency domain (F13 to F26) | Features extracted from wavelet packet coefficients including energy of $\alpha$, $\delta$, $\beta 1$, $\beta 2$, $\theta$ and spindle bands, total energy of all bands, energy ratio of $(\frac{\alpha}{\delta+\theta}, \frac{\delta}{\alpha+\theta}, \frac{\theta}{\alpha+\delta}, \frac{\delta}{\theta}, \frac{\alpha}{\theta})$, statistical features (mean and standard deviation of coefficients in all of the bands) |
| | Entropy (F27 to F30) | Spectral entropy, Rényi entropy, approximate entropy, permutation entropy [20] |
| | Non-linear (F31 to F36) | Petrosian fractal dimension, mean Teager energy, mean energy, mean curve length, hurst exponent [20], ISD |
| EOG | Time domain (F37 to F41) | Mean, maximum, standard deviation, skewness, kurtosis [21] |
| | Non-linear (F42) | Energy [21] |
| EMG | Frequency domain (F43 to F46) | Total power in the EMG frequency spectrum, statistical features of EMG frequency spectrum (maximum, mean, standard deviation) [21] |
| | Non-linear (F47 to F49) | Energy, ratio of the EMG signal energy for the current epoch and previous epoch, ratio of the EMG signal energy for the current epoch and next epoch [21] |

This feature set includes 49 features that can be considered as time, frequency, joint time-frequency domain, entropy-based and nonlinear types. Table 1 shows a summary of these features. For more detail see [19].

### 4.3 Normalization

In order to standardize the range of Features, scaling normalization method was applied. Each feature ($x_{ij}$) is independently normalized by applying the following equation:

$$x'_{ij} = \frac{x_{ij} - \min(\mathbf{x}_i)}{\max(\mathbf{x}_i) - \min(\mathbf{x}_i)} \tag{1}$$

where $\mathbf{x}_i$ is the vector of $i^{th}$ feature.

### 4.4 Discriminative Feature Selection

There are many potential advantages in removing the features before final modelling and classification. Fewer features mean lower computational complexity. Also, some features may reduce the performance by their corrupt distributions. Consider a feature that a single value for all of the samples. According to [22], this feature is called "zero-variance predictor". Even if it has little effect on the next step, this feature should be discarded from feature set, because it has no information. Similarly, some features may have few unique values that occur with low frequency. These features are called "near-zero variance predictors". Kuhn et al. [22] defines two criteria for detecting near-zero variance features as follows:

(a) The ratio of unique values to the number of samples is low, for example 10%.
(b) The ratio of the frequency of the most dominant value to the frequency of the second dominant value is high, for example 20.

Using these two criteria, we applied Discriminative Feature Selection (DFS) to remove the features that didn't have enough discriminative power. As a result, the dimension of feature set reduced to 37.

### 4.5 Stacked Sparse Autoencoder (SSAE)

An autoencoder is a special type of neural network whose output values are equal to the inputs. An autoencoder typically consists of an encoder and a decoder and it is trained in an unsupervised manner using backpropagation. During training, a cost function that measures the error between input and output of the autoencoder is optimized, in other words, the autoencoder tries to learn the identity function. By applying special constraints on the network such as the number of hidden units, an autoencoder can learn new representation or coding of the data [23]. Suppose the input vector to the autoencoder is a set of un-labelled data $\mathbf{x} \in \mathbb{R}^{D_x}$. This vector is encoded to another vector $\mathbf{z} \in \mathbb{R}^{D_1}$ in the hidden layer as follows:

$$\mathbf{z} = h^1\left(\mathbf{W}^1\mathbf{x} + \mathbf{b}^1\right) \tag{2}$$

where $h^1$ is the transfer function of the encoder, $\mathbf{W}^1$ is the weight matrix and $\mathbf{b}^1$ is the bias vector of the encoder. Then, the autoencoder tries to decode this new representation back to the original input vector as follows:

$$\hat{\mathbf{x}} = h^2\mathbf{z} = h^2\left(\mathbf{W}^1\mathbf{x} + \mathbf{b}^1\right) \tag{3}$$

where $h^2$ is the transfer function of decoder, $\mathbf{W}^2$ is weight matrix and $\mathbf{b}^2$ is bias vector of the decoder. Sparse autoencoder is a specific type of autoencoder in which in order to encourage the sparsity of the output of the hidden layer, a constraint is imposed on the number of active hidden neurons. The cost function of sparse autoencoder is slightly different from the original autoencoder as follows:

$$E = \underbrace{\frac{1}{N}\sum\sum(\mathbf{x}-\hat{\mathbf{x}})^2}_{\text{mean squared error}} + \underbrace{\lambda\,\Omega_{weights}}_{\text{weight regularization}} + \underbrace{\beta\,\Omega_{sparsity}}_{\text{sparsity regularization}} \tag{4}$$

where $N$ is length of the input vector, $\lambda$ is the weight regularization parameter $\beta$ is the sparsity regularization parameter [24].

A Stacked Sparse Autoencoder (SSAE) is a neural network with several sparse autoencoders. In this architecture, the output of each autoencoder is fully connected to the inputs of the next autoencoder. Greedy layer-wise training strategy is usually used for training SSAE. After the training of each layer is complete, a fine tuning is usually performed for enhancing the learned weights using backpropagation algorithm. Fine tuning can greatly improve the performance of the stacked autoencoder [23].

### 4.6 Softmax Classifier

After SSAE, Softmax classifier is stacked to the network as the output layer. Softmax classifier has a probabilistic interpretation of each. It is the generalization of binary Logistic Regression classifier to multiple classes. In sleep stage classification, the number of output classes is equal to the number of sleep stages.

## 5 Experimental Results and Discussion

For evaluating the performance of the proposed sleep stage classification algorithm, we used all data from 10 healthy subjects as described in Sect. 4.1. The EEG, EOG and EMG signals of each subject were divided into 30-second epochs. After feature extraction and normalization, the feature sets were fed to DFS block to eliminate the near-zero variance features.

According to the criteria mentioned in Sect. 4.3, 12 features were recognized as near-zero variance feature and removed from our sleep data model, as follows: maximum value (F1), minimum value (F2), variation (F5), median (F8), Petrosian fractal dimension

(F31), permutation entropy (F30), Hjorth parameter (Activity) (F10), zero crossing number (F9), total power in the EMG frequency spectrum (F43), mean of power in the EMG frequency spectrum (F45), absolute energy of the time domain EMG signal (F47), maximum value of time domain EOG signal (F38).

After the feature vector was set, data were divided into two parts, training, testing, using 10-fold cross validation method. For fine tuning step of SSAE, part of training data was utilized. Our deep learning consists of three layers, a two-layer SSAE and a Softmax layer. The number of hidden units for the first and second layer of SSAE was 20 and 12, respectively.

For finding the best hyper-parameters for the autoencoders, we tried several models by adjusting sparsity regularization parameter, weight regularization parameter and the number of iterations. We used autoencoders with logistic sigmoid activation function for both of the layers. The performance of the proposed algorithm was compared with two other classifiers, Softmax and *k*-Nearest Neighbor (*k*-NN) classifier. The number of neighbors was set to 18 and Euclidean distance was used as a measure of distance for *k*-NN. To evaluate our systems' performance we used classification accuracy as the evaluation criterion.

Table 2 shows the individual sleep stage and overall classification accuracy extracted from confusion matrix for three different classifiers. The boldface numbers indicates the best performance.

**Table 2.** Results of the statistical analysis for comparison of each stage and overall accuracy.

| Classifiers | Wake (%) | REM (%) | S1 (%) | S2 (%) | SWS (%) | Overall accuracy (%) |
|---|---|---|---|---|---|---|
| Softmax | 80 | 61.66 | 65 | 90 | 78.33 | 74.9 |
| *k*-NN | 85 | 66.66 | 61.66 | 70 | 83.33 | 73.33 |
| SSAE | 91 | 77 | 69 | 87 | 87 | 82.2 |

It is noticeable that SSAE method outperforms the other two classifiers in terms of overall accuracy. Also, for the individual sleep stages, in most of the cases SSAE discriminates the stages better. In addition to the higher performance, SSAE provides a considerable reduction in the dimension of the feature vector.

Considering that the second layer of SSAE had 12 hidden units, SSAE succeeded to decrease the dimension from 37 to 12, which means 67% reduction. Therefore, SSAE is a powerful tool to generate more descriptive features from original feature vector. In order to confirm the advantage of DFS block, the performance of SSAE-based sleep stage classification with and without this step was investigated. Without using DFS block, 49 original features were fed to SSAE.

The classification accuracy achieved in this way was 74.1% which is almost 8% less than accuracy with DFS block.

## 6 Conclusions and Future Works

Although feature transformation based on deep learning has already used in several machine learning applications, the advantages and potentials of applying these methods in sleep stage classification problem have not explored yet. This paper is a contribution in this regard. We proposed a method for dimension reduction and feature transformation based on SSAEs. The results show that SSAE can be considered as an appropriate tool for decreasing the complexity of sleep scoring issues. Future works will include comparing the performance of other conventional classifiers such as SVM and RF with SSAE in sleep stage classification.

**Acknowledgment.** This work was partially funded by FCT Strategic Program UID/EEA/00066/203 of UNINOVA, CTS.

## References

1. Berry, R.B., Brooks, R., Gamaldo, C.E., Harding, S.M., Lloyd, R.M., Marcus, C.L., et al.: AASM - Manual for the Scoring of Sleep and Associated Events version 2.1 (2014)
2. Hamida, S.T.-B., Ahmed, B.: Computer based sleep staging: challenges for the future. In: 2013 7th IEEE GCC Conference and Exhibition, pp. 280–285. IEEE (2013)
3. Koley, B., Dey, D.: An ensemble system for automatic sleep stage classification using single channel EEG signal. Comput. Biol. Med. **42**, 1186–1195 (2012). Elsevier
4. Şen, B., Peker, M., Çavuşoğlu, A., Çelebi, F.V.: A comparative study on classification of sleep stage based on EEG signals using feature selection and classification algorithms. J. Med. Syst. **38**, 18 (2014)
5. Radha, M., Garcia-Molina, G., Poel, M., Tononi, G.: Comparison of feature and classifier algorithms for online automatic sleep staging based on a single EEG signal. In: 36th Annual International Conference of the IEEE Engineering in Medicine and Biology Society, pp. 1876–1880. IEEE (2014)
6. Šušmáková, K., Krakovská, A.: Discrimination ability of individual measures used in sleep stages classification. Artif. Intell. Med. **44**, 261–277 (2008)
7. Hsu, Y.-L., Yang, Y.-T., Wang, J.-S., Hsu, C.-Y.: Automatic sleep stage recurrent neural classifier using energy features of EEG signals. Neurocomputing **104**, 105–114 (2013). Elsevier
8. Dursun, M., Gunes, S., Ozsen, S., Yosunkaya, S.: Comparison of artificial immune clustering with fuzzy c-means clustering in the sleep stage classification problem. In: 2012 International Symposium on Innovations in Intelligent Systems and Applications, pp. 1–4. IEEE (2012)
9. Rempe, M., Clegern, W., Wisor, J.: An automated sleep-state classification algorithm for quantifying sleep timing and sleep-dependent dynamics of electroencephalographic and cerebral metabolic parameters. Nat. Sci. Sleep. **7**, 85 (2015)
10. Fukumizu, K., Bach, F.R., Jordan, M.I.: Kernel dimensionality reduction for supervised learning. J. Mach. Learn. Res. **5**, 73–99 (2004)
11. CORDIS Archive: European Commission: CORDIS: FP7: ICT: Micro/nanosystems: Calls [Internet]. http://cordis.europa.eu/fp7/ict/micro-nanosystems/calls_en.html. Accessed 17 Dec 2016
12. Baig, M.M., Gholamhosseini, H.: Smart health monitoring systems: an overview of design and modeling. J. Med. Syst. **37**, 9898 (2013). Springer US

13. Tsinalis, O., Matthews, P.M., Guo, Y.: Automatic sleep stage scoring using time-frequency analysis and stacked sparse autoencoders. Ann. Biomed. Eng. **44**, 1587–1597 (2016)
14. Giri, E.P., Fanany, M.I., Arymurthy, A.M.: Combining Generative and Discriminative Neural Networks for Sleep Stages Classification (2016)
15. Tsinalis, O., Matthews, P.M., Guo, Y.: Automatic sleep stage scoring using time-frequency analysis and stacked sparse autoencoders. Ann. Biomed. Eng. **44**, 1587–1597 (2016). Springer US
16. Längkvist, M., Karlsson, L., Loutfi, A., Loutfi, A.: Sleep stage classification using unsupervised feature learning. Adv. Artif. Neural Syst. **2012**, 1–9 (2012). Hindawi Publishing Corporation
17. Dong, H., Supratak, A., Pan, W., Wu, C., Matthews, P.M., Guo, Y.: Mixed Neural Network Approach for Temporal Sleep Stage Classification (2016)
18. Khalighi, S., Sousa, T., Santos, J.M., Nunes, U.: ISRUC-sleep: a comprehensive public dataset for sleep researchers. Comput. Methods Programs Biomed. **124**, 180–192 (2016). Elsevier Ireland Ltd
19. Najdi, S., Gharbali, A.A., Fonseca, J.M.: A comparison of feature ranking and rank aggregation techniques in automatic sleep stage classification based on polysomnographic signals. In: Ortuño, F., Rojas, I. (eds.) IWBBIO 2016. LNCS, vol. 9656, pp. 230–241. Springer, Heidelberg (2016). doi:10.1007/978-3-319-31744-1_21
20. Şen, B., Peker, M., Çavuşoğlu, A., Çelebi, F.V.: A comparative study on classification of sleep stage based on EEG signals using feature selection and classification algorithms. J. Med. Syst. **38**, 18 (2014)
21. Dursun, M., Gunes, S., Ozsen, S., Yosunkaya, S.: Comparison of artificial immune clustering with fuzzy c-means clustering in the sleep stage classification problem. In: 2012 International Symposium on Innovations in Intelligent Systems and Applications, pp. 1–4 (2012)
22. Kuhn, M., Johnson, K.: Applied Predictive Modeling. Springer, New York (2013)
23. Ng, A., Ngiam, J., Foo, C., Mai, Y., Suen, C.: UFLDL Tutorial (2010). http://ufldl.stanford.edu/wiki/index.php/UFLDL_Tutorial
24. Olshausen, B.A., Field, D.J.: Sparse coding with an overcomplete basis set: a strategy employed by V1? Vis. Res. **37**, 3311–3325 (1997)

# Embedded and Real Time Systems

# Quality Evaluation Strategies for Approximate Computing in Embedded Systems

Olaf Neugebauer[1(✉)], Peter Marwedel[1], Roland Kühn[1], and Michael Engel[2]

[1] TU Dortmund University, Dortmund, Germany
{olaf.neugebauer,peter.marwedel,roland.kuehn}@tu-dortmund.de
[2] Coburg University of Applied Sciences and Arts, Coburg, Germany
engel@multicores.org

**Abstract.** The quest for increased performance at lower energy consumption rates, especially in embedded systems used in smart systems, has reached physical limits that can no longer be exploited using traditional optimization techniques. One popular way to achieve additional gains is to intentionally perform inaccurate computations. Our framework eases evaluation, analysis and comparison of approximation techniques in terms of energy consumption, run time, quality and user-defined criteria. Applied to a set of benchmarks, we obtain valuable insights into related side effects, including increased file sizes indicating that a careless utilization of approximate computing threatens its usefulness.

**Keywords:** Approximate computing · Energy consumption · Quality metrics

## 1 Introduction

Growing demands for processing power, especially in embedded systems bundled in smart systems such as autonomous cars and media signal processing, increase the pressure on hardware and software developers to create well performing systems which are also power and energy efficient. Traditionally, hardware manufacturers solved the challenge by increasing the frequencies of processors and memories, whereas the reduction of power and energy consumption was mostly achieved by shrinking semiconductor structure sizes. More recently, multiple, potentially heterogeneous, cores are being combined onto a single chip to tackle the needs of versatile performance requirements. However, frequency increase is limited by thermal and energy constraints and parallelism is restricted by synchronization overhead.

One increasingly popular method to mitigate the rising impact of these performance and power walls is to intentionally perform inaccurate computations but still satisfy the quality of service (QoS) requirements of a given application. Similar ideas have been exploited especially in the media processing domain to reduce file sizes and throughput requirements by the introduction of *lossy encoding*.

To determine the effectiveness and efficiency of approximate computing approaches, it becomes obvious that it is required to ensure that results stay within an acceptable range. As a consequence, suitable quality metrics are required. The selection of such


L.M. Camarinha-Matos et al. (Eds.): DoCEIS 2017, IFIP AICT 499, pp. 203–210, 2017.
DOI: 10.1007/978-3-319-56077-9_19

metrics is complex, error-prone, and highly application dependent. Especially if metrics lead to contrary or insufficient conclusions, it is necessary to take multiple metrics into account. Our assessment framework gives indications of an application's performance with respect to run time, energy consumption and quality. An extensible implementation enables introduction of user-defined metrics and objectives. We found that the careless utilization of approximate computing threatens its usefulness. We observed increased output file sizes which can render the benefits of approximate computing useless in later processing steps.

## 2 Contribution to Smart Systems

Today's smart systems expose new requirements to embedded systems. Coupled closely with sensors and actuators, embedded systems play an important role in the whole processing chain of modern smart systems. We think, approximate computing offers new opportunities for the embedded system domain. By either applying approximation techniques to the source code or accepting inaccurate results, applications not designed for this domain can be executed on embedded systems. We showed that pattern recognition can be performed within certain limits on mobile devices. We ported an image-based virus detection application [12] from a dedicated computer to a high performance embedded system by using approximation techniques.

The next step to an aerial virus detection system with dozens of sensors is to move at least the preprocessing to a low-power embedded system closer to the sensor. We believe that new enhanced approximation techniques are key to achieve this goal. In this paper we present our assessment framework which enables us to pursue our overall goal of an aerial virus detection as an example of modern smart systems. The framework is not restricted to our application; a universal approach allows its usage in other domains.

In this paper we start with well-known approximation techniques and applications. One important result of our analyses is that focusing solely on a single metric or category can lead to significantly incorrect conclusions in the evaluation of approximation techniques; in addition, we discovered mutual influences between approximation techniques and lossy compression which can have negative side effects on objectives such as the file size. Since it is not obvious which approximation approach is suitable for a given application, the framework enables a design space exploration (DSE) of approximation techniques to assess the related gains and impacts.

In summary, the key contributions of this paper are:

- An analysis of the impact of exemplary approximation techniques on representative applications showing the difficulty of using single-metrics QoS analyses,
- An insight into conflicts between unfocused use of approximation and application-inherent lossy compression techniques,
- Enabling DSE for approximation techniques through an automatic evaluation of multiple metrics,
- Enabling runtime and energy assessment on real embedded hardware,
- Easy integration of user-defined metrics and objectives, and
- Determining the impact of approximation techniques in early design stages.

## 3 On Approximate Computation and Metric Selection

Approximate computing promises to be a worthwhile approach to overcome or at least shift the influence of performance and power walls. In recent years, related techniques have seen significant adoption [11]. On the hardware side, approximate computing can be applied to almost all components. For example, approximate adders [7], logic complexity reduction at the transistor level [4], and reduced refresh rates of memories [9]. Software approaches are usually more flexible than hardware ones since they can be adapted to specific scenarios or dynamically adjusted at run-time. Application knowledge can be introduced, exploring the specific requirements and capabilities of a given piece of software. Software approximation techniques can be executed on commodity hardware. Also, these techniques avoid the problem of obtaining unexpected results from inaccurate hardware computations and the problem of testing inaccurate hardware. Omitting iterations is the basic idea behind loop perforation [10, 15]. Using approximate data types and operations are the key features of EnerJ [14]. The Green system [2] enables programmers to express approximate functions and loops. Parallel applications can suffer from large synchronization overhead. Relaxing the synchronization by keeping the output quality in an acceptable range can improve the performance [13].

The impact of approximations on accuracy or QoS of applications can be quantified with metrics. Numerous metrics for different domains have been developed in the last decades. The presented approaches either rely on user-defined or heavily application-dependent quality metrics. In most cases, only one metric is considered. Akturk et al. [1] presented hints on metric selection for different domains. However, choosing the appropriate metric during the development process is a complicated task as we show in the following. First, we present some of the most important metrics used in today's approximate computing research. This paper focuses on applications from the image processing domain. Thus, metrics concerning this domain are discussed.

Metrics can be at least classified into two categories. The first category focuses on the signal itself where mathematical properties are used to quantify the quality. The second category takes the context information and recipient of the result into account. For image processing applications with humans as recipients, this category is called *Perceptual Visual Quality Metrics* (PVQM) and according to Lin et al. [8] the first category is called *Signal Fidelity Metrics* (SFM).

SFM are widespread and often used due to their simplicity. Prominent examples are the Mean-Squared Error (MSE), Root-Mean-Squared Error (RMSE), Mean-Absolute Error (MAE) and Peak-Signal-to-Noise Ratio (PSNR). In contrast to typical SFM, PQVM try to involve human visual system characteristics in the evaluation. Here, we present two metrics from this domain, the Universal Image Quality Index (UIQI) [16] and its successor the Structural Similarity Index (SSIM) [17]. Since the UIQI tends to be unstable in some cases, the SSIM uses some improvements to obtain stable results.

During the development and analysis of applications using approximate computing, a selection of proper quality metrics is of high importance. Figure 1 shows three versions of the same picture with Fig. 1a being the reference picture. In Fig. 1b the image has been brightened, and in Fig. 1c is distorted with Gaussian blur. Applying a signal fidelity metric like MSE results in almost identical values. Thus, judging which picture is *better*

**Fig. 1.** Quality comparison according to MSE and SSIM.

is complicated. In contrast to SFM, PVQM consider the human as receiver of the results. The resulting SSIM value indicates that Fig. 1b is much *better* than Fig. 1c. Not only for humans, also for pattern recognition applications these results are relevant. For example, an edge detection algorithm would usually perform better on images with sharp edges such as the image from Fig. 1b.

This example highlights the importance of quality metrics and their selection for approximate computing. Selecting proper metrics is complex, error-prone and usually requires deep application knowledge.

## 4 Framework

The *Quality Comparison for Approximate Programs on Embedded Systems* (QCAPES) framework supports developers to introduce approximate computing into their applications taking the special requirements of embedded systems into account. As described in the previous sections, this paper focuses on applications from the multimedia domain. However, as this section shows, the framework can be applied to nearly all domains. QCAPES is available under an open source license[1].

The user provides an executable without modifications which is considered as the reference and one or several inaccurate versions of the application to the framework. By passing a configuration file, the user selects metrics, objectives and data sets which are considered for the evaluation. Several metrics and some testing data from the multimedia domain are already included in the framework. Due to the open structure of QCAPES, new metrics and objectives can be added easily. Evaluation results are stored in a database and visualizations are generated.

Considering multiple metrics is crucial to evaluate the results generated by approximate versions of a given program in terms of output quality. This framework includes all previously discussed signal fidelity and perceptual-based metrics (cf. Sect. 3). For UIQI and SSIM, overlapping pixel by pixel or non-overlapping blockwise sliding window implementations are provided.

In addition, windows of block size $4 \times 4$, $8 \times 8$ or $16 \times 16$ pixels are used for the local quality evaluation. For the SSIM, a Gaussian filter can be used to reduce the influence of pixels at the edges. To summarize, a total of 13 PVQM settings are available

[1] http://sfb876.tu-dortmund.de/auto?self=Software.

in the QCAPES framework. Additional metrics included, e.g. numerical applications, are not covered in this paper. The user of the framework selects which metrics and objectives should be used. In this paper the Odroid-XU3, containing a modern heterogeneous MPSoC, is used as a target platform. It provides facilities to measure the energy consumption of key components.

## 5 Evaluation

To highlight the importance of considering multiple quality metrics and objectives simultaneously, this paper presents two use-case studies from the multimedia and image processing domain typical for embedded systems. In these experiments, state of the art approximation techniques were applied.

In the first use-case essential loops of the x264[2] video encoder were perforated following the work from Sidiroglou et al. [10, 15]. The perforated applications were evaluated with two videos (640 × 360 pixel, 32 and 128 frames) from PARSEC [3] and two videos (352 × 288 pixel, both 300 frames) from Xiph.Org [18]. We used *perf*[3] to validate Sidiroglou's selection of run time critical loops to perforate since an older version of the x264-encoder was used. We identified loops to perforate in two functions (`x264_pixel_sad_x3_8x8` and `x264_pixel_sad_x3_16x16`). We choose to perforate the loops with a rate of 50% and 75% resulting in an execution of every second or forth iteration, respectively.

For each video, energy consumption and run time are measured on the Cortex-A15 and Cortex-A7 (cf. Fig. 2). As expected, run time and energy consumption decrease with higher perforation rates. The results from run time reduction matches the conclusion of the original work. However, energy consumption measured on a real embedded system extends the existing work.

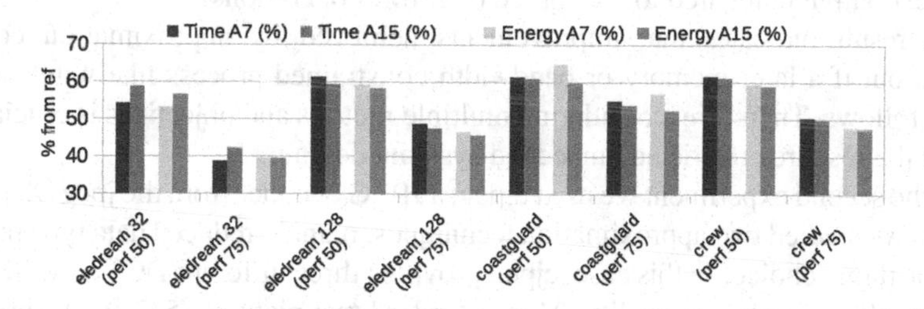


**Fig. 2.** Run time and energy consumption for the x264-encoder.

QCAPES enables analysis with multiple metrics. Figure 3 shows Y-component of the YCbCr signal encoded with a perforation rate of 50%. The blockwise calculation led to almost identical results as pixel-based, while the run time for blockwise calculation was significantly shorter than pixel-based calculation. This indicates a fast alternative to pixel-based evaluation. Compared to the encoding time, applying blockwise PVQM

---

[2] x264-snapshot-20160203-2245.
[3] Linux profiling with performance counters.

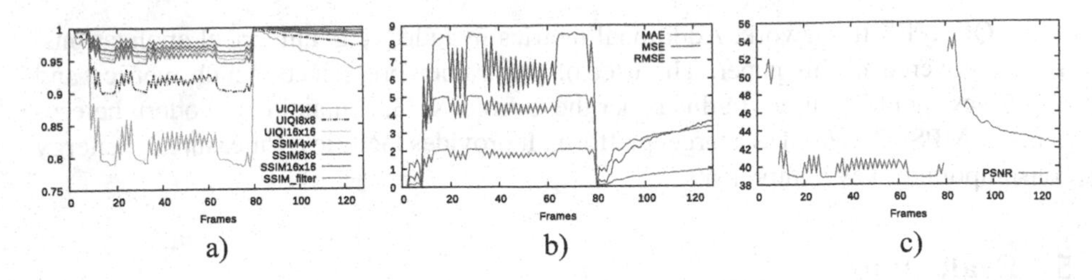


**Fig. 3.** Frame-by-frame analysis of the Y channel for x264 encoding of eledream 128. Higher values in 3a and 3c are better, and in 3b lower values are better.

seems feasible during run time, especially when idle cores are available. In general, the resulting videos were *good* with both perforation rates.

For memory limited or distributed systems, the file size is an additional important aspect. By usage of a user-defined objective, the file size was also measured. The results indicate that dropping iterations leads to an increased file size. At a perforation rate of 50%, the file size never deviated more than 5% from the reference, while a perforation rate of 75% could lead to an increase of 20%. A detailed analysis of the frame types for the eledream 128 video shows that increasing the perforation rate increases the amount of P- and decreases B-frames.

At this point it should be emphasized again, that QCAPES compares the output of a reference implementation with the output of one or more approximate implementations. Due to this fact, the quality between the different implementations is examined. For example, in an approximate version of a video encoder, a larger fraction of P-frames could be generated, which would result in better output quality of the approximate version than the reference implementation. We have chosen this method, since it is not always possible to obtain access to the source material or it is not even desirable to get the closest output compared to the source (e.g. filter operations).

This result indicates that improvements achieved by approximation could be canceled out if a later memory or bandwidth constrained process like wireless transmission follows. Therefore, considering multiple metrics and objectives is crucial especially in the resource restricted embedded systems domain.

For the second experiment we used cjpeg, a JPEG encoder from the jpeg-9b package [6]. Here we studied two approximation techniques, namely reduced data type precision and algorithmic choice. In this case, cjpeg provides three different DCT algorithms and an adjustable compression quality. Three standard test pictures ($512 \times 512$ pixel) [5], and a large picture ($4000 \times 4500$ pixel) [18] were analyzed.

Run time and energy consumption in relation to the different algorithms are very similar. For the small images, only minor differences are measured. On the Odroid-XU3, the float implementation is the fastest and most energy efficient choice. The file size differs too, but not as much as for the videos.

**Table 1.** Impact of different quality settings for lena_color.

| Benchmark | MSE | RMSE | MAE | PSNR |
|---|---|---|---|---|
| lena_int_100 | 0.12 | 0.34 | 0.12 | 57.48 |
| lena_int_80 | 0.24 | 0.49 | 0.15 | 54.40 |
| lena_int_50 | 0.27 | 0.52 | 0.08 | 53.76 |
| lena_int_30 | 0.22 | 0.47 | 0.06 | 54.69 |
| lena_int_10 | 0.28 | 0.53 | 0.02 | 53.67 |

| Benchmark | MSE | RMSE | MAE | PSNR |
|---|---|---|---|---|
| lena_fast_100 | 0.72 | 0.85 | 0.56 | 49.55 |
| lena_fast_80 | 1.05 | 1.02 | 0.55 | 47.92 |
| lena_fast_50 | 1.09 | 1.04 | 0.40 | 47.75 |
| lena_fast_30 | 1.21 | 1.10 | 0.33 | 47.32 |
| lena_fast_10 | 1.48 | 1.22 | 0.18 | 46.41 |

To highlight some insights, SFM results for lena_color are listed in Table 1. For this image we have evaluated five different quality settings to assess if the quality between the different implementations change. In the comparison of the *fast* and the *float* implementation we could observe very interesting results. As shown in Table 1 the MAE decreases with lower quality settings, while the MSE and RMSE increase. A possible explanation could be that small errors vanish, while larger errors increase. By using the square in MSE to avoid averaging effects, larger errors can have a greater effect on the MSE and all other metrics based upon it, even if the absolute error is getting smaller. This assumption is supported by the cjpeg documentation; the fast implementation tends to produce worse results on higher quality settings than on lower settings. This shows that it is always advisable and recommended not to rely on just one specific metric or metric family, but to involve multiple metrics in the evaluation.

## 6 Conclusion and Future Work

Evaluating a small set of complex, real-world benchmarks using QCAPES has already demonstrated the usefulness of an automatic, multi-metric assessment framework to discover previously unexpected effects of the naïve application of approximation techniques. In cases where approximation techniques interfere with application-specific inherent optimizations, e.g., the application of psychoacoustic or visual models for lossy data compression, the consideration of additional metrics, such as file size (and, consequently, energy consumption for transfer or storage) is necessary to efficiently apply approximation techniques without introducing undesirable side effects. We expect similar effects to show up especially in applications which are supposed to benefit most from approximation, thus multi-criterial assessment will be required for holistically building of efficient approximate systems. In addition, we propose to involve application domain experts in the QoS assessment process. Our JPEG example shows that using multiple metrics can be used to identify the structure of errors. In addition, contradicting metrics evaluating the same approximation approach demonstrate that a careful selection and combination of metrics is required for real-world use cases. Here, QCAPES provides an easy approach to integrate the simultaneous evaluation of different metrics. This is becoming especially relevant since we expect a large amount of the output of the approximate applications to be consumed by subsequent pieces of software, e.g., in autonomously driving cars, instead of a human end user who can detect and complain about insufficient quality. A dynamic QoS adaptation at runtime is a worthwhile goal, e.g. to adapt a system's output to changing requirements or to cope with previously unknown input data. Our evaluation has shown that the overhead of QoS assessment is small compared to the computational requirements for the application itself. This motivates to investigate the evolution of QCAPES into an online QoS assessment and control tool, enabling the easy establishment of *software control circuits*. The discussed metrics rely on comparing the "lossy" QoS-reduced application with a reference version. Integrating reduced reference metrics into the framework would increase the usage and flexibility.

**Acknowledgements.** The authors like to thank the German Research Foundation (DFG) for supporting part of this work within the Collaborative Research Center SFB 876 "Providing Information by Resource-Constrained Data Analysis", projects A3 and B2, http://sfb876.tu-dortmund.de/.

## References

1. Akturk, I., Khatamifard, K., Karpuzcu, U.R.: On quantification of accuracy loss in approximate computing. In: Proceedings of WDDD (2015)
2. Baek, W., Chilimbi, T.M.: Green: A framework for supporting energy-conscious programming using controlled approximation. In: Proceedings of PLDI (2010)
3. Bienia, C., Kumar, S., Singh, J.P., Li, K.: The PARSEC benchmark suite: Characterization and architectural implications. In: Proceedings of PACT (2008)
4. Gupta, V., Mohapatra, D., Park, S.P., Raghunathan, A., Roy, K.: IMPACT: IMPrecise adders for low-power approximate computing. In: Proceedings of ISLPED (2011)
5. Image Processing Place, December 2016. http://www.imageprocessingplace.com
6. Independent JPEG Group, December 2016. http://www.ijp.org
7. Kedem, Z., Mooney, V.J., Muntimadugu, K.K., Palem, K.V., Devarasetty, A., Parasuramuni, P.D.: Optimizing energy to minimize errors in dataflow graphs using approximate adders. In: Proceedings of CASES (2010)
8. Lin, W., Jay Kuo, C.C.: Perceptual visual quality metrics: A survey. J. Vis. Commun. Image Represent. **22**(4), 297–312 (2011)
9. Liu, J., Jaiyen, B., Veras, R., Mutlu, O.: RAIDR: Retention-aware intelligent DRAM refresh. In: Proceedings of ISCA, Vol. 40 (2012)
10. Misailovic, S., Sidiroglou, S., Hoffmann, H., Rinard, M.: Quality of service profiling. In: Proceedings of ICSE, vol. 1 (2010)
11. Mittal, S.: A survey of techniques for approximate computing. ACM Comput. **48**(4), 62 (2016)
12. Neugebauer, O., Libuschewski, P., Engel, M., Müller, H., Marwedel, P.: Plasmon-based virus detection on heterogeneous embedded systems. In: Proceedings of SCOPES (2015)
13. Renganarayana, L., Srinivasan, V., Nair, R., Prener, D.: Programming with relaxed synchronization. In: Proceedings of RACES (2012)
14. Sampson, A., Dietl, W., Fortuna, E., Gnanapragasam, D., Ceze, L.: EnerJ: approximate data types for safe and general low-power computation. In: Proceedings of PLDI (2011)
15. Sidiroglou, S., Misailovic, S., Hoffmann, H., Rinard, M.: Managing performance vs. accuracy trade-offs with loop perforation. In: Proceedings of ESCE/FSE (2011)
16. Wang, Z., Bovik, A.C.: A universal image quality index. IEEE Sig. Process. Lett. **9**(3), 81–84 (2002)
17. Wang, Z., Bovik, A.C., Sheikh, H.R., Simoncelli, E.P.: Image quality assessment: From error visibility to structural similarity. IEEE Trans. Image Process. **13**(4), 600–612 (2004)
18. Xiph.Org Foundation, December 2016. http://www.xiph.org

# Configurable Reprogramming Methodology for Embedded Low-Power Devices

Ondrej Kachman(✉) and Marcel Balaz

Institute of Informatics, Slovak Academy of Sciences, Dubravska cesta 9, 84507 Bratislava, Slovakia
{ondrej.kachman,marcel.balaz}@savba.sk

**Abstract.** The embedded low-power devices are very important part of any smart system. With the large amounts of sensors and actuators used, it is a good practice to implement remote reprogramming capabilities into the firmwares of these devices. This paper presents a new configurable reprogramming methodology that can be applied to various platforms. It is built on the best reprogramming practices while giving developers more control over firmware outline, updated functions and modules. It also refers energy efficiency, as the data shared over the network and memory operations on the devices are minimal. The multi-platform capabilities make this scheme ideal for smart systems.

**Keywords:** Reprogramming · Embedded · Low-power · Configurable · Over-the-air

## 1 Introduction

The area of remote and efficient reprogramming of the embedded devices has been researched since the introduction of the low-power devices and the wireless sensor networks. Various reprogramming methods have been developed over the past 15 years. The main goal of these methods is to replace the old firmware version with the new one while keeping the procedure energy efficient and secure. Recent progress in the internet of things technologies, cyber-physical systems and smart systems requires these methods to be evaluated and adjusted to the new trends in these areas.

The research topic of this paper is a configurable reprogramming methodology for low-power devices, as these devices are widely used in the modern intelligent systems, including smart systems. The main advantage of the solution that this paper will present is its platform independency. The configurability (see Sect. 4.2) of our solution will provide programmers with more control over the outline of the updated firmware and we present experiments that show how it can help to share less update data on the network and spare program memory with less operations executed.


L.M. Camarinha-Matos et al. (Eds.): DoCEIS 2017, IFIP AICT 499, pp. 211–219, 2017.
DOI: 10.1007/978-3-319-56077-9_20

## 2 Relationship to Smart Systems

Smart system is a general term for a system that collects information, processes it and acts based on the results of the data analysis. Smart systems evolved from the simpler systems capable of sensing and basic actuation into the systems with perception superior to a human being. The design of a smart system includes cloud services and networking technologies, choices of the appropriate hardware and software components and their implementation [1]. The sizes of smart systems can be very different depending on their function. They range from nano-scales to meter-scales. Smart systems can be applied in many sectors, for example healthcare, automotive industry, retail, building management, military, etc. [2]. Low-power embedded devices are usually used for sensing, actuation a control.

An embedded device used in a smart system is preloaded with a default firmware. Even the thorough testing may not detect all the errors that this initial firmware has. Once the devices are used in the real environment in great numbers, an identified firmware error may cause a long system downtime until the problem is fixed and the system can be considered reliable again. The over-the-air update methods are developed to update the faulty firmware as fast as possible. These methods can be focused on various sub problems, closely analyzed in the Sect. 3.1. The new challenge for these methods regarding the smart systems is their adaptability to the devices based on various platforms that form a smart system. Most works in the area of remote reprogramming are focused on the wireless sensor networks (WSNs) and the TinyOS operating system [3, 4]. Most WSNs are homogenous, using devices based on the same platform for data collection. Smart systems and may use different devices for sensing, different devices for networking and different devices for presentation of the information to the users of the system.

To create a multiplatform reprogramming method suitable for the smart systems, there should be no alterations to the compilers or the instructions of the linked firmware images, as these alterations are platform specific. Jump, branch and relative call instructions, that are often altered by these methods, may have different operation codes for each platform. Also, not every platform may support the same instruction set and relocation types. The ideal target for a platform independent solution are the object files, the product of compiler, and the executable file, the product of the linker, later transformed into the binary representation of the firmware.

Once a firmware fault is discovered, the developers may estimate if the fix will require one or more updates. In the case of a single update, it is not required for a device to change the outline of its firmware functions, it can just apply a simple update script. If the problem fix requires incremental updates and debugging, it is good to prepare the device for these patches. An example is to put every function that requires an update to its own memory section [5]. The amount of the data shared on the system's network must be as small as possible so it will not affect the other parts of system. Different approaches may be the most suitable for the different parts of the smart system.

# 3 State of the Art and Related Work

This section analyzes the most important principles and works in the area of the remote reprogramming for embedded devices. Analyzed papers serve as a base for our methodology that we present in the Sect. 4 and evaluate in the Sect. 5.

## 3.1 Main Challenges for the Remote Reprogramming Methods

The methods for remote reprogramming of the low power devices are mainly focused on the 4 main sub problems [6]:

**Improving Firmware Similarity.** Previous works used altered compilers [7], object file alterations [8] and custom linker directives [5] to make the firmware images as similar as possible.

**Generating Delta Files.** Many differencing algorithms have been developed for various networks of embedded devices [9, 10]. These algorithms are designed to generate very small delta files (deltas, patches). Smaller size improves the energy consumption of the wireless interfaces and prevents the network from congestion or desynchronization.

**Network Dissemination.** Network protocols are responsible for the delivery of the delta files to the target devices. In the past, many protocols have been designed specifically for over-the-air updates [3]. Smart systems may take advantage of the newer and standard protocols used for the low-power devices. The integrity and security are also a concern in this area. Delta files are split into the number of packets protected by CRC codes and error correcting codes [11]. Network security of the smart systems and the dissemination of the delta files is out of scope of this paper.

**Update Execution.** The last problem deals with the implementation of the update procedure on the target devices. Reprogramming code is often embedded into the bootloader [4] or resides in its own reserved space [10]. The updates of this code are very rare and critical.

## 3.2 Taking Advantage of the Standard ELF File Format and Relocatable Entries

Object files are the product of compilers. Probably the most popular format of these files is the executable and linkable format, ELF [12]. This format is supported by the variety of compilers, including the well-known GCC and its ports for embedded platforms. The file in this format is also the output of the linker and is later used to create a binary image of a firmware. Embedded systems use the following formats:

- Relocatable file – created by a compiler, relocatable entries are not resolved
- Executable file – created by a linker, all relocations are resolved

The process is shown in the Fig. 1. Relocatable files include unresolved relocations, for example calls, jumps, branches, etc., and these entries are resolved during linking. The authors of [4] managed to reduce the data shared on the network by setting all relocations to the same value for each firmware, making different versions more similar. This allows the differencing algorithm to generate very small deltas. The real relocation values are propagated as a metadata with the delta files and written to their positions by a loader on the target device.

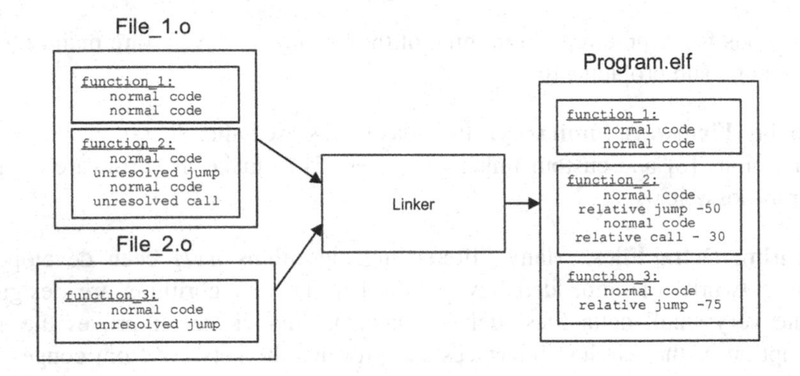


**Fig. 1.** Linking relocatable ELF files into the executable ELF file

The authors of [8] alter ELF files in various ways. New functions are placed at the end of the program memory and calls to them are edited. This approach also changes the way sections for initialized and uninitialized variables are allocated in the RAM, so there are fewer shifts for these entries. Some instructions are changed directly, so the algorithm understands operation codes for the selected platform. However, it may not work for a completely different microcontroller.

**Relaxation.** During development of our own solution, we found some problems with the relocations that changed sizes after linking. This is called relaxation. For example, a compiler may allocate 4 bytes for a *call* instruction to some function. Linker discovers, that it placed the called function within 128 bytes of the relocation and thus reduces the relocation to a 2-byte *relative call*, shrinking the calling code. It is therefore important to check the resolved relocations in the executable ELF file.

### 3.3 Fragmented Firmware

Linkers can place code segments to any free address in the program memory. These segments are also called *.text* segments or sections. If there is a *.text* section generated for each function, linker can provide these functions with the slop region, a free space in which the function can grow or shrink without causing any other functions to shift [5]. This approach helps to generate small deltas and prolong the lifespan of the flash memories. Some works criticized this solution for an inefficient use of the program memory [4]. More jumps between pages in the memory cause more energy to be consumed, as activation of a different page consumes more power than activation of a different block within the same page [13].

# 4 Configurable Method for the Remote Reprogramming

This section presents our remote reprogramming methodology. It consists of methods that form a complete, platform independent solution capable of learning about the changes in any firmware and generating as small delta files as possible. Our methods are applied in a software tool we developed. Our methodology includes methods responsible for the following tasks:

- Analyzing the object files, listing *.text* sections and relocations
- Tracking changes for each function, detecting new or deleted functions
- Assigning addresses to the functions based on their analysis
- Invoking linker, checking for relaxed entries in the final files
- Generating delta files

Our tool includes a wrapper for the GCC compiler and linker. We aim to test our method and tool on the three different platforms – 8-bit microcontroller from the AVR family, 16-bit microcontroller from the MSP430 family and a system-on-chip with the 32-bit ARM Cortex-M0 core. Each platform has its version of the GCC available and supports ELF files. We do not alter the instructions generated by the compiler. The flowchart of our method is in the Fig. 2.

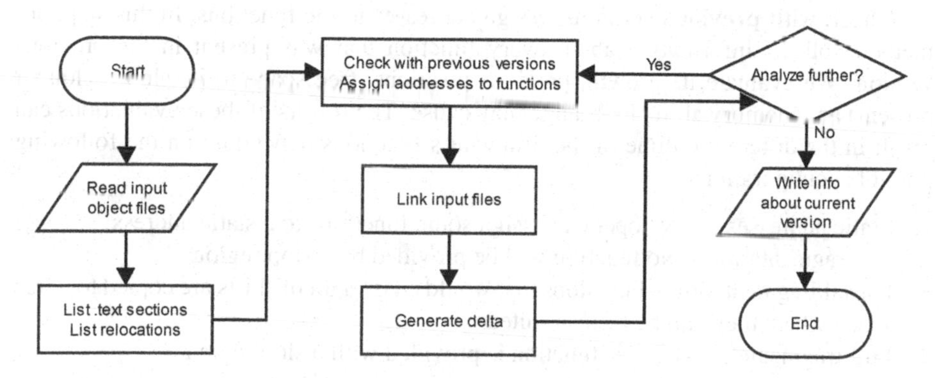


**Fig. 2.** Flowchart of the proposed method

## 4.1 Research Contribution and Innovation

None of the methods used in our methodology alter any ELF sections that hold instructions or relocations. There is no need to specify instruction formats or relocation types for any platform. This makes our solution platform independent.

Our methodology provides developers with various configurable options described in the following subsection. Produced output files can be reverse analyzed and provide feedback on how the current configuration can be improved. This is a new approach. Various configurations may be the best for different scenarios.

Our methodology generates smaller deltas than basic differencing algorithms. Less data shared on the network improves the energy efficiency of our solution.

### 4.2 Description of the Processes and Configurability Options

This subsection describes the configurable processes from the Fig. 2. These are the processes within our proposed method that contribute to its good performance.

**List text sections,** list relocations. Compilers enable developers to generate a *.text* section for each function, for example *.text.function1*, *.text.function2*. However, both will be linked into the *.text* section in the executable file. To prevent this, programmer must define a target section for each function in the source code. Our method includes two configurations:

1. Split functions into the sections defined explicitly by the developer
2. Edit string tables of the object files, place each function in its own section

The second option will simply iterate through the ELF string tables and add the number to each *.text* section, putting previously mentioned functions into sections *.text1* and *.text2*. This does not affect any firmware instructions.

Relocation entries can be read at this point, even though they are not resolved. List of these relocations may be later used before the effective delta generation by setting them to zero and making the firmware images more similar – inspired by [4].

**Check with previous versions,** assign addresses to the functions. In this step, our method collects information about every function that was present in the previous version. We evaluate its growth, previous positions, free space in its slop region (if present) and memory shifts its change may cause. The results of these evaluations can result in the different outline of the firmware's functions depending on the following possible configurations:

1. Static addresses – developer can assign some functions to a static address
2. No fragmentation – No function will be provided by a slop region
3. Partial fragmentation – Functions that would cause a lot of shifts are copied to a free space where they can be further edited
4. Full fragmentation – Every function is provided with a slop region.

**Generate Delta.** After linking, when the relocations are resolved, there are two possible configurations:

1. Set all the relocations to zero, generate INSERT operations for the delta
2. Do not alter the relocations, run the differencing algorithm directly

Differencing algorithm also supports two different modes. These modes are relevant when the firmware is being fragmented or defragmented:

1. Dirty mode – unused sections that previously contained data are not erased
2. Clean mode – unused sections are erased and filled with 0xFF symbols.

## 5 Experimental Results and Discussion

This section provides experimental results that show, how various configurations of our methodology may be the most suitable for the different scenarios of over-the-air updates. In [10], we presented our differencing algorithm that outperforms existing solutions. Therefore, we use it for the following experiments executed on the ATmega32u4 microcontroller. We consider three scenarios:

1. In the first scenario, the devices on the network require one big update of the functions that sense, store and send data – the sensor module
2. The second scenario updates these three sensor module functions incrementally, one after another
3. The third scenario involves 9 incremental updates to various functions in the firmware code – the sensor module (3 functions), the communication module (3 functions), the flash storage module (3 functions)

We configure our tool to edit string tables of the ELF files, every function is placed in its own section. We do not set relocations to zero. For each scenario, we evaluate three different configurations:

1. No fragmentation; All deltas are generated in clean mode
2. Partial fragmentation; Changed functions are provided with a slop region (dirty mode), final delta generated in the clean mode defragments the memory
3. Full fragmentation; All functions are provided with slop regions before any update (clean mode), functions are then updated (dirty mode) and finally, the firmware is defragmented (clean mode)

**Results.** The results of our experiments are shown in the Table 1. Column "Delta files" shows, how many delta files were used for the update. Column "Total bytes" shows the sum of the delta file sizes used for reprogramming. Column "Frag. overhead" shows, how many of those bytes were used for memory fragmentation, defragmentation and cleanup. Table 2 shows the sizes of every delta file from the 3rd scenario. The sizes are in bytes.

**Table 1.** Experimental results for the three different scenarios and configurations

| Config | Scenario 1 - single update | | | Scenario 2 - 3 updates | | | Scenario 3 - 9 updates | | |
|---|---|---|---|---|---|---|---|---|---|
| | Delta files | Total bytes | Frag. overhead | Delta files | Total bytes | Frag. overhead | Delta files | Total bytes | Frag. overhead |
| No frag. | 1 | 98 (best) | 0 | 3 | 216 (+7%) | 0 | 9 | 1398 (+65%) | 0 |
| Partial frag. | 2 | 164 (+64%) | 112 | 4 | 202 (best) | 102 | 10 | 844 (best) | 260 |
| Full frag. | 3 | 680 (+593%) | 634 | 5 | 638 (+215%) | 576 | 11 | 1036 (+22%) | 568 |

**Table 2.** Delta file sizes in bytes for the Scenario 3

| Config | $\Delta_{frag}$ | Δ1 | Δ2 | Δ3 | Δ4 | Δ5 | Δ6 | Δ7 | Δ8 | Δ9 | $\Delta_{defrag}$ | Total |
|---|---|---|---|---|---|---|---|---|---|---|---|---|
| No frag. | – | 74 | 84 | 58 | 106 | 162 | 150 | 244 | 102 | 418 | – | 1398 |
| Partial frag. | – | 12 | 40 | 48 | 26 | 126 | 50 | 200 | 58 | 24 | 260 | 844 |
| Full. frag. | 302 | 14 | 14 | 34 | 26 | 126 | 20 | 200 | 20 | 14 | 266 | 1036 |

**Discussion.** The experiment shows, that the single firmware update does not require any memory fragmentation in order to be efficient. With more incremental updates required, configurations with memory fragmentation become more efficient – partial fragmentation performs the best for both 3 and 9 incremental updates. Full fragmentation configuration has the penalty of the big deltas that are used to fragment the memory at the beginning and to defragment it at the end. It is performing significantly better for more updates and it can be expected, that for a great number of incremental updates, this configuration will perform the best. Table 2. shows, that the delta sizes for the full fragmentation approach are mostly the smallest for the single function updates (Δ1–Δ9).

## 6 Conclusion

This paper presents a configurable reprogramming methodology that is built on the best practices in the area, and introduces configurations that make remote firmware updates more efficient for the different reprogramming scenarios. Experiments have shown, that various configurations may reduce total amount of the data shared on the network. The solution is multiplatform, which makes it ideal for the smart systems.

**Acknowledgement.** This work has been supported by Slovak national project VEGA 2/0192/15.

## References

1. Arsan, T.: Smart systems: from design to implementation of embedded smart systems. In: 13th HONET-ICT International Symposium on Smart Microgrids for Sustainable Energy Sources Enabled by Photonics and IoT Sensors, Haspolat, pp. 59–64 (2016)
2. Moshnyaga, V.: Guidelines for developers of smart systems. In: IEEE 8th International Conference on Intelligent Systems (IS), Sofia, pp. 455–460 (2016)
3. Hui, J.W., Culler, D.: The dynamic behavior of a data dissemination protocol for network programming at scale. In: Proceedings of the 2nd International Conference on Information Processing in Sensor Networks (IPSN 2008), pp. 81–94. ACM Press, New York (2004)
4. Dong, W., Mo, B., Huang, C., Liu, Y., Chen, C.: R3: optimizing relocatable code for efficient reprogramming in networked embedded systems. In: IEEE INFOCOM Proceedings, Turin, pp. 315–319 (2013)
5. Koshy, J., Pandey, R.: Remote incremental linking for energy-efficient reprogramming for sensor networks. In: Proceedings of the Second European Workshop on Wireless Sensor Networks, Istanbul, pp. 354–365 (2005)

6. Kachman, O., Balaz, M.: Effective over-the-air reprogramming for low-power devices in cyber-physical systems. In: Camarinha-Matos, Luis, M., Falcão, António, J., Vafaei, N., Najdi, S. (eds.) DoCEIS 2016. IAICT, vol. 470, pp. 284–292. Springer, Heidelberg (2016). doi:10.1007/978-3-319-31165-4_28
7. Huang, Y., Zhao, M., Xue, C.J.: WUCC: joint WCET and update conscious compilation for cyber-physical systems. In: 18th Asia and South Pacific Design Automation Conference (ASP-DAC), Yokohama, pp. 65–70 (2013)
8. Shafi, N.B., Ali, K., Hassanein, S.: No-reboot and zero-flash over-the-air programming for wireless sensor networks. In: 9th Annual IEEE Communications Society Conference on Sensor, Mesh and Ad Hoc Comm. and Networks (SECON), Seoul, pp. 371–379 (2012)
9. Hui, U.P., Jongsoo, J., Pyeongsoo, M.: Non-invasive rapid and efficient firmware update for wireless sensor networks. In: Proceedings of the 2014 ACM International Joint Conference on Pervasive and Ubiquitous Computing, Seattle, pp. 147–150 (2014)
10. Kachman, O., Balaz, M.: Optimized differencing algorithm for firmware updates of low-power devices. In: IEEE 19th International Symposium on Design and Diagnostics of Electronic Circuits & Systems (DDECS), Kosice (2016)
11. Unterschütz, S., Turau, V.: Fail-safe over-the-air programming and error recovery in wireless networks. In: Proceedings of the 10th International Workshop on Intelligent Solutions in Embedded Systems (WISES), Klagenfurt, pp. 27–32 (2012)
12. TIS Committee: Executable and Linking Format (ELF) Specification Version 1.2 (1995)
13. Pallister, J., Eder, K., Hollis, S. J., Bennet, J.: A high-level model of embedded flash energy consumption. In: International Conference on Compilers, Architecture and Synthesis for Embedded Systems (CASES). ACM Press, New York (2014)

# Upper Bounds Prediction of the Execution Time of Programs Running on ARM Cortex-A Systems

Irina Fedotova(✉), Bernd Krause, and Eduard Siemens

Department of Electrical, Mechanical and Industrial Engineering,
Anhalt University of Applied Sciences, Bernburger Str. 57, 06366 Köthen, Germany
{i.fedotova,e.siemens}@emw.hs-anahlt.de,
bernd.krause@hs-anhalt.de

**Abstract.** This paper describes the application of statistical analysis of the timing behavior for a generic real-time task model. Using specific processor of ARM Cortex-A series and an empirical approach of time values retrieval, the algorithm to predict the upper bounds for the task of the time acquisition operation has been formulated. For the experimental verification of the algorithm, we have used the robust Measurement-Based Probabilistic Timing Analysis based on the Extreme value theory (EVT). In the ongoing work, we provide a systematic method for safe worst-case execution time estimations of programs running on the specific hardware. We focus here primary on the consequences of EVT assumptions and their correct interpretation for the upper bounds prediction.

**Keywords:** Extreme Value Theory · Worst-case execution time · Probabilistic Timing Analysis · Timing verification

## 1 Introduction

The implementation of most smart systems nowadays presumes the availability of high computing, memory and IO-resources on very small and low-power consuming hardware platforms - so called SoC devices. In this area, the ARM Cortex-A platform plays a central role. On such systems, the time acquisition process is one of the most critical steps in meeting real-time requirements. This presumes certain guarantees that the probability of the system failing to meet its timing constraints is below an acceptable threshold. However, the CPU of said system includes many features, which imposes correspondent limitations to the aimed timing analysis. One way to get suitable values for meeting real-time requirements is the development of statistical methods, predicting the probability distribution of the timing operations delays for particular tasks.

Nowadays, deterministic as well as probabilistic approaches pursue the goal of providing a safe upper bound of execution time for a particular task [1]. The difference between these approaches is mainly that the deterministic method produces a single worst-case execution time (WCET) estimation, while probabilistic - multiple WCET estimates with their respective probabilities. Both approaches have their static and measurement-based variants. Classical static timing analysis operates on deterministic


L.M. Camarinha-Matos et al. (Eds.): DoCEIS 2017, IFIP AICT 499, pp. 220–229, 2017.
DOI: 10.1007/978-3-319-56077-9_21

processor architectures and provide safe WCET estimates as they are proven to be the worst ones [2]. However, most real-time operating systems have become increasingly complex and it makes the applicability of static timing analysis is very challenging. On the contrary, measurement-based methods provide an estimation based on the derived maximal and minimal observed execution times or their distributions. In fact, with a probabilistic hardware architecture and measurement-based approaches, it is possible to guarantee an accurate probabilistic pWCET, what is solved by applying the Measurement-Based Probabilistic Timing Analysis (MBPTA).

The objective of the MBPTA is to derive the timing behavior of a task by means of statistical modeling. It aims at modeling extreme execution times for characterizing the worst-case, relied on both execution time measurements and the application of the Extreme Value Theory (EVT). The EVT deals with the extreme deviations from the median of probability distributions and estimates the tails, where the worst case should lie. However, hardware effects in real-time systems make EVT applicability difficult with regard to its required theoretical hypotheses. In this work, we focus on a particular hardware presented on the Atmel SAMA5D4 board with one ARM Cortex-A5 core, which can be a common example of realistic embedded applications. The platform based on these series of processors employs a cache of random-replacement policy. It facilitates upper bounding the probability of pathological worst-cases behaviors to extremely low levels (failure probability of $10^{-9}$) [3]. For our investigations, the failure probability at the level $10^{-9} < P < 10^{-7}$ (from hazardous class) was chosen. This level is proposed by PED certification [4] and usually applied in flight control systems.

In the related work, a number of methods for application of EVT theory for WCET estimations has been proposed [2, 5, 6]; nevertheless, a considerable uncertainty due to the complexity of the problem still exists. Results of statistical tests are often fuzzy and it is hard to make a correct decision on their basis to fulfill EVT requirements. To overcome difficulties with EVT, it is necessary to perform a complete set of diagnostic tests, check hypotheses for generalizing the EVT applicability and assume the validity of different methods. In this article, we intent to significantly reduce the existing ambiguities and suggest a systematic step-by-step method. Moreover, while previous studies have investigated time bounds for the whole application using corresponding benchmarks, we focus on the particular task for obtaining time values. The MBPTA approach is then applied only for this most critical part of the system, as other components can be assessed with more deterministic methods. The rest of the paper is organized as follows: the relation of the ongoing research to smart systems is discussed in Sect. 2, followed by discussion on related work in Sect. 3. The description of the problem of the real-time probabilistic modeling, as well as the main steps of proposed algorithm are presented in Sect. 4. Section 5 describes the experiment setup. The proof of fitting the target distribution and obtaining WCET estimators are provided in Sect. 6. Finally, Sect. 7 sums up and collects the conclusions and future work.

## 2 ARM Based SoC and Smart Systems

What principally differentiates the ARM Cortex-A series from others, is their wide use in performance-intensive systems. However, for our research the choice of these processors is based not only on the high performance and low power capabilities of the hardware, but also on the extensive application in control smart systems with timing constraints. The term "smartness" refers to the ability of the system performing very complex set of heuristic operations, based on a rich data basis, often gathered from external cloud environments within some given time boundaries, networking and real-time capabilities. Here selected embedded devices play a dominant role. A wide range of applications for Cortex-A is possible also due to the support of rich OSs such as Linux or Android. However, using them in embedded systems brings additional challenges to the development process. While one of the primary aim of any smart system is making decisions more predictive under certain conditions, in fact, the standard Linux OS is not productive enough for that. It follows therefore the necessity to better address some challenges related to the reaction of the system, especially where highly precise timing must be kept at one microsecond level.

The specific feature of these series of processor is, that CPU cycle counter is not directly available from user-space and wraps around in every 6.5 min. Thereupon, the investigations of timing capabilities are being performed within the high performance *HighPerTimer* library [7], which main idea is to simplify the process of timestamps acquisition. It allows avoiding the invocation of any system calls, since the main timing mechanism (supporting the procedure of handling overflows) is processed in user space. While microsecond precision of time stamping becomes standard for the development of remote monitoring and control systems, the underlying OS (particularly due to invoking system calls) is not able to meet the real-time requirements. Moreover, exactly the task for obtaining time values from the hardware can often take up to 20% of the whole program execution time. Therefore, it is important to minimize the cost of time retrieval and provide the precise time bounds of this critical task. Therefore, based on the used implementation of timing, it is possible to move forward in the direction of calculating worst-case execution time and predict the upper bound. These predictions allow a verification if the timing constraints will be met at run-time and so help to lead to a faster and more secure design process of the system.

## 3 Related Work

One of the first comprehensive comparisons among timing analysis techniques was presented by J. Abella et al. in [2]. Authors compare static and MBPTA methods in qualitative and quantitative terms under different cache configurations. Besides this, their previous work [8] establishes requirements to EVT with the MBPTA method to derive WCET estimates. Thereby they address the WCET problem by introducing randomization into the timing behavior of the system hardware and software.

The recent work by F. Guet et al. [6] proposes a DIAGnostic tool, which applies the MBPTA method without human intervention. Depending on the certain theoretical hypotheses of the EVT, the logical work flow of the framework derives its probabilistic pWCET estimate from traces of execution times. Concerning the DIAGnostic tool, K. Berezovskyi et al. in [9, 10] also investigate different methods inside the EVT theory for Graphical Processor Units (GPUs). The main results showed that hardware time-randomization is not that essential for the applicability of EVT and can be applied even to some non-time-randomized systems as GPUs.

An approach based on EVT theory has also been used for optimal performance analysis. Radojković et al. [5] presented a new method for predicting the performance of the thread assignment in multi-core processors. Using statistical inference of each thread assignment in a random sample, the authors estimates the optimal one.

Despite the intensive research in this area over the past few years, open problems in applying MBPTA for computing probabilistic WCET estimate still remain. In particular, said approaches give little information about the sequences of checking statistical hypotheses and making safe decision on their basis. Ongoing work intents to overcome these challenges by developing systematic and more coherent approach of analyzing the timing behavior. Moreover, the innovation novelty of this work is to cluster tasks exactly for time acquisition and apply the MBPTA method only for this critical non-deterministic part of the system.

## 4 The Probabilistic Modeling of Timing Behavior

Measurement-based methods produce estimates (for parameters of some distributions) by executing the given task on the given hardware and measuring the execution time of the task or of its parts. In particular, MBPTA approaches are aimed at modeling extreme execution times and characterizing its worst-case. The probabilistic theory that focuses on extreme values and large deviations from the average values is the EVT theory [11]. It estimates the probability of occurrence of extremely large values, which are known to be rare events. More precisely, the EVT predicts the distribution function for the max (or min) values of a set of $n$ observations, which are modeled with random variables. Its main result is provided in the Fisher-Tippett-Gnedenko theorem, which characterizes the max-stable distribution functions for a single family of continuous cumulative distribution functions (CDFs), known as the generalized extreme value (GEV) distribution. The result implies that an average discrete cumulative distribution function belongs to the Maximum Domain of Attraction (MDA) of the GEV.

Furthermore, the GEV distribution is characterized by three parameters: location $\mu$, scale $\sigma$, and shape $\xi$ and by their estimation, we can prove the resulting GEV distribution. If the shape parameter is $\xi > 0$, than the measured values in trace belong to Fréchet distribution, whereas if $\xi < 0$, than this is a "reversed" Weibull case. If the shape parameter $\xi = 0$, than values belong to a Gumbel distribution, which in most previous works [12, 13], has been assumed as applied to the pWCET distribution. Nonetheless, there is no restriction on the values that $\xi$ can take and the resulting distribution, we intent to check all three distributions, in order to be close to the accurate estimation of the parameter. Therefore, for applying the EVT, the following hypotheses are required to be verified: that $n$ random variables of execution time measurements are: (i) independent,

(ii) identically distributed and (iii) the existence of a distribution in the MDA of the GEV distribution is given [6]. These three elements are checked for guaranteeing the safety of the reliable probabilistic pWCET estimates.

### 4.1 Selecting Extreme Values

Within the EVT context, there are several approaches to measure the extreme values. The Peak over Threshold (PoT) method is preferred and presents a good model for the upper tail, providing reliable extrapolation for exceedances over a sufficiently high threshold. The Pickands-Balkema-de Haan theorem [14], on which the PoT method is based, corresponds to the basic principle of extracting the execution time measurements from the trace above a threshold $u$ and fitting of the Generalized Pareto Distribution (GPD) to the exceedances. The PoT allows to use data more efficient than other approaches, though the evident disadvantage is the selection of the suitable threshold value. Moreover, in the single-path case the PoT appears to be more accurate (with respect to the measurements), but the increase of the threshold $u$ can result into more pessimistic pWCET estimations [14, 15].

### 4.2 Proposed WCET Estimation Algorithm

Considering the probabilistic modelling briefly described above, the following algorithm for WCET estimation is suggested: **Step 1.** Selecting extreme values. The objective of this step is to collect from the original distribution the values, which fit into the tail, and hence can be modeled with the GEV distribution. For estimation of extremes, the PoT method was chosen: Step 1.1. The choice of the best-fitted threshold. It is based on the method of graphical diagnostic, which includes: mean residual life plot and parameter stability plot. They allow finding the lowest possible threshold, for which the extreme value model provides a reasonable fit to exceedances; Step 1.2. Filtering values above the threshold. **Step 2.** Fitting the GEV distribution: Gumbel, Fréchet and Weibull. If none of three types distributions fits, then going back to Step 1.1 and increase the threshold value. **Step 3.** Estimating the remaining parameters of the fitted distribution: $\mu$, $\sigma$ and $\xi$. **Step 4.** Verification of EVT hypotheses of independence and identical distribution: Step 4.1. Checking that the data are identically distributed; Step 4.2. Proving that samples are independent. **Step 5.** Return WCET estimation based on $\mu$, $\sigma$ and $\xi$ parameter. Due to the space constrains, a more detailed description of the Step 4 is omitted here. However, the statistical analysis shows that EVT can be applied to the obtained results with high confidence (95%). At the same time, it must be considered that recent work [10] shows that independence is not a necessary hypothesis for EVT applicability and the theory can be applied even for stationary weakly dependent time series, even if the identical distribution of random variables does not represent a limiting hypothesis to EVT applicability.

## 5 Experimental Setup

There are multiple ways, in which measurements can be performed for the probabilistic approach. The most evident method is by extra instrumentation of the program code that collects a timestamp or CPU cycle counter. All experiments for this research have been carried out on ARM Cortex-A5 (see Sect. 2). We have evaluated the overhead of setting two consecutive timers of *HighPerTimer* library [7], which practically gives the estimate of the timer cost; the job, which is required to be performed in order to measure executions times of any program or the piece of code. The measured process is scheduled by SCHED_OTHER - the default scheduling policy of the Linux kernel version 4.4.11, used in these investigations. The influence of the real-time scheduling policy and kernel extensions are out of scope of this work. We rather concern about measuring the time acquisition process under the standard conditions. Table 1 lists basic statistics of the 27 min estimation part of a representative trace. The representation of complementary cumulative distribution function (CCDF) for this execution time is shown in Fig. 1(a). As can be seen from the graph, the distribution peaks near the mean and falls with rapidly decreasing probability density below 1 μsec. The Fig. 1(b) gives the general picture of the time behavior showing additionally sigma values: $\sigma$ and $3\sigma$.

**Table 1.** Statistical properties of the original data set.

| Mean execution time | Std. deviation | Max value | Min value |
|---|---|---|---|
| 0.519 μsec | 2.217 μsec | 5.995 ms | 0.455 μsec |

## 6 Fitting the GPD Distribution and Estimation the Upper Bounds

Having determined the threshold value, the parameters of the GPD can be calculated by the maximum likelihood estimation (MLE). The appropriate goodness-of-fit statistics compares observed data to quantiles of the specified distribution. We estimate three distributions by the critical value of Chi-square test, the Bayesian (BIC) and Akaike (AIC) information criteria and QQ-plot. Firstly, since preliminary test series show that the shape parameter $\xi$ can't be negative, the case of Weibull distribution was excluded. Secondly, based on the Chi-square test the hypotheses that data follow Gumbel or Fréchet distributions can't be rejected. Thirdly, comparing models according to the AIC and BIC criteria, the hypothesis that data fit Fréchet distribution is slightly preferred against the Gumbel case. Fourthly, the straight diagonal line of a QQ-plot indicates that Gumbel distribution is a relatively good fit to the tail. However, since the difference of most statistics results for Gumbel and Fréchet models is not significantly differed, it makes sense to retrieve the WCET estimate for both cases.

The final step is to use the computed and verified GPD parameters and the exceedance probability of failure $p$ to estimate the WCET. The WCET estimate is derived based on parameters $\xi$ and $\alpha_u$ for each respective scenario as following [16]:

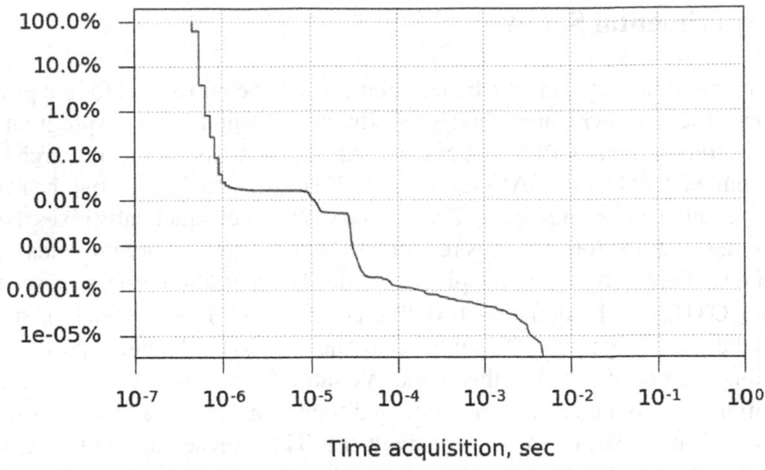


(a)

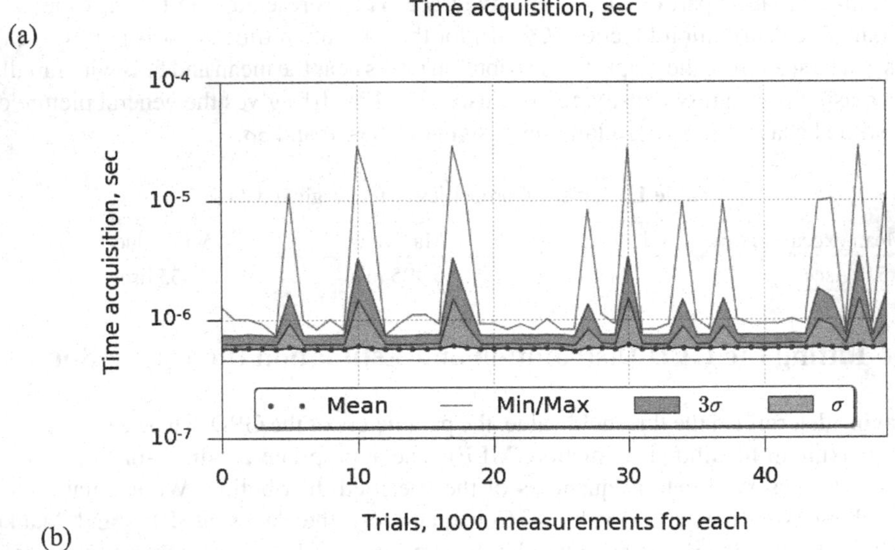


(b)

**Fig. 1.** The original dataset of time execution: (a) its CCDF representation; (b) samples with statistical properties (mean, min/max, 3σ, σ) grouped into 50 clusters.

$$WCET = \begin{cases} u + \frac{\alpha_u}{\xi}\left(\left(\frac{n}{k}p\right)^{-\xi} - 1\right) & if\ \xi > 0, \\ u - \alpha_u \log\left(\frac{n}{k}p\right) & if\ \xi = 0, \end{cases} \quad (1)$$

where $\alpha_u$ is an estimated scale parameter, $k$ - the number of peaks over the threshold $u$. Table 2 gives modeling results of the extreme execution times and Fig. 2 (a) shows the distribution convergence for Gumbel and Fréchet scenarios. The x-axis shows the pWCET estimation, the y-axis shows the respective probabilities. We use a CCDF of GPD to predict the potential values. According to the Table 2 the WCET estimates for both distributions differ significantly. Formally, too pessimistic results of Fréchet are quite tight and practically less useful: for failure probability $p = 10^{-9}$ the bound is 5.921 s. The Gumbel distribution converges to 0 faster than the Fréchet and it decreases the pessimism of the WCET thresholds. Moreover, shape parameter stability plot (on Fig. 2 (b)) estimates maximum likelihood of $\xi$ over a range of threshold and confirms the hypothesis that data fit Gumbel distribution with $\xi = 0$ respectively.

**Table 2.** EVT results for the hazardous class of p considering GPD expressed in cycles and msec.

| | $\sigma$ | $\mu$ | $\xi$ | (WCET; $10^{-7}$) | (WCET; $10^{-8}$) | (WCET; $10^{-9}$) |
|---|---|---|---|---|---|---|
| Gumbel | 12 247.56 | 12 968.35 | 0 | 27 968 = 2.54 ms | 56 170 = 5.11 ms | 84 370 = 7.67 ms |
| Fréchet | 10 542.529 | −1 539.657 | 1.351 | 124 185.2 = 11 ms | 2 898 825 = 263 ms | 65 126 941 = 5.921 s |

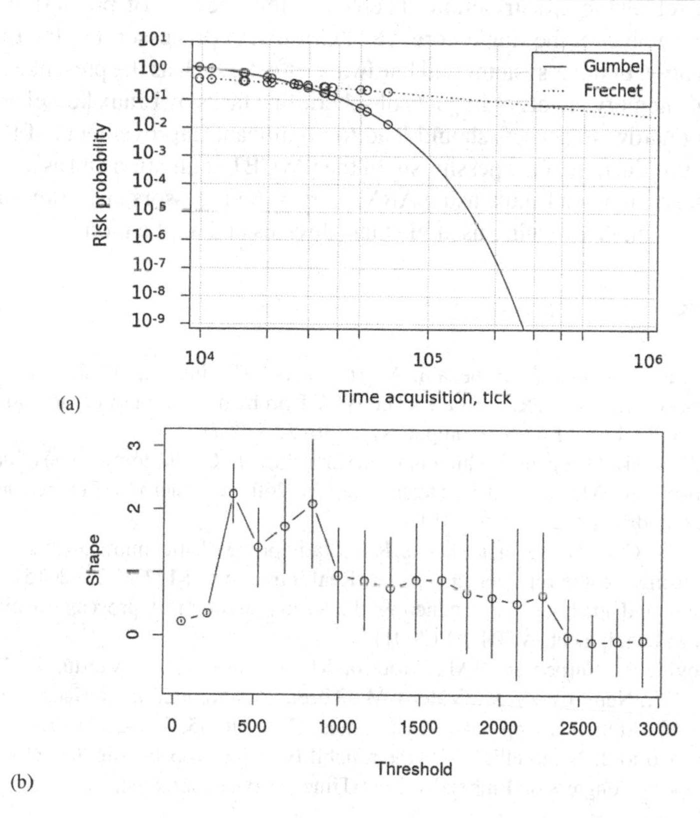


**Fig. 2.** EVT results: (a) estimation of upper bounds; (b) shape parameter stability plot.

Having possibility to run the experiment including the target risk probabilities, we can compare the real max value and predicted WCET estimates. Therefore, the max value of the original dataset is 5.99 ms (Table 1) and the predicated upper bound using parameters for Gumbel distribution for failure probability $p = 10^{-9}$ gives 7.67 ms, which fits the assumption. Considering any real-time system (soft or firm), these results allow to verify the timing constraints specifically for the application component, which includes operations for time acquisition. This can help to design more stable and secure system running on platforms based on the ARM Cortex-A core.

## 7 Conclusion

The ability to predict the exceedance probability and reduce the cost to perform trustworthy timing analysis makes the MBPTA approach along with the EVT theory very attractive. The main contribution of this paper is to propose a systematics for the estimation of probabilistic WCET of a time acquisition task. Assuming several cases of GPD distributions, we compare WCET estimates for them. Our results show the least pessimistic estimation (about 7.67 ms for failure probability levels of $10^{-9}$), using parameters for Gumbel distribution. Therefore, this method of predicting the upper bound is acceptable for the single-core ARM Cortex-A5 processor. For the future work, the investigation of other systems and hardware effects, such as the presence of multiple cores, cache memory effects or impacts of scheduling tasks by Linux kernel are planned. Considering hardware aspects should lead to significant improvements of the dependence metrics, reduction of the pessimism of the pWCET estimation and as a result, make applications, running on Linux, using ARM Cortex-A processors more time-predictable and so more suitable for being used in smart devices and applications.

## References

1. Wilhelm, R., Engblom, J., Ermedahl, A., Holsti, N., Thesing, S., Whalley, D., Bernat, G., Ferdinand, C., Heckman, R., Mitra, T.: The WCET problem - overview of methods and survey of tools. ACM Trans. Embed. Comput. Syst., 36–53 (2007)
2. Abella, J., Hardy, D., Puaut, I., Quinones, E., Cazorla, F.J.: On the comparison of deterministic and probabilistic WCET estimation techniques. In: 26th Euromicro Conference on Real-Time Systems, Madrid, pp. 266–275 (2014)
3. Altmeyer, S., Cucu-Grosjean, L., Davis, R.I.: Static probabilistic timing analysis for real-time systems using random replacement caches. Real-Time Syst. **51**, 77–123 (2015)
4. Guidelines and methods for conducting the safety assessment process on civil airborne systems and equipment. ARP4761 (2001)
5. Radojković, P., Carpenter, P.M., Moretó, M., Čakarević, V., Verdú, J., Pajuelo, A., Cazorla, F.J., Nemirovsky, M., Valero, M.: Thread assignment in multicore/multithreaded processors: a statistical approach. IEEE Trans. Comput. **65**, 256–269 (2016)
6. Guet, F., Morio, J., Santinelli, L.: On the reliability of the probabilistic WCET estimates. In: 8th European Congress on Embedded Real-Time Software and Syst, Toulouse, pp. 758–767 (2016)
7. Fedotova, I., Siemens, E., Hu, H.: A high-precision time handling library. J. Commun. Comput. **10**, 1076–1086 (2013)
8. Cazorla, F., Vardanega, T., Quinones, E., Abella, J.: Upper-bounding program execution time with extreme value theory. In: the Proceedings of WCET workshop (2013)
9. Berezovskyi, K., Santinelli, L., Bletsas, K., Tovar, E.: WCET measurement-based and extreme value theory characterization of CUDA kernels. In: 22nd International Conference on Real-Time Networks and Systems, New York, pp. 279–288 (2014)

10. Berezovskyi, K., Guet, F., Santinelli, L., Bletsas, K., Tovar, E.: MBPTA is for graphics process or units. In 29th International Conference, Nuremberg, pp. 223–236 (2016)
11. Coles, S.: An Introduction to Statistical Modeling of Extreme Values. Springer, Heidelberg (2001)
12. Cucu-Grosjean, L., Santinelli, L., Houston, M., Lo, C., Vardanega, T., Kosmidis, L., Abella, J., Mezzetti, E., Quinones, E., Cazorla, J.F.: MBPTA for multi-path programs. In: 23nd Euromicro Conference on Real-Time Systems. IEEE Press (2012)
13. Hansen, J., Hissam, S.A., Moreno, G.A.: Statistical-Based WCET Estimation and Validation - viewcontent.cgi. In 9th International Workshop on WCET Analysis (2009)
14. Balkema, A., Haan, L.: Residual life time at great age. Ann. Probab. **2**, 792–804 (1974)
15. Santinelli, L., Morio, J., Dufour, G., Jacquemart, D.: On the sustainability of the EVT for WCET estimation. In: 14th International Workshop on WCET Analysis, p. 21 (2014)
16. Embrechts, P., Klueuppelberg, C., Mikosch, T.: Modelling Extremal Events for Insurance and Finance. Springer Science & Business Media, Berlin (1996)

# Energy: Management

# Assessment of Ancillary Service Demand Response and Time of Use in a Market-Based Power System Through a Stochastic Security Constrained Unit Commitment

Saber Talari[1], Miadreza Shafie-khah[1], Neda Hajibandeh[1], and João P.S. Catalão[1,2,3(✉)]

[1] C-MAST, University of Beira Interior, 6201-001 Covilhã, Portugal
{saber.talari,miadreza,neda.hajibandeh}@ubi.pt, catalao@fe.up.pt
[2] INESC TEC and The Faculty of Engineering of the University of Porto, 4200-465 Porto, Portugal
[3] INESC-ID, Instituto Superior Técnico, University of Lisbon, 1049-001 Lisbon, Portugal

**Abstract.** In this paper, the impacts of an incentive-based Demand Response, i.e., Ancillary Service DR (ASDR), and a price-based DR, i.e., Time of Use (ToU), are revealed in a restructured power system which has some wind farms. This network is designed based on the pre-emptive market which is a day-ahead market with a balancing market prognosis. It is a proper mechanism to deal with the stochastic nature of non-dispatchable and outage of all units of the network. With Monte Carlo Simulation (MCS) method, several scenarios are generated in order to tackle the variability and uncertainties of the wind farms generation. The impacts of merging ASDR and ToU are investigated through running a two-stage stochastic security constrained unit commitment (SCUC), separately .

**Keywords:** Ancillary service demand response · Security constrained unit commitment · Time of Use · Two-stage stochastic programming

## 1 Introduction

With the increasing of these renewable energy resources in the power systems, Independent System Operators (ISO) have been faced with new challenges, mainly related to the random and uncertain nature of wind speed [1, 2].

On the other hand, with growing the share of wind energy in power system's energy, novel methods have been proposed to improve the power consumption pattern of consumers [3]. Demand Response Programs (DRPs) are one of the most practical methods for this purpose.

With the expansion of smart meters, like Advanced Metering Infrastructure (AMI) and Internet of Things (IoT), in demand side of power systems, implementation of different methods of Demand Response (DR) is going to be much more applicable [4, 5]. A suitable DR method can not only decrease total operation cost but also provide security and safety of the network operation [6].


L.M. Camarinha-Matos et al. (Eds.): DoCEIS 2017, IFIP AICT 499, pp. 233–241, 2017.
DOI: 10.1007/978-3-319-56077-9_22

In [7], an approach is proposed for shifting suitable amount of load from peak hours to off-peak hours to facilitate effective use of wind generators, reduce possible contingencies in the network and nonnegative local marginal prices (LMP).

Transferring and load shifting has been performed for those loads that can be moved from peak hours to off-peak hours under the load control of Independent System Operator (ISO), while the total load over the planning horizon is completely fixed. Load shifting of the consumers from peak hours to off-peak hours has been conducted by implementing Incentive Based Demand Response (IBDR) programs where only one of the incentive demand response programs has been modeled. By implementing several programs of demand response programs and using time-based demand response programs, more reliable results will be presented.

In [8], wind resources have been used for the flexibility of supply side and their uncertainties have been considered. A flexible stochastic security-constrained shedding framework has been presented for simultaneous optimization of utilization in the supply and demand side aided a demand response plan or program named Time of Use (TOU) optimization tariff [9]. In [10–12], TOU tariffs have been optimized so that the flexibility potential of load side in confronting with the load uncertainty and large scale changes of generation side is maximized. In [13, 14], the impact of two important demand response strategies of Load curtailment and Load shifting on power systems with the wind generation has been investigated. In [15] the problem of constructing the bidding curve has been discussed for a retailer to offer in the pool market. The problem is formulated as a stochastic linear programming.

In this paper, the impacts of Ancillary Service DR (ASDR) and Time of Use (ToU) are studied. A day-ahead market with a balancing market is considered. With a Monte Carlo Simulation (MCS) method, several scenarios are generated in order to tackle the variability and uncertainties of wind generation. The impacts of merging ASDR and ToU are thoroughly investigated.

## 2 Relationship to Smart Systems

The various current facilities in smart grids either in terms of communication technology e.g. Home Area Networks (HAN) or smart meters e.g. advanced metering infrastructures (AMI) enable end-users to participate actively in electricity market in order to improve security, economy, efficiency and reliability of network operation [16]. The information flow in the smart grids can be classified based on power flow and power system sections. Accordingly, communication architecture for demand response is introduced. Home Area Networks/Business Area Networks/Industrial Area Networks (HANs/BANs/IANs) are communication technologies which are deployed in residential units, commercial buildings, and industrial plants in order for connecting multiple electrical appliances to smart meter through ZigBee, WiFi, or Power-line communication (PLC) [17].

On the other hands, Neighborhood Area Networks/Field Area networks (NANs/FANs) are designed for communication between different smart meters of power distribution system through Data aggregate Unit (DAU), WiFi, world-wide interoperability for microwave access (WiMax) or cellular networks like GPRS, 3G and LTE.

Meanwhile, for communication between bulk generation, transmission lines (Wide-Area Network) WANs is used through fiber-optic communication or microwave transmission.

## 3 Formulation

In this paper, two objective functions are introduced in order to deal with both priced-based demand response that is ToU and incentive-based demand response which is ASDR. In fact, for the earlier, a two-stage stochastic program is run for maximizing social welfare and the later, a two stage-stochastic program is run for minimizing total operation cost. Some amount of customers' consumption in ToU program are presented as load bidders. Hence, the objective is to maximize deference of end-user cost and operation cost. Accordingly, the objective function for ToU program and their constraints are shown as follows:

$$Max\sum_{t=1}^{T}\left\{\begin{array}{l}\sum_{b=1}^{B}\{(Bid_{b,t}DisL_{b,t})-\sum_{n=1}^{N}\left(P_{n,t}B_{n,t}+C_{n,t}^{Up}SR_{n,t}^{Up}+C_{n,t}^{Down}SR_{n,t}^{Down}+SU_{n}y_{n,t}+SD_{n}z_{n,t}\right)\\ +\sum_{s=1}^{S}(\Pr_{t,s}\times(\sum_{n=1}^{N}r_{s,n,t}^{up}\times resC_{n}^{up}+r_{s,n,t}^{down}\times resC_{n}^{down})+\sum_{b=1}^{B}VOLL\times Lshed_{s,b,t})\end{array}\right\} \tag{1}$$

In this formulation, the first line is regarding first stage where $Bid_{b,t}$, is load bidding at hour t and bus b, $DisL_{b,t}$, is the dispatched load at hour t and bus b. Moreover, $P_{n,t}B_{n,t}$ is cost of scheduled power production of unit n at hour t. $C_{n,t}^{Up}SR_{n,t}^{Up}$, $C_{n,t}^{Down}SR_{n,t}^{Down}$ are cost of scheduled up-spinning reserve and down-spinning reserve of unit n at hour t, respectively. $SU_n$, $SD_n$ are the start-up cost and shut-down cost of unit n, respectively. The second line is related to second stage which handles the balance market by different scenarios where $\Pr_{t,s}$, is the probability of scenario s at time t, $r_{s,n,t}^{up}\times resC_{n}^{up}$ is the cost of scheduled up-spinning reserve of unit n at scenario s, $r_{s,n,t}^{down}\times resC_{n}^{down}$ is the cost of scheduled down-spinning reserve of unit n at scenario s. The last term $VOLL\times Lshed_{s,n,t}$ is the cost of forced load curtailment at bus n and scenario s and hour t.

Subject to:

DC power flow equation

$$\sum_{n=1}^{N}P_{n,t}+W_{b,t}^{dispach}=\mu\sum_{b=1}^{B}Ld_{b,t}+\sum_{l=1}^{L}Pflow_{l,t} \tag{2}$$

where, $W_{b,t}^{dispach}$ is the real power usage from wind farm node n at hour t which should be less than $W_{b,t}^{expect}$(the expected produced power from wind farm of node n at hour t), $Ld_{b,t}$ is the whole demand at bus b and hour t, $\mu$ is the percentage of the load which must be feed and are not placed in ToU program, $Pflow_{l,t}$ is real power flow in line $l$ and hour t which is as follows:

$$Pflow_l = \frac{1}{X_l}(\delta_{ls} - \delta_{lr}) \tag{3}$$

where $X_l$ is reactance of line $l$, $\delta_{ls}$ is voltage angle of sending-end bus of line $l$ and $\delta_{lr}$ is voltage angle of receiving-end bus of line $l$ and its constraint is as follows:

$$-Pflow_{l,t}^{\max} \leq Pflow_{l,t} \leq Pflow_{l,t}^{\max} \tag{4}$$

where $Pflow_{l,t}^{\max}$ is the maximum capacity of line $l$. And the real power generation constraints are as follows:

$$p_n^{\min}.u_{n,t} + SR_{n,t}^{Down} \leq p_{n,t} \leq p_n^{\max}.u_{n,t} - SR_{n,t}^{Up} \tag{5}$$

where $p_n^{\min}, p_n^{\max}$ are the minimum and maximum capacity of unit n, $u_{n,t}$ is the commitment state of unit n at hour t. Moreover, start-up and shut-down constraint of units are as follows:

$$\begin{aligned} u_{n,t} - u_{n,t-1} &= y_{n,t} - z_{n,t} \\ y_{n,t} + z_{n,t} &\leq 1 \end{aligned} \tag{6}$$

Up and down spinning reserve constraints are as follows:

$$\begin{aligned} 0 \leq SR_{n,t}^{Up} &\leq R_n^{up} \\ 0 \leq SR_{n,t}^{Down} &\leq R_n^{down} \end{aligned} \tag{7}$$

where $R_n^{up}$ is ramp-up limitation for unit n and $R_n^{down}$ is ramp-down limitation for unit n. Plus, ramp-up and -down constraints are as follows:

$$\begin{aligned} P_{n,t} - P_{n,t-1} &\leq R_n^{down} \\ P_{n,t-1} - P_{n,t} &\leq R_n^{up} \end{aligned} \tag{8}$$

Constraints of second stage include DC power flow equation which is as follows:

$$\sum_{n=1}^{N} us_{s,n,t} P_{n,t} + r_{s,n,t}^{up} - r_{s,n,t}^{down} + W_{s,b,t}^{scen} - W_{s,b,t}^{spill} = \mu \sum_{b=1}^{B} Ld_{b,t} + \sum_{l=1}^{L} Pflow_{s,l,t} - Lshed_{s,b,t} \tag{9}$$

where $us_{s,n,t}$ is the given state of unit n at scenario s and hour t, $W_{s,n,t}^{scen}$ is the amount of produced power by wind farm node b in scenario s and hour t, $W_{s,b,t}^{spill}$ is the spillage power of wind farm node b in scenario s and hour t, $Pflow_{s,l,t}$ is power flow of line $l$ in scenario s and hour t which is as follows:

$$Pflow_{l,s} = \frac{1}{X_l}(\delta_{s,ls} - \delta_{s,lr}) \tag{10}$$

where $\delta_{s,ls}$ is voltage angle of sending-end bus of line $l$ in scenario s and $\delta_{s,lr}$ is voltage angle of receiving-end bus of line $l$ and its constraint is as follows:

$$-Pflow_{l,t}^{\max} \leq Pflow_{s,l,t} \leq Pflow_{l,t}^{\max} \tag{11}$$

The deployed up- and down-spinning reserve constraints are as follows:

$$\begin{aligned} 0 &\leq sr_{s,n,t}^{Up} \leq us_{s,n,t}.SR_{n,t}^{Up} \\ 0 &\leq sr_{s,n,t}^{Down} \leq us_{s,n,t}.SR_{n,t}^{Down} \end{aligned} \tag{12}$$

Finally, the maximum load that is participated in this demand response program should be specified as follows:

$$DisL_{b,t} \leq \lambda \times Ld_{b,t} \tag{13}$$

where $\lambda$ is the percentage of the load at bus b that should be joined in DRP.

On the other hands, the objective function for ASDR and some constraints have a few differences compared with ToU. In this program, total operation cost will be minimized and demand response is scheduled as a reserve for the loads which participate in this program. the objective function is as follows:

$$Min \sum_{t=1}^{T} \left\{ \begin{aligned} & \sum_{n=1}^{N} (P_{n,t}B_{n,t} + C_{n,t}^{Up}SR_{n,t}^{Up} + C_{n,t}^{Down}SR_{n,t}^{Down} + SU_n \times y_{n,t} + SD_n \times z_{n,t}) + \sum_{b=1}^{B} CDR_{b,t} \Bigg) \\ & + \sum_{s=1}^{S} (\Pr_{t,s} (\sum_{n=1}^{N} r_{s,n,t}^{up}.resC_n^{up} + r_{s,n,t}^{down}.resC_n^{down}) + \sum_{b=1}^{B} EDRs_{s,b,t} + \sum_{b=1}^{B} VOLL.Lshed_{s,b,t}) \end{aligned} \right\} \tag{14}$$

where $CDR_{b,t}$ is the capacity cost of scheduled demand response reserve for load of bus b at hour t and $EDRs_{s,b,t}$ is the energy cost of scheduled demand response reserve for load of bus b in scenario s and at hour t which are defined as follows:

$$\begin{aligned} DR_{b,t} &= q_{b,t}^0 u_{b,t}^0 + \sum_{k=1}^{K} \lambda_{b,t}^k u_{b,t}^k \\ CDR_{b,t} &= cc_{b,t}^0 q_{b,t}^0 u_{b,t}^0 + \sum_{k=1}^{K} cc_{b,t}^k \lambda_{b,t}^k u_{b,t}^k \\ \lambda_{b,t}^k &= q_{b,t}^k - q_{b,t}^{k-1} \end{aligned} \tag{15}$$

In this program, several demand response discrete points k is determined as $q_{b,t}^k$ for bus b and hour t and each discrete point k has a capacity cost $cc_{b,t}^k$. The first k is separated and the difference of next k and previous one is assigned as $\lambda_{b,t}^k$ with specific price of $cc_{b,t}^k$. Each level can be selected by the objective function based on profitability. The state of which level k at bus b and hour t is being deployed is determined as $u_{b,t}^k$. This program is almost the same in second stage which is as follows:

$$
\begin{aligned}
&0 \le dr_{s,b,t} \le DR_{b,t} \\
&dr_{s,b,t} = q^0_{b,t} uk^0_{s,b,t} + \sum_{k=1}^{K} \lambda^k_{s,b,t} uk^k_{s,b,t} \\
&EDRs_{s,b,t} = ec^0_{b,t} q^0_{b,t} uk^0_{s,b,t} + \sum_{k=1}^{K} ec^k_{b,t} \lambda^k_{b,t} uk^k_{s,b,t}
\end{aligned}
\tag{16}
$$

where $ec^k_{b,t}$ is the energy cost of demand response reserve in scenarios and $uk^k_{s,b,t}$ is the state of which level k at bus b and hour t in scenario s is being deployed. $dr_{s,b,t}$ is scheduled demand response reserve of bus b in scenario s at hour t. Some of the constraints are different from the ToU program that are as follows:

$$
\sum_{n=1}^{N} P_{n,t} + W^{dispach}_{b,t} = \sum_{b=1}^{B} Ld_{b,t} + \sum_{l=1}^{L} Pflow_{l,t}
$$

$$
\sum_{n=1}^{N} us_{s,n,t} P_{n,t} + r^{up}_{s,n,t} - r^{down}_{s,n,t} + W^{scen}_{s,b,t} - W^{spill}_{s,b,t} + \sum_{b=1}^{B} dr_{s,b,t} = \sum_{b=1}^{B} Ld_{b,t} + \sum_{l=1}^{L} Pflow_{s,l,t} - Lshed_{s,b,t} \tag{17}
$$

## 4 Case Study

The 14-bus IEEE power system is used in this study. The slack bus of system is node1. Two wind farms are connected to the power system at Node 5 and Node8. The capacity of each one is equal to 100 MW. Here are several scenarios that each trajectory is the sum of power production of two nodes. There are four cases to assess running ToU and ASDR and compare them in terms of efficiency. First case is running a stochastic SCUC to maximize social welfare without participating loads as ToU DRP and with a constant tariff for load bidding $Bid_{b,t}$. The second case is similar to first one with difference of considering 20% of loads for joining ToU DRP with different tariffs $Bid_{b,t}$. The third case is running a stochastic SCUC to minimize total operation cost without considering ASDR. The fourth case is similar the third one with considering 20% of loads to join ASDR program. In the first case, it is assumed that load bidding is 50$/MWh for all hours. In the second case, two tariff levels are defined for ToU. For peak hours e.g. 13–20 and 28–45, tariff is 40 $/MWh and for others tariff is 70 $/MWh. Accordingly, social welfare for each hour is obtained and it is shown in Fig. 1. As can be seen, in case 2 where the ToU is used the social welfare is mostly higher than case 1, especially in peak times. Moreover, in Fig. 2 which shows the amount of DR participated loads that are supplied. As can be seen, some of loads at peak hours are not supplied in order to get higher social welfare. For the third and fourth cases, demand response is applied just for node 3 and 4. Furthermore, information of ASDR is shown in Table 1. As can be seen in Fig. 3, the operation cost of forth case at hours 34–38 and hour 42 is lower than the case without ASDR. Because, demand response is just scheduled for load in node 4 at hour 34–38 and hour 42. Therefore, the operation cost at these hours are less than operation cost without considering ASDR. Table 2 shows cost of reserve-up, reserve-down, power production units, and security cost which is related to balancing market for all four case studies. The total cost case 1 which is social welfare is lower than case 2 because of using ToU in second one. Finally, total operation cost for case 4 is lower than case 3 because of using ASDR in fourth case.

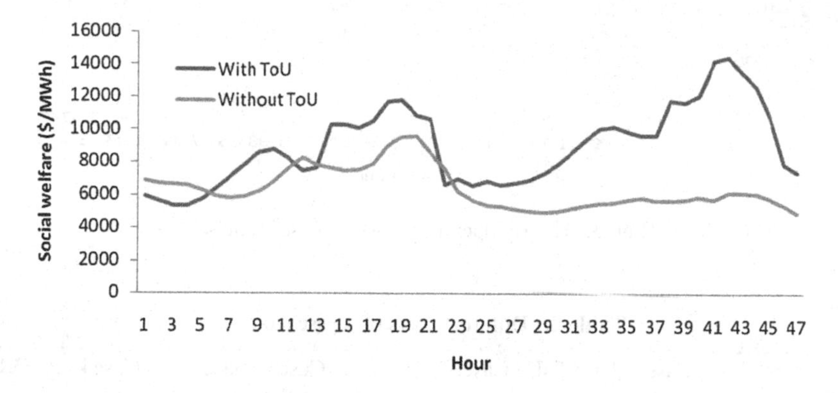


**Fig. 1.** social welfare for case 1 and 2 at each hour

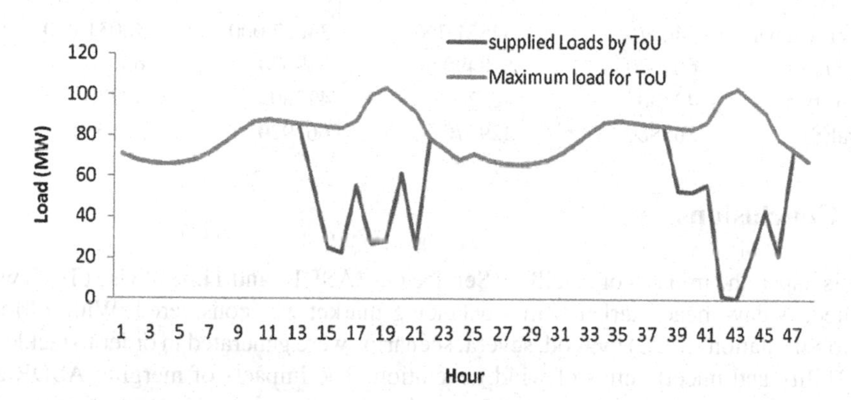


**Fig. 2.** Maximum load for joining ToU and the amount of supplied loads within ToU program

**Table 1.** ASDR data

| k | 0 | 1 | 2 | 3 |
|---|---|---|---|---|
| $q_t^k(MW)$ | 20 | 30 | 40 | 50 |
| $cc_t^k(\$/MWh)$ | 10 | 12.5 | 15 | 17.5 |
| $ec_t^k(\$/MWh)$ | 20 | 22.5 | 25 | 27.5 |

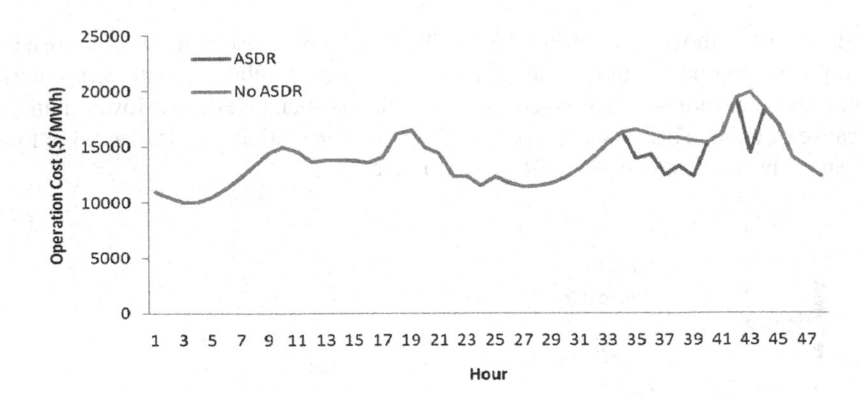


**Fig. 3.** Hourly operation cost for case 3 and 4

**Table 2.** Different costs of 4 case studies

| | Case1 - base ToU | Case2 - ToU | Case3 - base ASDR | Case4 - ASDR |
|---|---|---|---|---|
| Reserve up($) | 0 | 0 | 0 | 0 |
| Reserve down($) | 34627 | 34521.700 | 34627.000 | 40031.820 |
| Production($) | 669670 | 628490 | 669670 | 651300 |
| Security($) | 42380 | 42270 | 42380 | 48270 |
| Total($) | 306880 | 429770 | 661920 | 651080 |

## 5 Conclusions

In this paper, the impacts of Ancillary Service DR (ASDR) and Time of Use (ToU) were studied. A day-ahead market with a balancing market was considered. With a Monte Carlo Simulation (MCS) method, several scenarios were generated in order to tackle the variability and uncertainties of wind generation. The impacts of merging ASDR and ToU were thoroughly investigated. The results of this program demonstrate that using demand response program leads to not only compensate wind generators uncertainties but also increase operator benefits. In peak hours, volume of supplied loads will be dropped in both DR programs and social welfare will be higher which means both of operator and customer are more satisfied.

**Acknowledgment.** This work was supported by FEDER funds through COMPETE 2020 and by Portuguese funds through FCT, under Projects SAICT-PAC/0004/2015 - POCI-01-0145-FEDER-016434, POCI-01-0145-FEDER-006961, UID/EEA/50014/2013, UID/CEC/50021/2013, and UID/EMS/00151/2013. Also, the research leading to these results has received funding from the EU Seventh Framework Programme FP7/2007-2013 under grant agreement no. 309048.

## References

1. Jannati, M., Hosseinian, S.H., Vahidi, B., Li, G.J.: A survey on energy storage resources configurations in order to propose an optimum configuration for smoothing fluctuations of future large wind power plants. Renew. Sustain. Energy Rev. **29**, 158–172 (2014)
2. Zafirakis, D., Kaldellis, J.K.: Autonomous dual-mode CAES systems for maximum wind energy contribution in remote island networks. Energy Convers. Manag. **51**, 2150–2161 (2010)
3. De Vos, K., Petoussis, A.G., Driesen, J., Belmans, R.: Revision of reserve requirements following wind power integration in island power systems. Renew. Energy **50**, 268–279 (2013)
4. Arasteh, H., et al.: Iot-based smart cities: a survey. In: 2016 IEEE 16th International Conference on Environment and Electrical Engineering (EEEIC), Florence, pp. 1–6 (2016). doi:10.1109/EEEIC.2016.7555867
5. Kamyab, F., Amini, M.H., Sheykhha, S., Hasanpour, M., Jalali, M.M.: Demand response program in smart grid using supply function bidding mechanism. IEEE Trans. Smart Grid **7**(2), 1277–1284 (2016)
6. Bahrami, S., Sheikhi, A.: From demand response in smart grid toward integrated demand response in smart energy hub. IEEE Trans. Smart Grid **7**(2), 650–658 (2016)
7. Sioshansi, R., Short, W.: Evaluating the impacts of real-time pricing on the usage of wind generation. IEEE Trans. Power Syst. **24**, 516–524 (2010)
8. Papavasiliou, A., Oren, S.O., Neill, R.P.: Reserve requirements for wind power integration: a scenario based stochastic programming framework. IEEE Trans. Power Syst. **26**, 2197–2206 (2011)
9. Heydarian-Forushani, E., Golshan, M.E.H., Moghaddam, M.P., Shafie-khah, M., Catalão, J.P.S.: Robust scheduling of variable wind generation by coordination of bulk energy storages and demand response. Energy Convers. Manag. **106**, 941–950 (2015)
10. Liu, G., Tomsovic, K.: Quantifying spinning reserve in systems with significant wind power penetration. IEEE Trans. on Power Syst **27**, 2385–2393 (2012)
11. Pina, A., Silva, C., Ferrão, P.: The impact of demand side management strategies in the penetration of renewable electricity. Energy **41**, 128–137 (2012)
12. Heydarian-Forushani, E., Golshan, M.E.H., Shafie-khah, M.: Flexible security-constrained scheduling of wind power enabling time of use pricing scheme. Energy **90**, 1887–1900 (2015)
13. Critz, D.K., Busche, S., Connors, S.: Power systems balancing with high penetration renewables: The potential of demand response in Hawaii. Energy Convers. Manag. **76**, 609–619 (2013)
14. Dietrich, K., Latorre, J.M., Olmos, L., Ramos, A.: Demand response in an isolated system with high wind integration. IEEE Trans. Power Syst. **27**, 20–29 (2012)
15. Zhechong, Z., Wu, L.: Impacts of high penetration wind generation and demand response on LMPs in day-ahead market. IEEE Trans. Smart Grid **5**(1), 220–229 (2014)
16. Boroojeni, K.G., Amini, M.H., Iyengar, S.S.: Overview of the security and privacy issues in smart grids. In: Boroojeni, Kianoosh G., Amini, M.Hadi, Iyengar, S.S. (eds.) Smart Grids: Security and Privacy Issues, pp. 1–16. Springer, Cham (2017). doi: 10.1007/978-3-319-45050-6_1
17. Sheikhi, A., et al.: A cloud computing framework on demand side management game in smart energy hubs. Int. J. Electr. Power Energy Syst. **64**, 1007–1016 (2015)

# Self-scheduling of Wind-Thermal Systems Using a Stochastic MILP Approach

Rui Laia[1,2], Isaias L.R. Gomes[1,2], Hugo M.I. Pousinho[1], Rui Melício[1,2(✉)], and Victor M.F. Mendes[2,3]

[1] Instituto Superior Técnico, IDMEC, Universidade de Lisboa, Lisbon, Portugal
ruimelicio@gmail.com
[2] Departamento de Física, Universidade de Évora, Escola de Ciências e Tecnologia, Évora, Portugal
[3] Instituto Superior de Engenharia de Lisboa, Lisbon, Portugal

**Abstract.** In this work a stochastic (Stoc) mixed-integer linear programming (MILP) approach for the coordinated trading of a price-taker thermal (Ther) and wind power (WP) producer taking part in a day-ahead market (DAM) electricity market (EMar) is presented. Uncertainty (Uncer) on electricity price (EPr) and WP is considered through established scenarios. Thermal units (TU) are modelled by variable costs, start-up (ST-UP) technical operating constraints and costs, such as: forbidden operating zones, minimum (Min) up/down time limits and ramp up/down limits. The goal is to obtain the optimal bidding strategy (OBS) and the maximization of profit (MPro). The wind-Ther coordinated configuration (CoConf) is modelled and compared with the unCoConf. The CoConf and unCoConf are compared and relevant conclusions are drawn from a case study.

**Keywords:** Wind thermal coordination · Stochastic programming · MILP

## 1 Introduction

The emissions derived of the use of nonrenewable fuels and the aspiration to attain independence of energy [1] lead a considerable European countries to promote generation of the electricity from renewable (Rnew) resources by adopting some instruments of support for Rnew energy production, namely investments incentives, green certificates, soft balancing costs and feed-in-tariffs [2].

At the end of 2014, 43.70% of all novel Rnew farms were based on WP and was the 7th year consecutively that over 55.0% of added capacity of power in the EU was Rnew [3]. In the face of the increasing Rnew energy incorporation in the last years, supply of energy still depending on nonrenewable fuels since more than 60% of the electricity generated all over the world in 2012 was based on nonrenewable fuel Ther plants [4].

In a restructured EMar, power resources owners' operate under competition level due to the nodal variations of EPr [5] in order to obtain the best revenue bidding in the DAM [6]. For the WP producers (WPP), WP and the market-clearing EPr Uncer are to be addressed in order to know the amount of energy to produce in order to present optimal


L.M. Camarinha-Matos et al. (Eds.): DoCEIS 2017, IFIP AICT 499, pp. 242–250, 2017.
DOI: 10.1007/978-3-319-56077-9_23

offers. In absence of conformity, i.e., there is a deviation (Dev), economic penalizations is due to happen [7]. For Ther power producers, only market-clearing EPr Uncer has to be addressed.

## 2 Relationship to Smart Systems

A smart system can be stated as an embedded system that incorporates advanced systems and provide the inhabitants with sophisticated monitoring and control over how something happens in the system [8], for example a wind farm or a TU. Smart systems are capable of sensing, making diagnosis, describing, qualifying and managing how something happens in the system, incorporating both technical intelligence and cognitive functions. In smart systems, electronic devices will be communicating with software base system, allowing the user to access information about the functionality of the system [8]. These systems are highly reliable, often miniaturized, networked, predictive and energy autonomous [9]. Future power systems should ensure security, reliability and efficiency in energy management. Using the abilities of smart systems to monitoring the energy demand and the energy production of other units can play a vital role in what regards the unit commitment of TU. Particularly, monitoring and high quality real-time data of the exploitation of Rnew energy sources, namely WP, that usually requires a certain amount of spinning reserve due to their intermittent nature may represent additional information at the moment of unit commitment of TU. With this information, the Wind-Ther Power Producer (WTPP) can make a more accurate decision concerning the participation in EMars and therefore foremost revenue [10, 11]. Also, benefits of environmental are predictable with the increase in the capability of discovery offers able to be satisfied with a high level of being pleased and less needed of spinning reserve, less TU are needed and less nonrenewable fuel is used.

## 3 State of the Art

For Ther conversion of energy into electricity, several methods of optimization to resolve the problem of unit commitment (UC) have been used in the literature, including a technique of primacies list, classical mathematical programming techniques, like Lagrangian Relaxation (LR) and dynamic programming (DP) and more newly artificial intelligence (AI) techniques [12]. Although, requiring small computation time and easy to implement, the priority list technique does not guarantee an opportune resolution near the global optimal one, which implies an operation of higher cost [13, 14]. DP methods are flexible but these methods are characterized by a known limitation by the "curse of dimensionality". Although the LR can overcome the previous limitation, does not necessarily lead to a viable resolution, implying further processing for satisfying the infringed constraints in order to find a viable resolution, which does not guarantee solution optimal. Although, AI techniques based on simulating annealing and ANN have been applied, the major limitation of the AI techniques concerning with the possibility to obtain a resolution near the global optimum one is a disadvantage. The MILP method has been useful with success for solving the problem of UC [15]. MILP is suitable for the formulation of bidding strategies due to its rigorousness and extensive capability of modeling [16]. WPP usually have significant difficulties to predict their power output accurately. In addition, WPP have to face Uncer on EPr. These Uncer have to be expediently considered, i.e., treated into the variables of the problems [17] to be addressed by a WPP in order to know how much to produce and the price for bidding. The technical literature presents methods for WP bidding strategies solving using different approaches: the first one is the use of WP with technologies of storage of energy [18]; the use of economic options as a tool for WPP to hedge against WP Uncer [19]; another approach is the design of Stoc models in order to obtain OBS for WPP participating in an EMar [20], without the aforementioned policies. The 3rd line of action is a Stoc formulation explicitly modelling the Uncer faced by a WPP [21], using indeterminate measures and an established of scenarios built by WP forecast and market-clearing EPr forecast [22] requests.

Hence, this paper provides an effective approach based on Stoc MILP to find out the optimal bidding strategies of a single entity having to manage a coordinated wind-Ther system, so as to maximize the expected revenue in the Iberian day-ahead EMar.

# 4 Problem Formulation

## 4.1 WPP

Considering the variability and intermittent nature of WP the physical delivering usually differs from the offer submitted by WPP to the DAM. The revenue $RV_h$ of a WPP proposing a power of $E_h^{offer}$, but actually producing $E_h^{act}$ for period $h$ is stated as:

$$RV_h = \pi_h^D E_h^{offer} + IC_h \tag{1}$$

In (1), the DAM price is $\pi_h^D$, the imbalance (Imb) cost is $IC_h$. The total Dev for period $h$ is stated as:

$$D_h = E_h^{act} - E_h^{offer} \tag{2}$$

The price that WPP will pay for excess of production is $\pi_h^+$, the price to be charged for deficit of production is $\pi_t^-$. The Imb prices can be given by means of price ratios stated as:

$$pr_h^+ = \frac{\pi_h^+}{\pi_h^D},\ pr_h^+ \leq 1 \quad \text{and} \quad pr_h^- = \frac{\pi_h^-}{\pi_h^D},\ pr_h^- \geq 1 \tag{3}$$

In (3), the $\pi_h^+$ is never greater than 1. The $\pi_t^-$ is never lower than 1.
The power producer
The operating cost, $T_{sih}$, for a TU can is stated as:

$$T_{sih} = B_i b_{sih} + g_{sih} + u_{sih} + A_i z_{sih} \qquad \forall s, \quad \forall i, \quad \forall h \tag{4}$$

In (4), the fixed production cost is $B_i$, the added variable cost is $g_{sih}$, the ST-UP and shut-down (Sh-Down) costs are $u_{sih}$ and $A_i$, of the unit. The last three costs are in general described by nonlinear function (Func) and worse than that some of the functions are non-convex and non-differentiable functions, but some kind of smoothness is expected and required to use MILP, for instance, as being sub differentiable functions.

The ST-UP and Sh-Down costs of units in (4) are considered to be such that is possible to approximate those Func by a piecewise linear. Hence, the $g_{sih}$, is:

$$g_{sih} = \sum_{l=1}^{L} T_i^l \Delta_{sih}^l \qquad \forall s, \quad \forall i, \quad \forall h \tag{5}$$

$$E_{sih} = E_i^{\min} b_{sih} + \sum_{l=1}^{L} \Delta_{sih}^l \qquad \forall s, \quad \forall i, \quad \forall h \tag{6}$$

$$(M_i^1 - E_i^{\min}) j_{sih}^1 \leq \Delta_{sih}^1 \qquad \forall s, \quad \forall i, \quad \forall h \tag{7}$$

$$\Delta_{sih}^1 \leq (M_i^1 - \mathrm{E}_i^{\min}) b_{sih} \qquad \forall s, \quad \forall i, \quad \forall h \tag{8}$$

$$(M_i^l - M_i^{l-1}) j_{sih}^l \leq \Delta_{sih}^l \qquad \forall s, \quad \forall i, \quad \forall h, \quad \forall l = 2, \ldots, L-1 \tag{9}$$

$$\Delta_{sih}^l \leq (M_i^l - M_i^{l-1}) j_{sih}^{l-1} \qquad \forall s, \quad \forall i, \quad \forall h, \quad \forall l = 2, \ldots, L-1 \tag{10}$$

$$0 \leq \Delta_{sih}^L \leq (E_i^{\max} - M_{sih}^{L-1}) j_{sih}^{L-1} \qquad \forall s, \quad \forall i, \quad \forall h \tag{11}$$

In (5), the slope of each segment is $T_i^l$, the segment power is $\Delta_{sih}^l$. In (6), the binary variable $b_{sih}$ guarantee that the power production is equal to 0 if the unit is in the state offline. In (7), if the binary variable $j_{sih}^l$ has a null value, then the segment power $\Delta_{sih}^1$ can be lower than the segment *1* maximum power (MaxPow); otherwise and in conjunction with (8), if the unit is in the state on, then $\Delta_{sih}^1$ is equal to the segment *1* MaxPow. In (9), if the binary variable $j_{sih}^l$ has a null value, then the segment power $\Delta_{sih}^l$ can be lower than the segment *l* MaxPow; otherwise and in conjunction with (10), if the unit is in the state on, then $\Delta_{sih}^l$ is equal to the segment *l* MaxPow. The exponential nature of the ST-UP costs functions, $u_{sih}$ is approached by a linear formulation [21] is:

$$u_{sih} \geq K_i^{\alpha} \left( b_{sih} - \sum_{r=1}^{\alpha} b_{sih-r} \right) \qquad \forall s, \quad \forall i, \quad \forall h \tag{12}$$

The constraints to limit the power produced by the unit are:

$$E_i^{\min} b_{sih} \leq E_{sih} \leq E_{sih}^{\max} \quad \forall s, \quad \forall i, \quad \forall h \tag{13}$$

$$E_{sih}^{\max} \leq E_i^{\max}(b_{sih} - z_{sih+1}) + SD\, z_{sih+1} \quad \forall s, \quad \forall i, \quad \forall h \tag{14}$$

$$E^{\max}_{sih} \leq E^{\max}_{sih-1} + RU\,b_{sih-1} + SU\,y_{sih} \quad \forall s, \quad \forall i, \quad \forall h \tag{15}$$

$$E_{sih-1} - E_{sih} \leq RD\,b_{sih} + SD\,z_{sih} \quad \forall s, \quad \forall i, \quad \forall h \tag{16}$$

In (13) and (14), the upper bound of $E^{\max}_{sih}$ is established, which is the maximum available power of the unit. The minimum down time (MDT) constraint is imposed by a formulation:

$$\sum_{h=1}^{F_i} b_{sih} = 0 \quad \forall s, \quad \forall i \tag{17}$$

$$\sum_{h=k}^{k+DH_i-1} (1 - b_{sih}) \geq DH_i z_{sih} \quad \forall s, \quad \forall i, \quad \forall k = F_i + 1 \ldots H - DH_i + 1 \tag{18}$$

$$\sum_{h=k}^{H} (1 - b_{sih} - z_{sih}) \geq 0 \quad \forall s, \quad \forall i, \quad \forall k = H - DH_i + 2 \ldots H \tag{19}$$

$$F_i = \min\{H, (DH_i - t_{si0})(1 - b_{si0})\}$$

The MUT constraint is also imposed:

$$\sum_{h=1}^{N_i} (1 - b_{sih}) = 0 \quad \forall s, \quad \forall i \tag{20}$$

$$\sum_{h=k}^{k+UH_i-1} b_{sih} \geq UH_i y_{sih} \quad \forall s, \quad \forall i, \quad \forall k = N_i + 1 \ldots H - UH_i + 1 \tag{21}$$

$$\sum_{h=k}^{H} (b_{sih} - z_{sih}) \geq 0 \quad \forall s, \quad \forall i, \quad \forall k = H - UH_i + 2 \ldots H \tag{22}$$

$$N_i = \min\{H, (UH_i - U_{si0})\, b_{si0}\}$$

The relation between the binary variables to identify start-up, shutdown and forbidden operating zones is:

$$y_{sih} - z_{sih} = t_{sih} - b_{sih-1} \quad \forall s, \quad \forall i, \quad \forall h \tag{23}$$

$$y_{sih} + z_{sih} \leq 1 \quad \forall s, \quad \forall i, \quad \forall h \tag{24}$$

The total power produced by the TU is:

$$E^{g}_{st} = \sum_{i=1}^{I} E_{sih} \quad \forall s, \quad \forall h \tag{25}$$

*Objective function*: The total offer is:

$$E^{offer}_{sh} = E^{th}_{sh} + E^{D}_{sh} \quad \forall s, \ \forall h \tag{26}$$

The physical delivering is:

$$E^{act}_{sh} = E^{g}_{sh} + E^{\omega d}_{sh} \quad \forall s, \quad \forall h \tag{27}$$

In (27), $E^{g}_{sh}$ is the physical delivering by the TU and $E^{\omega d}_{sh}$ is the physical delivering by the wind farm for scenario $s$. The expected revenue of the GNCO is:

$$\sum_{s=1}^{N_S} \sum_{h=1}^{N_H} p_s \left[ \left( \pi^{D}_{sh} E^{offer}_{sh} + \pi^{D}_{sh} pr^{+}_{sh} D^{+}_{sh} - \pi^{D}_{sh} pr^{-}_{sh} D^{-}_{sh} \right) - \sum_{i=1}^{I} T_{sih} \right] \quad \forall s, \ \forall h \tag{28}$$

Subject to:

$$0 \le E^{offer}_{sh} \le E^{M}_{sh} \quad \forall s, \quad \forall h \tag{29}$$

$$D_{hs} = \left( E^{act}_{sh} - E^{offer}_{sh} \right) \quad \forall s, \ \forall h \tag{30}$$

$$D_{hs} = D^{+}_{hs} - D^{-}_{hs} \quad \forall s, \ \forall h \tag{31}$$

$$0 \le D^{+}_{hs} \le E_{hs} x_s \quad \forall \omega, \ \forall h \tag{32}$$

The maximum Ther generation is:

$$E^{M}_{sh} = \sum_{i=1}^{I} E^{\max}_{sih} + E^{w\max} \quad \forall s, \ \forall h \tag{33}$$

An additional constraint for (28) appears:

$$(E^{offer}_{sh} - E^{offer}_{s'h})(\pi^{D}_{sh} - \pi^{D}_{s'h}) \ge 0 \quad \forall s, \ s', \quad \forall h \tag{34}$$

## 5 Case Study

The case study is from a GNCO with a WTPP, with 1440 MW of installed capacity. The used data is available in [6]. The energy prices are from the Iberic Market of electricity and available in [23], considering 10 days of June. The EPr and the energy generated from wind are displayed in Fig. 1.

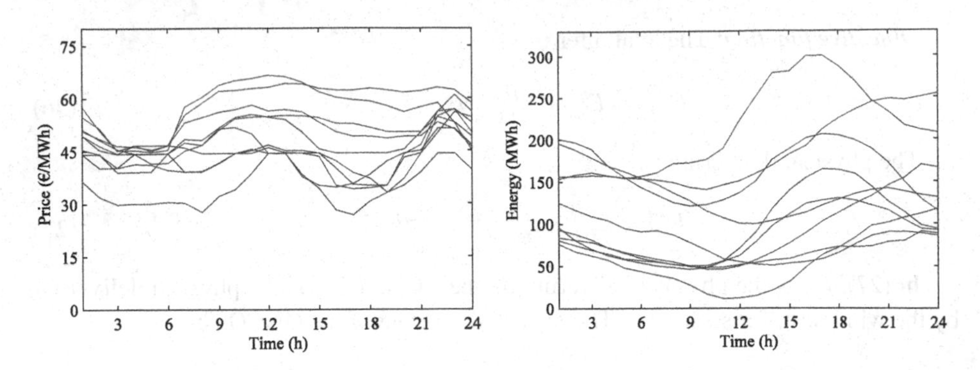


**Fig. 1.** Market Iberic: June 2014; left: EPr, energy from wind: right.

The energy generated is obtained using the total energy generated from the wind farm having 360 MW of rated power. The expected revenue for CoConf and unCoConf are displayed in Table 1.

**Table 1.** Expected revenue for CoConf and uncoordinated configurations

| Case | Expected revenue |
|---|---|
| Wind uncoordinated (€) | 119 200 |
| Ther uncoordinated (€) | 516 848 |
| Coordinated Wind and Ther (€) | 642 326 |
| Gain (%) | 0,99 |

The non-decreasing energy bid for the unCoConf approach is displayed in Fig. 2.

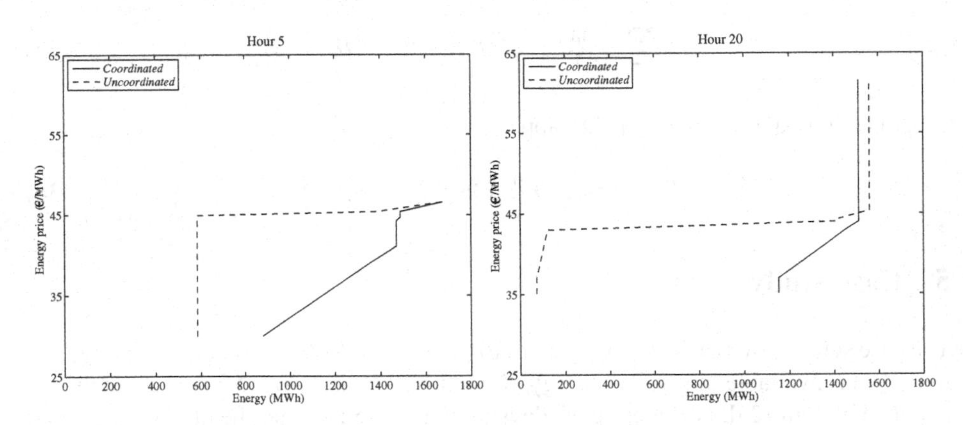


**Fig. 2.** Bids of energy.

In Fig. 2 the CoConf permits for a Min value of offered power upper than the one offered in the unCoConf and permits for a lesser price of the offering, which is a possible operation benefit.

## 6 Conclusion

Smart Systems can play an important role for a Ther and WP producer since the operation till the bidding in day-ahead EMars. The ability to provide real-time data from the wind production may result in foremost decisions for the decision-maker and therefore higher revenues. As result of the proposed approach for uncoordinated and coordinated operations optimal schedule of the TU and the short-term bidding strategies are obtained. The presented approach is appropriate for the GNCO involvement with TU and a wind farm. The offer coordinated of Ther with WP power permits providing foremost outcomes than the sum of the lonely offers. The Uncer are modelled using established scenarios for the prices of the energy and power production. In the literature of all trading problems and management involving production by wind prove to be optimization problems under Uncer.

**Acknowledgments.** Portuguese Funds partially support this work through the FCT-Foundation for Science and Technology, LAETA 2015/2020, UID-EMS-50022-2013.

## References

1. Al-Awami, A.T., El-Sharkawi, M.A.: Coordinated trading of wind and thermal energy. IEEE Trans. Sustain. Energy **2**(3), 277–287 (2011)
2. Pineda, S., Bock, A.: Renewable-based generation expansion under a green certificate market. Renew. Energy **91**, 53–63 (2016)
3. Wind in power: 2014 European Statistics, February 2015. http://www.ewea.org/fileadmin/files/library/publications/statistics/EWEA-Annual-Statistics-2014.pdf
4. Key World Energy Statistics 2014: International Energy Agency (2014). http://www.iea.org/newsroomandevents/agencyannouncements/key-world-energy-statistics-2014-now-available-for-free.html
5. Laia, R., Pousinho, H.M.I., Melício, R., Mendes, V.M.F.: Bidding strategy of wind-thermal energy producers. Renew. Energy **99**, 673–681 (2016)
6. Laia, R., Pousinho, H.M.I., Melício, R., Mendes, V.M.F.: Self-scheduling and bidding strategies of thermal units with stochastic emission constraints. Energy Convers. Manag. **89**, 975–984 (2015)
7. Laia, R., Pousinho, H.M.I., Melício, R., Mendes, V.M.F.: Optimal scheduling of joint wind-thermal systems. Adv. Intell. Syst. Comput. **527**, 136–146 (2017). Springer
8. Arsan, T.: Smart systems: from design to implementation of embedded smart systems. In: 13th International Symposium on Smart MicroGrids for Sustainable Energy Sources Enabled by Photonics and IoT Sensors, Nicosia, Cyprus, pp. 59–64 (2016)
9. Topham, D., Gouze, N.: Smart Systems Integration Knowledge Gateway. SSI, Copenhagen (2015)
10. Laia, R., Pousinho, H.M.I., Melício, R., Mendes, V.M.F.: Optimal bidding strategies of wind-thermal power producers. In: Camarinha-Matos, Luis M., Falcão, António J., Vafaei, N., Najdi, S. (eds.) DoCEIS 2016. IAICT, vol. 470, pp. 494–503. Springer, Cham (2016). doi:10.1007/978-3-319-31165-4_46
11. Laia, R., Pousinho, H.M.I., Melício, R., Mendes, V.M.F.: Bidding decision of wind-thermal GenCo in day-ahead market. Energy Procedia **106**, 87–96 (2016)

12. Trivedi, A., Srinivasan, D., Biswas, S., Reindl, T.: Hybridizing genetic algorithm with differential evolution for solving the unit commitment scheduling problem. Swarm Evol. Comput. **23**, 50–64 (2015)
13. Laia, R., Pousinho, H.M.I., Melício, R., Mendes, V.M.F.: Stochastic emission constraints on unit commitment. Procedia Technol. **17**, 437–444 (2014)
14. Gomes, I.L.R., Pousinho, H.M.I., Melício, R., Mendes, V.M.F.: Bidding and optimization strategies for wind-pv systems in electricity markets assisted by CPS. Energy Procedia **106**, 111–121 (2016)
15. Ostrowski, J., Anjos, M.F., Vannelli, A.: Tight mixed integer linear programming formulations for the unit commitment problem. IEEE Trans. Power Syst. **27**(1), 39–46 (2012)
16. Floudas, C., Lin, X.: Mixed integer linear programming in process scheduling: modeling, algorithms, and applications. Annals of Operations Res. **139**, 131–162 (2005)
17. El-Fouly, T.H.M., Zeineldin, H.H., El-Saadany, E.F., Salama, M.M.A.: Impact of wind generation control strategies, penetration level and installation location on electricity market prices. IET Renew. Power Gener. **2**, 162–169 (2008)
18. Angarita, J.L., Usaola, J., Martinez-Crespo, J.: Combined hydro-wind generation bids in a pool-based electricity market. Electr. Power Syst. Res. **79**(7), 1038–1046 (2009)
19. Hedman, K.W., Sheble, G.B.: Comparing hedging methods for wind power: using pumped storage hydro units vs options purchasing. In: Proceedings of International Conference on Probabilistic Methods Applied to Power Systems, Stockholm, Sweden, pp. 1–6 (2006)
20. Matevosyan, J., Soder, L.: Minimization of imbalance cost trading wind power on the short-term power market. IEEE Trans. Power Syst. **21**(3), 1396–1404 (2006)
21. Ruiz, P.A., Philbrick, C.R., Sauer, P.W.: Wind power day-ahead uncertainty management through stochastic unit commitment policies. In: Proceedings IEEE Power Systems Conference and Exposition, Seattle, USA, pp. 1–9 (2009)
22. Coelho, L.S., Santos, A.A.P.: A RBF neural network model with GARCH errors: application to electricity price forecasting. Electr. Power Syst. Res. **81**(1), 74–83 (2011)
23. Red Electrica de España (2016). http://www.esios.ree.es/web-publica/

# Impact of Distributed Generation on the Thermal Ageing of Low Voltage Distribution Cables

Gergely Márk Csányi(✉), Zoltán Ádám Tamus, and Árpád Varga

Department of Electric Power Engineering, Budapest University of Technology and Economics, Egry József Str. 18, Budapest 1111, Hungary
{csanyi.gergely,tamus.adam}@vet.bme.hu, varpi950128@gmail.com

**Abstract.** The low voltage (LV) distribution cable networks were installed some decades ago but the new paradigm in electric power engineering generates new requirements from these old assets. The distributed generation, storage and new appliances can cause high variation of load and reverse power flow, nevertheless the LV cable grid was not designed to these new stresses. The aggregate load and generation can surpass the rated capacity of the cable lines causing short term temperature increasing. This temperature stress can decrease the expected lifetime of the cable lines. In this study the short term thermal overload of LV distribution cables was investigated. The experiments were executed on PVC insulated LV cable samples and electrical and mechanical properties of the cable jacket were investigated. The effect of these short-term overloads on the expected lifetime of cables is introduced and non-destructive measurement for tracking the effect of the short term thermal overloads on the cable is suggested.

**Keywords:** Distributed generation · Low voltage cables · Distribution network · Cable ageing · PVC cable · EVR · Extended voltage response method · Smart

## 1 Introduction

Low voltage underground cables play an important role in electricity distribution all across Europe and are valuable assets. Hence, preventing outages, assessing the cables' condition and optimizing replacement strategies are usually the main goals of an electricity provider [1]. Low voltage distribution cables were installed some decades ago, however due to the new paradigm in electric power engineering they have to withstand new stresses. The increasing number of novel appliances connected to the grid (e.g. renewable energy sources, electric vehicles, heat pumps etc.) can easily lead the load to surpass the cable's maximum capacity increasing its temperature [2]. Higher temperature has harmful effects on cable insulation's expected lifetime. However, overloads can be often remain undetected without blowing the fuses that are responsible for the protection of the given cable section while they have a certain thermal effect on the cable, reducing its lifetime. This kind of overload is more likely to happen in the near future due to the aggregated loads generated by the new appliances. In this study the short term thermal overload of LV distribution cables was investigated. The experiments were


L.M. Camarinha-Matos et al. (Eds.): DoCEIS 2017, IFIP AICT 499, pp. 251–258, 2017.
DOI: 10.1007/978-3-319-56077-9_24

executed on new, PVC insulated LV cable samples and short intensive thermal stresses due to the overloads were simulated by laboratory ageing tests. Electrical (tan δ, voltage responses) and mechanical (Shore D hardness) properties of the cable jacket were measured after the ageing cycles and changing of these parameters was evaluated.

## 2 Relationship to Smart Systems

As a result of the new paradigm in electric power engineering, low voltage (LV) distribution cables will more often operate in the near future under harmful conditions. The distributed generation, energy storage systems and new appliances can cause high variation of load and reverse power flow, in extreme cases, mostly the photovoltaic systems on hot summer days can cause overloads on low voltage distribution cable lines. These overloads appear when the temperature of the soil is also elevated therefore it can increase the temperature of the core insulation and the jacket over the maximum operation temperature. The increasing temperature decreases the LV distribution cables' expected lifetime. Nevertheless, the LV cable grid was not designed to these new stresses. However, rebuilding the LV cable network would be very expensive therefore the protection of LV assets offers a better solution. Hence, the attention increases on examining how these stresses may affect the LV cables' insulation and how smart solutions can prolong their expected lifetime [2–6]. By means of smart measurements the temperature of the cable sections could be estimated and therefore a more efficient replacement strategy could be developed helping the operators. In order to reach this goal it is essential to examine how LV cable insulations react to the thermal effect of short term overloads.

## 3 Experimental

### 3.1 Samples and Thermal Ageing

The samples were prepared from NYCWY 0.6/1 kV 4 × 10/10 $mm^2$ low voltage cables. The construction of the cable can be seen in Fig. 1. The cable builds up from a PVC jacket (5), wire and tape screens made of copper (4), filling material between (3) the grounding screen and the PVC core insulations (2) and four copper conductors (1).

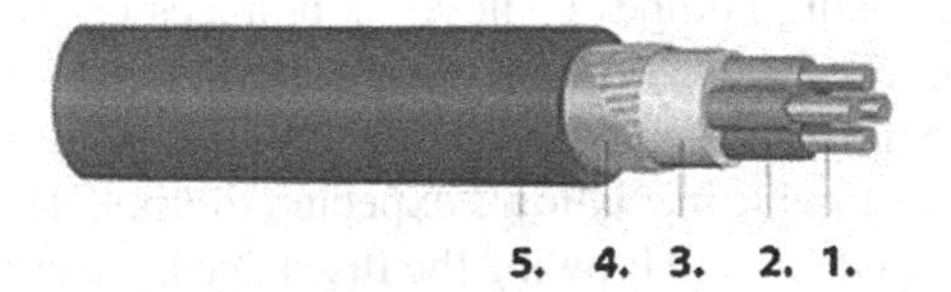


**Fig. 1.** The construction of the cable

Each cable sample was half-meter long and the wire and tape screens of the cables were connected to all of the core conductors providing an inner electrode while an outer electrode was created by wrapping an aluminum foil around each cable jacket.

To simulate the thermal effects of short time overloading periods the samples were subjected to accelerated thermal ageing. From the cable specimens two groups were created and were aged for 11 h at 125 °C and 140 °C, respectively.

After that a 4-hour-long ageing were carried out at the same temperatures. All in all, the mechanical and electrical properties were measured after 11 and 15 h of thermal ageing. The correlations between the applied and equivalent ageing durations were calculated by means of Arrhenius equation; the activation energy used for the calculation was 80 kJ/mol [7]. Results can be seen in Table 1.

**Table 1.** Equivalent ageing duration at different temperatures [hour]

| 80 kJ/mol | 11 h @ 125 °C | 15 h @ 125 °C | 11 h @ 140 °C | 15 h @ 140 °C |
|---|---|---|---|---|
| 80 °C | 239 | 326 | 575 | 784 |
| 100 °C | 56 | 76 | 133 | 182 |

### 3.2 Measurement

Both mechanical and electrical properties were measured during the investigation. The mechanical parameters were measured by Shore D hardness tester and the dielectric properties were measured by tan δ measurement and Extended Voltage Response (EVR) measurement [8]. The hardness was measured 11 times on each specimen's cable jacket's uncovered part. The measurements can be used only for comparison purposes since the thickness of the jacket is only 2 mm however, the standards (ASTM etc.) require at least 4 mm thickness of the material. The tan δ values were measured 14 times in the 20 Hz…500 kHz range at 5 V. The test equipment was a Wayne-Kerr 6430a impedance analyzer. The EVR method measures the initial slopes of the decay ($S_d$) and return ($S_r$) voltages. The decay voltage can be measured after the relatively long duration (100…1000 s) charging period of the insulation. The return voltage can be measured on the charged insulation after a few seconds of short circuiting [9]. By changing the shorting times different time constant polarizations can be investigated. During the EVR measurement the inner electrode (inner cores and wire and tape screens) was connected to 1000 V while the outer (aluminum foil) was grounded. The charging time was set to 4000 s the charging voltage to 1000 V DC and return voltages were measured after 20 different discharging times in the 1…2000 s range. During the investigation the specimens were measured at 23.2 ± 0.5 °C. Proper connections were examined before each electric measurement.

## 4 Results

Shore D hardness measurement, that is used for characterizing polymers, gives a dimensionless value between 0 (soft) and 100 (hard). The Shore D measurement results in Table 2 show that hardness of the jacket is increasing by the thermal ageing and the degradation at 140 °C is more intensive than at 125 °C.

**Table 2.** The results of the Shore D hardness measurement of the jacket

| Ageing: | New | 11 h @ 125 °C | 15 h @ 125 °C | 11 h @ 140 °C | 15 h @ 140 °C |
|---|---|---|---|---|---|
| Average | 36.95 | 39.12 | 40.04 | 43.46 | 45.67 |
| Deviation | 1.49 | 1.68 | 1.57 | 1.69 | 1.65 |

The dielectric parameters were measured by means of Extended Voltage Response method. This measurement provides information about the changing of the conduction by the value of slope of decay voltage ($S_d$) [10, 11]. The results can be seen in Table 3. As the data of Table 3 suggest the specific conductivity of the jacket material is decreasing by ageing. This suggests the decreasing of the plasticizer content of the jacket.

**Table 3.** The slopes of decay voltages ($S_d$) measured on the jacket [V/s]

| Ageing: | New | 11 h @ 125 °C | 15 h @ 125 °C | 11 h @ 140 °C | 15 h @ 140 °C |
|---|---|---|---|---|---|
| Average | 76.16 | 84.42 | 37.98 | 50.45 | 15.60 |
| Deviation | 9.14 | 4.47 | 9.17 | 5.55 | 0.48 |

The slopes of return voltages as the function of shorting times are in Fig. 2.

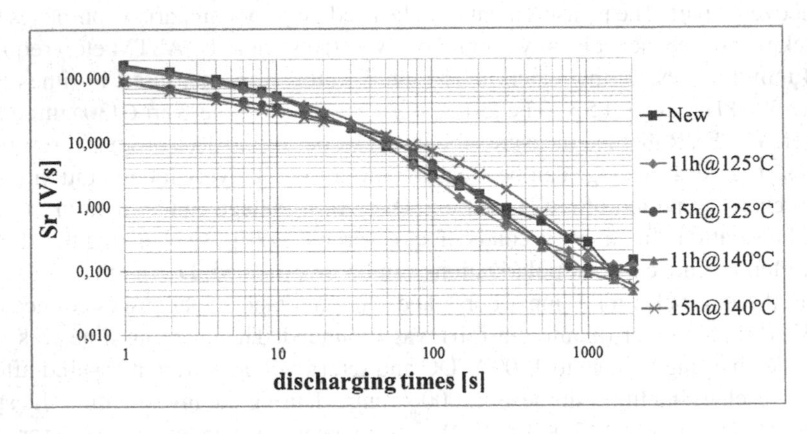


**Fig. 2.** The slopes of the return voltages ($S_r$) as the function of shorting times

The result shows that each curve has a certain cut-off point that shifts to the right by ageing. The order from smaller time constants to the faster: New, 11 h @ 125 °C, 11 h @ 140 °C, 15 h @ 125 °C, 15 h @ 140 °C. This also suggests the reduction of the plasticizer content of the jacket.

The tan δ as the function of the frequency can be seen in Fig. 3. The curve of 'New' cable suggests that the peak of the tan δ is over 500 kHz. By ageing, the peaks shift to the lower frequency range. This also suggests the changing of the material structure of the jacket.

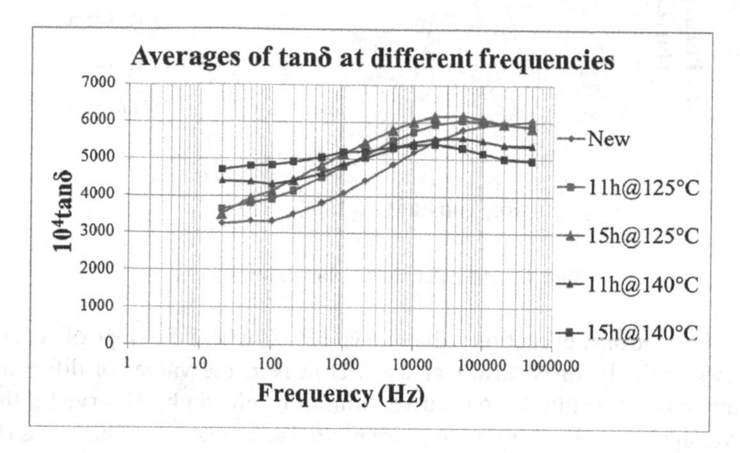


**Fig. 3.** Tan δ as the function of frequency

## 5 Discussion of the Results

Previous studies have shown that conductivity and the slow polarization processes above 1 s time constant are strongly influenced by the plasticizer content of the PVC [12]. The slow polarization processes are related to conduction processes e.g. space charge polarization [13]. The thermal degradation of PVC increases [14], however the loss of plasticizer content reduces the conductivity that is directly proportional to the initial steepness of decay voltage ($S_d$). The slope of return voltage after short (1…3 s) shorting times is directly proportional to the polarization conductivity of slow polarization processes [10].

All of the measurements indicate that the short term but relatively high temperature ageing cycles initiate deterioration processes in the material structure of the jacket.

The shifting of the cut-off point to longer shorting times (Fig. 2) suggests that the most intensive polarization processes become slower. In order to prove this assumption, using an iteration algorithm, the polarization spectra have been calculated from the results of the EVR measurements [8, 9]. Figure 4 shows the calculated spectra.

The result shows that in case of new cables the peak of the most intensive polarization process can be found at 12.6 s. By ageing, the peaks shift towards the higher time constants. The highest time constant peak can be observed at 79.4 s on the samples that were exposed to most intensive ageing (15 h @ 140 °C). Moreover, the shifting of the peaks in the polarization spectrum follows the same order as the changing of the cut-off points in Fig. 2.

The relation between the hardness and the slope of decay voltage is plotted in Fig. 5a. The result shows lower slope of decay voltage (conductivity) correspond to increased

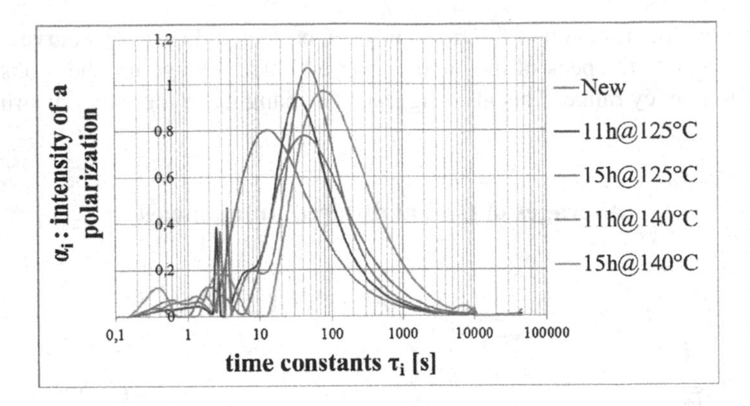


**Fig. 4.** The calculated polarization spectra

hardness values the only exception is the 11 h @ 125 °C where the slope of decay voltage increased while the hardness also increased. However, the values of different ageing temperatures cannot be fitted to one curve. Similar result can be observed if the slopes of return voltages after 1 s shorting time are plotted as a function of hardness (Fig. 5b).

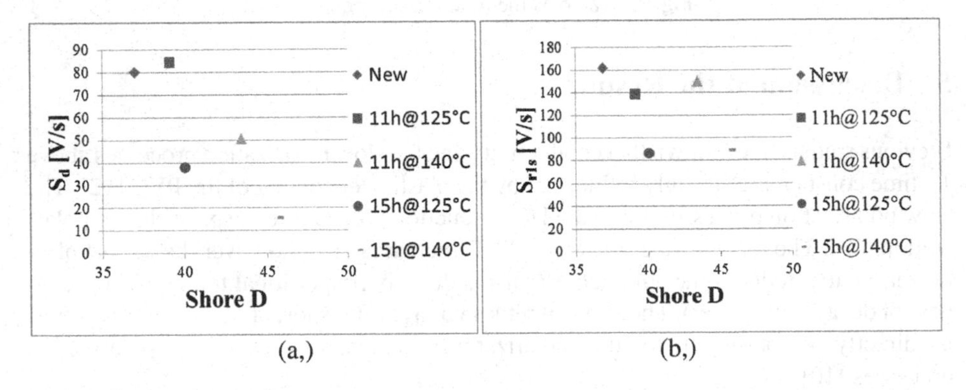


**Fig. 5.** Slope of decay voltage ($S_d$) versus Shore D hardness (a,) Slope of return voltage after 1 s shorting time (Sr1) versus Shore D hardness (b,)

The general trend of measuring lower slope of decay voltages by increasing degree of thermal ageing indicates that the dominant process is not the dehydrochlorination but the loss of plasticizer from the PVC. In case of the 11 h @ 125 °C result probably another, faster chemical process (e.g. decomposition of compound) has an effect on the conductivity and later the effect of plasticizer loss becomes dominant, however proving this assumption needs further investigation.

Summing up the results of the study, it can be stated that after the ageing the jacket was not in seriously degraded condition, nevertheless both the mechanical and electrical tests prove the evaporation of plasticizer from the jacket. Moreover, the correlation between the mechanical and dielectric measurements proposes the application of non-destructive dielectric measurements for testing both mechanical and electrical condition of the cable jacket. This would be really useful in practice since the cables are laid under the ground therefore determining the mechanical condition is practically impossible, however by means of EVR method both mechanical and electrical degradations could be investigated.

Although the results are promising, to determine a threshold value for increased probability of failure more investigation is required involving the bending test which is generally accepted in nuclear power plant cable testing [15].

## 6 Conclusions

The distributed generation in extreme cases, mostly the photovoltaic systems on hot summer days can cause overloads on low voltage distribution cable lines. These overloads appear when the temperature of the soil is also elevated therefore it can increase the temperature of the core insulation and the jacket over the maximum operation temperature. These temperature peaks can initiate degradation processes in these polymeric components of the cable. The effect of the short term thermal overloads on the jacket of PVC insulated low voltage cables was investigated on cable samples. The samples were exposed to periodic thermal ageing test consisting of short but high temperature cycles. After the ageing, the mechanical and dielectric properties of the jacket were tested by Shore D hardness and voltage response measurements, respectively. The results show increasing hardness and decreasing conductive properties of the jacket moreover these properties show high correlation. The relation between these properties enables tracking the effect of the short term thermal overloads by non-destructive dielectric measurements. These results suggest the loss of the plasticizer additive from the jacket material deteriorating the mechanical properties of the jacket. In this condition the jacket is harder and ruptures and cracks can easily appear on it. Therefore, it will not be able to protect the inside of the cable from the environmental stresses e.g. water ingression increasing the probability of the cable failures, which increases the operation risk of low voltage distribution network.

**Acknowledgement.** This research work was supported by the ÚNKP-16-3-II. New National Excellence Program of the Ministry of Human Capacities.

## References

1. Kruizinga, B., Wouters, P.A.A.F., Steennis, E.F.: Fault development on water ingress in damaged underground low voltage cables with plastic insulation. In: 2015 IEEE Electrical Insulation Conference (EIC), pp. 309–312, June 2015
2. Höning, N., Jong, E.D., Bloemhof, G., Poutré, H.L.: Thermal behaviour of low voltage cables in smart grid-related environments. In: IEEE PES Innovative Smart Grid Technologies, Europe, pp. 1–6, October 2014
3. Armendariz, M., Babazadeh, D., Nordstrom, L., Barchiesi, M.: A method to place meters in active low voltage distribution networks using BPSO algorithm. In: 2016 Power Systems Computation Conference (PSCC), pp. 1–7, June 2016

4. Ramezani, S., Höning, N., Poutré, H.L.: Fast and revenue-oriented protection of radial LV cables with smart battery operation. In: 2013 IEEE Computational Intelligence Applications in Smart Grid (CIASG), pp. 107–114, April 2013
5. Kadurek, P., Cobben, J.F.G., Kling, W.L.: Overloading protection of future low voltage distribution networks. In: 2011 IEEE Trondheim PowerTech, pp. 1–6, June 2011
6. Trichakis, P., Taylor, P.C., Lyons, P.F., Hair, R.: Predicting the technical impacts of high levels of small-scale embedded generators on low-voltage networks. IET Renew. Power Gener. **2**(4), 249–262 (2008)
7. Ekelund, M., Edin, H., Gedde, U.: Long-term performance of poly (vinyl chloride) cables. part 1: mechanical and electrical performances. Polym. Degrad. Stab. **92**(4), 617–629 (2007)
8. Tamus, Z.Á., Csábi, D., Csányi, G.M.: Characterization of dielectric materials by the extension of voltage response method. J. Phys: Conf. Ser. **646**(1), 012043 (2015)
9. Csányi, G.M., Tamus, Z.Á., Iváncsy, T.: Investigation of dielectric properties of cable insulation by the extended voltage response method. In: 2016 Conference on Diagnostics in Electrical Engineering (Diagnostika), pp. 1–4, September 2016
10. Németh, E.: Proposed fundamental characteristics describing dielectric processes in dielectrics. Periodica Polytech. Electr. Eng. **15**(4), 305–322 (1971)
11. Németh, E.: Measuring voltage response: a non-destructive diagnostic test method of hv insulation. IEE Proc. Sci. Measur. Technol. **146**(5), 249–252 (1999)
12. Nagy, A., Tamus, Z.Á.: Effect of dioctyl phthalate (DOP) plasticizing agent on the dielectric properties of PVC insulation. In: 2016 Conference on Diagnostics in Electrical Engineering (Diagnostika), pp. 1–4, September 2016
13. Raju, G.G.: Dielectrics in Electric Fields. 19. CRC Press, Boca Raton (2003)
14. Tamus, Z.Á., Németh, E.: Condition assessment of pvc insulated low voltage cables by voltage response method. In: International Conference on Condition Monitoring and Diagnosis, CMD 2010, pp. 721–724 (2010)
15. IEEE standard for qualifying electric cables and splices for nuclear facilities - red- line. IEEE Std 383-2015 (Revision of IEEE Std 383-2003) - Redline, pp. 1–50, October 2015

# A Hybrid Anti-islanding Method for Inverter-Based Distributed Generation

Ebrahim Rokrok[1], Miadreza Shafie-khah[1], Hamid Reza Karshenas[2], Esmail Rokrok[3], and João P.S. Catalão[1,4,5](✉)

[1] C-MAST, University of Beira Interior, 6201-001 Covilhã, Portugal
{ebrahim.rokrok,miadreza}@ubi.pt, catalao@fe.up.pt
[2] Isfahan University of Technology, Isfahan, Iran
karshen@cc.iut.ac.ir
[3] University of Lorestan, Khorramabad, Iran
rokrok.e@lu.ac.ir
[4] INESC TEC and the Faculty of Engineering of the University of Porto, 4200-465 Porto, Portugal
[5] INESC-ID, Instituto Superior Técnico, University of Lisbon, 1049-001 Lisbon, Portugal

**Abstract.** Nowadays, high penetration of Distributed Generations (DG)s in power systems caused some protection issues. One of these issues is unintentional islanding. As regards IEEE 1547 standard, this situation must be recognized immediately, and DG must be separated from the load in less than 2 s. In this paper, to detection of islanding in an inverter-based distributed generation, a new hybrid method with high performance is proposed. In the proposed method, a primary detection of islanding is conducted by measuring the voltage harmonic distortion at the Point of Common Coupling (PCC), as well as comparing the variations to a specified threshold level. After this primary detection, a temporary reactive current signal is injected to the PCC by the inverter of DG, and its terminal voltage and frequency are measured. In the case of deviation of voltage or frequency from permissible range, definitive detection of islanding is determined. Simulation results indicate the efficiency and accuracy of the proposed detection method in different circumstances, especially for loads with the different quality factors.

**Keywords:** Inverter-based distributed generation · Islanding · Non-detection zone · Quality factor · Total harmonic distortion

## 1 Introduction

Due to the technical, economic and environmental advantages, integration of distributed generation units to power systems is growing increasingly. It seems that in the future, we will face with complex grids with a significant number of DGs [1].

In this regard, analysis of the issues and problems related to the connection of DGs is very important. One of these issues is unintentional islanding. Unintentional islanding happens when the switch of grid side is disconnected without a prior plan and in an unforeseen way, and subsequently, the DG has to feed the local loads alone.


L.M. Camarinha-Matos et al. (Eds.): DoCEIS 2017, IFIP AICT 499, pp. 259–266, 2017.
DOI: 10.1007/978-3-319-56077-9_25

According to IEEE 929-1988 standard, the DG must be disconnected immediately from the load once it islanded [2]. So, it is crucial to immediately detect islanding state and then separate the DG from the load.

Generally, there is two main categories for detection methods: (1) local methods (2) remote methods. Local methods are classified to passive methods, active methods and hybrid methods. Remote methods are also divided into two general methods: at the utility level and communication-based methods.

In passive methods, the detection procedure is based on the measurement of variables such as frequency, voltage amplitude and harmonic distortion at the PCC. Passive techniques are low cost and have short operating time. These methods have no effect on grid's power quality. Various passive methods including 1- Voltage and frequency protection [3], 2- voltage and current harmonics detection [4], 3- Voltage phase jump detection [5], 4- Comparing the rate of change of frequency [6], 5- Voltage unbalance [7], 6- Output power change rate [4], 7- Rate of change of frequency over power [8] have been studied. The main problem of these methods is that they have Non-Detection Zones (NDZs). NDZ is a region in which the islanding detection technique cannot detect islanding state [9]. This limitation of passive schemes is solved by using active schemes.

The main idea of an active scheme is that when a DG is islanded, a small disturbance will lead to a substantial change in system parameters, while the change are small when the DG is connected to the grid [10]. The active methods including: 1- Impedance measurement [6], 2- Detection of impedance at specific frequency [6], 3- Slip mode frequency shift [6], 4- Frequency bias or active frequency drift [6], 5- Sandia frequency and voltage shift [11], 6- Frequency jump [6] can be noticed. Active methods have small NDZ. Instead, they have longer operating time than passive methods and have a negative effect on grid's power quality [9].

For a more accurate detection of islanding, the hybrid methods (both active and passive method) are used. These methods take advantage of the two-stage detection process. First, a passive method can be used as the primary diagnoses. After that, an active method is applied as the passive scheme became suspects to the islanding [12]. Hybrid methods including 1- Method based on voltage unbalance and positive feedback, 2- Method based on the shift of the voltage and reactive power [4, 10] can be cited.

In this paper, a hybrid detection method with high accuracy is proposed to detect the islanding state in an inverter-based DG. In this method, the primary detection is done by measuring the voltage harmonic distortion at the PCC and comparing its variation with a threshold level. Then, for a definitive detection of islanding, a reactive current signal is injected by the inverter of the DG to the PCC, and its voltage and frequency are measured. If the deviation of voltage and frequency passes the allowable limits, islanding is detected. Simulation results indicate the efficiency of the suggested method in different conditions.

## 2 Relationship to Smart Systems

The environmental issues besides increasing concern for traditional energy resources lead to increasing concentrations on distributed generation based on renewables. The grow of DG penetration level in the future smart grids leads to increasing need for high-performance anti-islanding protection methods [12]. Therefore, anti-islanding protection is so important that specific capabilities and specifications for anti-islanding are required in the countries with a developed power grid system [13].

To connect the DGs to the grid, there are some conditions that must be considered. Standard institutions have published these conditions. On of the very significant conditions which are required to distributed generators is detecting the islanding state.

Unintentional islanding, because of loss of synchronism, can have undesirable effects on power system stability [14]. It is likely to damage the electrical appliances and equipment in the island. It is also dangerous for personnel or people. So, in modern smart grids, it is necessary to use effective and reliable methods to detect islanding conditions [15]. Numerous methods have been proposed to detect unintentional islanding so far. These methods are different in terms of cost, efficiency, execution speed and power quality issues. This paper, proposes an effective anti-islanding detection method to the goal of reliable operation in future smart grids.

## 3 Islanding Detection by Reactive Current Injection

Figure 1 depicts the inverter-based DG and the local load at the PCC. When islanding occurs, the Circuit Breaker (CB) opens, and DG alone feeds the local load.

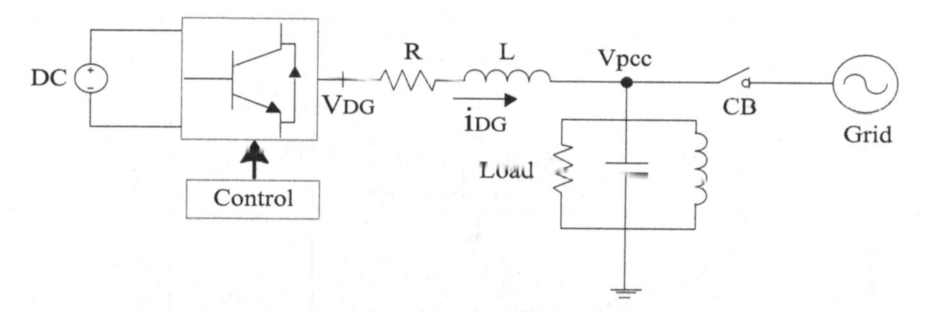


**Fig. 1.** DG with the local load connected to the grid.

The basis of islanding detection in reactive current injection method is making the reactive flow component in DG's output current through the control system and measuring the frequency and voltage at the PCC. To this end, a reactive triangular current signal with the frequency of 20 Hz can be used. The amplitude of this signal is considered as 10% of DG's rated current.

The voltage and frequency deviation lay outside of the standard ranges if the local load and DG's output power is not matched with each other. In this case, islanding can be diagnoses easily by voltage and frequency relays. However, if the DG's output power matches with the local load, frequency and voltage deviation are so small that relays can't detect islanding. In this case, injection of reactive current causes the frequency or voltage will be outside of the standard ranges and islanding is detected by voltage and frequency relays.

## 4 Proposed Method

For improving power quality in reactive current injection method, it can be combined with a passive scheme. To detect islanding, first the passive scheme is used, and if it suspects of islanding, reactive current is injected. The reactive current injection will continue for a short time. If voltage or frequency was out of range during this period, the islanding has been detected, and DG will be separated. Otherwise injecting reactive current is interrupted and passive method is re-applied. So, DG does not have to inject reactive current continuously, and thus power quality is improved.

The passive method that used in this paper is measuring of the voltage Total Harmonic Distortion (THD) at the PCC. In the control system of inverter, the d-q components of the voltage are available and the voltage THD is obtained from these components [16].

In the suggested technique, voltage THD is sampled and is stored. The number of samples is fixed, and by adding a new sample in every cycle, the oldest sample is discarded. An average of samples is calculated by a moving window. The new sample will be compared with the average of previous samples. If the difference between these two is greater than a defined threshold level, the primary diagnosis of islanding is made, and then DG injects the reactive current. The flowchart of Fig. 2 shows islanding detection procedure by proposed method.

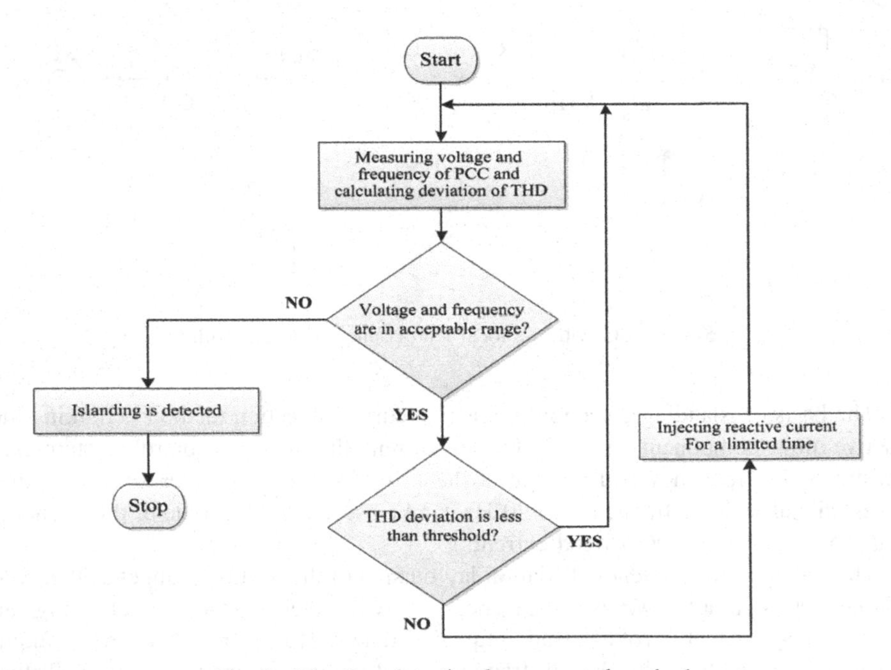


**Fig. 2.** Islanding detection by suggested method

## 5 Simulation Results

To demonstrate the usefulness of the suggested method, a DG is connected to the local load and the utility grid (Fig. 1) is simulated in the Matlab-Simulink. The inverter is controlled in the PQ control mode. The modeling detail is presented in [3]. The characteristics of load, system and DG are specified in Table 1.

**Table 1.** Test system characteristics.

| | | |
|---|---|---|
| Grid | Nominal voltage | 381 V |
| | Frequency | 60 Hz |
| DG | Rated power | 60 kW |
| | Rated voltage | 381 V |
| | Switching frequency | 5 kHz |
| | DC link | 800 V |
| | Coupling reactance | 2.5 mH |
| Load | High-quality factor | Q = 2.3 |
| | Low-quality factor | Q = 0.2 |

Simulations results are given for three different states. In these simulations, the DG's output power is considered equal to the local load. In the other words, the worst loading conditions are considered. According to IEEE 929-2000 standard, allowable range of frequency is $59.3\ \text{Hz} < f_{pcc} < 60.5\ \text{Hz}$ and allowable voltage range is $0.88$ p. u. $< V_{pcc} < 1.1$ p.u. that by taking effective amount 220 V for phase voltage, it will be $193.6\ \text{V} < V_{pcc} < 242\ \text{V}$.

### 5.1 Case 1: Islanding Detection by Permanent Injection of Reactive Current for a Load with a Low-Quality Factor

In the case of accordance of DG's real power and loads with high/low-quality factor, islanding is not detected based on voltage and frequency measurement. So, the triangular reactive current with the frequency of 20 Hz and the amplitude of 10% of the DG rated current is continuously injected. Figure 3 shows the frequency and voltage quantity at PCC. At t = 1 (s), the grid is disconnected, and islanding occurs.

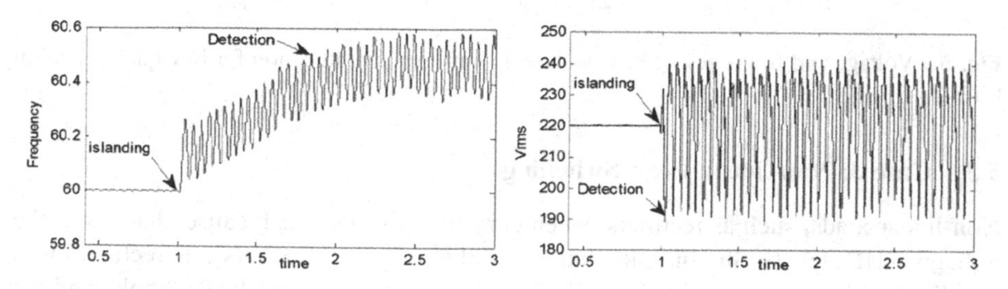


**Fig. 3.** Voltage and frequency of PCC with permanent reactive injection for low-quality load

As can be seen from these figures, frequency and voltage range are excluded from standard limit and islanding is detected. For the high-quality factor load, by decreasing the amplitude of injected reactive current up to 5% of the DG nominal current, detection can be done. But in any case, continuous injection of reactive current can reduce the system's power quality.

## 5.2 Case 2: Evaluation of Islanding Detection with the Proposed Method

In Fig. 4, the average of THD of the voltage at PCC (average of 50 samples with 1 kHz sampling rate) and reactive current injected are depicted. At t = 1 (s), the grid is disconnected, and islanding happens. At t = 1.02 (s), THD changes will be higher than the threshold level (3%), and the primary detection is done. Then the reactive current is injected. Figure 5 illustrates the voltage and frequency of PCC. It is observed that the voltage is out of range in a short time and islanding is detected. Current injection is stopped, and DG must be separated.

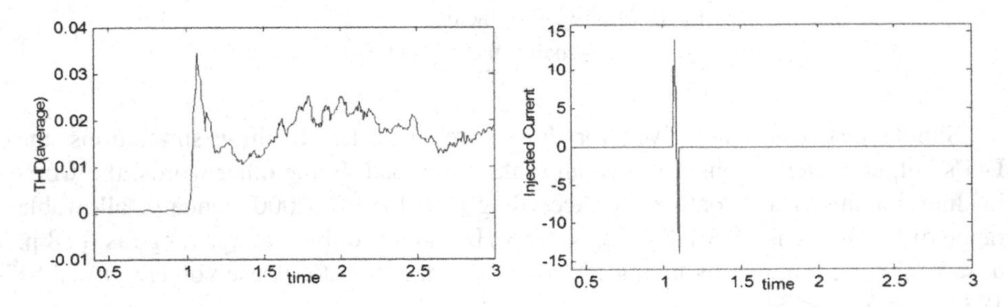


**Fig. 4.** The average of THD of the voltage at PCC and injected reactive current in Case 2.

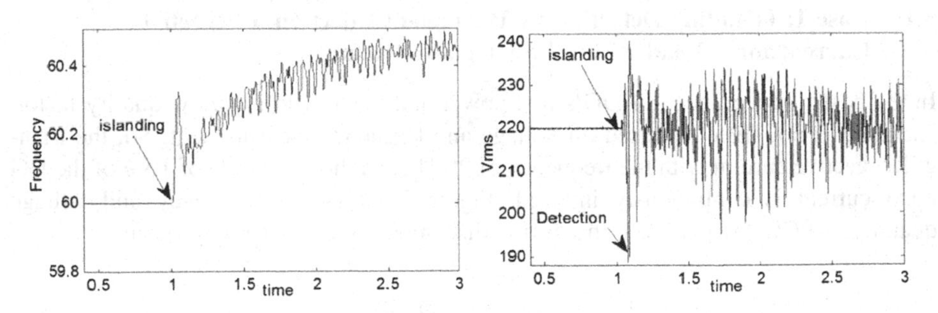


**Fig. 5.** Voltage and frequency of PCC with temporary reactive injection for low-quality load in Case 2.

## 5.3 Case 3: Non-linear Load Switching

Non-linear loads, such as rectifiers, when connecting to PCC can cause changes in the voltage THD levels. In this simulation, at the moment t = 1 (s), a rectifier as a non-linear load is connected to PCC. THD changes are more than the threshold, and the primary detection of islanding is done. Then the reactive current is injected within 0.1 s (Fig. 6). Figure 7 shows the frequency and voltage of PCC. It is observed that the voltage and frequency will not be out of allowable range. So DG is not islanded, and injection of reactive current will be stopped, and DG will continue to work.

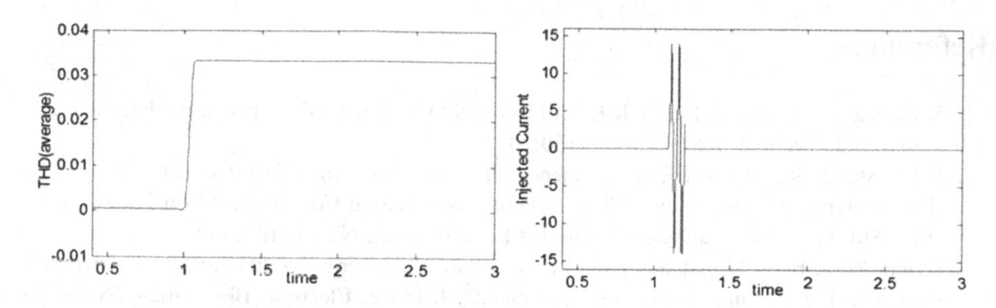


**Fig. 6.** The average of THD of the voltage at PCC and injected reactive current in Case 3.

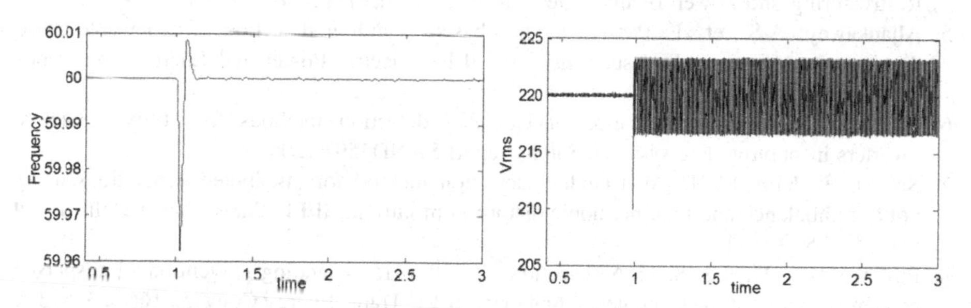


**Fig. 7.** Voltage and frequency of PCC with temporary reactive injection for low-quality load in Case 3.

# 6 Conclusions

In connection of DGs to power systems, unintentional islanding is one of the important problems that must be investigated. According to existing standards, this phenomenon should be immediately diagnosed. Grid and load characteristics can determine factors for selection of an islanding detection method. This paper suggested a hybrid approach to detection of islanding. In the beginning, the primary detection of the islanding was done by measuring the THD of voltage at the PCC and comparing its variation to a threshold level. After confirmation of primary detection, for definite detection, a reactive current signal was injected to the PCC and voltage, and frequency deviations were measured at this point. Simulation results showed the efficiency of the suggested technique especially for the loads with non equal quality factor. Likewise, the switching of non-linear loads and correct diagnosis of non-islanding was simulated.

**Acknowledgment.** This work was supported by FEDER funds through COMPETE 2020 and by Portuguese funds through FCT, under Projects SAICT-PAC/0004/2015 -

POCI-01-0145-FEDER-016434, POCI-01-0145-FEDER-006961, UID/EEA/50014/2013, UID/CEC/50021/2013, and UID/EMS/00151/2013. Also, the research leading to these results has received funding from the EU Seventh Framework Programme FP7/2007-2013 under grant agreement no. 309048.

## References

1. Shrivastava, S., Jain, S., Nema, R.K.: A proposed hybrid method for islanding detection. Int. J. Eng. Sci. Technol. **2**(5), 813–817 (2010)
2. IEEE Std. 929-2000, IEEE Recommended Practice for Utility Interface of Photovoltaic (PV) Systems, Sponsored by IEEE Standards Coordinating Committee 32 on Photovoltaic, IEEE Std. 929-2000, Published by the IEEE, New York, NY, April 2000
3. Byung-Yeol, B., et al.: Islanding detection method for inverter-based distributed generation systems using a signal cross-correlation scheme. J. Power Electron. **10**(6), 762–768 (2010)
4. Pukar, M., Chen, Z., Bak-Jensen, B.: Review of islanding detection methods for distributed generation. In: Third International Conference on Electric Utility Deregulation and Restructuring and Power Technologies DRPT 2008. IEEE (2008)
5. Aljankawey, A.S., et al.: Passive method-based islanding detection of renewable-based distributed generation: the issues. In: 2010 IEEE Electric Power and Energy Conference (EPEC). IEEE (2010)
6. Ward, B. Ropp, M.: Evaluation of islanding detection methods for utility-interactive inverters in photovoltaic systems. Sandia report SAND3591 (2002)
7. Sung-Il, J., Kim, K.-H.: An islanding detection method for distributed generations using voltage unbalance and total harmonic distortion of current. IEEE Trans. Power Delivery **19** (2), 745–752 (2004)
8. Fu-Sheng, P., Huang, S.-J.: A detection algorithm for islanding-prevention of dispersed consumer-owned storage and generating units. IEEE Trans. Energy Convers. **16**(4), 346–351 (2001)
9. Etxegarai, A., Eguía, P., Zamora, I.: Analysis of remote islanding detection methods for distributed resources. In: International Conference Renewable Energies Power Quality (2011)
10. Chandrakar, C.S., Dewani, B., Chandrakar, D.: An assessment of distributed generation islanding detection methods. Int. J. Adv. Eng. Technol. **5**(1), 218–226 (2012)
11. Kunte, R.S., Wenzhong, G.: Comparison and review of islanding detection techniques for distributed energy resources. In: Power Symposium (2008)
12. Teoh, W.Y., Chee, W.T.: An overview of islanding detection methods in photovoltaic systems. World Acad. Sci. Eng. Technol. **58**, 674–682 (2011)
13. Boroojeni, K.G., Hadi Amini, M., Iyengar, S.S.: Overview of the Security and Privacy Issues in Smart Grids. Smart Grids: Security and Privacy Issues, pp. 1–16. Springer International Publishing, Heidelberg (2017)
14. Shahab, B., Sheikhi, A.: From demand response in smart grid toward integrated demand response in smart energy hub. IEEE Trans. Smart Grid **7**(2), 650–658 (2016)
15. Arasteh, H., et al.: Iot-based smart cities: a survey. In: 2016 IEEE 16th International Conference on Environment and Electrical Engineering (EEEIC). IEEE (2016)
16. Salem, R., et al.: A combination of shunt hybrid power filter and thyristor-controlled reactor for power quality. IEEE Trans. Industr. Electron. **61**(5), 2152–2164 (2014)

# Energy: Optimization

# A New DG Planning Approach to Maximize Renewable - Based DG Penetration Level and Minimize Annual Loss

Soroush Najafi[1], Miadreza Shafie-khah[1], Neda Hajibandeh[1], Gerardo J. Osório[1], and João P.S. Catalão[1,2,3](✉)

[1] C-MAST, University of Beira Interior, 6201-001 Covilhã, Portugal
{soroush.najafi,miadreza,neda.hajibandeh}@ubi.pt,
gjosilva@gmail.com

[2] INESC TEC and the Faculty of Engineering of the University of Porto, 4200-465 Porto, Portugal
catalao@fe.up.pt

[3] INESC-ID, Instituto Superior Técnico, University of Lisbon, 1049-001 Lisbon, Portugal

**Abstract.** Distributed Generation (DG) using renewable technologies is increasing due to their benefits including energy security and emission reduction. However, installing new DGs in distributed networks is limited due to network constraints such as feeder capacity and short circuit level, as well as higher investment costs. In this paper, network reconfiguration and reactive power planning are used to maximize DG penetration level and to minimize annual loss for DGs with biomass technologies. In order to model the problem uncertainties, 96 scenarios considering ten different network load levels are studied. A multi-objective method is applied for solving this optimization problem by using Pareto front. The numerical results indicate the positive impacts of the proposed approach on improving the network security.

**Keywords:** Annual loss · Biomass · DG penetration · Network reconfiguration · Pareto front · Reactive power planning · Renewable energy

## 1 Introduction

Environmental concerns, such as global warming causes that renewable energy, e.g., biomass, play an important role in supplying energy [1]. The planning of distribution systems is a critical issue that the system planners faced with [2]. In one hand, appropriate placement of DGs in the distribution systems has an essential role in the enhancement of the system efficiency; consequently finding the optimum placement of DGs is critical [3]. On the other hand, a large number of researchers have focused on the reactive power planning in the distribution system, especially when it has have high-penetration of renewable resources DG [4].

In this sense, network reconfiguration is a method to change the topology of branches through altering the open and close statuses of sectionalizing and tie switches [3]. This can be carried out in different seasons of a year to reduce the total loss and increase the system reliability.


L.M. Camarinha-Matos et al. (Eds.): DoCEIS 2017, IFIP AICT 499, pp. 269–276, 2017.
DOI: 10.1007/978-3-319-56077-9_26

For instance, as reported in [4], a DG planning was proposed for the coordination of reconfiguration of feeders and voltage controlling with the goal of maximizing the DG capacity at a specific bus.

In the current work, it is planned biomass DG for four different scenarios. In the first scenario, the annual loss is minimized without considering network penetration. The second scenario represents minimized annual loss considering network penetration. In the third scenario, network penetration is maximized, and finally, in the fourth scenario, a multi-objective problem is provided that minimizes the annual loss and maximizes the network penetration.

## 2 Relationship to Smart Systems

Distributed Generation (DG) of electricity using renewable technologies is increasing due to their benefits including energy security and emission reduction [5]. Smart grid can be defined as a contemporary structure of electricity systems to improve effectiveness, reliability, and safety, via integration of sustainable energies, as well as automated and intelligent controlling and advanced communications technologies [6, 7]. It affects entire parts of power systems, and consequently it needs a widespread data communication structure [8].

Importantly, smart grid allows modern network controlling schemes and provides operative integrated DGs in the demand side [9, 10]. Smart Grid will have a significant role to support high penetration level of DGs; nevertheless, the existing standard/code leading DGs' interconnection does not allow implementing various applications that can be advantageous for the system [11]. Novel approaches for utilizing the data in the supervisory controllers would be required, as well as an environment in which tools could be emerged in online stream event processing.

## 3 PSO and MOPSO Algorithm

The particle swarm optimization (PSO) applied in optimization techniques; it was inspired from the movement migration of population species that lives grouped [12]. The elementary philosophy behind PSO is that each individual from the swarm, hereafter called as particle, uses the information from the swarm, based on the knowledge, sharing, and historical information to find its survival means. Computationally, PSO process starts with a random swarm of particles delimited from delimited universe, where each particle, from the movement process, will be a momentary solution of the problem with tuned fitness rate. The aforementioned fitness is computed and optimized in each iteration. The movement rule, defined by a position and velocity, defines the direction, speed, and position of the swarm to reach the optimal and are updated in each iteration. In this process, the particles interact with each other and them they move from the delimited universe till reach the best solution (the optimal from the optimization process). Both, position and velocity are updated them from the experience from the past position and velocity and the best position found from the swarm. Computationally, the aforementioned description can be expressed:

$$X_p^{iter+1} = X_p^{iter} + V_p^{iter} \tag{1}$$

$$V_P^{iter+1} = V_p^{iter} + Rand_1\left(X_{p,best}^{iter} - X_p^{iter}\right) + Rand_2\left(X_{g,best}^{iter} - X_p^{iter}\right) \quad p = 1, 2, \ldots, N_p \tag{2}$$

Where $X_p^{iter}$ is the last position of a particle *p* from iteration *iter* that may express the best result to the optimization problem. By other words, and specifically in DG optimization problem, $X_p^{iter}$ is a set of DG decision variables and $V_p^{iter}$ is the velocity from particle *p* from iteration *iter*. Moreover, $X_{g,best}^{iter}$ is the best global position *g* found from the swarm in iteration *iter*. So, the movement direction of the swarm is dependent by the influence, respectively, from $V_p^{iter}$, and $X_{g,best}^{iter}$ [13]. By other words, the swarm movement is strictly influenced by the inertia, memory and cooperation of the swarm, expressed by the three term of (2). However, for the multi-objective problem of DG optimization proposed in this work the original PSO strategy is modified. The original goal to find a unique solution is then turned in one new array of multiple solutions, based on Pareto strategy to solve the multi-objective DG problem [14, 15].

## 4 Formulation

The planning model aims the minimization from the yearly energy loss and maximizes network penetration in the distribution system for all possible loads. Time segments are represents in sets of 90 h (i.e., the number of days for each season). Moreover, in (3), *g* represents as load states, and $P_{loss}(g)$ represents energy loss at $(g)$ load state. Furthermore, in (4), $C_m$ represents biomass capacity installed at each candid bus $(CB)$.

$$Minimize\ E_{Loss} = \sum_{g=1}^{N} P_{loss}(g) * 90 \tag{3}$$

$$Maximize\ Penetration = \sum_{m} C_m m \in CB \tag{4}$$

Subject to:
Active and Reactive power balance:

$$P_{DGB,i} - LP(g) * P_{D,i} = \sum_{j=1}^{n} V_{g,i} * V_{g,j} * Y_{ij} * Cos(\theta_{ij} + \delta_{g,j} - \delta_{g,i}), \quad i \in NBus \tag{5}$$

$$-LP(g) * Q_{D,i} = -\sum_{j=1}^{n} V_{g,i} * V_{g,j} * Y_{ij} * Sin(\theta_{ij} + \delta_{g,j} - \delta_{g,i}), i \in NBus \tag{6}$$

In (5) and (6), *LP*(*g*) represent load peak level that shown in Table 1. $V_{g,i}$ represents voltage magnitude at load level *g*. $P_{D,i}$ and $Q_{D,i}$ represent active and reactive demands, respectively.

**Table 1.** Load profile for different hours

| Time | Winter | Spring | Summer | Autumn |
|---|---|---|---|---|
| 12:00–1:00 am | 0.4756 | 0.3970 | 0.64 | 0.3717 |
| 1:00–2:00 | 0.4473 | 0.3906 | 0.60 | 0.3558 |
| 2:00–3:00 | 0.4260 | 0.3780 | 0.58 | 0.3540 |
| 3:00–4:00 | 0.4189 | 0.3654 | 0.56 | 0.3422 |
| 4:00–5:00 | 0.4189 | 0.3717 | 0.56 | 0.3481 |
| 5:00–6:00 | 0.4260 | 0.4095 | 0.58 | 0.3835 |
| 6:00–7:00 | 0.5254 | 0.4536 | 0.64 | 0.4248 |
| 7:00–8:00 | 0.6106 | 0.5355 | 0.76 | 0.5015 |
| 8:00–9:00 | 0.6745 | 0.5958 | 0.87 | 0.5605 |
| 9:00–10:00 | 0.6816 | 0.6237 | 0.95 | 0.5841 |
| 10:00–11:00 | 0.6816 | 0.6300 | 0.99 | 0.5900 |
| 11:00–12:00 | 0.6745 | 0.6237 | 1.00 | 0.5841 |
| 12:00–1:00 am | 0.6745 | 0.5859 | 0.99 | 0.5487 |
| 1:00–2:00 | 0.6745 | 0.5796 | 1.00 | 0.5428 |
| 2:00–3:00 | 0.6603 | 0.5670 | 1.00 | 0.5310 |
| 3:00–4:00 | 0.6674 | 0.5544 | 0.97 | 0.5192 |
| 4:00–5:00 | 0.7029 | 0.5670 | 0.96 | 0.5310 |
| 5:00–6:00 | 0.7100 | 0.5796 | 0.96 | 0.5428 |
| 6:00–7:00 | 0.7100 | 0.6048 | 0.93 | 0.5664 |
| 7:00–8:00 | 0.6816 | 0.6174 | 0.92 | 0.5782 |
| 8:00–9:00 | 0.6461 | 0.6048 | 0.92 | 0.5664 |
| 9:00–10:00 | 0.5893 | 0.5670 | 0.93 | 0.5310 |
| 10:00–11:00 | 0.5183 | 0.5040 | 0.87 | 0.4720 |
| 11:00–12:00 am | 0.4473 | 0.4410 | 0.72 | 0.4130 |

Slack Bus Constraints:

$$V_{g,1} = 0 \tag{7}$$

$$\delta_{g,1} = 0 \tag{8}$$

Voltage Constraints at the Other Buses:

$$V_{min} \leq V_{g,i} \leq V_{max} \tag{9}$$

where $V_{min}$ and $V_{max}$ are 0.95p.u and 1.05 p.u respectively.

Feeder Capacity Limits:

$$0 \leq I_{g,ij} \leq I_{ij,max} \tag{10}$$

DG Penetration. To this end, it has 2 different type of DG penetration in DG planning. (11) represent DG penetration for candidate bus. This limitation depends on the land and protection level of the candidate bus. In this paper, it is assume this limitation at 1200 kW. (12) limit installed DG capacity in the whole network. This limitation assumes to be 1200 kW too. (12) just consider in the 3rd scenario.

$$P_{DGB,m} \leq 1200\ kWm \in CB \tag{11}$$

$$\sum_{m} C_m m \in CB \tag{12}$$

## 5 Numerical Results of the Model

The PG&E 69bus distribution test case [16] has been considered to implement the proposed model. In normal operation, disregarding DGs, the yearly energy loss is 199.5 MWh. The slack bus of the system is node 1. The candidate bus for installing DG is CB = {20, 46, 49, 50, 53}. Total active and reactive demands are 3802 kW and 2695kVAr, respectively. The 69bus case study is also presented in Fig. 1. Implemented DG in the network is biomass resources with the controllable output. The hourly demand is represented in Table 1 as a proportion of the yearly demand peak. The proposed planning problem is applied to 4 scenarios with different constraint and objective function. Outcomes are represented in Table 2. Based on the results, if it is added DG, penetration constraint to an optimization problem, annual loss would increase.

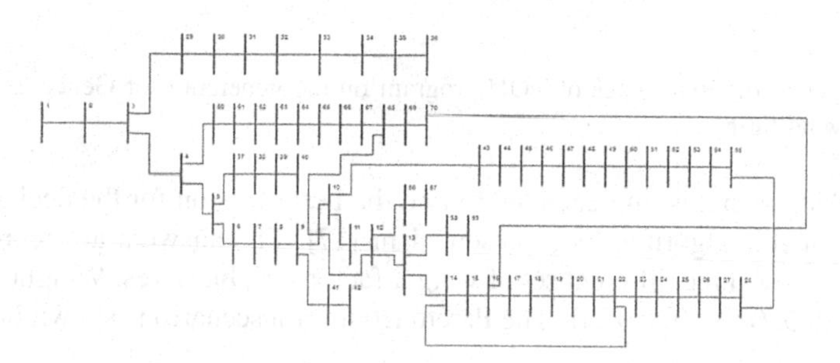

**Fig. 1.** 69bus PG&E single diagram

Another effect of adding this constraint is that it is not possible to control voltage bus within 0.95 to 1.05 per-units (p.u). In this scenario, the buses voltage is varied from 0.9 to 1.1. In the third scenario, it is maximized DG penetration without violation any constraints. In this case voltage limit is $0.95 < V < 1.05$. In the fourth scenario is planning biomass DG with two objective functions. Pareto optimal front found by the algorithm is shown in Fig. 2. In this scenario voltage limit will be in 0.95 to 1.05 p.u.

**Table 2.** DG Planning results for different scenarios

| Candidate bus | Annual loss minimization TA(with consider DG Penetration) | Annual loss minimization (without consider DG penetration) | DG penetration maximization |
|---|---|---|---|
| 20 | 560 | 0 | 1600 |
| 46 | 90 | 0 | 0 |
| 49 | 0 | 0 | 3020 |
| 50 | 480 | 0 | 1160 |
| 53 | 1235 | 1200 | 0 |
| Annual loss (KWh) | 2365 | 1200 | 5760 |
| Total installed capacity | 252120 | 313240 | 388510 |

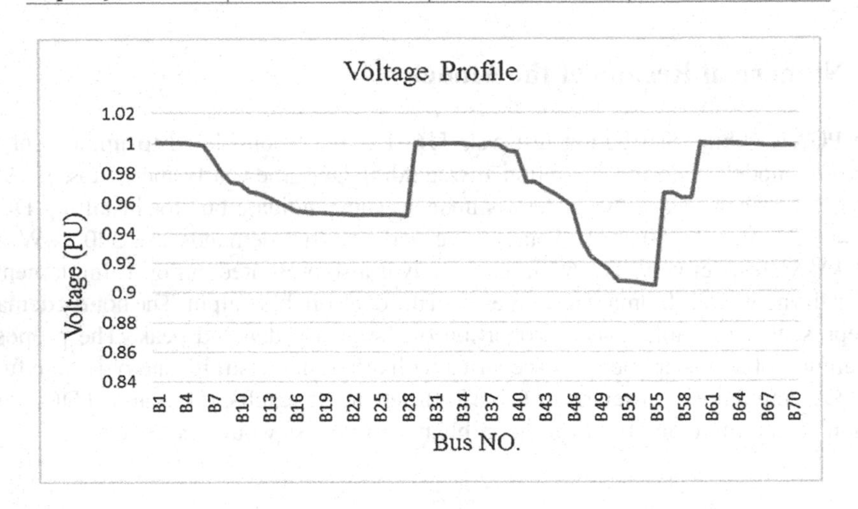


**Fig. 2.** Impact of different types of TOU program on the generation of Genco 1 without the presence of wind farm.

TOPSIS algorithm is implemented to find the best decision for the decision maker. The theory of this algorithm is represented in [17]. The answers are represented in Table 3. It is used three different set of weight factor for objectives. Weight factors are {(0.2, 0.8), (0.5, 0.5), (0.8, 0.2)}. The Pareto front for a scenario is shown in Fig. 3.

**Table 3.** PLANNER best decision based on different weight factors

| Weight factor | Annual loss (kWh) | DG penetration |
|---|---|---|
| (0.2, 0.8) | 258310 | 2770 |
| (0.5, 0.5) | 423140 | 6239 |
| (0.8, 0.2) | 423140 | 6239 |

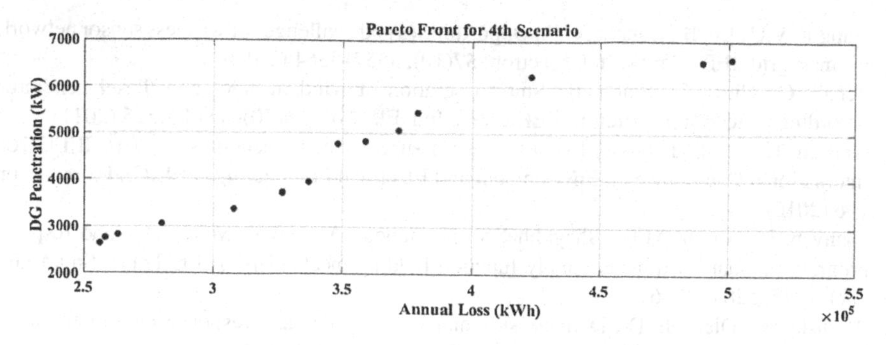


**Fig. 3.** Pareto front for 4th scenarios.

# 6 Conclusions

In this work a multi-objective model was presented for the distributed generation planning and solved the formulated problem with PSO algorithm. DGs were biomass with controllable output, although the proposed strategy could just tackle several types of uncertainty from wind or solar energy, for instance. Annual loss minimization and DG penetration maximization were considered in this paper. After determining the Pareto front, it was used TOPSIS method to choose the best decision for the decision maker. The numerical results showed that considering only the DG penetration constraint increased the annual loss. In addition, by adding this constraint, the voltage bus could not be constrained between desired amounts. However, the proposed approach considering both DG penetration constraint and annual loss could significantly improve the effectiveness of the DG planning.

**Acknowledgment.** This work was supported by FEDER funds through COMPETE 2020 and by Portuguese funds through FCT, under Projects SAICT-PAC/0004/2015 - POCI-01-0145-FEDER-016434, POCI-01-0145-FEDER-006961, UID/EEA/50014/2013, UID/CEC/50021/2013, and UID/EMS/00151/2013. Also, the research leading to these results has received funding from the EU Seventh Framework Programme FP7/2007–2013 under grant agreement no. 309048.

# References

1. Atwa, Y.M., El Saadany, E.F.: Probabilistic approach for optimal allocation of wind based distributed generation in distribution systems. IET Renew. Power Gener. **5**, 79–88 (2011)
2. Arefifar, S.A., Abdel-Rady, Y.I.M.: Probabilistic optimal reactive power planning in distribution systems with renewable resources in grid-connected and islanded modes. IEEE Trans. Ind. Electron. **61**(11), 5830–5839 (2014)
3. Atwa, Y.M., El Sadaany, E.F., Salama, M.M.A., Seethapathy, R.: Optimal renewable resource mix for distribution system energy loss minimization. IEEE Trans. Power Syst. **25**(1), 360–370 (2010)
4. Su, S.-Y., Lu, C.-N., Chang, R.-F., Gutiérrez-Alcaraz, G.: Distributed generation interconnection planning: a wind power case study. IEEE Trans. Smart Grid **2**(1), 181–189 (2011)

5. Gungor, V.C., Lu, B., Hancke, G.P.: Opportunities and challenges of wireless sensor networks in smart grid. IEEE Trans. Ind. Electron. **57**(10), 3557–3564 (2010)
6. Cecati, C., Citro, C., Siano, P.: Smart operation of wind turbines and diesel generators according to economic criteria. IEEE Trans. Ind. Electron. **58**(10), 4514–4525 (2011)
7. Arasteh, H., et al.: IoT-based smart cities: a survey. In: Proceedings of 2016 IEEE 16th International Conference on Environment and Electrical Engineering (EEEIC), Florence, pp. 1–6 (2016)
8. Kamyab, F., Amini, M.H., Sheykhha, S., Hasanpour, M., Jalali, M.M.: Demand response program in smart grid using supply function bidding mechanism. IEEE Trans. Smart Grid **7**(2), 1277–1284 (2016)
9. Palensky, P., Dietrich, D.: Demand side management: demand response intelligent energy systems and smart loads. IEEE Trans. Ind. Inform. **7**(3), 381–388 (2011)
10. Calderaro, V., Hadjicostis, C., Piccolo, A., Siano, P.: Failure identification in smart grids based on Petri net modeling. IEEE Trans. Ind. Electron. **58**(10), 4613–4623 (2011)
11. Bahrami, S., Sheikhi, A.: From demand response in smart grid toward integrated demand response in smart energy hub. IEEE Trans. Smart Grid **7**(2), 650–658 (2016)
12. Couceiro, M., Ghamisi, P.: Fractional Order Darwinian Particle Swarm Optimization Applications and Evaluation of an Evolutionary Algorithm. SpringerBriefs in Applied Sciences and Technology, 2nd edn. Springer, Heidelberg (2016). p. 75
13. Soroudi, A., Afrasiab, M.: Binary PSO-based dynamic multi-objective model for distributed generation planning under uncertainty. IET Renew. Power Gener. **6**(2), 67–78 (2012)
14. Devi, S., Agade, A.K., Dehuri, S.: Comparison of various approaches in multi-objective particle swarm optimization (MOPSO): empirical study. Stud. Comput. Intell. **592**, 75–103 (2015)
15. Santos, S.F., et al.: Novel multi-stage stochastic DG investment planning with recourse. IEEE Trans. Sust. Energy **8**(1), 164–178 (2016)
16. Baran, M.E., Wu, F.F.: Optimal capacitor placement on radial distribution systems. IEEE Trans. Power Deliv. **4**(1), 725–734 (1989)
17. Eldurssi, A.M., O'Connell, R.M.: A fast nondominated sorting guided genetic algorithm for multi-objective power distribution system reconfiguration problem. IEEE Trans. Power Syst. **30**(2), 593–601 (2015)

# Stochastic Optimization for the Daily Joint Operation of Wind/PV and Energy Storage

Isaías L.R. Gomes[1], Hugo M.I. Pousinho[2], Rui Melício[1,2(✉)], and Vitor M.F. Mendes[1,3]

[1] Departamento de Física, Escola de Ciências e Tecnologia, Universidade de Évora, Évora, Portugal
ruimelicio@gmail.com

[2] IDMEC, Instituto Superior Técnico, Universidade de Lisboa, Lisbon, Portugal

[3] Instituto Superior of Engenharia de Lisboa, Lisbon, Portugal

**Abstract.** This paper deals with the problem of optimal bidding in a day-ahead market of electricity for a power producer having joint operation of wind with photovoltaic power systems and storage of energy. Uncertainty, not only on electricity market prices, but also on wind and photovoltaic powers, has to be faced in order to achieve optimal bidding. The problem is viewed as a sort of a two-stage stochastic optimization problem formulated by mix-integer linear programming. A case study with data from the Iberian Peninsula is presented and a comparison between joint and disjoint operations is discussed, allowing concluding that the joint operation attenuates the economic impact of disjoint operation volatility.

**Keywords:** Day-ahead electricity market · Energy storage · Mixed-integer linear programming · Stochastic optimization · Wind with PV powers

## 1 Introduction

Although of the uncertainty on the availability of power, renewable energy sources (RES) exploitations are the main announced measures to reduce anthropogenic greenhouse gases emissions and the way of future in what regards a sustainable development [1]. The exploitation of RES has worldwide grown rapidly, benefiting from sustenance schemes and incentive policies [2]. A main characteristic of the RES exploitation of wind power (WP) and of photovoltaic power (PVP) is uncertainty, which have to be considered in the scheduling of these powers. WP or PVP systems standing alone are not capable of warranting power generation during all day due to the uncertainty in the availability of the source of energy at different time periods of day or different periods of the year. The mismatch between periods of low availability of these sources lead to a question about if complementarity between WP and PVP has a better appropriateness for presenting a single offer in a day-ahead market (DAM). Also, energy storage is a way to harness RES, allowing the store and discharge of energy at adequate conditions. A report of the NREL states that the challenges associated with meeting the variation


L.M. Camarinha-Matos et al. (Eds.): DoCEIS 2017, IFIP AICT 499, pp. 277–286, 2017.
DOI: 10.1007/978-3-319-56077-9_27

in demand while providing reliable services has motivated historical development of the storage of energy and that large penetrations of variable generation rises the need for flexibility options [3]. Large scale RES integration without energy storage is said to be a challenge for future power systems [4]. Hence, storage technologies play an important role in the actual deregulated markets, providing capability for arbitrage, increasing the economic value of RES [5]. Joint operation with the aim of submitting in a DAM a single bid of a producer having WP, PVP and a battery for energy storage is the line of research perused in this paper. Which has as a contribution the application of MILP approaches to two-stage stochastic optimization problems for optimal bid of: WP; PVP and joint operation of WP, PVP and energy storage device.

## 2 Relationship to Smart Systems

Intelligence in the form of algorithms, new computing models, and advanced analytics prone to allows enhancement in decision making [6]. Power system can benefit from this intelligence. A smart system can be stated as an embedded system that incorporates advanced systems and provide the inhabitants with sophisticated monitoring and control over how something happens in the system [7], for example, the management of a wind system, a PV system or a storage device. Smart systems are capable of sensing, making diagnosis, describing, qualifying and managing how something happens in the system, incorporating both technical intelligence and cognitive functions. This incorporation is labeled as the third-generation of smart systems, offering an interface between the virtual and the physical world. Smart systems are highly reliable, often miniaturized, networked, predictive and energy autonomous [8]. The exponential increase in energy demand, that is forecasted to be around 45% by 2030, and the increase integration of RES in power systems are facts driving research and development about managing how things happen in the system [9]. Particularly, energy management will be one of the most urgent topics of the century and a significant driver for research and developments [9]. Important issues in energy management are reliability and efficiency. These issues are implied by the change of paradigm from the traditional electric grid to a smart grid one [9]. A smart grid in power system can lead to better results in decision-making under uncertainty. Particularly, monitoring and high quality real-time data for the exploitation of WP or PVP are a valuable contribution to mitigate the implications of uncertainty, mitigating the loss of a producer's profit due to bad decisions of bidding in a day-ahead electricity market. So, smart systems in an electric grid are expected to be: crucial for increase of power reliability; favorable for achievement of better bidding in electricity markets and an overall enhancement in power system.

## 3 State of the Art

One of the greatest challenges for exploitation of WP or PVP is to face the intermittence and variability of those powers in order to mitigate the uncertainty in power generation. Exploitation of WP or PVP by a power producer bidding in DAM at day d-1 has a menace. Which is due to the power output delivered at day d being predisposed to differ from the clearing accepted level of power in the market. Consequently, the power producer enters in imbalance of delivering and is subjected to economic penalty. Therefore, uncertainty in power output has to be properly accommodated into the decision of bidding. A complementarity of WP with PVP has been shown on the Iberian Peninsula, leading to a question about if the joint exploitation of the operation of WP with PVP is able to mitigate the uncertainty in the power output [10]. Literature dealing with WP producer (WPP) mitigation of WP uncertainty favors the implementation of combining strategy with the aim of reducing uncertainty on power generation. This combining strategy for a WPP includes the use of energy storage technology, namely pump-storage plants, compressed air facilities or vanadium redox flow batteries [11–13]. Also, the use of: purchase call/put options to pumped-storage facility is proposed for wind producers to hedge against WP uncertainty [14]; joint operation of WP, PVP, pump-storage power plant and energy storage devices is proposed for reserve market [15].

Stochastic nonlinear programming is an approach suitable for the bidding strategy of a WPP to achieve convenient decision with the aim of considering uncertainty [16]. Even more suitable, if an approach supported by MILP is conceivable. MILP either deterministic [17] or stochastic [18] have been proposed to convenient address the appropriated decision of bidding strategy a WPP needing to consider uncertainty.

## 4 Profit Description

A power producer exploiting WP or PVP needs to consider the uncertainty on the availability of power impacts in the worth of bidding in a DAM. If joint operation of the exploitation of these powers with energy storage is possible is important to further research the worth of profiting from the mismatch between periods of low availability of these powers. This research is the subject of the following problem description, having the objective of enquiring if this joint operation can improve profit of the power producer in DAM. A prerequisite is that the market rules permit that joint operation participates in the aforementioned electricity market. The concept presented in this paper is the same of the concept of a Virtual Power Plant (VPP). A VPP is defined as a power plant that aggregates different distributed generation units, such as WP and PVP systems, in order to participate in the market as a single entity. This participation is expected to not only facilitate trading of energy in electricity markets, but also being prone to achieve better profits.

The power system balance between the total amount of energy converted into electric energy and the usage of electric energy has to be null in real time. This balance has to be guaranteed with a convenient anticipation relatively to the real time delivering. Usually, the balance is guaranteed by production/consumption programs established in advance that guarantees system balance and by a real-time balancing market. The producer imbalance or the system imbalance may be negative, null or positive in the market, but as long as there is producer imbalance the producer is subjected to a market penalty procedure. The procedure in the Iberian electricity market is to subject the producer to a price for the positive energy imbalance and another price for negative energy imbalance. In the Iberian electricity market the procedures for computing these prices are reported in [19]. The imbalance incurred by a power producer with a power plant $k$ in a disjoint assessment in time $t$ is stated as follows:

$$d_t^k = P_{ts}^k - P_t^k. \quad (1)$$

In (1) $P_{ts}^k$ is the physical delivering of energy in time $t$ and $P_t^k$ is the bid submitted in the DAM. An imbalanced of a physical energy delivering is associated with a price ratio: positive and negative imbalances are associated with price ratios $pr_t^+$ and $pr_t^-$, respectively. The positive price ratio is stated as follows:

$$pr_t^+ = \frac{\lambda_t^+}{\lambda_t}, \quad pr_t^+ \leq 1. \quad (2)$$

The negative price ratio is defined as follows:

$$pr_t^- = \frac{\lambda_t^-}{\lambda_t}, \quad pr_t^- \geq 1. \quad (3)$$

In (2) and (3) $\lambda_t$ is the DAM price and $\lambda_t^+$ and $\lambda_t^-$ are respectively the positive and negative imbalance prices. Consequently, the power plant $k$ at period $t$ in a disjoint assessment has a profit given as follows:

$$PR_t^k = \lambda_t P_t^k + (\lambda_t pr_t^+ d_t^{k+} - \lambda_t pr_t^- d_t^{k-}). \quad (4)$$

In (4) the first term represents the profit associated with the DAM price $\lambda_t$ from an accepted production $P_t^k$. The term in brackets is associated with the income derived from the procedure of economic penalty. The positive and negative deviations are respectively quantified by $d_t^{k+}$ and $d_t^{k-}$ subjected to the constraint of at least one of the deviations is always null at a period t.

## 5 Problem Formulation

The goal of the optimization problem is the profit maximization. Maximization of profit implies a convenient management of the economic impact of energy delivering deviations. This management has to face uncertainty on market prices and on availability of energy in due time to be converted into electric energy in order to satisfy delivering. If there is a mismatch between the energy traded and the physical delivering, then eventual losses on profit can occur. The uncertainties on WP and on PVP availabilities are modelled by a convenient selection of scenarios from a set $S$ said to be the set of appropriate scenarios for the next day. A scenario $s$ has the probability $\pi_s$. The MILP formulation for the disjoint operation of WP or PVP systems are equivalent problems stated as follows:

$$\max \sum_{s=1}^{S}\sum_{t=1}^{T} \pi_s\left(\lambda_{ts} P_t^k + \lambda_{ts} pr_{ts}^{+} d_{ts}^{k+} - \lambda_{ts} pr_{ts}^{-} d_{ts}^{k-}\right). \tag{5}$$

Energy offer constraints

$$0 \le P_t^k \le P^{k\max}, \ \forall t. \tag{6}$$

Imbalance constraints

$$d_{ts}^k = P_{ts}^k - P_t^k, \ \forall t, \forall s \tag{7}$$

$$d_{ts}^k = d_{ts}^{k+} - d_{ts}^{k-}, \ \forall t, \forall s \tag{8}$$

$$0 \le d_{ts}^{k+} \le P_{ts}^k u_{ts}^k, \ \forall t, \forall s \tag{9}$$

$$0 \le d_{ts}^{k-} \le P^{k\max}(1 - u_{ts}^k), \ \forall t, \forall s. \tag{10}$$

In (6) the limit on the bid is set to be the rated power of the $k$ unit (wind or PV system). In (7) to (10) the imbalance is decomposed into a difference of positive values, i.e., the difference between the positive $d_{ts}^{k+}$ and the negative $d_{ts}^{k-}$ imbalances. If the imbalance is negative, the term $\lambda_{ts} pr_{ts}^{+} d_{ts}^{k+}$ is null and the term $\lambda_{ts} pr_{ts}^{-} d_{ts}^{k-}$ is subtracted. If the imbalance is positive the term $\lambda_{ts} pr_{ts}^{-} d_{ts}^{k-}$ is null and the term $\lambda_{ts} pr_{ts}^{-} d_{ts}^{k-}$ is added. The problem for joint operation of WP, PVP and energy storage is formulated as a sort of a two-stage stochastic optimization problem where the hourly bids and imbalances are first and second-stage variables, respectively. The goal of the problem is to find a single optimal bid in DAM for the wind power, PVP generation and energy storage. The stochastic MILP formulation is specified by the maximization problem stated as follows.

$$\max \sum_{s=1}^{S}\sum_{t=1}^{T} \pi_s\left(\lambda_{ts} P_t + \lambda_{ts} pr_{ts}^{+} d_{ts}^{+} - \lambda_{ts} pr_{ts}^{-} d_{ts}^{-}\right). \tag{11}$$

General constraints

(a) Energy offer constraint

$$0 \le P_t \le P^{W\max} + P^{PV\max} + P_t^{De\max}. \tag{12}$$

(b) Output power of combined wind power, PV power and energy storage device

$$P_{ts} = P_{ts}^{W} + P_{ts}^{PV} - P_t^{Chbat} + P_t^{Debat}. \tag{13}$$

Imbalance constraints

$$d_{ts} = P_{ts} - P_t, \forall t, \forall s \tag{14}$$

$$d_{ts} = d_{ts}^{+} - d_{ts}^{-}, \forall t, \forall s \tag{15}$$

$$0 \leq d_{ts}^{+} \leq P_{ts} u_{ts}, \forall t, \forall s \tag{16}$$

$$0 \leq d_{ts}^{-} \leq (P^{PV}\max + P^{PV}\max + P^{De}\max)(1 - u_{ts}), \forall t, \forall s. \tag{17}$$

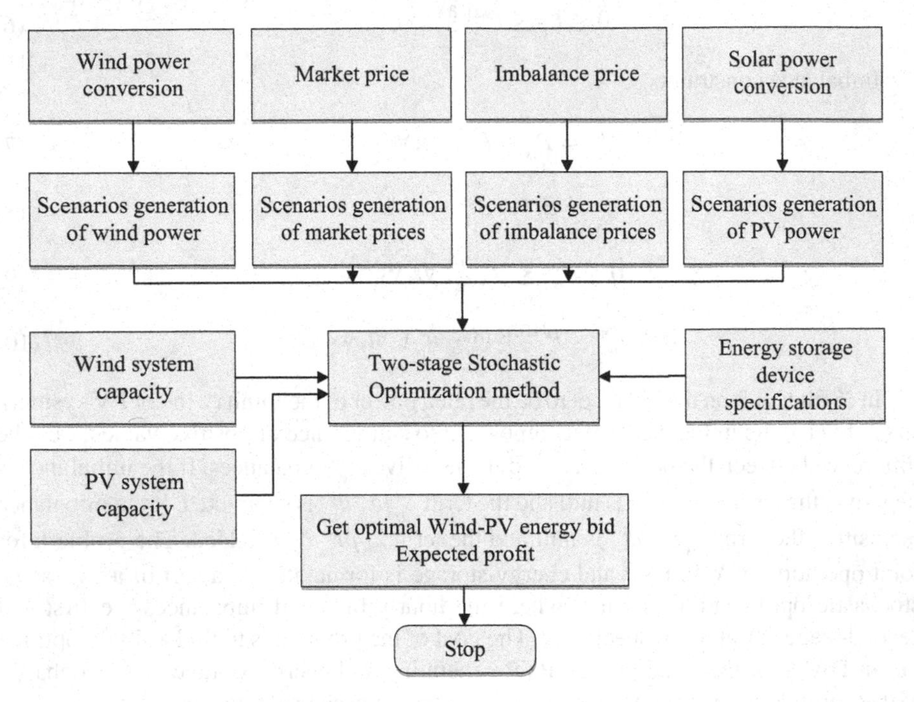


**Fig. 1.** Flowchart of joint bidding strategy.

Constraints of energy storage device

(a) Energy storage equation

$$E_t^{bat} = E_{t-1}^{bat} + \eta^{Chbat} P_t^{Chbat} - \frac{1}{\eta^{Debat}} P_t^{Debat}. \tag{18}$$

(b) Energy storage limits

$$0 \leq E_t^{bat} \leq E^{bat}\max. \tag{19}$$

(c) Storage power limits

$$0 \leq P_t^{Chbat} \leq P_t^{Chbat}\max k_t \tag{20}$$

$$0 \leq P_t^{Debat} \leq P_t^{Debat}\max (1 - k_t) \tag{21}$$

In (18) is the equation for the model of a Vanadium redox flow battery, assuming a null depth of discharge and considering the efficiency of charging and of discharging. This type of energy storage is one of the most promising technology for energy management in order to alleviate the variation and intermittence of WP or PVP.

The procedure of the joint bid strategy is shown in Fig. 1.

In Fig. 1 the upper blocks are for scenario generation procedures obtained via historical data of wind speed, solar irradiance and historical data of the Iberian electricity market. After the blocks for scenario generation procedures, technical data blocks access the wind system, PV system capacities and energy storage speciation to deliver the information into the block implementing the MILP approach for the two-sage stochastic optimization method in GAMS/CPLEX.

## 6 Case Study

The case study is based on a wind-PV system in the Iberian Peninsula with rated power of 200 MW and an energy storage rated power of 30 MW. The day-ahead energy prices and the

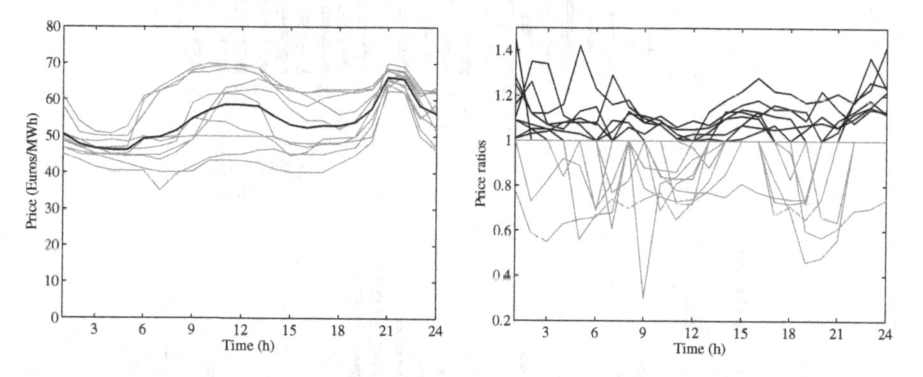


**Fig. 2.** Left, electricity price (average - black line). Rigth, price ratios: positive (blue lines), negative (black lines). (Color figure online)

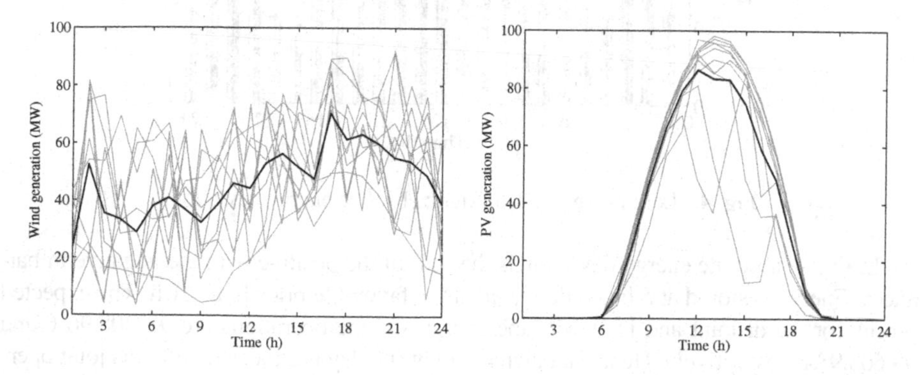


**Fig. 3.** Left, WP scenarios and average scenario (black line). Rigth, PV scenarios and average scenario (black line). (Color figure online)

price ratios $pr_t^+$ e $pr_t^-$ are reported in [20]. The bidding is for a 24 h horizon on an hourly basis. The scenarios are 10 for WP, 10 for PVP, 10 for DAM prices and 10 for price ratios. The DAM-clearing price scenarios and the price ratios scenarios $pr_t^+$ and $pr_t^-$ are shown in Fig. 2.

The scenarios for WP and for PVP are shown in Fig. 3.

The energy traded and the absolute value of the energy deviation by the disjoint and by the joint operations are shown in Fig. 4.

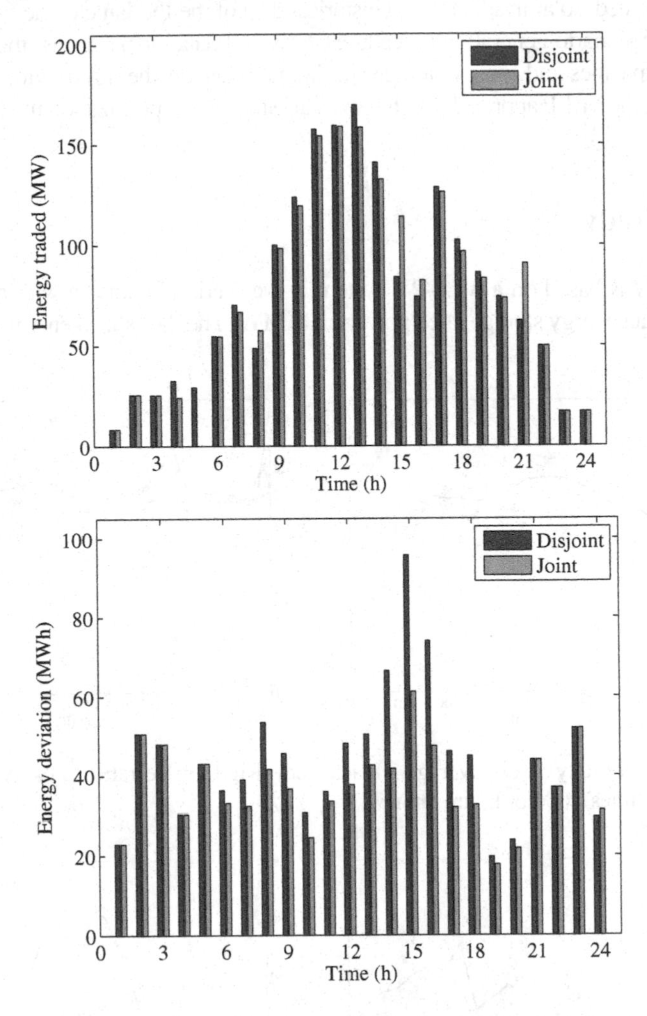


**Fig. 4.** Left: energy traded. Right: absolute energy deviation.

In Fig. 4, absolute energy deviation is the sum of the positive with the negative imbalances. Energy is stored at 5 h and delivered at the favorable price hour 21 h. The expected profits for the disjoint and joint with the energy storage operations are 98 308.96 € and 98 608.98 €, respectively. Hence, an increase of 300 €/day is expected with this joint operation. A higher increase occurs at 15 h, a period of favorable prices, low negative imbalance price ratios and high positive imbalance price ratios. Although, the numeric values change in function of availability of WP and PVP, the energy deviation is conveniently managed in what regards economic penalties in order to achieve better profit, i.e., this joint operation is never worse than the disjoint one.

## 7 Conclusion

In this paper, an optimization problem of bidding in DAM for a power producer having joint operation of WP with PVP and energy storage of the type Vanadium redox flow battery is presented. The problem is stated as sort of a two-stage stochastic optimization problem addressed as MILP problem. The joint bidding is envisaged as a favorable one when the mismatch of uncertainty due to the WP and the PVP is partial disabled by one another and an energy storage device allows flexibility of storing energy and discharging at hours of convenient DAM prices. A future work can include the participation in intraday-markets.

**Acknowledgments.** To thank the Millennium BCP Foundation for the financial support; and Foundation for Science and Technology-FCT: LAETA 2015/2020, ref: UID-EMS-50022-2013.

## References

1. Gutierrez, M.F., Silva, R.A., Montoro, P.P.: Effects of wind intermittency on reduction of CO2 emissions: the case of the Spanish power system. Energy **61**, 108–117 (2013)
2. Wang, T., Gong, Y., Jiang, C.: A review on promoting share of renewable energy by green-trading mechanisms in power system. Renew. Sustain. Energy Rev. **40**, 923–929 (2014)
3. Denholm, P., Ela, E., Kirby, B., Milligan, M.: The role of energy storage with renewable electricity generation. National Renewable Energy Laboratory, Colorado (2010)
4. Beaudin, M., Zareipour, H., Schellenberglabe, A., Rosehart, W.: Energy storage for mitigating the variability of renewable electricity sources: an updated review. Energy. Sustain. Dev. **14**(4), 302–314 (2010)
5. Divya, K.C., Østergaard, J.: Battery energy storage technology for power systems–an overview. Electr. Power Syst. Res. **79**(4), 511–520 (2009)
6. Gomes, I.L.R., Pousinho, H.M.I., Melício, R., Mendes, V.M.F.: Bidding and optimization strategies for wind-PV systems in electricity markets assisted by CPS. Energy Procedia **106**, 111–121 (2016)
7. Arsan, T.: Smart systems: from design to implementation of embedded smart systems. In: 13th International Symposium on Smart MicroGrids for Sustainable Energy Sources enabled by Photonics and IoT Sensors, Nicosia, Cyprus, pp. 59–64 (2016)
8. Topham, D., Gouze, N.: Smart systems integration knowledge gateway. In: European Technology Platform on Smart Systems Integration, Turin, Italy (2015)
9. Glesner, M., Philip, F.: Embedded systems design for smart system integration. In: IEEE Computer Society Annual Symposium on VLSI, Natal, Brazil 32–33 (2013)
10. Jerez, S., Trigo, R.M., Sarsa, A., Lorente-Plazas, R., Pozo-Vásquez, D., Montávez, J.P.: Spatio-temporal complementarity between solar and wind power in Iberian Peninsula. Energy Procedia **40**, 48–57 (2013)

11. González-Garcia, J., Muela, R.M.R., Santos, L.M., González, A.M.: Stochastic joint optimization of wind generation and pumped-storage units in an electricity market. IEEE Trans. Power Syst. **23**(2), 460–468 (2008)
12. Angarita, J.L., Usaola, J., Martínez-Crespo, J.: Combined hydro-wind generation bids in a pool-based electricity market. Electr. Power Syst. Res. **79**(7), 1038–1046 (2009)
13. Sundararagavan, S., Baker, E.: Evaluating energy storage technologies for wind power integration. Sol. Energy **86**(9), 2707–2717 (2012)
14. Hedman, K.W., Sheble, G.B.: Comparing hedging methods for wind power: using pumped storage hydro units vs. options purchasing. In: Proceedings of International Conference on Probabilistic Methods Applied to Power Systems – PMAPS, Stockholm, Sweden, pp. 1–6 (2006)
15. Parastegari, M., Hooshmand, R.A., Khodabakhshian, A., Zare, A.H.: Joint operation of a wind farm, photovoltaic, pump-storage and energy storage devices in energy and reserve markets. Electr. Power Energy Syst. **64**, 275–284 (2015)
16. Matevosyan, J., Söder, L.: Minimization of imbalance cost trading wind power on the short-term power market. IEEE Trans. Power Syst. **21**(3), 1396–1404 (2006)
17. Pousinho, H.M.I., Mendes, V.M.F., Catalão, J.P.S.: Investigation on the development of bidding strategies for a wind farm owner. Int. Rev. Electr. Eng. **5**(3), 1324–1329 (2010)
18. Laia, R., Pousinho, H.M.I., Melício, R., Mendes, V.M.F.: Self-scheduling and bidding strategies of thermal units with stochastic emission constraints. Energy Convers. Manag. **89**, 975–984 (2015)
19. Gomes, I.L.R., Pousinho, H.M.I., Melício, R., Mendes, V.M.F.: Optimal wind bidding strategies in day-ahead markets. In: Camarinha-Matos, Luis, M., Falcão, António, J., Vafaei, Nazanin, Najdi, Shirin (eds.) DoCEIS 2016. IAICT, vol. 470, pp. 475–484. Springer, Heidelberg (2016). doi:10.1007/978-3-319-31165-4_44
20. http://www.esios.ree.es/web-publica/

# The Impacts of Demand Response on the Efficiency of Energy Markets in the Presence of Wind Farms

Neda Hajibandeh[1], Miadreza Shafie-khah[1], Saber Talari[1], and João P.S. Catalão[1,2,3](✉)

[1] C-MAST, University of Beira Interior, 6201-001 Covilhã, Portugal
{neda.hajibandeh,miadreza, Saber.talari}@ubi.pt, catalao@fe.up.pt
[2] INESC TEC and the Faculty of Engineering of the University of Porto, 4200-465 Porto, Portugal
[3] INESC-ID, Instituto Superior Técnico, University of Lisbon, 1049-001 Lisbon, Portugal

**Abstract.** In this paper, an optimal scheduling of thermal and wind power plants is presented by using a stochastic programming approach to cover the uncertainties of the forecasted generation of wind farms. Uncertainties related to wind forecast error, consequently wind generation outage power and also system load demand are modeled through scenario generation. Then, with regard to day-ahead and real-time energy markets and taking into account the relevant constraints, the thermal unit commitment problem is solved considering wind energy injection into the system. Besides, in order to assess impacts of Demand Response (DR) on the problem, a load reduction demand response model has been applied in the base model. In this approach, self and cross elasticity is used for modeling the customers' behavior modeling. The results indicate that the DR Programs (DRPs) improves the market efficiency especially in peak hours when the thermal Gencos become critical suppliers and the combination of DRPs and wind farm can be so efficient.

**Keywords:** Demand response · Electricity market · Stochastic programming · Wind production

## 1 Introduction

Different countries have increased contribution of renewable energy resources by planning and applying protectionist policies of renewable energies [1]. In some countries, wind sources participate in the electricity markets disregarding imbalance penalties [2]. Imbalance penalties are defined as power planned by supplier minus generated power. It is used for security and appropriate utilization of system and avoids to be trifled with market. In some power markets, wind resources will be allowed to offer energy in the market, but some of supportive methods are used to increase their income. These methods usually increase power imbalance in the time of use and thereupon, will follow increasing the cost for the system [3].


L.M. Camarinha-Matos et al. (Eds.): DoCEIS 2017, IFIP AICT 499, pp. 287–296, 2017.
DOI: 10.1007/978-3-319-56077-9_28

In [4], a changeable tariff method has been used to specify income of wind power unit in the power market. In the current method, the consumption cost rises in proportion to its increase and therefore, more consumption is not associated with more subsidy and high consumption customers are in the focus of rising cost instead of whole customers. In [5, 6], the method of profit maximization is used to participate wind power plants. Modelling of uncertainties is necessary for these methods. Also, since the interval between time of bidding and time of use may be high, wind power plant's income will be reduced in these methods. A supportive method for contribution of wind power plant to the power markets has been discussed in [7]. However, if the content of wind resources increases, use of supportive methods will be inefficient.

In this paper, two-stage stochastic bidding considering uncertainty in the market price is carried out. Uncertainty in the market price is modelled as probability density function and finally, each renewable generating unit presents a set of its strategies as probability density function. Among the presented models for estimation of wind speed, Weibull distribution has the best fitting for statistical data, so in most articles it is used as a suitable distribution to estimate the wind speed [8, 9]. Appropriate values for parameters estimation of this probability function accepted for wind speed have been given in [10, 11].

In the current paper, the aim is to model players' behavior in the imperfect market and obtain the optimum point to tariff the DRPs, while the scope of research is from the perspective of independent system operator (ISO). A model is proposed for day-ahead and real-time energy markets with implementation of DRPs in presence of the wind turbines. In fact, system uncertainties including outage of wind farms are merged in a two-stage stochastic model.

The remainder of this paper is structured as follows. Relationship to smart systems is provided in Sect. 2. In Sect. 3, modelling the wind power units in the market is introduced and their mathematical model is presented. Numeric results of the model are presented in Sect. 4, and the last section concludes the paper.

## 2 Relationship to Smart Systems

With the expansion of smart meters, like Advanced Metering Infrastructure (AMI) and Internet of Things (IoT), in demand side of power systems, implementation of different methods of DR is going to be much more applicable [11]. A suitable DR method can not only decrease total operation cost but also provide security and safety of the network operation [12].

In [13], DR is discussed considering wind units connected to a smart grid. They have tried to maximize the social welfare while the intermittent nature of wind may impose some real challenges. It is performed a comparison with the predictable scheduling and with a non-smart grid. The proposed approach balances the wind power fluctuations in the grid due to DR management.

In continuous we proposed DR programs as well to mitigate the challenges of grid including wind units.

## 3 The Proposed Model

In the current paper, the price-based DRP, Time of Use (TOU) [14, 15], is considered. Equation (1) denotes the optimal customer's demand in 24 h while taking part in price-based DRPs [16].

$$d(t) = d_0(t)\left\{1 + \sum_{t'=1}^{24} E(t,t').\frac{[\rho(t') - \rho_0(t')]}{\rho_0(t')}\right\} \tag{1}$$

where $d(t)$ and $d_0(t)$ are the final and initial electricity demand, respectively. $\rho_0(t)$ and $\rho_0(t)$ are the final and initial electricity tariffs, respectively. $E(t,t')$ represents the elasticity of demand. Considering to a guaranteed purchase of electricity from the renewable resources including wind and solar, in many studies and planning, renewable loads are considered as negative loads when setting the market prices based on offer for sale of all power plants. In addition to, in some markets, these renewable resources can also bid their price in the power market and sell their energy as thermal and other power plants.

But, according to the policy makings of the governments to encourage and attract investor toward generating electricity from the renewable energies and developing clear energies in the countries, they usually purchase the electricity of clear resources from the vendors at prices several times the market clearing price to make an incentive to attract investors to this industry and guarantee to purchase the electricity at high and fixed price [17].

The unit commitment program and its aggregate constraints can be modeled as below:

*Minimize*

$$\sum_{t=1}^{NT}\sum_{i=1}^{NG}\left(SUC_{i,t} + MPC_iU_{i,t} + \sum_{m=1}^{NM}(P^e_{i,t,m}C^{G\_Eng}_{i,t,m})\right) + \sum_{t=1}^{NT}\sum_{i=1}^{NG} C^{G\_UC}_{i,t}R^{G\_UC}_{i,t} + C^{G\_DC}_{i,t}R^{G\_DC}_{i,t}$$
$$+ \sum_{t=1}^{NT}\sum_{w=1}^{NW}\rho_w\left(\sum_{i=1}^{NG} C^{G\_UE}_{i,t}r^{G\_up}_{i,t,w} - C^{G\_DE}_{i,t}r^{G\_dn}_{i,t,w}\right) \tag{2}$$

where $P^e_{i,t,m}$ denotes the production of segment $m$ in a linear form of cost function. $C^{G\_Eng}_{i,t,m}$ is the slope of each segment. $C^{G\_DC}_{i,t}$ and $C^{G\_UC}_{i,t}$ are the offered price of down- and up-reserves of power producers. $C^{G\_UE}_{i,t}$ and $C^{G\_DE}_{i,t}$ represent the offered price of up- and down- deployed reserves of each unit. $R^{G\_UC}_{i,t}$ and $R^{G\_DC}_{i,t}$ the scheduled up- and down-reserve capacities of each unit, respectively. $r^{G\_up}_{i,t,w}$ and $r^{G\_dn}_{i,t,w}$ represent the real-time up- and down- deployed reserves of each unit. $SUC_{i,t}$ denotes the start-up cost. $MPC_iU_{i,t}$ is the minimum production cost. The costs of start-up, minimum generation, and the electricity of producer plants are included in the objective function. The second part of the objective function is associated with the offer price of up/down reserve from production plants. The next term consists of the scenarios consideration.

The prices with regard to the expected delivered up/down reserve through production plants are formulated in the next term. The scenarios of wind power are generated based on a roulette wheel mechanism presented in [18]. The objective function is subjected to following constraints.

### 3.1 Day-Ahead Equations

Equation (3) guarantees the demand-supply balance. As regard with modelling the linearized transmission limits, DC load flow is used as formulated through (4). Furthermore, the branch limitations are modelled in (5). Wind units considered in the day-ahead session are constrained due to the predicted wind production in (6). Electricity and capacity equations of thermal plants are proposed in (7)–(11). Block electricity output of thermal plants and the limits are described through (7). Equations (8)–(9) restrain the generation of thermal plants. The reserve up/down limits are described in (10)–(11) as well. The inequality (12) denotes the start-up limit of thermal plants. Constraints (13) and (14) formulate the minimum up/down times of thermal plants, respectively [19].

$$\sum_{i \in G_b} P_{it} + \sum_{wf \in WF_b} P_{wf,t}^{WP,S} - \sum_{j \in J_b} L_{j,t}^{C} = \sum_{l \in L_b} F_{l,t}^{0} \quad \forall b,\ t \tag{3}$$

$$F_{l,t}^{0} = \left(\delta_{b,t}^{0} - \delta_{b',t}^{0}\right) \Big/ X_l \quad \forall l,\ t \tag{4}$$

$$-F_l^{\max} \le F_{l,t}^{0} \le F_l^{\max} \quad \forall l, t \tag{5}$$

$$0 \le P_{wf,t}^{WP,S} \le P_{wf,t}^{WP,\max} \quad \forall wf, t \tag{6}$$

$$P_{i,t} = \sum_{m=1}^{NM} P_{i,t,m}^{e}\ ,\ 0 \le P_{i,t,m}^{e} \le P_{i,m}^{\max} \quad \forall i, t \tag{7}$$

$$P_{i,t} + R_{i,t}^{G\_UC} \le P_i^{\max} \quad \forall i, t \tag{8}$$

$$P_{i,t} - R_{i,t}^{G\_DC} \ge 0 \quad \forall i, t \tag{9}$$

$$0 \le R_{i,t}^{G\_UC} \le RU_i \tau \quad \forall i, t \tag{10}$$

$$0 \le R_{i,t}^{G\_DC} \le RD_i \tau \quad \forall i, t \tag{11}$$

$$SUC_{i,t} \ge SC_i(U_{i,t} - U_{i,t-1}) \quad \forall i, t \tag{12}$$

$$\sum_{t'=t+2}^{t+MUT_i} \left(1 - U_{i,t'}\right) + MUT_i \left(U_{i,t} - U_{i,t-1}\right) \leq MUT_i \quad \forall i, t \tag{13}$$

$$\sum_{t'=t+2}^{t+MDT_i} U_{i,t'} + MDT_i \left(U_{i,t-1} - U_{i,t}\right) \leq MDT_i \quad \forall i, t \tag{14}$$

### 3.2 Real-Time Equations

This stage is dealing with fulfilling the limits in the realization of all scenarios. The demand-supply balance is taken into account for all the scenarios in (15) by considering the updates of wind generation. Constraints (16)–(17) are similar to (4)–(5) that consider for all the scenarios. The delivered up/down reserve in all scenarios needs to be limited to the planned reserve capacity formerly submitted to the markets (18)–(19). Constraint (20) characterizes the net generation of thermal plants in real-time. Equation (21) represents the technical constraints of thermal generators. Ramp up/down limit is given in (22)–(23).

$$\sum_{i \in G_b} \left(r_{i,w,t}^{G\_up} - r_{i,w,t}^{G\_dn}\right) + \sum_{wf \in WF_b} \left(P_{wf,w,t}^{W} - P_{wf,t}^{WP,S}\right) = \sum_{l \in L_b} F_{l,w,t} - F_{l,t}^{0} \quad \forall b, w, t \tag{15}$$

$$F_{l,w,t} = \left(\delta_{b,w,t} - \delta_{b',w,t}\right) / X_l \quad \forall l, w, t \tag{16}$$

$$-F_l^{\max} \leq F_{l,w,t} \leq F_l^{\max} \quad \forall l, w, t \tag{17}$$

$$0 \leq r_{i,w,t}^{G\_up} \leq R_{i,t}^{G\_UC} \quad \forall i, w, t \tag{18}$$

$$0 \leq r_{i,w,t}^{G\_dn} \leq R_{i,t}^{G\_DC} \quad \forall i, w, t \tag{19}$$

$$P_{i,w,t} = P_{i,t} + r_{i,w,t}^{G\_up} - r_{i,w,t}^{G\_dn} \quad \forall i, w, t \tag{20}$$

$$P_i^{\min} U_{i,t} \leq P_{i,w,t} \leq P_i^{\max} U_{i,t} \quad \forall i, w, t \tag{21}$$

$$P_{i,w,t} - P_{i,w,t-1} \leq RU_i\, U_{it} + SUR_i \left(1 - U_{i,t-1}\right) \quad \forall i, w, t \tag{22}$$

$$P_{i,w,t-1} - P_{i,w,t} \leq RD_i\, U_{i,t-1} + SDR_i \left(1 - U_{i,t}\right) \quad \forall i, w, t \tag{23}$$

## 4 Numerical Results of the Model

In order to indicate the impact of DRPs on the oligopolistic behavior of electricity market in presence of wind farm, the IEEE six-bus test system is employed [20]. The mixed-integer linear programming has been modeled in GAMS 24.0 and solved by

CPLEX 12.0. In addition different types of TOU programs are studied as presented in Table 1. It is assumed that 20% of consumers are responsive demand except the TOU-3 tariff that its price tariff is considered as the same as TOU-1 but at this program 40% of consumers are responsive demand.

**Table 1.** Tariffs/incentives of considered DRPs (€/MWH)

| Case | Valley (1 to 8) | Off-peak. (9–11,22–24) | Peak. (12–14,19–21) | Critical peak (15 to 18) |
|---|---|---|---|---|
| Base case (fixed-rate) | 63.2 | 63.2 | 63.2 | 63.2 |
| TOU-1 | 31.6 | 63.2 | 94.8 | 94.8 |
| TOU-2 | 15.8 | 63.2 | 126.4 | 126.4 |
| TOU-3 | 31.6 | 63.2 | 94.8 | 189.6 |

The impact of different types of TOU programs on Genco 1 is compared with the generation in base case without the presence of wind farm and in presence of the wind units, the power output of Genco 1 in these cases are presented in Figs. 1 and 2, respectively. According to Fig. 1, Genco 1 is affected by the load shifting arisen from TOU tariffs. On this basis, the generation in valley period is increased based on the low tariff, while the generation reduces in the peak period. Figure 2 depicts the same great affection but the very considerable and beneficial changes in base case in presence of wind units. As it is shown there is a comprehensive reform in the curve and chart. It should be noted that the prices in base case program is obtained from the simulation of the electricity market without implementation of DRPs. In the base case, the average of market prices is considered as electricity tariff in all hours. In TOU programs, the mentioned tariff is considered as the tariff in off-peak hours. The self- and cross-elasticities are extracted from [21].

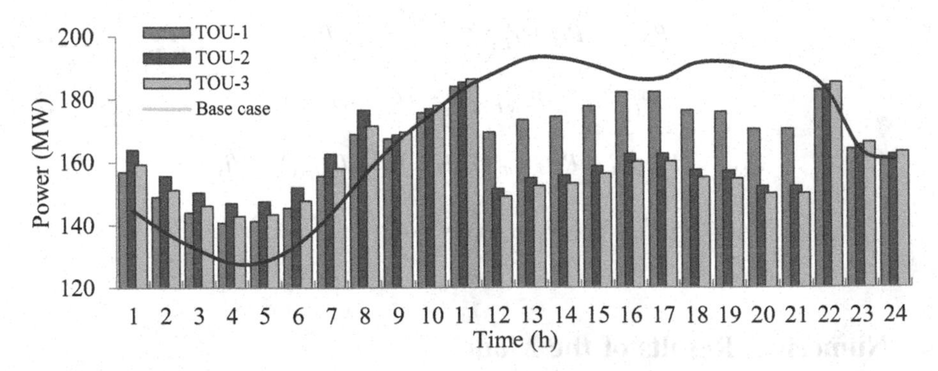


**Fig. 1.** Impact of different types of TOU program on the generation of Genco 1 without the presence of wind farm.

Developing the investigation of the impact of different types of TOU programs on the behaviour of market players Figs. 3 and 4 are presented. These figures depict the power output of Genco 2 without the presence of wind farm and in presence of the wind units respectively. These results show the different tariffs of TOU demand response in presence of wind farm can more considerably effect on shifting peak loads to Valley. Figure 3 shows the significant impact of TOU programs on the electricity market prices in the peak period, because of the reduction of offered prices of the system's Gencos. The impact of different types of TOU programs on objective function is presented in Table 2. As it is expected, it shows that total price after implementation of DRPs with presence of wind farm is less than that in the system with no renewable units.

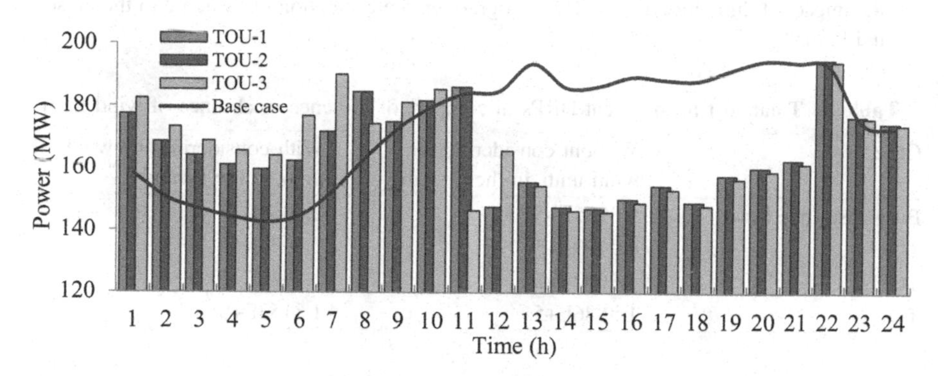


**Fig. 2.** Impact of different types of TOU program on the generation of Genco 1 in the presence of wind farm.

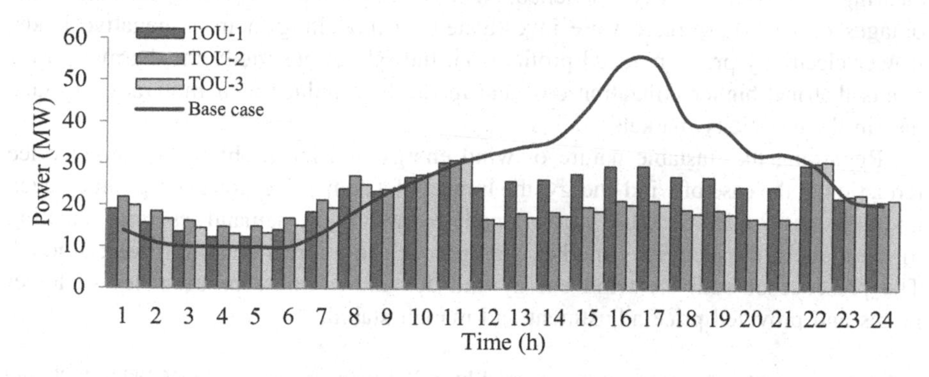


**Fig. 3.** Impact of different types of TOU program on the generation of Genco 2 without the presence of wind farm.

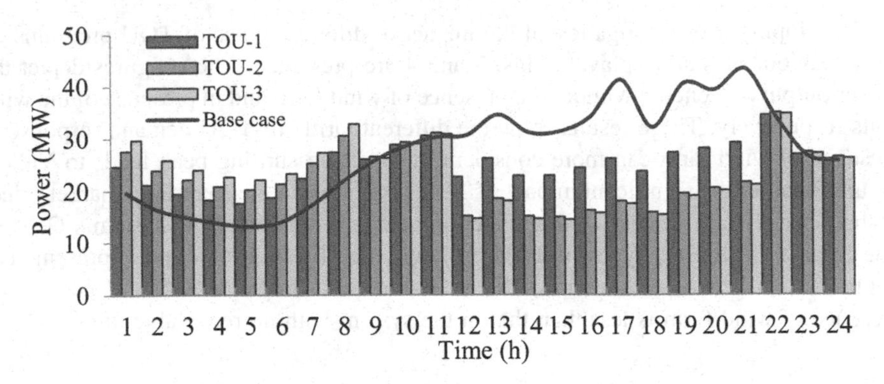


**Fig. 4.** Impact of different types of TOU program on the generation of Genco 2 in the presence of wind farm.

**Table 2.** Total cost for different DRPs in compere of presence or absence of wind farm

| Case | Without considering the wind units in the system | With considering the wind units in the system |
|---|---|---|
| Base case (fixed-rate) | 2.0935E+5 | 2.0036E+5 |
| TOU-1 | 1.9799E+5 | 1.8994E+5 |
| TOU-2 | 1.8692E+5 | 1.7940E+5 |
| TOU-3 | 1.8886E+5 | 1.8153E+5 |

## 5 Conclusions

In this paper, modelling the power market was conducted and bidding of power plants and making the supply curve and then its intersection with the demand curve to find clearing price of market were modelled. Next, the impacts of uncertainty of wind farms outages on electricity price were investigated by modelling them as negative loads. Lower electricity prices reduced profits from the sale of electricity for thermal power plants that had higher utilization cost and made them reduce their motivation to take part in the electricity markets.

Regarding the unstable nature of wind energy and the problem of market price reduction in the case of wind energy, the impact of a demand response program on such a system was studied. In other words, finally by adding the demand response planning to the models, the electricity market and approach to the real condition were studied. The presence of both the wind energy and demand response program led to lower prices and provided price adjustment and market stability.

**Acknowledgment.** This work was supported by FEDER funds through COMPETE 2020 and by Portuguese funds through FCT, under Projects SAICT-PAC/0004/2015 - POCI-01-0145-FEDER-016434, POCI-01-0145-FEDER-006961, UID/EEA/50014/2013, UID/CEC/50021/2013, and UID/EMS/00151/2013. Also, the research leading to these results has received funding from the EU Seventh Framework Programme FP7/2007-2013 under grant agreement no. 309048.

## References

1. Kennedy, J., Fox, B., Flynn, D.: Use of electricity price to match heat load with wind power generation. In: Proceedings of the IEEE Conference on Sustainable Power Generation and Supply (2009)
2. Bathurst, G.N., Weatherrill, J., Strbac, G.: Trading wind generation in short-term energy markets. IEEE Trans. Power Syst. **17**, 782–789 (2002)
3. Matevosyan, J., Söder, L.: Minimization of imbalance cost trading wind power on the short-term power market. IEEE Trans. Power Syst. **21**, 1396–1404 (2006)
4. Yu, W., et al.: Valuation of switchable tariff for wind energy. Electr. Power Syst. Res. **76**, 382–388 (2006)
5. Pinson, P., Chevallier, C., Kariniotakis, G.: Trading wind generation from short-term probabilistic forecasts of wind power. IEEE Trans. Power Syst. **22**, 1148–1156 (2007)
6. Morales, J.M., Conejo, A.J., Perez-Ruiz, J.: Short-term trading for a wind power producer. IEEE Trans. Power Syst. **25**, 554–564 (2010)
7. Parsa Moghadm, M., Talebi, E., Mohamadian, M.: A new method for pricing of wind power in short-term power markets. In: Presented at 16th Power Systems Computation Conference (PSCC) (2008)
8. Moeini-Aghtaie, M., Abbaspour, A., Fotuhi-Firuzabad, M.: Incorporating large-scale distant wind farms in probabilistic transmission expansion planning—part i: theory and algorithm. IEEE Trans. Power Syst. **27**, 1585–1593 (2012)
9. Kianoosh, G., Boroojeni, M., Amini, H., Bahrami, H., Iyengar, S.S., Arif, I., Karabasoglu, O.: A novel multi-time-scale modeling for electric power demand forecasting: from short-term to medium-term horizon. Electr. Power Syst. Res. **142**, 58–73 (2017)
10. Yu, H., et al.: Probabilistic load flow evaluation with hybrid latin hypercube sampling and cholesky decomposition. IEEE Trans. Power Syst. **24**, 661–667 (2009)
11. Verbic, G., Schellenberg, A., Rosehart, W., Canizares, C.A.: Probabilistic optimal power flow applications to electricity markets. In: Probabilistic Methods Applied to Power Systems (PMAPS) (2006)
12. Boroojeni, K.G., Amini, M.H., Iyengar, S.S.: Overview of the Security and Privacy Issues in Smart Grids. Smart Grids. Security and Privacy Issues. Springer International Publishing, Heidelberg (2017)
13. Çiçek, N., Deliç, H.: Demand response management for smart grids with wind power. IEEE Trans. Sustain. Energ. **6**, 625–634 (2015)
14. Heydarian-Forushani, E., Golshan, M.E.H., Moghaddam, M.P., Shafie-khah, M., Catalão, J.P.S.: Robust scheduling of variable wind generation by coordination of bulk energy storages and demand response. Energ. Convers. Manag. **106**, 941–950 (2015)
15. Heydarian-Forushani, E., Golshan, M.E.H., Shafie-khah, M.: Flexible security-constrained scheduling of wind power enabling time of use pricing scheme. Energy. **90**, 1887–1900 (2015)
16. Aalami, H.A., Moghaddam, M.P., Yousefi, G.R.: Demand response modeling considering interruptible/curtailable loads and capacity market programs. Appl. Energ. **87**, 243–250 (2010)
17. Arasteh, H., et al.: Iot-based smart cities: a survey. In: IEEE 16th International Conference on Environment and Electrical Engineering (EEEIC) (2016)
18. Shafie-khah, M., et al.: Strategic offering for a price-maker wind power producer in oligopoly markets considering demand response exchange. IEEE Trans. Industr. Inform. **11** (6), 1542–1553 (2015)

19. Shafie-khah, M., Moghaddam, M.P., Sheikh-El-Eslami, M.K.: Unified solution of a non-convex SCUC problem using combination of modified branch-and-bound method with quadratic programming. Energ. Convers. Manag. **52**(12), 3425–3432 (2011)
20. Fu, Y., Shahidehpour, M., Li, Z.: AC contingency dispatch based on security-constrained unit commitment. IEEE Trans. Power Syst. **21**, 897–908 (2006)
21. Zhao, C., Wang, J., Watson, J.P., Guan, Y.: Multi-stage robust unit commitment considering wind and demand response uncertainties. IEEE Trans. Power Syst. **28**, 2708–2717 (2013)

# Implementing an Integer Linear Approach to Multi-objective Phasor Measurement Unit Placement

Amir Baharvandi[1], Miadreza Shafie-khah[1], Saber Talari[1], and João P.S. Catalão[1,2,3](✉)

[1] C-MAST, University of Beira Interior, 6201-001 Covilhã, Portugal
amir.baharvandi@gmail.com, {miadreza,saber.talari}@ubi.pt
[2] INESC TEC and the Faculty of Engineering of the University of Porto, 4200-465 Porto, Portugal
catalao@fe.up.pt
[3] INESC-ID, Instituto Superior Técnico, University of Lisbon, 1049-001 Lisbon, Portugal

**Abstract.** In this paper, an Integer Linear Programming (ILP) problem for a model of Multi-Objective Optimal PMU Placement (MOPP) is proposed. The proposed approach concurrently deals with two objectives. The first objective is the number of phasor measurement units (PMUs) which should be minimized. The second objective function is measurement redundancy which is the number of observable buses in the case of PMU outage. In fact, whatever the amount of second objective increases, the system would be more reliable. Furthermore, some linearized formulations are defined for each nonlinear formula. In fact, the nonlinear nature of formulation related to redundancy is substituted by linear inequality and so there is no nonlinear formula such that the calculation of the problem would be simplified. Finally, a modified 9-bus test system is implemented to show how the proposed method is effective.

**Keywords:** Phasor Measurement Unit (PMU) · Multi-objective mathematical programming · Integer linear programming · Optimal PMU Placement

## 1 Introduction

The key element of the wide-area monitoring system and application of smart systems is to attain a instantaneous measurement of state variables which Phasor Measurement Units (PMUs) are able to deliver it [1–3]. To increase the security of the system, the future state of the entire network should be estimated [4]. Accordingly, devices known as PMU are used to measure the system's phasors (voltage and current) [5, 6].

In this paper, two previously mentioned objective functions will be described as a Multi-Objective problem in the optimum PMU placement area. Also, to prove the usefulness of the method a case study is tested.

## 2 Relationship to Smart Systems

Since the metering of the smart systems is of paramount importance to the operator. There are several devices equipping smart cities. These devices are in the kind of


L.M. Camarinha-Matos et al. (Eds.): DoCEIS 2017, IFIP AICT 499, pp. 297–304, 2017.
DOI: 10.1007/978-3-319-56077-9_29

electronic devices base of Internet of Things (IoT) [7]. Additionally, the internet facilitates interconnection between people and objects and it is known as the main revolution of the internet infrastructure.

One of these devices using Global Positioning System (GPS) to increase the reliability of the system is Phasor Measurement Unit. It is absolutely certain that implementation of the technological devices has increased drastically in smart cities. In this condition, Phasor Measurement Units can be very applicable in the power electrical systems of smart cities.

To make a clear relationship between the proposed research area and smart system, it should be noted that in the basis of IoT, there is a huge communication between devices using the internet. Likewise, in the smart systems, consumers can manage their consumption in order to decrease their cost of electricity [8].

First relationship between PMU and smart systems is about using the internet. PMUs implement GPS to calculate the voltage phasor of buses and in this process gathers data from buses and uses an internet interconnection with satellite subsequently sends data to a central server. This process is shown in Fig. 1. Another relationship between PMU device and smart system is about managing consumption in smart grids [9].

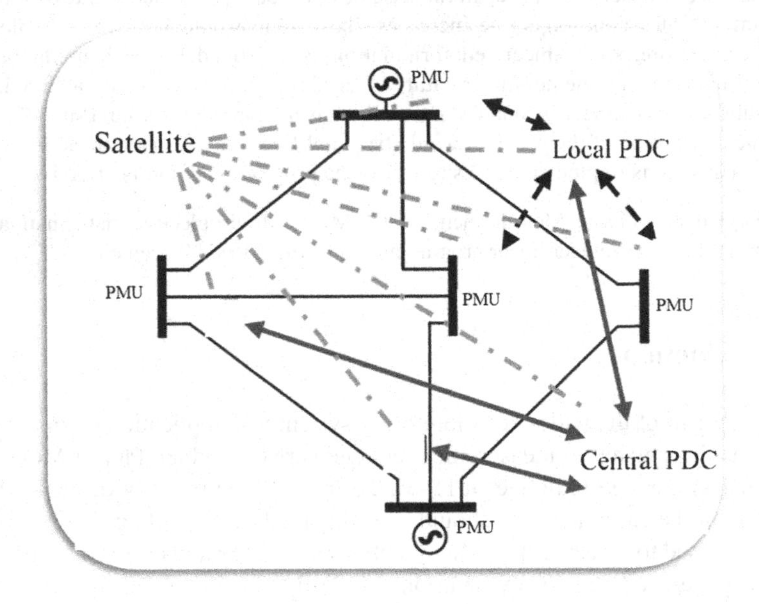


**Fig. 1.** PMU process

The calculation of phasors by PMU is utilized to estimate the future states of the power systems and this help to anticipate several parameters such as future consumption in order to supply future loads [10]. State estimating helps the generators and consumers to manage their supply and consumption.

## 3 Problem Formulation

In the MOPP problem the best condition is making the entire network observable while the redundancy of the system is in a suitable range. It goes without saying that this observability is better to achieve using the minimum PMUs and at the same time all the system constraints would be satisfied.

In the following objective functions and constraints are introduced.

### 3.1 Objective Function

$F_1$: First Objective Function As mentioned before we want to increase the observability of the system (all buses) by minimum cost. Thus, minimum number of PMU which is equivalent the minimum cost is the first objective function:

$$\min F_1 = \sum_{\forall i \in I} x_i. \tag{1}$$

where $F_I$ is the first objective function, $x_i$ is a binary variable (which equals to 0 if the bus has not PMU and 1 otherwise) and $I$ is a set for all buses from 1 to $N_b$ ($N_b$ is the number of buses).

$F_2$: Second Objective Function The second objective is redundancy which is associated with contingency occurrence in the system. This contingency may be PMU. It means when a contingency occurs how many buses lose their observability. The desire condition is that we will have the minimum number of unobservable bus. In other word it is absolutely certain that whatever the PMUs increases the number of unobservable buses decreases as well as the redundancy increases. On the contrary, whatever the number of PMUs would be more the cost of installation related to the first objective function increases. So, we have the limitation of PMU and cost and should be aware about it.

Redundancy parameter ($r$) is equal to the number of observable buses while a contingency occurs. In this study we suppose that this contingency could be PMU loss. The second objective function is introduced as the number of set of buses which are observable when all the contingencies occur separately. This set is defined as $R$ and the member of this set ($r$) should be maximized. Furthermore, maximizing of $r$ is equivalent to the following formulation:

$$\min F_2 = N_b - r. \tag{2}$$

If a PMU is placed on especial bus, that bus and buses connected to that bus are observable. For Example, in Fig. 2, assume that 3 PMUs are located at buses 1, 2, and 7 and all buses are observable. To explain the redundancy criterion, we should disregard PMUs one by one and after each contingency the observability of buses should be analyzed. Now suppose that PMU installed at bus 1 is failed. In this state, all the buses are observable too. Accordingly, the PMU located at bus 7 would be failed. In this contingency bus 7 would be unobservable. Similarly, in the case of PMU outage at bus

2, buses 3, 4 and 5 are unobservable. Thus, the number of entire unobservable buses in all PMU outages is 4 and this is the value of second objective function.

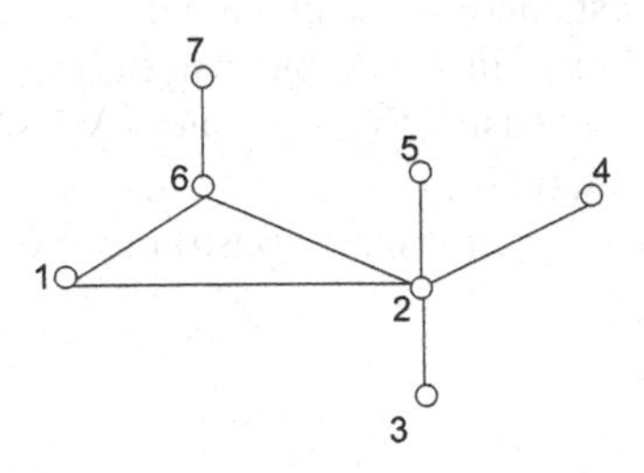


**Fig. 2.** Example for illustration

### 3.2 Constraints

As mentioned before, an installed PMU results in observability for he installed bus and its adjacent buses. A function ($f_i$) is introduced as sum of $u_i$ for bus $i$ and its neighbors.

$$f_i = GX = \sum_{j \in I} g_{ij} \times u_j, \quad \forall i \in I. \tag{3}$$

where,

$$G = \left[g_{ij}\right]_{N_b \times N_b} \tag{4}$$

$$g_{ij} = \begin{cases} 1, & \text{if bus } i \text{ and } j \text{ are connected or } i = j \\ 0, & \text{otherwise} \end{cases} \tag{5}$$

$$U = \left[u_1\, u_2\, \ldots\, u_{Nb}\right]^T \tag{6}$$

To avoid any unobservable bus, all the functions of $f_i$ should be nonzero:

$$f_i \geq 1, \quad \forall i \in I. \tag{7}$$

There is another issue related to zero injection buses. These buses may reduce the PMU number using KCL. In fact, when all buses associated with a zero injection bus are observable but one of them is unobservable, the KCL law leads to observability of that unobservable bus. Therefore, $f_i$ in (3) would be as follows [11]:

$$f_i = \sum_{j \in I} g_{ij} \times u_j + \sum_{j \in I} g_{ij} \times t_j \times q_{ij}, \quad \forall i \in I. \tag{8}$$

$$t_j = \sum_{i \in I} u_{ij} \times q_{ij}, \quad \forall j \in I. \tag{9}$$

where, $t_j$ is another binary variable (1 for zero injection bus and zero otherwise) and $q_{ij}$ represents an auxiliary binary variable. Suppose a zero injection bus $j$ with $t_j = 1$. In this state in (9) just one of the $q_{ij}$ equals to 1 which means zero injection bus $j$ results in bus $i$ to be observable and the equality (9) would be correct. Hence, the impact of zero injection buses is taken into account as the second part of (8) that is increased if a zero injection bus causes that another bus is observable. The zero injection buses, $t_j$ and $q_{ij}$ are explained completely in [11]. Similarly, when a power flow is placed on one line, if one of the sending-end or receiving-end buses of that line is observable, there is no need for other end-bus to have PMU for observability and it can be observable due to other bus [12]. Thus, constraint (7) for line $ij$ which is a power flow line is substituted by:

$$f_i + f_j \geq 1 \tag{10}$$

To formulate redundancy mathematically the following formula is presented using another binary variable i.e. $\beta_i$ [13]:

$$\frac{1}{2 \times N_b} + \frac{f_i - 3}{N_b} \leq \beta_i \leq 1 + \frac{f_i - 3}{N_b}. \tag{11}$$

In inequality (11), when $f_i$ is higher than 2 the left hand side of (11) will be between zero and one, and the other side would be greater than one so $\beta_i$ will be 1. Also, it is be equal to 0 if $f_i < 2$, then $\beta_i = 0$. Thus, sum of $\beta_i$ denotes the redundant buses with $f_i > 2$.

$$F_2 = N_b - \sum_{\forall i \in I} \beta_i . \tag{12}$$

This is the constraint of redundancy which should be greater than a specific value according to necessity of system operator.

$$r \geq p_{\min}. \tag{13}$$

where the required minimum value of redundancy is $p_{min}$.

### 3.3 Multi-objective Solution Methodology

The MOPP problem is as follows:

$$\begin{aligned} & \min\{F_1(x), F_2(x)\} \\ & s.t.: g(x) \leq 0, \ \ h(x) = 0. \end{aligned} \tag{14}$$

## 4 Simulation Results

A nine-bus test system has been implemented to simulate the proposed multi-objective optimization problem. The proposed approach has been employed through a Pentium IV, 1-GHz Core i5 with 8 GB RAM using mixed integer linear programming solver

CPLEX 9.0 in the GAMS [14]. Figure 3 shows the 9-bus test system. Except buses 1, 2 and 4 all the buses have load and also there are 2 generation units located at buses 1 and 2. Thus, bus 4 has no load and generator and is considered as a zero injection bus.

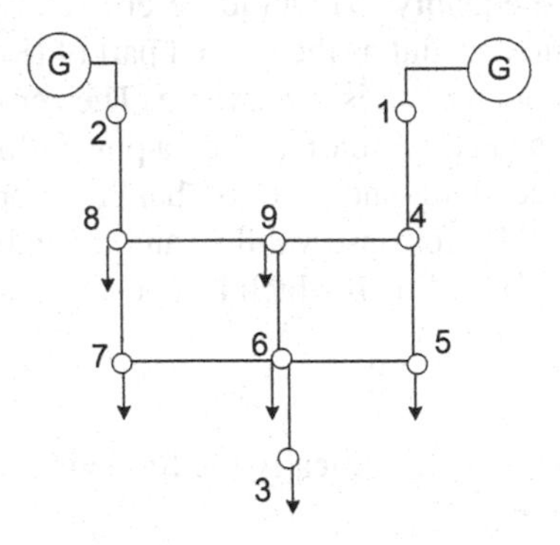


**Fig. 3.** Modified 9-bus test system

If the first objective function is considered as the main objective, we can have an observable system using lowest number of PMUs that is 3 PMUs at buses 4, 6 and 8. Subsequently, the amount of second objective stays in the worst state and there is no redundant bus with $f_i > 2$.

$$F_2 = N_b - \sum_{\forall i \in I} \beta_i = 9 - 0 = 9.$$

In the flip side, if we solve the single objective problem so that the second objective function is the main objective, there would be 6 PMU at buses 4-9. In this condition the second objective is equal to 3. The amount of $F_2$ cannot be less than 3, because there are 3 buses in the system which cannot have $f_i > 2$.

The Pareto points of the proposed Multi-objective method are shown in Fig. 4 and Table 1. As we can see from these Figure and Table, two objective functions have conflict

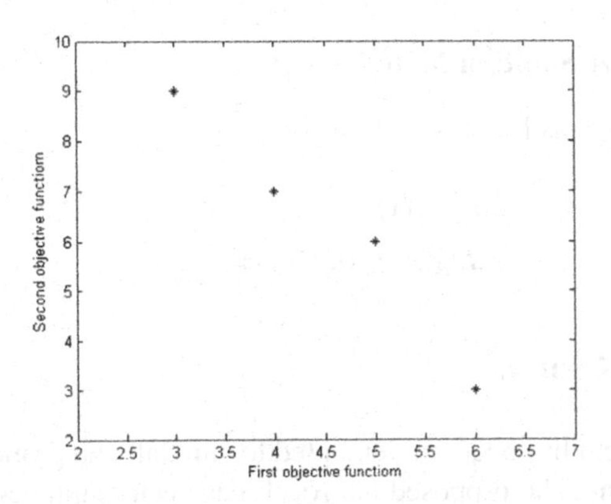


**Fig. 4.** Pareto set related to test system

manner and they cannot be optimum simultaneously. Whatever the first objective function would be minimized the second one would be worse and conversely.

**Table 1.** Pareto points for 9-bus test system according to MOPP

| Pareto # | $F_1$ | $F_2$ | Number of PMU buses | Number of redundancy buses |
|---|---|---|---|---|
| 1 | 3 | 9 | 4, 6, 8 | – |
| 2 | 4 | 7 | 6, 7, 8, 9 | 7, 8 |
| 3 | 5 | 6 | 2, 5, 6, 7, 8 | 6, 7, 8 |
| 4 | 6 | 3 | 4, 5, 6, 7, 8, 9 | 4, 5, 6, 7, 8, 9 |

Also, constraint (13) can decrease the number of points. For example in the case of $p_{min} = 1$ (the system needs 1 redundant bus) the first row of Pareto set in Table 1 should be disregarded. It is noteworthy that, the proposed approach can be implemented on larger case studies such as 118-bus or more because the formulation is linear and not complex.

## 5 Conclusion

This paper proposed a multi-objective PMU placement taking into account the redundancy as an extra objective. The redundancy criterion was introduced when a PMU was not in a suitable operation mode. Moreover, the formulation is linear for redundancy which decreases the complexity of the calculation. Two conflict objective functions are minimized simultaneously and the system operator can make a trade-off between points in the Pareto set and choose the best solution point according to the system requirements.

**Acknowledgment.** This work was supported by FEDER funds through COMPETE 2020 and by Portuguese funds through FCT, under Projects SAICT-PAC/0004/2015 - POCI-01-0145-FEDER-016434, POCI-01-0145-FEDER-006961, UID/EEA/50014/2013, UID/CEC/50021/2013, and UID/EMS/00151/2013. Also, the research leading to these results has received funding from the EU Seventh Framework Programme FP7/2007-2013 under grant agreement no. 309048.

## References

1. Khajeh, K.J., Bashar, E., Rad, A.M., Gharehpetian, G.B.: Integrated model considering effects of zero injection buses and conventional measurements on optimal PMU placement. IEEE Transactions on Smart Grid, in press. doi:10.1109/TSG.2015.2461558
2. Manousakis, N., Korres, G.: Optimal allocation of PMUs in the presence of conventional measurements considering contingencies. IEEE Transactions on Power Delivery, in press. doi:10.1109/TPWRD.2016.2524658
3. Korres, G.N., Manousakis, N.M., Xygkis, T.C., Löfberg, J.: Optimal phasor measurement unit placement for numerical observability in the presence of conventional measurements using semi-definite programming. IET Gener. Trans. Distri. **9**(15), 2427–2436 (2015)
4. Manousakis, N.M., Korres, G.N., Georgilakis, P.S.: Taxonomy of PMU placement methodologies. IEEE Trans. Power Syst. **27**(2), 1070–1077 (2012)

5. Abiri, E., Rashidi, F., Niknam, T.: An optimal PMU placement method for power system observability under various contingencies. Int. Trans. Electr. Energy Syst. **25**(4), 589–606 (2015)
6. Korres, G.N., Manousakis, N.M.: State estimation and observability analysis for phasor measurement unit measured systems. IET Gener. Trans. Distrib. **6**, 902–913 (2012)
7. Arasteh, H., et al.: Iot-based smart cities: a survey. In: 2016 IEEE 16th International Conference on Environment and Electrical Engineering (EEEIC), Florence, pp. 1–6 (2016). doi:10.1109/EEEIC.2016.7555867
8. Boroojeni, K.G., Amini, M.H., Bahrami, S., Iyengar, S.S., Sarwat, A.I., Karabasoglu, O.: A novel multi-time- scale modeling for electric power demand forecasting: From short-term to medium-term horizon. Electr. Power Syst. Res. **142**, 58–73 (2017)
9. Boroojeni, K.G., Amini, M.H., Iyengar, S.S.: Overview of the security and privacy issues in smart grids. In: Boroojeni, K.G., Amini, M.H., Iyengar, S.S. (eds.) Smart Grids: Security and Privacy Issues, pp. 1–16. Springer, Cham (2017). doi:10.1007/978-3-319-45050-6_1
10. Bahrami, S., Sheikhi, A.: From demand response in smart grid toward integrated demand response in smart energy hub. IEEE Trans. Smart Grid **7**(2), 650–658 (2016)
11. Aminifar, F., Khodaei, A., Fotuhi-Firuzabad, M., Shahidehpour, M.: Contingency-constrained PMU placement in power networks. IEEE Trans. Power Syst. **25**(1), 516–523 (2010)
12. Xu, B., Abur, A.: Observability analysis and measurement placement for systems with PMUs. IEEE Power Syst. Conf. Expos. **2**, 943–946 (2004)
13. Chen, J., Abur, A.: Placement of PMUs to enable bad data detection in state estimation. IEEE Trans. Power Syst. **21**(4), 1608–1615 (2006)
14. Generalized Algebraic Modeling Systems (GAMS). http://www.gams.com

# Distributed Infrastructure

# RELOAD/CoAP P2P Overlays for Network Coding Based Constrained Environments

Eman Al-Hawri, Noelia Correia, and Alvaro Barradas(✉)

CEOT, University of Algarve, 8005-139 Faro, Portugal
{esalhawri,ncorreia,abarra}@ualg.pt

**Abstract.** The Internet of Things will bring into the Internet all kinds of smart systems, which will be able to interact with each other. Therefore, applications relying on data sharing for collaboration will increase, and effective distributed solutions for data storage become necessary. This need led to the proposal of a CoAP Usage for RELOAD, a generic P2P protocol that accepts pluggable application layers (Usages). This allows P2P overlay networks to be built where constrained systems store their data and clients are able to retrieve it. Since many smart systems rely on wireless networks to communicate, where network coding can be used to reduce packet error rate, P2P overlays should be prepared to store data from network coding based networks. More specifically, encoding vectors and encoded data must be stored, and a decoding service is required. In this article, we propose a CoAP Usage extension so that network coding based constrained networks can use RELOAD/CoAP P2P distributed storage.

**Keywords:** P2P · RELOAD · CoAP Usage · Network coding

## 1 Introduction

Smart systems appear everywhere and this is expected to increase in the near future [1]. Such trend, combined with the potential for the Internet of Things (IoT) to connect tens of billions of objects to the Internet, will allow geographically distributed systems to collaborate and smart services to emerge. In such scenarios, peer-to-peer (P2P) networks may have a key role allowing federated peers to collaborate and improve their businesses.

In the context of federated constrained systems/networks some open standards become important. Within Internet Engineering Task Force (IETF), the Constrained RESTful Environments (CoRE) working group has focused on the development of Constraint Application Protocol (CoAP), a data messaging/transfer protocol providing a request/response interaction model between application endpoints [2, 3]. The "coap://" Uniform Resource Identifier (URI) scheme is used to identify CoAP resources, while "/.well-known" is defined as a default entry point for resource discovery. The Internet media type "application/link-format" is assigned for CoRE Link Format payloads [4, 5].

More recently, CoAP Usage for REsource LOcation And Discovery (RELOAD), a base protocol that provides a generic self-organizing P2P overlay network service, was proposed [6]. The use of pluggable application layers, called Usages, allows RELOAD to fit any purpose [7]. In a RELOAD/CoAP architecture the proxy nodes form a


L.M. Camarinha-Matos et al. (Eds.): DoCEIS 2017, IFIP AICT 499, pp. 307–315, 2017.
DOI: 10.1007/978-3-319-56077-9_30

distributed P2P overlay network to announce resources, allowing clients to discover them. More specifically, the overlay can be used: as a lookup service, to store existing resources, and as cache for data.

Many applications (e.g. industrial, environmental) require smart systems to communicate using wireless constrained networks. Such constrained environments usually have energy efficiency and end-to-end packet error rate concerns, which are competing goals (e.g. a packet may be sent through multiple paths to reduce error rate but this increases energy consumption). An elegant way of achieving a balance between these two goals is to use Network Coding (NC) [8]. When NC is used, nodes can perform some linear combination on the received packets and route encoded packets. At the destination, the decoding process can be applied to recover the original packets. Since linear combinations of packets are being done, the sink will be able to recover lost packets from a sufficient number of other arriving packets.

In this article, we propose an extension of CoAP Usage so that NC based constrained networks can benefit from RELOAD/CoAP P2P distributed storage. That is, although packets will travel from sources toward sinks/gateways, their final destination will be the P2P overlay where storage is done. When compared with the work in [6] a new data Kind structure is proposed for encoded data and encoding vectors to be stored. For the decoding process, required when original data packets are not received, a decoding service is required at the P2P overlay network. Such architecture allows constrained networks to reduce packet error rate while benefiting from an effective distributed solution for data storage. This work serves as a basis for future proposal of algorithms that determine the most effective routing trees, for packet forwarding toward sink/gateway nodes, and best placement of coding nodes.

The remainder of this article is organized as follows. Section 2 explains how smart systems can benefit from our proposal. Section 3 discusses network coding, while Sect. 4 presents the proposed architecture and extension of CoAP Usage. Finally, in Sect. 5 some conclusions are drawn.

## 2 Relationship to Smart Systems

A smart system is a collection of sensing, actuation, and controlling functions, allowing situations to be analyzed and decision making. Such systems mainly rely on sensor information sources, which might be geographically distributed [1]. With the increasing presence of smart systems in the Internet, applications relying on data sharing will increase and effective solutions for storage/discovery of data become necessary.

Although Cloud storage solutions have emerged, extra storage space always requires extra physical disks and processing, meaning that connecting hundreds of millions of sensors to the Cloud will be extremely demanding. Besides this, such places are known as storage places meaning that they are exposed to attacks. P2P approaches can overcome both these issues because participating peers/proxies also contribute to storage and processing, peers communicate directly posing no burden on a specific set of servers, and bandwidth bottlenecks are avoided. Encrypted communication tunnels can also be built between peers.

CoAP is a Web application transfer protocol that was developed for constrained systems to provide RESTful services. Although devices will be heterogeneous regarding their radio layer, CoAP is expected to become a common application layer protocol. For this reason, the CoAP Usage for RELOAD has been proposed, avoiding the use of centralized servers. The use of P2P overlays is particularly useful for power-constrained systems that are expected to explore sleeping modes for higher energy saving. This is so because clients can retrieve information directly from the overlay without performing either a connection to the proxy or the power-constrained system. Thus, RELOAD/CoAP P2P overlays are an effective response to many of the challenges that smart systems face.

Many smart systems rely on wireless constrained networks to communicate. Therefore, it makes sense to prepare P2P overlays to store data arriving from NC based wireless networks. More specifically, encoding vectors and encoded data must be stored, and a decoding service supplied. NC improves packet delivery rate and provides some degree of protection because an attacker cannot control the outcome of the decoding process without knowing all other coded packets [9]. These issues are critical for smart systems.

# 3 Network Coding

In network coding the nodes take several packets and combine them together for transmission, instead of simply relaying the data packets they receive [9]. This approach can be used to improve the throughput, efficiency, scalability, resilience to attacks and eavesdropping in networks. In the case of sensor Networks it can be used to reduce packet error rate [8].

When linear global coding is used the encoded packets are transmitted along the network, where encoding can be performed recursively. When packets reach their destination the decoding process can be applied on these packets for the original ones to be extracted. In the following discussion it is assumed that:

- Each packet consists of $l$ bits, small packets are padded with 0's;
- s consecutive bits form a symbol over finite field $F_{2^s}$;
- Each packet is a vector of $\frac{l}{s}$ symbols;
- The resulting linearly combined packets also have length $l$.

## 3.1 Encoding

Let us assume the original packets $(k^1, \ldots, k^n)$. An encoded packet $\omega$ will be a vector of $\frac{l}{s}$ symbols, $\omega = [\omega_1, \ldots, \omega_u, \ldots, \omega_{\frac{l}{s}}]$, where the $u^{th}$ symbol will be:

$$\omega_u = \sum\nolimits_{i=1}^{n} \alpha_i k_u^i \quad (1)$$

where coefficients $\alpha_1, \ldots, \alpha_n$ are in $F_{2^s}$, while $k_u^i$ and $\omega_u$ are the $u^{th}$ symbols of $k^i$ and $\omega$, respectively. The encoding vector $\alpha = (\alpha_1, \ldots, \alpha_n)$ is used by recipients to decode the data. Since the encoding process can occur recursively through the network, any intermediate node can receive the encoded packets $\omega^1, \ldots, \omega^m$ and encode them again to generate a linear combination of these packets. Although a different set of coefficients is used to make the new linear combinations, $h_1, \ldots, h_m$, the corresponding encoding vector is not simply $h = (h_1, \ldots, h_m)$ since coefficients are with respect to the original packets $k^1, \ldots, k^n$. That is, the re-encoded packet $\tilde{\omega}$ ends up being tagged with an encoding vector whose elements are:

$$\tilde{\alpha}_i = \sum_{j=1}^{m} h_j \alpha_i^j \qquad (2)$$

### 3.2 Decoding

When a node receives $m$ arrivals $(\alpha^1, \omega^1), \ldots, (\alpha^m, \omega^m)$, it needs to retrieve the original packets. This can be achieved by solving the system:

$$\omega^j = \sum_{i=1}^{n} \alpha_i^j k^i, \forall_j = 1, \ldots, m \qquad (3)$$

This linear system has $m$ equations that need to be solved to extract the $n$ unknowns (unknowns are $k^i$). To perform this, it requires that $m \geq n$ to be able to recover the original data, which means that the number of received packets should be at least equal to the number of the original packets.

## 4 Proposed P2P Architecture

### 4.1 Motivation

Network coding is an elegant way of finding a balance between energy efficiency and end-to-end packet error rate, which are competing goals. However, network coding based constrained wireless networks cannot benefit from RELOAD/CoAP P2P storage. For this reason we propose to extend CoAP Usage data Kind. A decoding service, to be provided by the P2P overlay, is required.

### 4.2 Overall Architecture

At the wireless section the nodes are organized in a way that packet flowing toward a sink/proxy node is ensured. Figure 1 shows the proposed architecture. Depending on factors like location, energy and functionality, nodes can be of type:

- Sensing: Data sources that forward their packets according to some routing table. Nodes can turn to sleep mode to reduce energy consumption.
- Encoding: Node able to perform encoding. Besides forwarding original packets, extra encoded packets are created by linearly combining received packets and packets overheard from neighbors. These might be required to have more energy, memory and processing capabilities than others.
- Relay: Work just as relays and forward any received packets, either original or encoded, towards the sink. Perform no encoding/decoding operations.
- Sink: Destination of data. It is connected to the RELOAD overlay network where packets can be stored and fetched by others.

It is assumed that encoding nodes propagate linear combinations of original packets and packets overheard from neighbors, besides original packets. Sinks/gateways store original and encoded packets at the overlay.

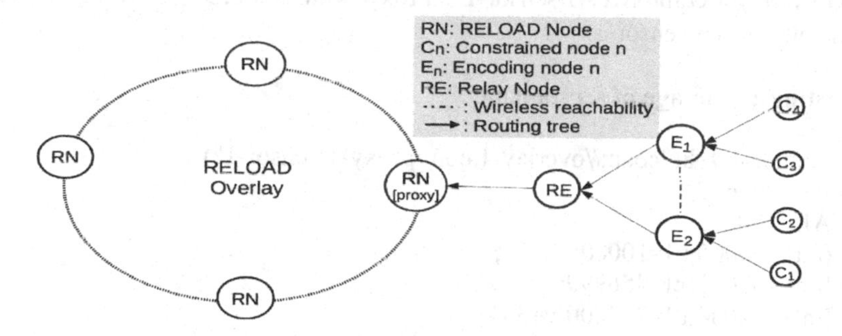


**Fig. 1.** Overall architecture.

## 4.3 Storing Data and Standard Extensions Needed

### 4.3.1 Data Kinds

The data Kinds defined by CoAP Usage include the CoAP-REGISTRATION, to announce resources available, and CoAP-CACHING for the storage of sensor measurements [6]. Although the CoAP protocol itself supports the caching of sensor measurements to reduce the response time and network bandwidth consumption for future requests (see [2]), the additional caching mechanism of CoAP Usage allows such data to be stored in the P2P overlay, improving even more the response time and bandwidth use because nodes can go to sleep mode for longer periods of time.

Regarding CoAP-CACHING, the possibility of storing proxy data and sensor data directly are both considered. Their resource names will be their CoAP URIs (e.g. "coap://overlay-1.com/proxy-1/", "coap://overlay-1.com/proxy-1/sensor-temp"). That is, the proxy data structure supports data from multiple sensor nodes forwarding their data to a proxy, while the sensor data structure stores measurements of a specific sensor node. Listing 1 shows the data stored for proxy with NodeID "9996172" and URI "coap://overlay-1.com/proxy-1", where "mt" is the measurement time of data, "ttl" is the time-to-live, and "v" is the measurement value.

**Listing 1.** Storage of proxy data.

```
Resource-ID=h("coap://overlay-1.com/proxy-1/")
KEY=9996172,
VALUE=[
 </sensor-1/>; {mt=100000;ttl=10000;v=38};
         {mt=100055;ttl=456990;v=42};
         {mt=134000;ttl=234000;v=30}
 </sensor-2/>; {mt=100000;ttl=10000;v=40};
                         {mt=400000;ttl=17000;v=25}]
```

The "h(…)" is the hash over the URI, because the key space for storage in the overlay must be numeric. Listing 2 shows a direct storage of data from sensor with URI "coap://overlay-1.com/proxy-1/sensor-1". Proxy with NodeID "9996172" is the one responsible for the sensor-1.

**Listing 2.** Storage of sensor data.

```
Resource-ID=h("coap://overlay-1.com/proxy-1/sensor-1")
KEY=9996172,
VALUE=[
 {mt=100000;ttl=10000;v=38};
 {mt=100055;ttl=456990;v=42};
 {mt=134000;ttl=234000;v=30}]
```

#### 4.3.2 Requirements Regarding Network Coding

Network coding based constrained networks forward original and encoded packets toward the proxy participating as peers in the P2P overlay network. Therefore, for network coding based constrained networks to benefit from such a distributed storage, the data Kind must also allow the storage of:

- *Encoding vectors:* Required for the decoding. For deterministic network coding these may be stored at the overlay at peer registration time (done once). A proxy stores encoding vectors of sensors for which it is responsible for.
- *Encoded packets:* Since decoding may not be done if a sufficient number of encoded packets has not arrived, storage of encoded packets is needed.

Therefore, the current CoAP Usage data Kind must be extended. Listing 3 shows an example when using such extended data Kind. The first value regards to the current data Kind structure and second value includes predefined encoding vectors and arriving encoded data.

**Listing 3.** Extended sensor data.

```
Resource-ID=h("coap://overlay-1.com/proxy-1/")
KEY=9996172,
VALUE=[
 </sensor-1/>; {mt=100000;ttl=10000;v=38};
              {mt=100055;ttl=456990;v=42};
              {mt=134000;ttl=234000;v=30}
 </sensor-2/>; {mt=100000;ttl=10000;v=40};
              {mt=400000;ttl=17000;v=25}]
ENCODING=[
  {enVector1;enVector2;...};
  {1stcodedData;2ndcodedData;...}]
```

### 4.3.3 Storing/Fetching Packets of Network Coding Based Constrained Network

Figure 2 illustrates how storage of packets would be done. In this example, encoding nodes $E_1, E_2$ receive packets from their children, namely $E_1$ receives $k_1, k_2$, and $E_2$ receives $k_3, k_4$. Moreover, these encoding nodes can hear each other. In this case, the linear combinations of packets from children plus packets they have heard is done $(k_1 + k_2 + k_3 + k_4)$. When a single packet gets lost, it can be retrieved when the decoding is performed. Please note that encoding can be performed at different areas of the wireless section, and nodes may forward packets toward different proxies/gateways. Therefore, decoding might require encoded packets stored by different proxies. Therefore, a relevant research issue is how to build routing trees that increase the effectiveness of encoding. Also, other encoding scenarios are possible (e.g. $E_1$ and $E_2$ could forward some overheard packets so that multiple packets can be recovered) with different impacts on bandwidth usage.

When a client needs to fetch sensor data from the overlay network, such data can be directly available (original data) or there might be a need to perform the decoding process, using stored encoded data and encoding vectors, in order to extract the original data. This decoding would be accomplished by RELOAD nodes able to provide such

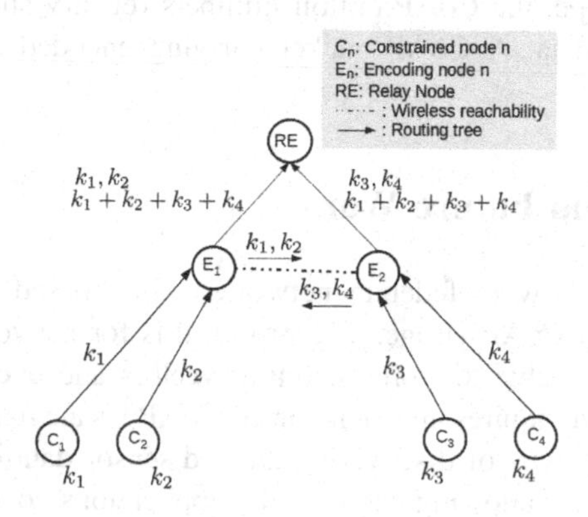


**Fig. 2.** Example of network coding at constrained network.

decoding service. Figure 3 illustrates a scenario where the decoding service is requested. A fetch request by some client may involve fetching original data and encoded data, which also requires fetching the encoding vectors, so that the highest amount of sensor measurements is returned to the client.

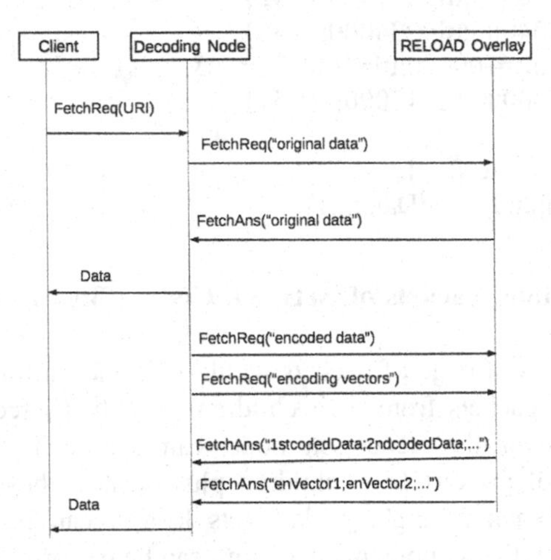


**Fig. 3.** Decoding at RELOAD/CoAP P2P overlay network.

Since peers face the problem of finding the peers providing the decoding service, the *Recursive Distributed Rendezvous (*ReDiR) service discovery mechanism can be used [10]. ReDiR ensures that the load is distributed among the nodes providing the service.

### 4.3.4 CoAP Option

CoAP allows options to be included in a message [2]. Each option instance specifies the option number, length of the option value, and the option value itself. For the proxy to be able to differentiate the type of payload at CoAP packets, which may be of original or encoded type, the CoAP option numbers registry should include an entry referring to encoded data. A CoAP packet carrying encoded data must carry such option.

## 5 Conclusions and Future Work

This article discusses how to federate network coding based constrained networks through RELOAD and CoAP Usage. The proposal is for network coding based constrained networks to be able to store encoding vectors and encoded data at the P2P overlay network, which requires an extension of CoAP Usage data Kind. This provides a scalable and efficient way of discovering cached sensor data of geographically dispersed sensor networks, allowing large scale applications to emerge, while taking advantage of network coding at the wireless section. This architecture serves as a basis for future proposal of algorithms that determine the most effective routing trees and placement of coding nodes.

**Acknowledgment.** This work was supported by FCT (Foundation for Science and Technology) from Portugal within CEOT Research Center and UID/MULTI/00631/2013 project. Eman AL-Hawri is supported by a grant from Thamar University - Yemen.

## References

1. Jones, A., Subrahmanian, E., Hamins, A., Grant, C.: Humans' critical role in smart systems: a smart firefighting example. IEEE Internet Comput. **19**(3), 28–31 (2015)
2. Shelby, Z., et al.: The Constrained Application Protocol (CoAP). RFC 7252 (2014)
3. Correia, N., Schütz, G., Mazayev, A., Martins, J., Barradas, A.: An energy aware resource design model for constrained networks. IEEE Commun. Lett. **20**(8), 1631–1634 (2016)
4. Shelby, Z.: Constrained RESTful environments (CoRE) link format. RFC 6690 (2012)
5. Nottingham, M., Hammer-Lahav, E.: Defining well-known Uniform Resource Identifiers (URIs). RFC 5785 (2010)
6. Jimenez, J., et al.: A Constrained Application Protocol (CoAP) usage for REsource LOcation And Discovery (RELOAD). RFC 7650 (2015)
7. Jennings, C., et al.: REsource LOcation And Discovery (RELOAD) base protocol. RFC 6940 (2014)
8. Keller, L., Atsan, E., Argyraki, K., Fragouli, C.: SenseCode: network coding for reliable networks. ACM Trans. Sens. Netw. **9**(2), 25 (2013)
9. Fragouli, C., Le Boudec, J., Widmer, J.: Network coding: an instant primer. ACM SIGCOMM Comput. Commun. Rev. **36**(1), 63–68 (2006)
10. Maenpaa, J., Camarillo, G.: Service discovery usage for REsource LOcation And Discovery (RELOAD). RFC 7374 (2014)

# PVInGrid: A Distributed Infrastructure for Evaluating the Integration of Photovoltaic Systems in Smart Grid

Lorenzo Bottaccioli(✉), Enrico Macii, Edoardo Patti, Abouzar Estebsari, Enrico Pons, and Andrea Acquaviva

Politecnico di Torino, Corso Duca Degli Abbruzzi 22, Turin, Italy
{lorenzo.bottaccioli,enrico.macci,edoardo.patti, abouzar.estebsari,enrico.pons,andrea.acquaviva}@polito.it

**Abstract.** Planning and developing the future Smart City is becoming mandatory due to the need of moving forward to a more sustainable society. To foster this transition an accurate simulation of energy production from renewable sources, such as Photovoltaic Panels (PV), is necessary to evaluate the impact on the grid. In this paper, we present a distributed infrastructure that simulates the PV production and evaluates the integration of such systems in the grid considering data provided by smart-meters. The proposed solution is able to model the behaviour of PV systems solution exploiting GIS representation of rooftops and real meteorological data. Finally, such information is used to feed a real-time distribution network simulator.

**Keywords:** Photovoltaic · GIS · Distributed infrastructure · Smart-grid · Real-time simulation

## 1 Introduction and Motivation

Thanks to the commitment of the participating countries to Kyoto Protocol and to Paris Agreement, the presence of distributed Renewable Energy Sources (RES) is increasing in our electricity systems. In this transition from a centralized to a distributed system, is important converting passive buildings rooftops of our cities into active Photovoltaic rooftops becomes crucial. However, the deployment of PV system cannot be left to chance but needs to be planed with specific tools able to *i)* estimate PV production in time and *ii)* to assess the capabilities and requirements of the networks [11]. This research aims at creating PVinGRID, a distributed infrastructure composed by software and hardware components. PVinGRID integrates: *i)* Geographic Information Systems (GIS) data and algorithms; *ii)* sensor, such as weather station and smart meter; *iii)* real-time hardware electricity grid simulator. Exploiting GIS tools and weather station data PVinGRID estimates the availability of deployable area for PV system on rooftops and simulates sub-hourly PV generation loads in real-sky conditions. BY


L.M. Camarinha-Matos et al. (Eds.): DoCEIS 2017, IFIP AICT 499, pp. 316–324, 2017.
DOI: 10.1007/978-3-319-56077-9_31

merging data coming from smart meters [1] with simulated PV generation loads PVinGRID evaluates the integration of PV system in the distribution network exploiting a real-time grid simulator. The rest of the paper is organized as follows. In Sect. 2 the relationship of this research with smart system is highlighted. Section 3 presents literature methodologies and services for PV energy simulation and integration into the grid. Section 4 presents the specifications and methodology to develop our solution. In Sect. 5 the results obtained with PVinGRID are presented. Finally, Sect. 6 provides the concluding remarks.

## 2 Relationship to Smart Systems

Information and Communication Technologies (ICTs) are a key factor in the development of smart systems for Smart Energy management. They can provide useful tools for planning and monitoring the deployment of RES systems. Furthermore, the spread diffusion of heterogeneous and pervasive sensors in our houses, district and cities is increasing the smartness of hour distribution grid. With such sensors, it is possible to monitor the electricity behaviours of the users. INformation on electricity consumption are needed in both planning and monitoring phases for evaluating: *i)* energy management actions; *ii)* management of electricity distribution networks; *iii)* integration of renewable sources in the city. PVinGRID is a tool useful for both planning and monitoring phases of PV systems, spanning all scales starting from single building up to block, district and city. Our solution can be used for different purposes depending on the needs of different stakeholders. For instance a ***Single citizen*** uses PVinGRID for estimating the economic benefits that he/she can achieve by installing a new PV system. End-user wants to be aware of the avoided $CO_2$ emission and virtuous actions to perform. ***Energy aggregators and Energy Communities*** can use our solution for scheduling the consumption pattern of their clients for maximizing self-consumption and minimizing energy bills. In particular ***Energy Communities*** can exploit such infrastructure to perform feasibility studies as proposed by [3]. ***Distribution system operators*** (DSO) can use PVinGRID for analysing the reliability of their grids after the deployment of PV system and plan, if needed, retrofits actions and/or extensions of existing distribution grids. They can exploit our solution to avoid network congestions exploiting load balancing. Lastly ***Energy and City planners*** can use PVinGRID for evaluating the benefits of large PV systems installations in the cities or monitoring the performance of already deployed PV systems. Furthermore, if accurate weather forecasts are available PVinGRID solution provides information on generation profiles of existing PV systems to: *i)* ***DSO*** for estimate possible unbalances on the grid and to evaluate load balancing actions (e.g. Demand Response and Energy Storage); *ii)* ***Market Operators*** for creating local or dynamic prices; *iii)* ***Smart Energy management systems*** for evaluating next actions in smart energy management (e.g. usage of storage and load scheduling).

## 3 State of the Art

Geographic Information Systems (GIS) tools have been widely applied in the estimation of renewable potential as reported in [5] in particular in urban contexts for solar applications [9]. New methodologies exploiting GIS for simulation with spatial and temporal domains of PV system energy production are recently emerging [4,10,12]. A spatio-temporal simulation of PV production integrated with network topology and users consumption load profiles allows to estimate the real PV potential by evaluating the effects of the penetration of PV systems into the grid [4]. Available web solution such as [13,18] are limited in the analysis of the time domain and do not integrate electricity consumption and network topology data. Hence, there is the need to develop frameworks that integrate consumption, production and network topology in order to evaluate the integration of PV system, or more in general RES.

Real-time simulator, such as Opal-RT®, are applied in the prototyping phase of the development of system or manufacture. This because real-time simulations allow testing solutions that are still not physical prototypes [2,6,19]. With respect to RES, real-time simulations have been applied for PV generation [15] and for wind-farms [14,20].

In this work we propose a simulation framework called PVinGRID that couples PV simulator with a real-time grid simulator, such as Opal-RT®. PVinGRID considers the electric behaviour of the distribution system by taking in to account both information on electricity consumption and network topology.

## 4 Contributions and Innovative Aspects

In this Section, we describe PVinGRID our distributed infrastructure for evaluating the integration of PV systems in a Smart Grid environment. PVinGRID is composed by both hardware and software components. The hardware components are: *i)* the real-time grid simulator, *ii)* weather station and *iii)* smart meters. The software component is a distributed infrastructure for estimating PV systems energy production (see Fig. 1(a)).

### 4.1 Real-Time Grid Simulator

Real time simulation (RTS) is a highly reliable method based on electromagnetic transient simulation which serves a platform to test new control strategies or technologies on a virtual environment emulating the real world system. It provides very reliable real-like information on impacts and benefits of new strategies or devices. RTS could support decision makings from real-time operation and control phase to long-term planning. Regarding electricity systems, RTS is being widely used in protection and control system development and testing. In particular for distributed generation modelling of RES integration (e.g. PV generation penetration), and intelligent grids development.

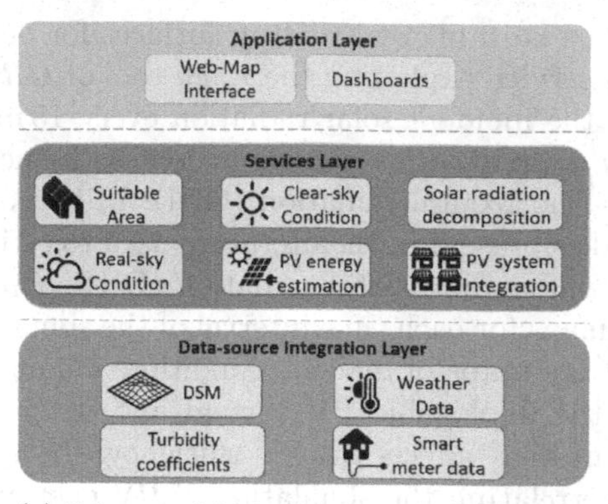


(a) Scheme of the software architecture for PV systems integration into Smart Grids

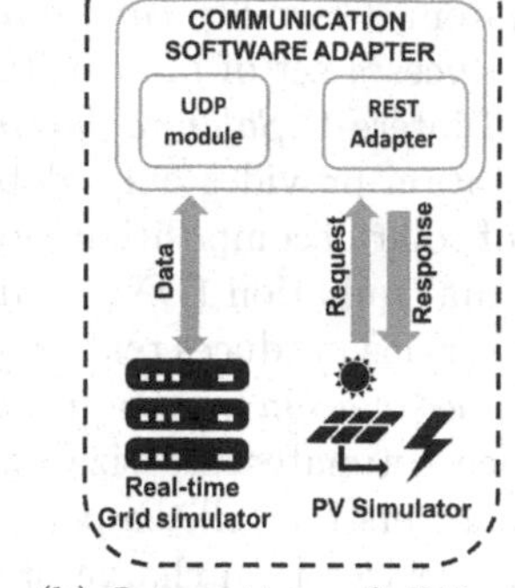


(b) Integration of PV and Real-Time grid simulators

**Fig. 1.** PVinGRID infrastructure

The purpose of using RTS in our work is to model a realistic distribution network to support investigations in terms of PV power penetration impacts in real-world situations. The objective is to simulate the behaviour of prosumers, and the set-up of a Software In-the-Loop (SIL) platform for laboratory validations of new control, operation, and planning algorithms for smart grids management [7].

### 4.2 Photovoltaic System Simulator

In this section we present the distributed software infrastructure for simulating PV generation profiles (see Fig. 1(a)). The development of this solution is needed to model and estimate PV system behaviours in both spatial and temporal domain. Thus, evaluations of their impact in the grid can be assessed. Our solution has been designed following both microservice[1] and REST [8] approaches. This due to the intention of developing a simulation tool with an easy maintenance and to allow integration of third-party software. Our solution is composed by three layers described in the following:

The **Data-source Integration Layer**, the lower layer, is in charge of integrating the following data-sources: *i) Digital Surface Model* (DSM) of the area in study, which is a raster image that represents terrain elevation considering the presence of manufactures; *ii) Linke Turbidity coefficients* that express the attenuation of solar radiation related to air pollution; *iii) Weather Data* coming from the nearest weather station provided by third-party services; *iv)* electricity consumption data provided by external software platforms(eg. [16,17]).

The **Services Layer** is the core of the simulation infrastructure and is composed by independent services that communicates trough REST-API: *Suitable*

[1] Fowler, M.: Microservices (2014), http://martinfowler.com/articles/microservices.html.

*Area* service is in charge of identifying suitable real rooftop surfaces for deploying PV system. *Clear-Sky condition* service exploiting the *r.sun* tool of *GRASS-GIS*[2], produces a set of maps with the incident solar radiation every 15 min in the area of interest. *Solar radiation decomposition* service is used if the nearest weather station provides only global horizontal radiation (GHI). In its core, it has a set of solar decomposition models present in the literature used to estimate direct normal radiation (DNI) and diffuse horizontal radiation (DHI). *Real-sky condition* service produces real-sky maps for each time interval of the simulation. *PV energy estimation* service using the maps of real-sky condition and suitable area service estimates the size of the deployable PV system and its production. It considers the cell temperature that affects the PV efficiency. *PV system Integration* service is in charge of correlating the simulation of PV energy production with data coming from the Smart Grid (e.g. energy consumption). This service provides information on the level of self-sufficiency (i.e. the share local demand fulfilled by PV energy production) and the level of self-consumption of the produced energy (i.e. the share of produced energy that is instantaneously consumed) for the area of interest. Furthermore, it enables the communication with the real-time grid simulator (see Sect. 4.1) as described in the following Sect. 4.3.

The **Application Layer** provides applications that allows end-users to interact with the simulation results exploiting *Web-Map interface* and *Dashboards*.

### 4.3 Smart-Grid Simulation Tool

As mentioned in Sect. 4.2, the *PV system Integration* service in Fig. 1(a) correlates the output of the PV system energy simulation with information coming from the Smart Grid with the same geographic area. In addition, it enables the communication with the real-time grid simulator through the *Communication Software Adapter* (see Fig. 1(b)). The adapter integrates two sub-modules: i) *REST adapter* and ii) *UDP module*. The *REST adapter* parses the simulation request from the RTS to the PV system simulator by translating them into REST calls to remote web-services. The UDP module pushes the simulated PV energy production data in to the RTS. Furthermore, it receives and processes the information coming from the RTS.

## 5 Results and Critical View

To demonstrate an application of the proposed integrated simulation platform in PV penetration assessment, we present some results of simulating the distribution grid behaviour in a real case-study considering a high PV penetration. The area under analysis is a city district with nearly 2200 residential buildings and 43 MV/LV substations. For this purpose a summer sunny day where there is the maximum production from PV system has been selected to show the maximum

[2] GRASS GIS, Open Source Geospatial Foundation, http://grass.osgeo.org.

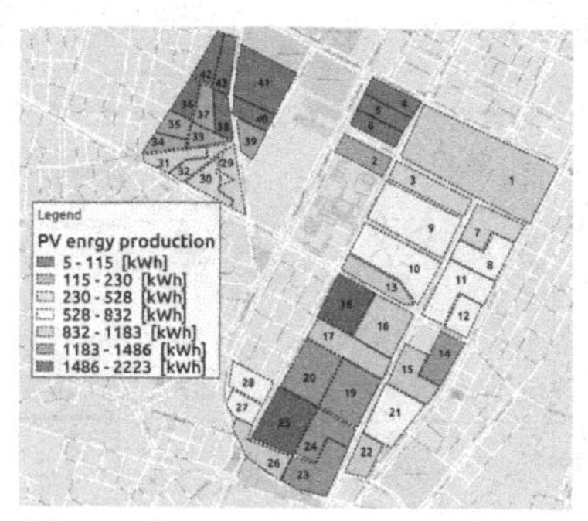


(a) Total PV energy generation in the summer sunny day

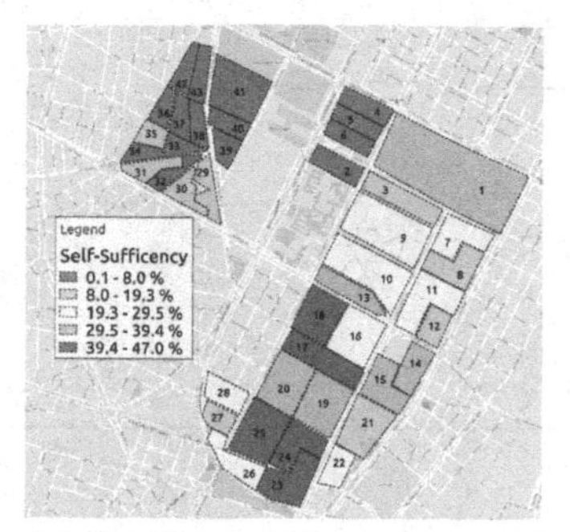


(b) Maximum Percentage of self-sufficency of each sub-station

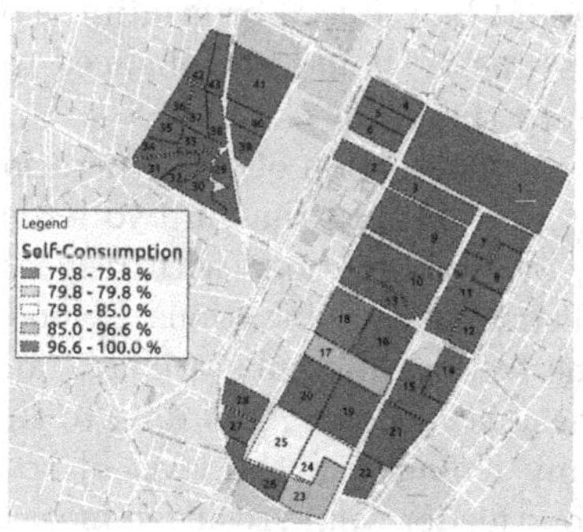


(c) Maximum Percentage of self-consumtion of each sub-station

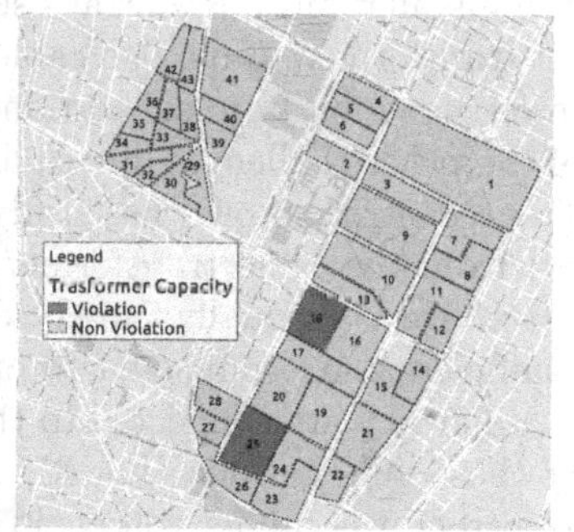


(d) Sub-station violating trasform capacity

**Fig. 2.** Mapped simulation results

level of both self-sufficiency and self-consumption and highlighting possible violation of MV/LV transformers. The simulation process identified 944 suitable areas in the building rooftops for deploying PV systems with a maximum nominal power equal to 14.21 $MW$. Figure 2(a) shows the distribution of energy production for each substation in a summer sunny day, that for the whole area is equal to 28.41 $MWh$. In Fig. 2(b) and (c) the levels of self-sufficiency and self-consumption are reported. Figure 2(b) shows that the sub-station with less production has less than 10% of self sufficiency. However, sub-stations 17, 18, 23, 24 and 25 have a level of self-sufficiency over 39%. Figure 2(c) revels that almost every sub-station absorbs all the produced energy from PV system. Only sub-stations 17, 18, 23, 24 and 25 feed the MV distribution grid and are the one with the highest level of self-sufficiency. Looking at the transformer capacity map (Fig. 2(d)) integrating PV generation would cause violations in sub-station 17 and 25. The maximum net consumption in these two sub-stations exceeds the transformers capacity due to an high amount of PV generation and low consumption. This highlights that in cases where PV generation is higher than local

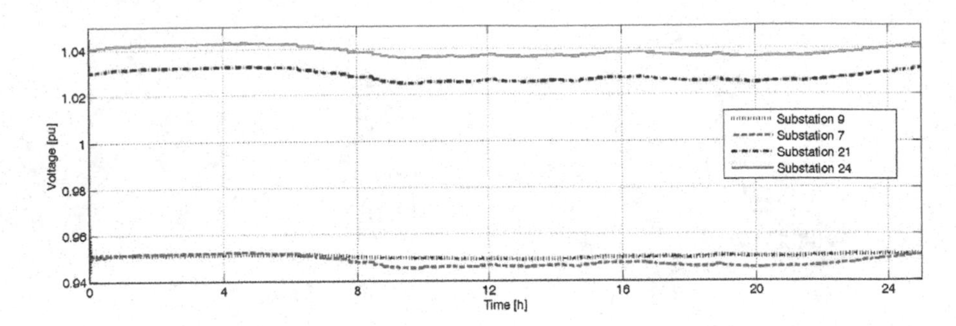


**Fig. 3.** Voltage profile for substations 7-9-22-24

demand the installation of PV arrays should be reduced or grid infrastructures, in terms of transformers (and also cables/lines), should be enhanced to tolerate reverse power injection from substations to the grid.

One of the other challenges due to high penetration of PV generation relates to the voltage control: in our case study, we observed that in two feeders which are derived from the same transformer, local generations with respect to local consumptions are not equally/closely distributed. Substations 22 and 24 with high self-sufficiency are connected to one feeder, and substations 7 and 9 with lower self-sufficiency are connected to the other feeder. According to the voltage profiles of these sub-stations (Fig. 3), any changes in the level of transformer voltage to correct over/under voltage in one feeder would result in more deviation in the other feeder.

## 6 Conclusions and Future Work

In this work PVinGRID a simulation infrastructure for evaluating PV integration in to smart grid has been presented. In addition motivations and relationships of the research with smart system has been highlighted. The results show how PVinGRID can be used for planning the deployment of PV systems and to evaluate the impact of such deployment in distribution grids. As future work a service for economic evaluation will be developed and included in the proposed solution. In particular, the simulation infrastructure will be integrated with a market simulator for evaluating dynamic prices and Demand Side Management policies. Furthermore the overall infrastructure will be used for evaluating Smart Energy actions as storage management, Demand Response events, Automatic Voltage Regulation, Network Reconfiguration and Load Balancing.

**Acknowledgement.** This work was supported by the EU project FLEXMETER, and by the Italian project "Edifici a Zero Consumo Energetico in Distretti Urbani Intelligenti".

## References


1. Bahmanyar, A., Jamali, S., Estebsari, A., Pons, E., Bompard, E., Patti, E., Acquaviva, A.: Emerging smart meters in electrical distribution systems: opportunities and challenges. In: Proceedings of ICEE, pp. 1082–1087 (2016)
2. Bompard, E., Monti, A., Tenconi, A., Estebsari, A., Huang, T., Pons, E., Stevic, M., Vaschetto, S., Vogel, S.: A multi-site real-time co-simulation platform for the testing of control strategies of distributed storage and V2G in distribution networks. In: Proceedings of EPE 2016 ECCE. IEEE (2016). Submitted for publication
3. Bottaccioli, L., Patti, E., Acquaviva, A., Macii, E., Jarre, M., Noussan, M.: A toolchain to foster a new business model for photovoltaic systems integration exploiting an energy community approach. In: Proceedings of ETFA2015. IEEE (2015)
4. Camargo, L.R., Zink, R., Dorner, W., Stoeglehner, G.: Spatio-temporal modeling of roof-top photovoltaic panels for improved technical potential assessment and electricity peak load offsetting at the municipal scale. Comput. Environ. Urban Syst. **52**, 58–69 (2015)
5. Domínguez, J., Amador, J.: Geographical information systems applied in the field of renewable energy sources. Comput. Ind. Eng. **52**(3), 322–326 (2007)
6. Dufour, C., Blanchette, H., Bélanger, J.: Very-high speed control of an FPGA-based finite-element-analysis permanent magnet synchronous virtual motor drive system. In: Proceedings of IECON 2008, pp. 2411–2416. IEEE (2008)
7. Estebsari, A., Pons, E., Patti, E., Mengistu, M., Bompard, E., Bahmanyar, A., Jamali, S.: An IOT realization in an interdepartmental real time simulation lab for distribution system control and management studies. In: Proceedings of EEEIC 2016. IEEE (2016). Submitted for publication
8. Fielding, R.T.: Architectural styles and the design of network-based software architectures. Open Source Geospatial Foundation (2000)
9. Freitas, S., Catita, C., Redweik, P., Brito, M.: Modelling solar potential in the urban environment: state-of-the-art review. Renew. Sustain. Energy Rev. **41**, 915–931 (2015)
10. Jakubiec, J.A., Reinhart, C.F.: A method for predicting city-wide electricity gains from photovoltaic panels based on LiDAR and GIS data combined with hourly Daysim simulations. Sol. Energy **93**, 127–143 (2013)
11. Keane, A., Ochoa, L.F., Borges, C.L., Ault, G.W., Alarcon-Rodriguez, A.D., Currie, R.A., Pilo, F., Dent, C., Harrison, G.P.: State-of-the-art techniques and challenges ahead for distributed generation planning and optimization. IEEE Trans. Power Syst. **28**(2), 1493–1502 (2013)
12. Lukac, N., Seme, S., Zlaus, D., Stumberger, G., Zalik, B.: Buildings roofs photovoltaic potential assessment based on lidar (light detection and ranging) data. Energy **66**, 598–609 (2014)
13. Mapdwell Solar System. http://www.mapdwell.com
14. Pak, L.F., Dinavahi, V.: Real-time simulation of a wind energy system based on the doubly-fed induction generator. IEEE Trans. Power Syst. **24**(3), 1301–1309 (2009)
15. Park, M., Yu, I.K.: A novel real-time simulation technique of photovoltaic generation systems using RTDS. IEEE Trans. Energy Convers. **19**(1), 164–169 (2004)
16. Patti, E., Syrri, A.L.A., Jahn, M., Mancarella, P., Acquaviva, A., Macii, E.: Distributed software infrastructure for general purpose services in smart grid. IEEE Trans. Smart Grid **7**(2), 1156–1163 (2016)

17. Patti, E., Pons, E., Martellacci, D., Castagnetti, F.B., Acquaviva, A., Macii, E.: multiflex: flexible multi-utility, multi-service smart metering architecture for energy vectors with active prosumers. In: Proceedings of SMARTGREENS, pp. 288–293 (2015)
18. Suri, M., Huld, T., Dunlop, E., Cebecauer, T.: Geographic aspects of photovoltaics in Europe: contribution of the PVGIS website. IEEE J. J-STARS **1**(1), 34–41 (2008)
19. Vamsidhar, S., Fernandes, B.: Hardware-in-the-loop simulation based design and experimental evaluation of DTC strategies. In: Proceedings of PESC 2004, vol. 5, pp. 3615–3621. IEEE (2004)
20. Wang, G., Gao, W.: Real time simulation for wind power generation system using RTDS. In: Proceedings of NAPS 2008, pp. 1–6. IEEE (2008)

# MAP Estimator for Target Tracking in Wireless Sensor Networks for Unknown Transmit Power

Slavisa Tomic[4,5](✉), Marko Beko[1,3], Rui Dinis[2,5], Milan Tuba[6], and Nebojsa Bacanin[6]

[1] CICANT-CIC.DIGITAL, Universidade Lusófona de Humanidades e Tecnologias, Lisbon, Portugal
mbeko@uninova.pt

[2] Instituto de Telecomunicações, Lisbon, Portugal
rdinis@fct.unl.pt

[3] CTS, UNINOVA – Campus FCT/UNL, Caparica, Portugal

[4] ISR/IST, LARSyS, Lisbon, Portugal
s.tomic@campus.fc.unl.pt

[5] DEE/FCT, NOVA University of Lisbon, Caparica, Portugal

[6] Faculty of Computer Science, John Naisbitt University, Belgrade, Serbia
tuba@ieee.org, nbacanin@naisbitt.edu.rs

**Abstract.** This paper addresses the target tracking problem, by extracting received signal strength (RSS) and angle of arrival (AoA) information from the received radio signal, in the case where the target transmit power is considered unknown. By combining the radio observations with prior knowledge given by the target transition state model, we apply the maximum a posteriori (MAP) criterion to the marginal posterior distribution function (PDF). However, the derived MAP estimator cannot be solved directly, so we tightly approximate it for small noise power. The target state estimate is then easily obtained at any time step by employing a recursive approach, typical for Bayesian methods. Our simulations confirm the effectiveness of the proposed algorithm, offering good estimation accuracy in all considered scenarios.

**Keywords:** Target tracking · Maximum a posteriori (MAP) estimator · Received signal strength (RSS) · Angle of arrival (AoA)

## 1 Introduction

### 1.1 Motivation

Locating a moving object in real-time has become an attractive topic in the academic and research community worldwide recently, owing to its great applicability potential in both military and commercial fields [1–5]. Taking advantage of the existing technologies, e.g., terrestrial radio frequency sources, to provide a solution to the real-time object localization problem is strongly encouraged, due to cost minimization. These include for example, extracting time of arrival, received signal strength (RSS), angle of arrival (AoA) information from the received radio signal, or a combination of them [5–11]. Which approach to use depends mainly on the available hardware. In this work,


L.M. Camarinha-Matos et al. (Eds.): DoCEIS 2017, IFIP AICT 499, pp. 325–334, 2017.
DOI: 10.1007/978-3-319-56077-9_32

we employ combined RSS and AoA measurements, because nowadays, practically all devices can measure the RSS information, and the AoA information can be extracted from RSS measurements by using rotating a directional antenna and choosing the direction from which the highest RSS measurement is obtained [12, 13].

## 1.2 Research Question

Designing real-time localization algorithms is a very difficult task, because of numerous challenges that have to be taken into consideration, such as accuracy, execution time and limited energy resources to name a few. Therefore, the research question of this work can be formulated as:

*How to design an efficient (highly accurate and computationally low complex) localization algorithm, robust to network topology and channel characteristics, applicable in real-time?*

In order to address the research question, we have begun with the following hypothesis:

*An efficient real-time localization algorithm can be developed by linearizing the considered non-linear measurement model and by following Bayesian approach, where prior knowledge is integrated with observations to enhance the estimation accuracy of an estimator.*

## 1.3 Related Work

Works [6–11] investigated the *classical* target localization problem, where the proposed estimators were based on radio measurements only, and prior knowledge was disregarded. The authors in [2, 3, 5] considered a problem of tracking of a moving target, and they combined the available radio measurements with prior knowledge given from the target state model. However, in [2, 3, 5] only pure RSS-based target tracking problem was investigated. In [4], the target tracking problem which employs hybrid, RSS and AoA, measurements was considered. The authors in [4] proposed a Kalman filter (KF) and a particle filter (PF) to solve the tracking problem. Furthermore, they proposed a generalized pattern search method for estimating the path loss exponent (PLE) for each link in every time step.

## 1.4 Contributions

In this paper, we investigate the RSS-AoA-based target tracking problem for unknown target transmit power. This setting is of practical interest for low-cost systems in which testing and calibration is not the priority. Also, due to sensor's battery drain during time, the true value transmit power becomes not perfectly known over time. By integrating prior knowledge given by the target transition model and by employing the maximum *a posteriori* (MAP) approach, we propose a tight approximation of the MAP estimator. Based on a well-known recursive approach, typical for Bayesian methods, we develop an iterative algorithm which updates the mean and covariance of the target state in each time step.

## 2 Relationship to Smart Systems

Smart systems embrace functions of sensing, actuation, and control with a goal of describing and analyzing an environment, and making decisions based on the available data in a predictive or adaptive manner. A good example of such a system is a wireless sensor network, composed of a number of scattered sensors that cooperate between themselves in order to respond in an attentive, adaptive and active way to the changes in the environment registered by sensors. However, in many applications, the information gathered by the sensors is meaningless if not correlated to accurate location of where the changes are occurring (e.g., a system might be set up to respond locally to changes in sensor data).

Therefore, determining accurate location of sensors is a very important task in forming a smart system. Moreover, accurate localization of objects and people in both indoor and outdoor environments enables new applications in emergency and commercial services that can improve safety and efficiency in everyday life (*e.g.*, smart parking, monitoring of storage conditions and goods, assistance for elderly or people with disabilities) [14].

Another important example where target localization helps building a smart system is video surveillance. It has been used to monitor security sensitive areas such as banks, highways, borders, *etc*. Owing to rapid advances high-speed network infrastructure and computing power, as well as large capacity storage devices, multi sensor video surveillance systems have been developed recently. However, traditional video outputs that were controlled by human operators became overwhelming both for operators and storage devices, due to the increased number of cameras. Therefore, in order to filter out redundant information and increase the response time to forensic events motivated the development of smart video surveillance systems. Such systems require fast, reliable and robust algorithms for moving object detection, classification, tracking and activity analysis [15].

## 3 Problem Formulation

Let $\boldsymbol{x}_t = [x_x, x_y]^T$ and $\boldsymbol{a}_i = [a_{ix}, a_{iy}]^T$, for $i = 1, \ldots, N$, denote the unknown location of a moving target at time instant $t$ and known location of the $i$-th static anchor, respectively[1]. For simplicity, we assume a constant velocity target state transition model [1–5, 16, 17], i.e.,

$$\boldsymbol{\theta}_t = \boldsymbol{S}\boldsymbol{\theta}_{t-1} + \boldsymbol{r}_t, \tag{1}$$

where $\boldsymbol{\theta}_t = [\boldsymbol{x}_t, \boldsymbol{v}_t]^T$ represents the target state at time $t$ (described by its location and velocity, $\boldsymbol{v}_t$, in a 2-dimensional plane) and $\boldsymbol{r}_t$ is the state process noise [1–5]. This process

[1] For simplicity and without loss of generality, here we focus on 2-dimensional scenario. The generalization to a 3-dimensional scenario is straightforward.

noise is assumed to be zero-mean Gaussian with a covariance matrix $\boldsymbol{Q}$, *i.e.*, $\boldsymbol{r}_t \sim \mathcal{N}(0, \boldsymbol{Q})$. Covariance $\boldsymbol{Q}$ and state transition matrix, $\boldsymbol{S}$, are defined as

$$\boldsymbol{Q} = q\begin{bmatrix} \Delta^3/3 & 0 & \Delta^2/2 & 0 \\ 0 & \Delta^3/3 & 0 & \Delta^2/2 \\ \Delta^2/2 & 0 & \Delta & 0 \\ 0 & \Delta^2/2 & 0 & \Delta \end{bmatrix},$$

$$\boldsymbol{S} = \begin{bmatrix} 1 & 0 & \Delta & 0 \\ 0 & 1 & 0 & \Delta \\ 0 & 0 & 1 & 0 \\ 0 & 0 & 0 & 1 \end{bmatrix},$$

with $q$ and $\Delta$ denoting the state process noise intensity and the sampling interval between two consecutive time steps [1, 3, 16, 17].

At each time step, $t$, the target emits a signal to the anchors which then withdraw the RSS and AoA information from it. Thus, the observation model can be written as

$$\boldsymbol{z}_t = \boldsymbol{h}(\boldsymbol{x}_t) + \boldsymbol{n}_t, \tag{2}$$

where $\boldsymbol{z}_t = \left[\boldsymbol{P}_t^T, \boldsymbol{\phi}_t^T\right](\boldsymbol{z}_t \in \mathbb{R}^{2N})$ is the observation vector composed of RSS, $\boldsymbol{P}_t = \left[P_i^{(t)}\right]^T$, and AoA, $\boldsymbol{\phi}_t = \left[\phi_i^{(t)}\right]^T$, measurements at time instant $t$. The function $\boldsymbol{h}(\boldsymbol{x}_t)$ in (2) is defined as $h_i(\boldsymbol{x}_t) = P_0 - 10\gamma \log_{10} \frac{\|\boldsymbol{x}_t - \boldsymbol{a}_i\|}{d_0}$ for $i = 1, \ldots, N$ [18], and $h_i(\boldsymbol{x}_t) = \tan^{-1}\left(\frac{x_y - a_{iy}}{x_x - a_{ix}}\right)$ for $i = N+1, \ldots, 2N$, [19], where $P_0$ is the reference power at a distance $d_0$, and $\gamma$ is the PLE. The measurement noise is modeled as $\boldsymbol{n}_t \in \mathcal{N}(0, \boldsymbol{C})$, where the noise covariance is defined as $\boldsymbol{C} = \text{diag}\left(\left[\sigma_{n_i}^2, \sigma_{m_i}^2\right]\right) \otimes \boldsymbol{I}_4$, with $\sigma_{n_i}^2$ (dB) and $\sigma_{m_i}^2$ (rad) being the variances of the RSS and AoA measurement noise, respectively, $\boldsymbol{I}_M$ representing the identity matrix of size $M$ and symbol $\otimes$. noting the Kronecker product. Often in practice, testing and calibration are not the priority in order to keep low network implementation costs; hence, the target transmit power, $P_T$, is often not calibrated, *i.e.*, not known. Not knowing $P_T$ corresponds to not knowing $P_0$ in (2) [18, 20].

We employ (1) and (2) to build the marginal posterior probability distribution function (PDF), $p(\boldsymbol{\theta}_t|z_{1:t})$, from which we can quantify the confidence we have in the values of the state $\boldsymbol{\theta}_t$ given all the past measurements $z_{1:t}$. From $p(\boldsymbol{\theta}_t|z_{1:t})$, we can obtain an estimate at any desired time step.

Below we show a recursive procedure, typical for Bayesian methods [1–5], for the evaluation of $p(\boldsymbol{\theta}_t|z_{1:t})$ at any time instant.

- *Initialization*: The marginal posterior PDF at $t = 0$ is set to the prior PDF $p(\boldsymbol{\theta}_0)$ of $\boldsymbol{\theta}_0$.
- *Prediction*: According to the state transition model (3), the predictive PDF of the state at $t$ is found as

$$p(\theta_t|z_{1:t-1}) = \int p(\theta_t|\theta_{t-1})p(\theta_{t-1}|z_{1:t.1})d\theta_{t-1}. \tag{3}$$

- *Update*: By following the Bayes' rule [1, 13], we have

$$p(\theta_t|z_{1:t}) = \frac{p(z_t|\theta_t)p(\theta_t|z_{1:t-1})}{\int p(z_t|\theta_t)p(\theta_t|z_{1:t-1})d\theta_t}. \tag{4}$$

It is worth noting that the denominator in (4) is just a normalizing constant. In general, the marginal PDF at $t-1$ cannot be calculated analytically, and the integral in (3) cannot be obtained analytically if the state model is non-linear. Therefore, some approximations are required in order to obtain $p(\theta_t|z_{1:t})$.

# 4 The Proposed MAP Algorithm

A state estimate, $\hat{\theta}_{t|t}$, of $\theta_t$ can be obtained from $p(\theta_t|z_{1:t})$ by employing the MAP criteria [13], i.e., by maximizing the marginal posterior PDF

$$\hat{\theta}_{t|t} = \arg\max_{\theta_t} p(\theta_t|z_{1:t}) \approx \arg\max_{\theta_t} p(z_t|\theta_t)p(\theta_t|z_{1:t-1}). \tag{5}$$

The problem in (5) resembles the maximum likelihood estimator, apart from the existence of the prior PDF. This problem is highly non-convex and its analytical solution cannot be obtained in general. Therefore, we approximate (5) by another estimator whose solution is readily available.

First, for small noise power we can write from the RSS and AoA model respectively

$$\lambda_i\|x_t - a_i\| \approx \eta d_0, \text{ for } i = 1, \ldots, N, \tag{6}$$

$$c_i^T(x_t - a_i) \approx 0, \text{ for } i = N+1, \ldots, 2N, \tag{7}$$

where, $\lambda_i = 10^{\frac{P_i^{(t)}}{10\gamma}}$, $\eta = 10^{\frac{P_0}{10\gamma}}$, and $c_i = [-\sin\phi_i^{(t)}, \cos\phi_i^{(t)}]^T$. Next, by transforming from Cartesian to polar coordinates we express $x_t - a_i = r_i u_i : r_i \geq 0$, $\|u_i\| = 1$, and make use of the available azimuth angle observations to define $u_i = [\cos\phi_i^{(t)}, \sin\phi_i^{(t)}]$, and rewrite (6) as

$$\lambda_i r_i \approx \eta d_0.$$

By multiplying the left hand side of the above equation by $u_i^T u_i$, we arrive at

$$\lambda_i u_i^T(x_t - a_i) \approx \eta d_0. \tag{8}$$

Hence, we can approximate the RSS and AoA measurement models by (8) and (7), respectively, which written in a vector form and applying the weighted least squares (WLS) criterion leads to

$$\hat{y}_{t|t} = \underset{y_t=[x_t,\eta]^T}{\arg\min} \|W(Ay_t - b)\|^2, \tag{9}$$

where $W = I_2 \otimes \text{diag}(w)$, $w = \left[\sqrt{w_i}\right]^T$ and $w_i = 1 - \frac{P_i^{(t)}}{\sum_{i=1}^N P_i^{(t)}}$

$$A = \begin{bmatrix} \vdots & \vdots \\ \lambda_i u_i^T & d_0 \\ \vdots & \vdots \\ c_i^T & 0 \\ \vdots & \vdots \end{bmatrix}, b = \begin{bmatrix} \vdots \\ \lambda_i u_i^T a_i \\ \vdots \\ c_i^T a_i \\ \vdots \end{bmatrix}.$$

The solution of (9) is given by $\hat{y}_{t|t} = \left(A^T W^T W A\right)^{-1} (A^T W^T b)$. Once we have an estimate of the target location, we can use it to find the maximum likelihood estimate of $P_0$, $\hat{P}_0$, from the RSS measurement model as

$$\hat{P}_0 = \frac{\sum_{i=1}^N P_i^{(t)} + 10\gamma \log_{10} \frac{\|x_t - a_i\|}{d_0}}{N}. \tag{10}$$

At this point, we can benefit from this estimated value by using it to resume the estimation process as if $P_0$ is known, *i.e.*, we can compute $\hat{\eta} = 10^{\frac{\hat{P}_0}{10\gamma}}$ in order to get

$$\hat{A} = \begin{bmatrix} \vdots & \vdots & \vdots \\ \lambda_i u_i^T & 0 & 0 \\ \vdots & \vdots & \vdots \\ c_i^T & 0 & 0 \\ \vdots & \vdots & \vdots \end{bmatrix}, \hat{b} = \begin{bmatrix} \vdots \\ \lambda_i u_i^T a_i^T + \hat{\eta} d_0 \\ \vdots \\ c_i^T a_i^T \\ \vdots \end{bmatrix}.$$

As far as the prior knowledge part in (5) is concern, we can assume that $p(\theta_{t-1}|z_{1:t-1})$ has Gaussian distribution with mean $\hat{\theta}_{t-1|t-1}$ and covariance $\hat{M}_{t-1|t-1}$ [17]. Then, according to (3), we obtain

$$p(\theta_t|z_{1:t-1}) \approx \frac{1}{k} \exp\left\{-\frac{1}{2}\left(\theta_t - \hat{\theta}_{t|t-1}\right)^T \hat{M}_{t|-1}^{-1} \left(\theta_t - \hat{\theta}_{t|t-1}\right)\right\}, \tag{11}$$

where $k$ is a constant, and $\hat{\theta}_{t|t-1}$ and $\hat{M}_{t|t-1}$ are respectively the mean and covariance of the one-step predicted state, obtained through (1) as

$$\hat{\theta}_{t|t-1} = S\hat{\theta}_{t-1|t-1},$$
$$\widehat{M}_{t|t-1} = SM_{t-1|t-1}S^T + Q. \tag{12}$$

Therefore, (5) can be written as

$$\hat{\theta}_{t|t} = \arg\min_{\theta_t} \left(z_t - h(x_t)\right)^T C^{-1}\left(z_t - h(x_t)\right) + \left(\theta_t - \hat{\theta}_{t|t-1}\right)^T \widehat{M}_{t|-1}^{-1}\left(\theta_t - \hat{\theta}_{t|t-1}\right). \tag{13}$$

Similarly as in (9), the problem in (13) can be rewritten as

$$\hat{\theta}_{t|t} = \arg\min_{\theta_t} \left\| \widehat{W}(H\theta_t - f) \right\|^2, \tag{14}$$

where $H = \begin{bmatrix} \hat{A}; \widehat{M}_{t|t-1}^{-\frac{1}{2}} \end{bmatrix}, f = \begin{bmatrix} \hat{b}; \widehat{M}_{t|t-1}^{-\frac{1}{2}} \theta_{t|t-1} \end{bmatrix}$, $\widehat{W} = I_2 \otimes \mathrm{diag}(\hat{w})$, $\hat{w} = [\hat{w}_i]^T, \hat{w}_i = 1 - \frac{\hat{d}_i}{\sum_{i=1}^N \hat{d}_i}$, $\hat{d}_i = d_0 10^{\frac{\hat{P}_0 - P_i^{(t)}}{10\gamma}}$. The solution to (14) is given by $\hat{\theta}_{t|t} = \left(H^T W^T W H\right)^{-1}(H^T W^T f)$.

The step by step proposed MAP algorithm for unknown $P_T$, is outlined in Algorithm 1.

**Algorithm 1.** The Proposed MAP Algorithm Description

Require: $z_t$, for $t = 1, \ldots, T$, $Q, S$

1: <u>Initialization</u>: $\hat{\theta}_{0|0} \leftarrow (9)$, $\widehat{M}_{0|0} \leftarrow I_4$, $\hat{P}_0 \leftarrow (10)$

2: for $t = 1, \ldots, T$ do

3: <u>Prediction</u>:

4: - $\hat{\theta}_{t|t-1} \leftarrow (12)$

5: - $\widehat{M}_{t|t-1} \leftarrow (12)$

6: <u>Update</u>:

7: - $\hat{\theta}_{t|t} \leftarrow (14)$

8: - $\widehat{M}_{t|t} \leftarrow I_4$

8: - $\hat{P}_0 \leftarrow (10)$

10: end for

## 5 Simulation Results

This section validates the performance of the proposed algorithm through computer simulation. All simulations in this work were performed in MATLAB.

### 5.1 Simulation Set-Up

Two scenarios, in which the target takes sharp maneuvers and a smoother one, are investigated. The state model (1) is used for the target state transition, and the radio measurements are acquired according to (2) at each time instant. Three anchors are fixed at $[[70, 10]^T, [40, 70]^T, [10, 40]^T]$, and a sample is taken every $\Delta = 1$ s during $T = 150$ s trajectory duration in each Monte Carlo, $M_c = 1000$, run. The reference distance is set to $d_0 = 1$ m, the reference power to $P_0 = -10$ dBm, and the PLE to $\gamma = 3$, $\sigma_{n_i} = 9$ dB, $\sigma_{m_i} = 4\pi/180$ rad, and $q = 2.5 \times 10^{-3} \text{m}^2/\text{s}^3$. The performance metric used here is the root mean square error (RMSE), defined as RMSE $= \sqrt{\sum_{i=1}^{M_c} \frac{\|\boldsymbol{x}_{i,t} - \hat{\boldsymbol{x}}_{i,t}\|^2}{M_c}}$, where $\hat{\boldsymbol{x}}_{i,t}$ represents the estimate of the true target location, $\boldsymbol{x}_{i,t}$, in the $i$-th Mc run at time instant $t$.

The performance of the proposed MAP algorithm is compared with a *classical* approach described by (9), which makes use of the radio measurements only and disregards the prior knowledge, labeled here as WLS.

### 5.2 Results

Figure 1 illustrates the true target trajectory in the two considered scenarios as well as the estimated trajectories given by the considered approaches. In the first scenario, the initial target's location was set to $[21, 20]^T$, whereas in the second scenario the starting point was set to $[35, 15]^T$. We can see that the proposed MAP estimator performs considerably better in both scenarios than the WLS one. However, it seems like there is still room for further improvement of the proposed approach, since the results are not particularly smooth. This might be done by *better* utilization of the prior knowledge, *i.e.*, in the update step of the state covariance matrix, and is left for future work. Figure 2 illustrates the RMSE (m) versus $t$ (s) performance comparison of the considered approaches. From it, one can see that the

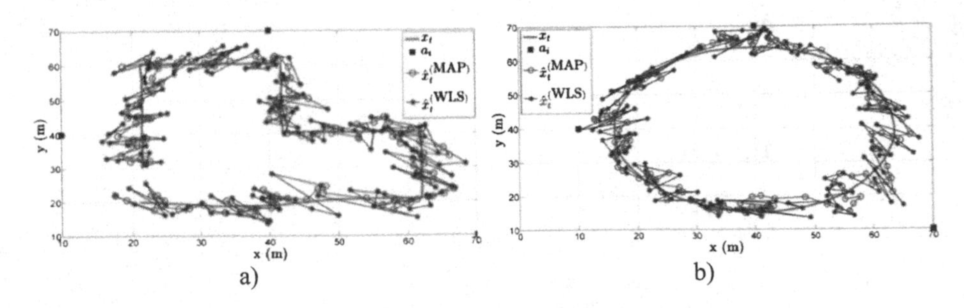


**Fig. 1.** The true target trajectory in: (a) the first and (b) the second considered scenario and the estimated ones.

proposed algorithm performs well in both scenarios, obtaining an average RMSE below 3 m. Furthermore, the superiority of the Bayesian approach, which combines the observations with prior knowledge, over the *classical* one, which utilizes the observations exclusively, is clearly observed in every time step.

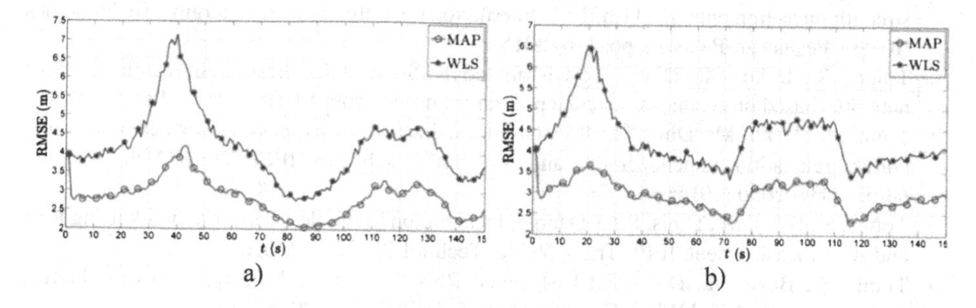


**Fig. 2.** RMSE versus t(s) comparison for (a) the first and (b) the second considered scenario.

# 6 Conclusions

In this work, we have investigated the target tracking problem which makes use of the combined RSS and AoA measurements, in which the target transmit power is considered not known. By resorting to Bayesian approach and relying on the MAP criterion, we proposed a tracking algorithm, which efficiently solves the target tracking in all considered scenarios and significantly outperforms the considered classical approach in every time instant. Although the proposed algorithm offers excellent estimation accuracy, it seems that there is still room for further improvement, which will be the topic of our future work.

This work is a part of our ongoing research, and we have validated the performance of our algorithms by means of simulations only. As a part of our future work, we plan to validate their performance using real measurements from [12, 13].

**Acknowledgments.** This work was partially supported by Fundação para a Ciência e a Tecnologia under Project PEst-OE/EEI/UI0066/2014 (UNINOVA), and Project UID/EEA/50008/2013 (Instituto de Telecomunicações), Program Investigador FCT under Grant IF/00325/2015 and Grant SFRH/BD/91126/2012.

# References

1. Dardari, D., Closas, P., Djuric, P.M.: Indoor tracking: theory, methods, and technologies. IEEE Trans. Vehic. Technol. **64**(4), 1263–1278 (2016)
2. Beaudeau, J.P., Bugallo, M.F., Djuric, P.M.: RSSI-based multitarget tracking by cooperative agents using fusion of cross-target information. IEEE Trans. Sign. Process. **63**(19), 5033–5044 (2015)
3. Masazade, E., Niu, R., Varshney, P.K.: Dynamic bit allocation for object tracking in wireless sensor networks. IEEE Trans. Sign. Process. **60**(10), 5048–5063 (2012)

4. Khan, M.W., Kemp, A.H., Salman, N., Mihaylova, L.S.: Tracking of wireless mobile nodes in the presence of unknown path-loss characteristics. In: Fusion, Washington DC, USA, pp. 104–111 (2015)
5. Khan, M.W., Salman, N., Ali, A., Khan, A.M., Kemp, A.H.: A comparative study of target tracking with Kalman filter, extended Kalman filter and particle filter using received signal strength measurements. In: 11th IEEE International Conference on Emerging Technologies (ICET), Peshawar, Pakistan, pp. 1–6 (2015)
6. Tomic, S., Beko, M., Dinis, R.: Distributed RSS-based localization in wireless sensor networks based on second-order cone programming. Sensors **14**(10), 18410–18432 (2014)
7. Tomic, S., Beko, M., Dinis, R.: RSS-based localization in wireless sensor networks using convex relaxation: noncooperative and cooperative schemes. IEEE Trans. Veh. Technol. **64**(5), 2037–2050 (2015)
8. Tomic, S., Beko, M., Dinis, R.: 3-D target localization in wireless sensor network using RSS and AoA measurement. IEEE Trans. Vehic. Technol **PP**(99), 1 (2016)
9. Tomic, S., Beko, M., Dinis, R.: Distributed RSS-AoA based localization with unknown transmit powers. IEEE Wirel. Commun. Lett. **5**(4), 392–395 (2016)
10. Tomic, S., Beko, M., Dinis, R., Montezuma, P.: Distributed algorithm for target localization in wireless sensor networks using RSS and AoA measurements. Pervasive Mob. Comput. **PP**(99), 1 (2016)
11. Tomic, S., Beko, M., Dinis, R., Montezuma, P.: A closed-form solution for RSS/AoA target localization by spherical coordinates conversion. IEEE Wirel. Commun. Lett. **PP**(99), 1 (2016)
12. Niculescu, D., Nath, B.: VOR base stations for indoor 802.11 positioning. In: the 10thAnnual International Conference on Mobile Computing and Networking (ACM MobiCom), Philadelphia, PA, USA, pp. 58–69 (2004)
13. Elnahrawy, E., Francisco, J.A., Martin, R.P.: Adding angle of arrival modality to basic RSS location management techniques. In: the 2nd International Symposium on Wireless Pervasive Computing (ISWPC), San Juan, PR, USA, pp. 1–6 (2007)
14. Yang, J., Portilla, J., Riesgo, T.: Smart parking service based on wireless sensor networks. In: the 38th Annual Conference on IEEE Industrial Electronics Society (IECON), Montreal, QC, Canada, pp. 6029–6034 (2012)
15. Dedeoglu, Y.: Moving object detection, tracking and classification for smart video surveillance. Master thesis, Bilkent University, Ankara, Turkey (2004)
16. Kay, S.M.: Fundamentals of Statistical Signal Processing: Estimation Theory. Prentice-Hall, Upper Saddle River (1993)
17. Wang, G., Li, Y., Jin, M.: On MAP-based target tracking using range-only measurements. In: 8th International Conference on Communications and Networking in China (CHINACOM), Guilin, China, pp. 1–6 (2013)
18. Rappaport, T.S.: Wireless Communications: Principles and Practice. Prentice-Hall, Upper Saddle River (1996)
19. Yu, K.: 3-D localization error analysis in wireless networks. IEEE Trans. Wirel. Commun. **6**(10), 3473–3481 (2007)
20. Patwari, N., Ash, J.N., Kyperountas, S., Hero III, A.O., Moses, R.L., Correal, N.S.: Locating the nodes: cooperative localization in wireless sensor networks. IEEE Sign. Process. Mag. **22**(4), 54–69 (2005)

# Solar Energy

# Performance Assessment of Tank Fluid Purging and Night Cooling as Overheating Prevention Techniques for Photovoltaic-Thermal (PV-T) Solar Water Heating Systems

Pedro M.L.P. Magalhães[1,2(✉)], João F.A. Martins[1,2], and António L.M. Joyce[3]

[1] Centre of Technology and Systems, UNINOVA, Quinta da Torre, 2829-516 Monte de Caparica, Portugal
pmlpm@campus.fct.unl.pt

[2] Faculty of Science and Technology, NOVA University of Lisbon, Quinta da Torre, 2829-516 Monte de Caparica, Portugal

[3] National Laboratory of Energy and Geology, Estrada do Paço do Lumiar 22, 1649-038 Lisbon, Portugal

**Abstract.** Tank fluid purging and night cooling are two overheating prevention techniques with potential to prevent photovoltaic-thermal collectors from experiencing temperatures capable of undermining their longevity and commercial appeal. Both techniques are readily available, inexpensive but inherently wasteful to use. Dynamic numerical simulations were conducted to determine the primary energy efficiency and the level of protection afforded by these techniques in active residential grid-connected solar domestic hot water systems. Also evaluated was the use of occupancy rate information, possible via so-called "smart systems", to complement the techniques. The results revealed better performances for systems using stagnation control schemes relative to those not using them. Also, night cooling was shown to be unable to prevent overheating reliably while tank fluid purging proved to be more apt but resulted in substantial waste of water annually, which was slightly reduced by combining it with night cooling, which in turn proved to be the most energy efficient solution.

**Keywords:** Photovoltaic · Thermal · Stagnation · Night · Cooling · Purging

## 1 Introduction

Solar water heating systems such as those for domestic hot water or space heating routinely experience stagnation events, particularly during the Summer months. During this period, ambient temperature and irradiation levels are at their highest and demand tends to be low by design (e.g., space heating) or due to absence periods (e.g., vacations). While standard collectors for low temperature applications (<100 °C) such as flat-plate and evacuated tube collectors are generally designed to withstand their worst case scenario stagnation temperatures (100–300 °C), hybrid photovoltaic-thermal (PV-T)


L.M. Camarinha-Matos et al. (Eds.): DoCEIS 2017, IFIP AICT 499, pp. 337–347, 2017.
DOI: 10.1007/978-3-319-56077-9_33

collectors are sensitive to temperatures in excess of 85 °C - in part due to the limited temperature stability of common photovoltaic (PV) array encapsulants, namely ethylene vinyl acetate (EVA) - which can occur during stagnation [1, 2]. As a result, PV-T systems require stagnation control methods to prevent overheating in addition to the protections typically required in standard solar heating systems.

Stagnation control methods commonly used in solar heating systems can be categorised as either stagnation handling or overheating prevention [3]. Stagnation handling methods do not prevent stagnation but instead mitigate its harmful consequences such as thermal stress to other components in the collector circuit due to heat carrier evaporation, heat carrier degradation, accelerated corrosion and plugging of pipes [4]. Examples of stagnation handling methods include the drainback, draindown and steamback systems or the use high system pressures in the collector loop [2]. In contrast, overheating prevention methods avoid the onset of temperatures capable of compromising a system's functional integrity by avoiding stagnation or limiting the temperatures reached. Examples include collector shading systems, night cooling of the storage tank, tank fluid purging, active heat dumping, passive venting and defocusing (for tracking systems). Of the two categories, overheating prevention methods are necessary for glazed PV-T systems, whose temperatures can reach around 150 °C under stagnation, even though some stagnation handling measures can be seen as desirable or complementary [1, 2, 5, 6].

The present endeavour focuses on the performance of two overheating prevention methods, namely tank fluid purging and night cooling of the storage tank, in active residential solar domestic hot water (SDHW) systems using glazed PV-T collectors producing electricity and feeding it to the local utility grid using grid-tie inverters. Both methods have a low initial cost – one of which (night cooling) is a common feature in many commercial controllers - and are designed to lower the storage tank temperature and in doing so indirectly prevent stagnation and high temperatures.

Tank fluid purging consists of disposing of hot tank fluid – generally water - using a single purge valve and replacing it with colder fluid to prevent the tank from being fully charged. In doing so, stagnation due to low temperature differences between the collector and tank or due to the maximum allowed collector temperature being reached can be avoided indirectly. However, the hottest fluid in the tank (i.e., from the top) is wasted and in the case of water predominantly when it is scarce (Summer months) while requiring parasitic energy to power the pump(s) and valve – assuming the latter is not thermostatic. On the other hand, the method is unsuitable for some fluids and the valve is used sparingly during the Winter months which may cause it to fail prematurely unless periodic discharges or careful maintenance is carried out [2].

In contrast, night cooling of the storage tank does so by running the collector circuit pump at night in order to prevent stagnation from taking place the following day. It does not require additional components other than possibly a controller, relies on the often-used collector loop, requiring parasitic energy to do so but does not waste water. On the other hand, the method can only be used effectively when solar energy collection is not possible - unlike tank fluid purging which can be used preventively or during the collection period - and must necessarily anticipate periods of mismatched supply and demand, either using predictive or conservative controls [2, 3].

The implicitly wasteful nature of both methods contrasts with their low implementation cost but can nevertheless penalise the performance of SDHW systems. At the same time, cooling the tank results in lower collector temperatures and higher PV yields for solar cells exhibiting negative cell efficiency temperature coefficients, although these are generally low in absolute value (e.g., –0.45%/K for monocrystalline silicon). In other words, these methods are likely to contribute to a degradation of the system's energy efficiency and cost-effectiveness, which is unlikely to be outweighed by the positive effect on the electrical performance of PV-T systems [1, 2, 5]. Nevertheless, this hypothesis needs to be tested in accordance with the scientific method and the magnitude of the performance variation quantified.

As such, the objective of the work described in this paper is to evaluate and compare implementations of tank fluid purging and night cooling with respect to the problem of stagnation and their impact on the performance of residential SDHW systems using glazed PV-T collectors. Although these methods have been to some extent discussed in the literature prior to this effort, the emphasis has not been on quantifying the effect they have on the performance of SDHW PV-T systems [7].

In order to do so, dynamic annual simulations were conducted. These focused on a reference system without any stagnation control method and systems employing each control method or both as a way to evaluate their potentially complementary nature and implications for systems with more frequent stagnation events such as space heating systems. Finally, the results were primarily analysed from the viewpoint of primary energy efficiency but other figures of merit were considered, namely overheating impact, pump cycling and likelihood of premature system failure.

## 2 Relationship to Smart Systems

The research efforts undertaken and described in this paper concern the study of two overheating prevention methods, namely tank fluid purging and night cooling, whose use can be enhanced by the features commonly associated with so-called "smart systems", namely sensing, (internet) connectivity, informed decision-making and actuation [8]. These features could render the preventive use of both methods more reliable and overall better performing than alternative control schemes. For instance, rather than cooling the tank according to a static temperature setpoint during a given time window at night, the control unit could hypothetically determine which flow rate to use if any, for how long and the timing of the operation dynamically.

Ideally, preventive use of both methods should consider dynamic weather forecasts, occupancy rates, electricity prices and other factors to select the most appropriate method (if more than one is available) and timing, which would require a combination of internet connectivity, sensors and control modules with predictive capabilities able to use such information in a timely manner. In practice, a predictive stagnation control scheme could be too complex and costly to implement relative to its potential benefits over simpler methods. Therefore, the current study focuses on the evaluation of simpler stagnation control solutions for SDHW PV-T systems yet consistent with the features associated with "smart systems".

Concretely, the use of occupancy rate information possibly provided by sensors, schedules or remote communication with the main unit to disable the backup heater and/or select different night cooling setpoints for normal load and load absence periods were evaluated since these are simple to implement and also have advantages from the standpoint of energy efficiency and user comfort when users are away.

## 3 System Overview

The solar heating systems under consideration for the purposes of this study are small active (i.e., forced-circulation) residential grid-connected SDHW systems featuring parallel-connected glazed PV-T collectors supplying the electricity generated to the local utility grid. These systems typically have a collector area in the range of 4–6 $m^2$, tanks sized according to specific storage volumes of 40–70 $L/m^2$ of collector area and featuring internal heat exchangers of the immersed coil variety, which have a propensity for low thermal stratification [9–11]. Moreover, the systems are prepared to perform night cooling of the tank whereas purging tank fluid requires an actuator valve on the demand loop. The generic diagram for all three systems evaluated (no stagnation control, night cooling and purging) is given in Fig. 1.

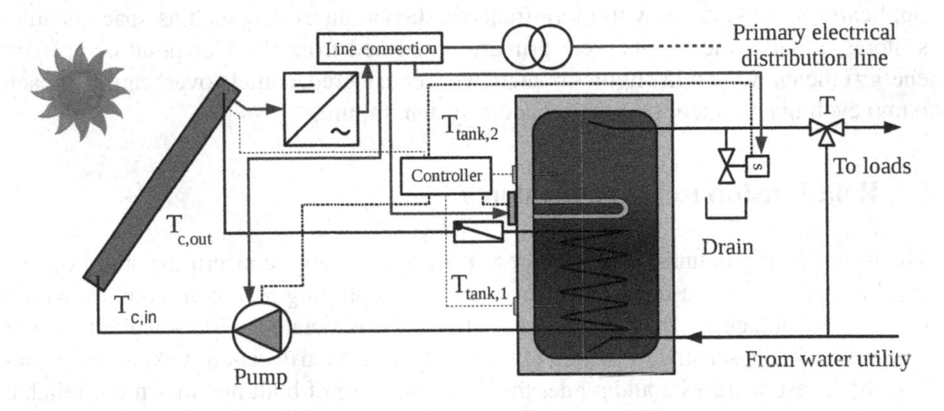


**Fig. 1.** Generic diagram for the solar domestic hot water (SDHW) systems studied.

Regular system operation is governed by simple controls, namely hysteretic controllers for the backup heater and pump as well as safety overrides designed to protect the system components other than the collector from high temperatures (>95 °C) by disabling heat carrier circulation. For the systems considered here, the backup heater is turned on once tank fluid temperatures drop below 55 °C and remains on until temperatures reach 60 °C (5 °C deadband) and in doing so prevents the growth of *legionella* while not significantly enhancing the formation of limestone deposits [10, 12]. Similarly, heat carrier circulation is initiated if the temperature difference ($\Delta T$) between the heat carrier at the collector outlet and the tank fluid near the heat exchanger increases beyond the turn on setpoint ($\Delta T_{on}$ = 10 K) and ends if it drops below the turn off setpoint ($\Delta T_{off}$ = 2 K) leading to stagnation in the collector loop.

On the other hand, the night cooling takes place when three conditions are met. The first condition is a negative temperature difference between the collector and tank to indicate cooling is possible while the second and third conditions concern a timer enable (in this case, active between midnight and 5 a.m. local time) and a minimum tank temperature setpoint to regulate the cooling period and how much cooling is to take place, respectively. Moreover, the tank temperature setpoint alternates between a value for normal occupancy periods ($T_n$) and another for periods of user absence ($T_{nl}$), according to occupancy information conveyed by the "smart" system elements.

Conversely, tank fluid purging is triggered by any one of three conditions. The first condition is met when the collector fluid outlet temperature is within a preset tolerance ($\Delta T_{f,tol} = 5$ °C) of its maximum allowed level ($T_{f,max} = 95$ °C) during fluid circulation. The second condition is met when the tank temperature exceeds the tank purge setpoint ($T_{purge}$, lower than the tank's maximum allowed temperature) and until the former drops to a safe level ($T_{safe}$, set to at least 5 °C below $T_{purge}$). The third and final condition is met when $\Delta T$ approaches $\Delta T_{off}$ during heat carrier circulation and the temperature difference between $T_{f,max}$ and the tank fluid is lower than $\Delta T_{on}$. Although unlikely to be frequently triggered, it concerns high temperature cycling and could prevent the system from being stuck in high temperature stagnation since if stagnation were to set in once $\Delta T$ dropped below $\Delta T_{off}$ while $T_{f,max} - T_{tank} < \Delta T_{on}$, the collector fluid temperature could exceed the maximum value for which circulation is allowed (95 °C) before it could resume ($\Delta T > \Delta T_{on}$) and after which fluid purging would not be effective since the system would be stuck until temperatures dropped.

## 4 Methodology

In order to address the proposed objectives, a set of annual dynamic simulations of the aforementioned SDHW PV-T systems were conducted. The choice of simulation work rather than physical experiments was based on the former's reasonable accuracy and the ability to compare systems under the same exact conditions in a time- and cost-effective manner. Annual simulation periods were selected due to the pertinence of evaluating the all-year round performance of the control methods and the system, even though stagnation events predominantly take place during Summer.

The simulations employed the models described in [2] except those for the stagnation controllers and the utility water temperature. The latter was modelled as 7 °C-amplitude 15 °C-average annual triangular wave, whose extrema were made to coincide (time-wise) with those of the outdoor temperature [12]. Typical meteorological year (TMY) data for Lisbon, Portugal (38°42'N, 9°8'W) was used, as represented in Fig. 2. However, the simulations assumed the lack of appreciable wind ($v_{wind} = 0$ m/s) as a worst case scenario and focused on common situations likely to lead to high stagnation temperatures in well-sized SDHW systems, namely the use of representative load patterns including Summer vacations and weekends off. A full load day corresponded to 200 L at 45 °C and no load periods included a weekend off every four weeks and four one-week Summer vacations spread a month apart.

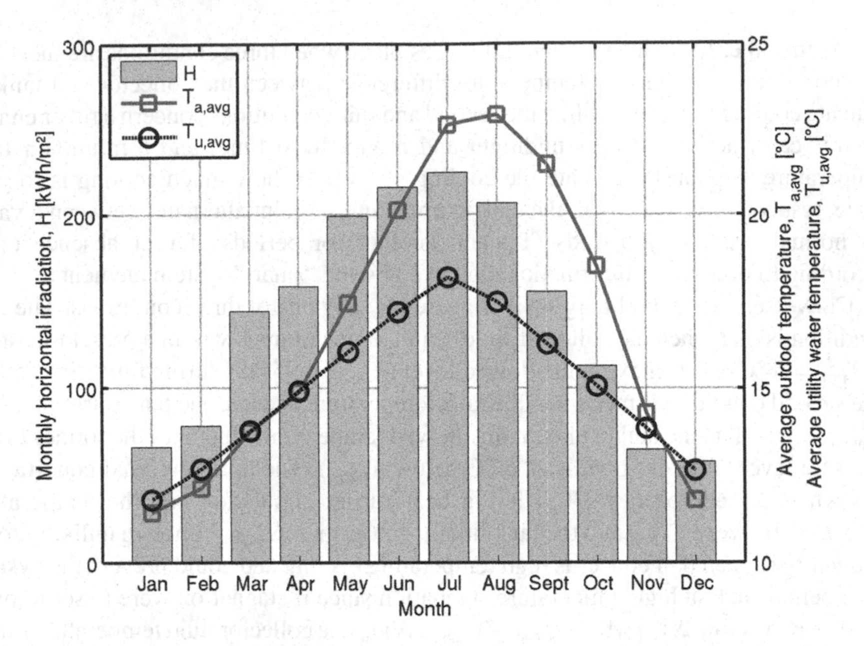


**Fig. 2.** Monthly horizontal irradiation (H), average outdoor (Ta, avg) and utility water temperatures (Tu, avg) according to the (Lisbon) TMY data used in the simulations conducted.

The simulations were run using MATLAB and most models implemented were validated against their respective counterparts from other well known simulation tools, namely TRNSYS, as described in [2]. Conversely, the dynamic PV-T collector model used is based on the equation for the quasi-dynamic test method featured in the standard EN 12975-2:2006 and reproduces the performance of the PV-T collector described in [1], which was measured according to the aforementioned standard. Moreover, the system's performance is in line with the results found in [13], which focused on the same location, using comparable PV-T technology and areas.

At the same time, the dynamic PV-T collector model used does not incorporate longwave radiative heat losses explicitly ($c_4$ or $\varepsilon = 0$), which has thermodynamic implications particularly for night cooling with PV-T collectors lacking low-emissivity coatings, as is the case [1]. Nevertheless, the implications are arguably minor for the present study since the temperature differences between the hot storage tank and the outdoors still allow for significant convective heat transfer (and thus a conservative cooling power assessment) whereas night cooling of cold storage tanks (e.g., used in space cooling systems) requires radiative heat transfer [14]. Additional information on the parameters used, unless stated otherwise, can be found in [2].

The simulations' primary results were used to assess the system's primary energy efficiency using a weighted primary energy savings adapted from [15] and defined as:

$$Q_{pes} = f_{pv} * E_{pv} - f_{par} * E_{par} - f_{aux} * \left(E_{aux} - E_{ref}\right). \quad (1)$$

where the primary energy factors for PV electricity ($f_{pv}$), parasitic energy ($f_{par}$) and auxiliary energy ($f_{aux}$) were all set to 2.5 (general purpose electricity) [16]. Moreover, the backup heater (i.e., an immersed electrical heating element) efficiency was set to 100%, the power converter efficiency to 95% and the pump power ($P_{pump}$) defined as a cubic function of the collector array mass flow rate ($m_c$) according to (2), where the pump power coefficient ($K_p$) was set to 1000 $W^1Kg^{-3}s^3$ after [17].

$$P_{pump} = K_p * m_c^3. \quad (2)$$

## 5 Simulations and Analysis

The simulations revealed marginal primary energy savings and cumulative overheating period variations as the result of disabling the backup heater during periods of low energy demand. In particular, the increase in $Q_{pes}$ from this measure was limited to no more than 2 kWh for all flow rates whereas the reduction in cumulative overheating was only observed at low flow rates and limited to less than 3 h. These results can be explained by the fact that the tank is usually charged during those periods and because overheating is more likely at low flow rates.

With regard to stagnation control schemes, the simulations revealed night cooling as having a limited ability to prevent collector overheating, unlike tank fluid purging.

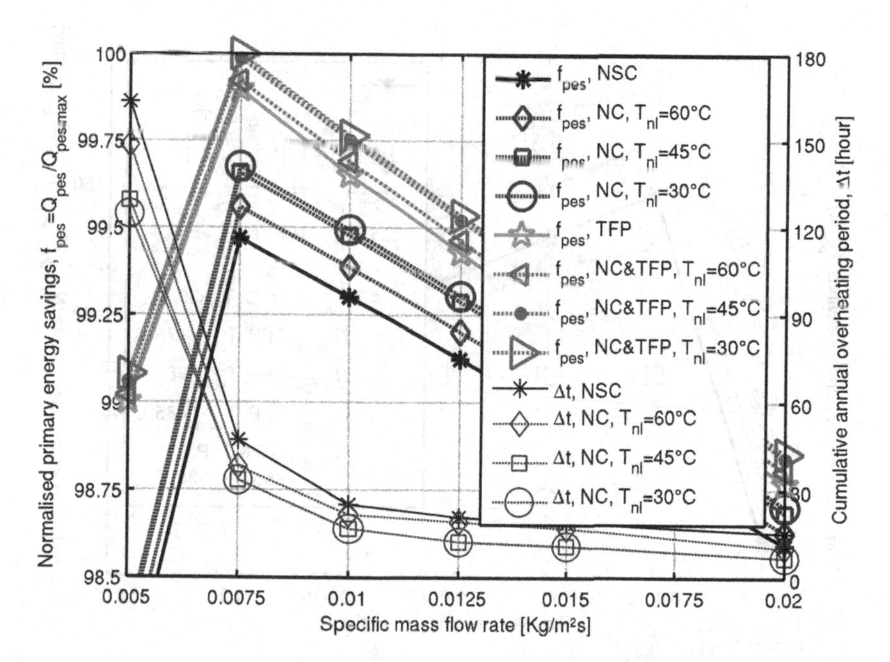


**Fig. 3.** Cumulative annual duration of overheating periods ($\Delta t$) and normalised primary energy savings ($f_{pes}$) versus the specific mass flow rate ($m_c/A_c$) for the SDHW PV-T systems simulated. Legend: NSC, no stagnation control; NC, night cooling ($T_n$ = 80 °C; no load mode limited to vacations, i.e., excludes weekends); TFP, tank fluid purging ($T_{purge}$ = 85 °C; $m_{purge} = m_c$).

Concretely, none of the systems simulated using tank fluid purging overheated – in contrast with results reported in [2] - while those relying exclusively on night cooling still overheated for at least a few hours annually although less than without a stagnation control method, as exemplified in Fig. 3. The inadequacy of night cooling for overheating prevention did not result from low collector loop cooling power but rather the inability to predict the need for cooling as such, as overheating was observed even during full load days in addition to low demand periods – either vacations or weekends off. Moreover, meeting the night cooling tank temperature setpoint while employing the nominal mass flow rate used during the daytime was routinely achieved in a fraction of the time available to do so (the average night cooling cycle lasted under 3 h but some lasted almost 5 h, particularly at the lowest flow rates) and overheating tended to drop as either of the setpoints was lowered, although 'no load' setpoints lower than 45 °C did not substantially reduce it.

At the same time, tank fluid purging either individually or combined with night cooling led to a substantial volume of water being flushed, mainly at low specific mass flow rates and predominantly during the Summer. While the water volume itself was not recorded, the energy wasted as the result of purging using tank purge setpoints from 75 °C to 85 °C ranged from 88 up to 927 kWh for the lowest flow rates, which is equivalent to approximately 13 and 135 full load days. Alternatively, this corresponds to between 1.2 and 14.3 m$^3$ of water at $T_{purge}$ being replaced by utility water at Summertime temperatures (18.5 °C) – all reasonable estimates for the estimation of the actual volume

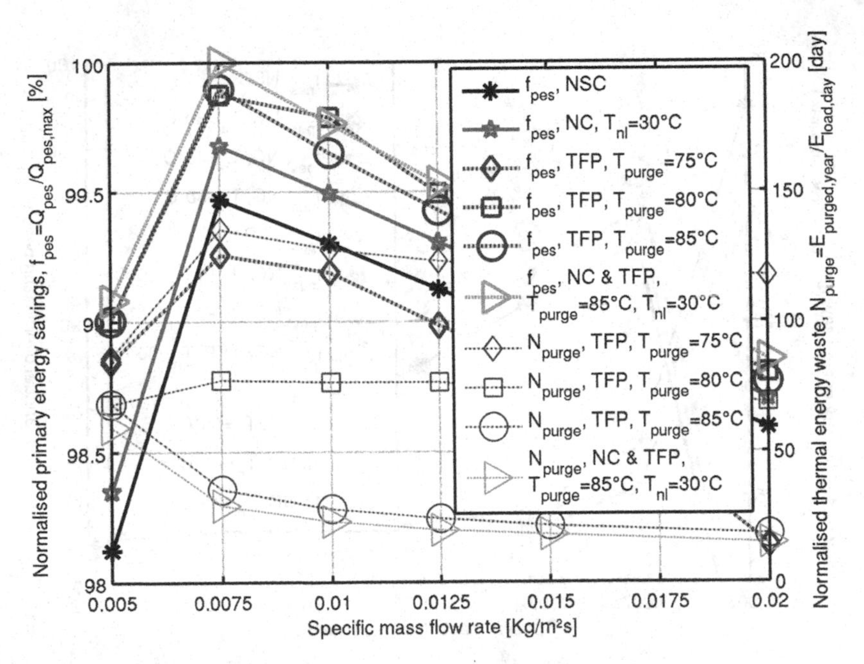


**Fig. 4.** Normalised primary energy savings ($f_{pes}$) and normalised thermal energy waste due to fluid purging ($N_{purge}$) versus the specific mass flow rate for the SDHW PV-T systems simulated. Legend: NSC, no stagnation control; NC, night cooling ($T_n$ = 80 °C; no load mode limited to vacations); TFP, tank fluid purging ($T_{safe} = T_{purge}$–5 °C; $m_{purge} = m_c$).

of water purged – which is equivalent volume-wise to between 6 and 72 normal load days annually. However, the minimum energy waste and equivalent volume corresponding to the use of tank fluid purging as stand-alone were 126 kWh (18 days) and 1.6 $m^3$ (8 days), respectively. Thus, combining both methods allowed for a reduction of the amount of purging and energy waste (see Fig. 4) and effectively constitutes a more environmentally viable solution.

Primary energy efficiency-wise, PV-T systems using night cooling, tank fluid purging or both were able to outperform those not using any stagnation control method, particularly at low and medium collector specific mass flow rates, high purge setpoints (80 and 85 °C) and if night cooling was limited to vacation periods, as shown in Figs. 3 and 4. Conversely, excessive purging brought on through the use of purge setpoints as low as 75 °C ($T_{safe}$ = 70 °C) inverted this trend while still reliably preventing overheating. Similarly, allowing night cooling during short 'no load' periods (i.e., weekends off) resulted in less overheating, mainly at low flow rates, but also slightly lower performances, predominantly at high flow rates. On the other hand, the range of performance variation itself was not significant: from – 16 kWh up to 35 kWh relative to a maximum $Q_{pes}$ of 3633 kWh. Nevertheless, the general outcome reflects higher PV yields (up to 22 kWh) due to cell cooling, despite increases in the parasitic (up to 1 kWh) and auxiliary (up to 55 kWh, and highest at high flow rates) energy consumptions, which ultimately stemmed from the methods' ability to cool the tank and the correlation between high irradiance periods and the need for cooling.

Moreover, the performance enhancement was highest for systems using tank fluid purging, either exclusively or combined with night cooling of the storage tank, which proved to be the most effective schemes in preventing overheating and in securing the highest primary energy efficiency. In this regard, the combined use of tank fluid purging and night cooling led to the highest observed primary energy efficiency.

## 6 Conclusions

Dynamic simulations of SDHW PV-T systems were conducted to evaluate implementations of tank fluid purging and night cooling of the storage tank as overheating prevention methods. While none of the methods evaluated can be useful in the event of blackouts, the results have shown tank fluid purging as a more effective method for overheating prevention than non-predictive night cooling of the storage tank. As such, the tank fluid purging implementation used for this study seemingly overcame the overheating problems reported in the literature.

Performance-wise, the PV-T systems simulated using tank fluid purging, night cooling or both combined surpassed the primary energy efficiency of the reference PV-T system not using any stagnation control method. In particular, the combined use of both individual methods led to the best performances of all, although the performance differences are arguably within the simulations' error range.

Moreover, combined use of both methods reduced the volume of water purged relative to the individual use of tank fluid purging, which proved significant. Thus, fluid purging is not environmentally sound vis-à-vis water scarcity, particularly if used in space heating systems, and can be complemented inexpensively by night cooling.

With regard to the implementation tweaks conceivably possible using features associated with "smart systems", the simulations did not reveal noticeable energy efficiency increases by using occupancy information to disable the auxiliary system. Conversely, using that information to alternate between normal and (lower) 'no load' night cooling setpoints reduced the cumulative duration of overheating events annually - even if insufficient to effectively prevent overheating as a whole - and improved the systems' primary energy efficiency. As such, night cooling has potential to improve and become the standard, low cost, effective and environmentally friendly overheating protection method for SDHW PV-T systems and others.

**Acknowledgments.** The current study was made possible by a doctoral grant (reference SFRH/BD/76747/2011) issued by the Foundation for Science and Technology (FCT). The authors would like to thank the National Laboratory of Energy and Geology (LNEG) and the Faculty of Science and Technology, New University of Lisbon (FCT-UNL) for the institutional support.

## References

1. Dupeyrat, P., Ménézo, C., Rommel, M., Henning, H.M.: Efficient single glazed flat plate photovoltaic–thermal hybrid collector for domestic hot water system. Sol. Energy **85**(7), 1457–1468 (2011)
2. Magalhães, P., Martins, J., Joyce, A.: Comparative analysis of overheating prevention and stagnation handling measures for photovoltaic-thermal (PV-T) systems. Energy Procedia **91**, 346–355 (2016)
3. Frank, E., Mauthner, F., Fischer, S.: Overheating prevention and stagnation handling in solar process heat applications. Technical report A.1.2, IEA SHC Task 49, Solar PACES Annex IV (2015)
4. Hillerns, F.: The behaviour of heat transfer media in solar active thermal systems in view of the stagnation conditions. In: IEA-SHC Task 26 Industry Workshop, Borlänge, Sweden, 3 April 2001
5. Lämmle, M., Thoma, C., Hermann, M.: A PVT collector concept with variable film insulation and low-emissivity coating. Energy Procedia **91**, 72–77 (2016)
6. Lämmle, M., Kroyer, T., Fortuin, S., Wiese, M., Hermann, M.: Development and modelling of highly-efficient PVT collectors with low-emissivity coatings. Sol. Energy **130**, 161–173 (2016)
7. Harrison, S., Cruickshank, C.A.: A review of strategies for the control of high temperature stagnation in solar collectors and systems. Energy Procedia **30**, 793–804 (2012)
8. Akhras, G.: Smart materials and smart systems for the future. Can. Mil. J. **1**, 25–31 (2000)
9. Eicker, U.: Solar Technologies for Buildings. Wiley, Hoboken (2003)
10. Kaltschmitt, M., Streicher, W., Wiese, A.: Renewable Energy: Technology, Economics and Environment. Springer, Heidelberg (2007)
11. Cruickshank, C.A., Baldwin, C.: Sensible thermal energy storage: diurnal and seasonal. In: Letcher, T.M. (eds.) Storing Energy, with Special Reference to Renewable Energy Sources, pp. 291–309. Elsevier (2016)
12. Dupeyrat, P., Ménezo, C., Fortuin, S.: Study of the thermal and electrical performances of PVT solar hot water system. Energy Build. **68**, 751–755 (2014)

13. Silva, R.M., Fernandes, J.L.M.: Hybrid photovoltaic/thermal (PV/T) solar systems simulation with Simulink/Matlab. Sol. Energy **84**, 1985–1996 (2010)
14. Eicker, U., Dalibard, A.: Photovoltaic-thermal collectors for night radiative cooling of buildings. Sol. Energy **84**, 1322–1335 (2011)
15. Lämmle, M., Fortuin, S., Hermann, M.: Thermisches management von pvt-kollektoren - ergebnisse aus systemsimulationen (2015)
16. DIN V 18599-1 Berichtigung 1, Ber 1:2013-05 (May 2013)
17. Nhut, L.M., Park, Y.C.: A study on automatic optimal operation of a pump for solar domestic hot water system. Sol. Energy **98**, 448–457 (2013)

# Stochastic Optimal Operation of Concentrating Solar Power Plants Based on Conditional Value-at-Risk

João A.R. Esteves[1,3(✉)], Hugo M.I. Pousinho[2], and Victor M.F. Mendes[3]

[1] R&D Nester – Centro de Investigação em Energia, REN – State Grid, S.A., Lisbon, Portugal
joao.esteves@rd.nester.com
[2] IDMEC/LAETA, Instituto Superior Técnico, Universidade de Lisboa, Lisbon, Portugal
[3] Instituto Superior de Engenharia de Lisboa, Lisbon, Portugal

**Abstract.** This paper presents a stochastic programming approach, using a risk measure defined by conditional value-at-risk, for trading solar energy in a market environment under uncertainty. Uncertainties on electricity price and solar irradiation are considered through a set of scenarios computed by simulation and scenario-reduction. The short-term operation problem of a concentrating solar power plant is formulated as a mixed-integer linear program, which allows modelling the discrete status of the plant. To improve the operational productivity of the plant during the non-insulation periods, energy storage systems are considered. The goal is to obtain the optimal operation planning that maximizes the total expected profits while evaluating trading risks. For risk evaluation, the conditional value-at-risk is used to maximize the expected profits of the least profitable scenarios. A case study is used to illustrate the usefulness and the proficiency of the proposed approach.

**Keywords:** Concentrating power plant · Conditional value-at-risk · Day-ahead market · Stochastic optimization

## 1 Introduction

Renewable energy grid integration increased in the E.U. to fulfil the Energy–2020 initiative. The global warming threat is a concern around the world and many policies and agreements are being created to mitigate it [1]. In 2016, the Global Energy Interconnection Development and Cooperation Organization promoted the first council meeting in order to improve the renewable energy efficient usage [2].

In 2016, the Portuguese energy consumption was totally fulfilled by renewable energy sources for almost 4 days. This proves that Portugal has a great potential to be a green energy country although more investments are needed, namely in solar energy [4]. In 2015, the Portuguese total annual energy consumption was fulfilled by 47% from renewable energy sources, but only 1.5% was from solar energy [3].


L.M. Camarinha-Matos et al. (Eds.): DoCEIS 2017, IFIP AICT 499, pp. 348–357, 2017.
DOI: 10.1007/978-3-319-56077-9_34

Since feed-in tariffs in many countries are coming to an end, the energy trade from renewable energy power plants must occur in electricity markets competing with non-renewable power plants.

Despite the large-scale wind power integration on power systems, other renewable energy technologies are emerging in several countries, such as: photovoltaics (PV) power plants and concentrated solar power (CSP) plants. The non-dispatchable characteristics of a solar energy power plant can be reduced through thermal energy storage (TES) systems. Moreover, TES systems allow the CSP power plants to participate in ancillary service markets [6].

This paper provides an optimization approach that maximizes the expected profits of a CSP producer taking part in the day-ahead electricity market. The main contribution of this paper is: (1) to use a stochastic mixed-integer linear programming (MILP) approach for determining the optimal self-scheduling of a CSP plant having TES systems; (2) to use the conditional value-at-risk so as to mitigate the effect of uncertainty; and (3) to use a scenario reduction algorithm so as to improve the computation time.

## 2 Relationship to Smart Systems

Smart systems represent a step in the development of a new era for the energy system. Regarding energy efficiency, the smart systems are a new advantage. In the electric power grids of the future, where the internet of things is connecting several measuring devices, are being optimized trough smart systems to increase the energy efficiency [20]. For CSP plants, the smart systems can measure different variables and provide them as inputs for an optimization algorithm. Moreover, the smart systems based on smart meters are vital for solar forecasting which allow to assess forecasts given by external providers. The smart meters allow storing variables in a database, helping power producers to develop the optimal offering strategies to submit in the electricity market [21].

## 3 State of the Art

The optimal self-scheduling of energy resources is growing in the research community. Different approaches for the short-term scheduling are in research, such as: a novel two-based approach for optimal self-scheduling of CSP plants where the MILP capabilities are combined with the accuracy of the detailed model showing improvements against the simple MILP approach [11], a robust optimization for the maximization of the profit in the day-ahead electricity market for a CSP plant with TES and a backup fossil system and considering bilateral contracts is presented with an acceptable computational time for an industrial application [12], a model-based predictive control approach for optimal generation scheduling of CSP plants is presented unveiling a significant improvement using short-term DNI forecasting and improvements reducing the deviation from the scheduled generation [13]. The location of the CSP plant has been studied due to is direct relation with the DNI [16]. Another research using a deterministic approach presented an optimal offering strategy for CSP plant with TES considering not only day-ahead markets, but also joint energy, reserve and regulations markets [17]. The value of TES is also being subject of high interest among the research community, it was estimated the capacity value of CSP plants with TES in the USA were it was optimized a CSP plant with TES using Solar Advisor Model (SAM) and a mix-integer program (MIP) that unveiled that only the critical hours of the year are necessary to estimate the capacity value of the CSP plant and unveiled that the TES allow extremely high capacities for CSP plant from around 79% to 92% [14], a dense study in TES options for CSP plants is shown in [15]. Coordination of renewable sources like solar and wind are also in the scope of the research community and this subject contributes to increase the renewable energy sources penetration, in [18] it is used robust optimization to integrate CSP plants with TES and wind power which unveiled an uncertainty generation reducing, in [19] a MILP approach is presented to schedule wind and CSP plant with TES revealing to be feasible.

## 4 Problem Formulation

The self-scheduling problem of a CPS plant having TES system is computed by the maximization of the objective function given by the profit, affected by the risk measure CVaR, subject to technical operation constraints and the risk management constraints.

### 4.1 Objective Function

The objective function is given by two parts, the first one is the profit which is equal to the revenues from the day-ahead electricity market sales during the time horizon of the schedule, and the second one is the equation resulting from the CVaR risk management application. The objective function is given as follows:

$$F = (1 - \beta)\left(\sum_{\omega=1}^{\Omega}\sum_{t=1}^{T}\left(\pi_{\omega,t}\lambda_{\omega,t}P^{s}_{\omega,t}\right)\right) + \beta\left(\eta - \frac{1}{1-\alpha}\sum_{\omega=1}^{\Omega}\left(\pi_{\omega,t}s_{\omega}\right)\right) \tag{1}$$

In (1), T is the set of hours in the time horizon, $\Omega$ is the set of scenarios resulting from the combination of solar resource and electricity market prices, $\pi_{\omega,t}$ is the probability for scenario $\omega$. hour t, $\lambda_{\omega,t}$ is the price of the electricity in the day-ahead market for scenario $\omega$. hour t, $P^{s}_{\omega,t}$ is the power output of the CSP plant having TES for scenario $\omega$. hour t, $\beta$ the level of risk that the decision maker is willing to take, $\alpha$. presents the level of confidence and is set to 0.95, $\eta$. It's a variable dependent of the first part of the objective function (maximization of the profit) and $s_{\omega}$ is a function given by:

$$s_{\omega} = \max\{\eta - f(x,\omega), 0\} \tag{2}$$

In (2), $f(x,\omega)$ represents the first part of the objective function.

## 4.2 Constraints

The constraints for the scheduling are due to operation, minimum up/down time, electricity market and risk management for the CSP plant having TES system.

**Operation Constraints.** The operation constraints are given as follows:

$$P^{s}_{\omega,t} = P^{FE}_{\omega,t} + P^{SE}_{\omega,t} \quad \forall\omega \in \Omega, \forall t \in T \tag{3}$$

$$P^{FE}_{\omega,t} = \eta_1 Q^{FE}_{\omega,t} \quad \forall\omega \in \Omega, \forall t \in T \tag{4}$$

$$P^{SE}_{\omega,t} = \eta_3 Q^{SE}_{\omega,t} \quad \forall\omega \in \Omega, \forall t \in T \tag{5}$$

$$Q^{FE}_{\omega,t} + Q^{FS}_{\omega,t} \leq E_{\omega,t} \quad \forall\omega \in \Omega, \forall t \in T \tag{6}$$

$$Q^{E}_{min} u_{\omega,t} \leq Q^{FE}_{\omega,t} + Q^{SE}_{\omega,t} \leq Q^{E}_{max} u_{\omega,t} \quad \forall\omega \in \Omega, \forall t \in T \tag{7}$$

$$P^{s}_{min} u_{\omega,t} \leq P^{s}_{\omega,t} \leq P^{s}_{max} u_{\omega,t} \quad \forall\omega \in \Omega, \forall t \in T \tag{8}$$

$$Q^{S}_{\omega,t} = Q^{S}_{\omega,t-1} + \left(\eta_2 Q^{FS}_{\omega,t}\right) - Q^{SE}_{\omega,t} \quad \forall\omega \in \Omega, \forall t \in T \tag{9}$$

$$Q^{S}_{min} \leq Q^{S}_{\omega,t} \leq Q^{S}_{max} \quad \forall\omega \in \Omega, \forall t = 0, 1, \ldots, T \tag{10}$$

$$\eta_2 \left(Q^{FS}_{\omega,t} - Q^{FS}_{\omega,t-1}\right) \leq R^{up} \quad \forall\omega \in \Omega, \forall t \subset T \tag{11}$$

$$P^{SE}_{\omega,t-1} - P^{SE}_{\omega,t} \leq R^{dn} \quad \forall\omega \in \Omega, \forall t \in T \tag{12}$$

$$P^{SE}_{\omega,t} \leq R e_{\omega,t} \quad \forall\omega \in \Omega, \forall t \in T \tag{13}$$

$$Q^{FS}_{\omega,t} \leq R(1 - e_{\omega,t}) \quad \forall\omega \in \Omega, \forall t \in T \tag{14}$$

$$P^{FE}_{\omega,t}, P^{SE}_{\omega,t}, P^{FS}_{\omega,t}, P^{s}_{\omega,t}, Q^{FE}_{\omega,t}, Q^{SE}_{\omega,t}, Q^{FS}_{\omega,t}, Q^{S}_{\omega,t} \geq 0 \quad \forall\omega \in \Omega, \forall t \in T \tag{15}$$

$$y_{\omega,t} - z_{\omega,t} = u_{\omega,t} - u_{\omega,t-1} \quad \forall\omega \in \Omega, \forall t \in T \tag{16}$$

$$y_{\omega,t} + z_{\omega,t} \leq 1 \quad \forall\omega \in \Omega, \forall t \in T \tag{17}$$

From (3) to (17), $P^{s}_{\omega,t}$ is the electric power produced by the CSP, $P^{FE}_{\omega,t}$ and $P^{SE}_{\omega,t}$ are the power produced by the power block (PB) from the solar field (SF) and from the TES for scenario ω. hour t, $\eta_1$ is the SF efficiency, $Q^{FE}_{\omega,t}$ is the thermal power from the SF for scenario ω. hour t, $\eta_3$ is the molten-salt tanks efficiency, $Q^{SE}_{\omega,t}$ is the storage power in TES to produce electricity for scenario ω. hour t, $E_{\omega,t}$ is the thermal energy available in the SF for scenario ω. hour t, $Q^E_{min}$ and $Q^E_{max}$ are the thermal power bounds of the PB for scenario ω. hour t, $u_{\omega,t}$ is a binary variable that represents the CSP plant commitment for scenario ω. hour t, $P^s_{min}$ and $P^s_{max}$ are the electrical power bounds of the PB, $Q^{FS}_{\omega,t}$ is the stored thermal power from the SF for scenario ω. hour t, $Q^S_{\omega,t}$ is the thermal energy stored in TES for scenario ω. hour t, $\eta_2$ is the TES efficiency, $Q^S_{min}$ and $Q^S_{max}$ are the TES thermal energy bounds, $R^{up}$ and $R^{dn}$ are the up and down ramp limitations of the TES, R. a sufficient large constant, $e_{\omega,t}$ is a binary variable that represents if TES is charging our discharging, $y_{\omega,t}$ is a binary variable that represents if the CSP plant started up for scenario ω. hour t, and $z_{\omega,t}$ is a binary variable that represents if the CSP plant shuts down for scenario ω. hour t. In (3), the electric power balance is obtained combining the electric power produced in the SF and the storage. In (4) and (5) are represented the efficiency between power energy and thermal energy for the SF and the TES, respectively. In (6) the bound of the thermal energy from the SF and TES are set. In (7) and (8) are the bounds of the SF and the TES. In (9) is given the thermal energy stored in the TES. In (10) the bounds of the TES are set. In (11) and (12) are represented the ramp-up and ramp-down limitations of the TES. In (13) and (14) guarantee that the SAE is not charging and discharging ate the same time. In (15) is guaranteed the positivity of energy. In (16) and (17) is guaranteed that the CSP plant don't start and shut down at the same time.

**Minimum Up/Down Time Constraints.** The minimum up/down constraints of the power block are given as follows:

$$\left(x_{\omega,t} - TMF\right)\left(u_{\omega,t-1} - u_{\omega,t}\right) \geq 0 \quad \forall\omega \in \Omega, \forall t \in T \tag{18}$$

$$\left(x_{\omega,t} - TMP\right)\left(u_{\omega,t} - u_{\omega,t-1}\right) \geq 0 \quad \forall\omega \in \Omega, \forall t \in T \tag{19}$$

In the Eqs. (18) and (19), $x_{\omega,t}$ represents the number of hours in which the CSP plant was not working (down time) for scenario ω. hour t, TMF and TMP represents the number of necessary hours of up time and down time, respectively.

**Electricity Market Constraints.** The electricity market constraints are given as follows:

$$\left(\lambda_{\omega,t} - \lambda_{\omega,t-1}\right)\left(P^s_{\omega,t} - P^s_{\omega,t-1}\right) \geq 0 \quad \forall\omega \in \Omega, \forall t \in T \tag{20}$$

In (20) the decreased monotony of the electricity market offer curve is guaranteed. By other others if the price of energy in the electricity market don't decrease the amount of energy to bid in the market also can't decrease.

**Risk Management Constraints.** The risk management constraints are given as follows:

$$\eta - \left(\sum_{t=1}^{T} \lambda_{t.\omega} P^{s}_{\omega,t}\right) \leq s_{\omega} \quad \forall \omega \in \Omega, \forall t \in T \tag{21}$$

$$s_{\omega} \geq 0 \quad \forall \omega \in \Omega \tag{22}$$

In (21) and (22) the risk management using CVaR is guaranteed. CVaR it is defined, for a certain level of confidence $\alpha \in \{0, 1\}$, as the average deviation of the worst scenarios.

## 5 Case Study

The case study intends to show the proficiency of the model developed for the scheduling of CSP plant having TES in the electricity market. The MILP approach has been solved using CPLEX 12.2 solver under GAMS environment [7]. A computer with 8 GB RAM with 2.40 GHz of CPU is used for the simulations of a realistic case study for the CSP plant having TES system carried out with technical data shown in Table 1.

**Table 1.** CSP having TES system data.

| $Q^{E}_{min}/Q^{E}_{max}$ (MWt) | $P^{s}_{min}/P^{s}_{max}$ (MWe) | $Q^{FE}_{min}/Q^{FE}_{max}$ (MWt) | $R^{dn}$ (MWe/h) | $R^{up}$ (MWe/h) |
|---|---|---|---|---|
| 50/125 | 0/50 | 0/150 | 35 | 80 |
| $Q^{S}_{min}/Q^{S}_{max}$ (MWht) | $Q^{S}_{0}$ (MWht) | R (MWe) | TMF (h) | TMP (h) |
| 45/700 | 350 | 150 | 2 | 2 |
| $K^{on}_{0}$ (h) | $K^{off}_{0}$ (h) | $u_0$ | $y_0$ | $z_0$ |
| 1 | 1 | 0 | 1 | 0 |

The time horizon in the simulations is a day on an hourly basis. The inputs considered within the time horizon are the solar power and the electricity market prices. The solar power profile was obtained using the System Advisor Model [8] and it was converted into available thermal power. The electricity market prices derived from the Iberian electricity market given in [9]. In this case study was used 250 scenarios composed by 25 electricity market prices, as shown in Fig. 1(a), and 10 scenarios for the available thermal power, as shown in Fig. 1(b).

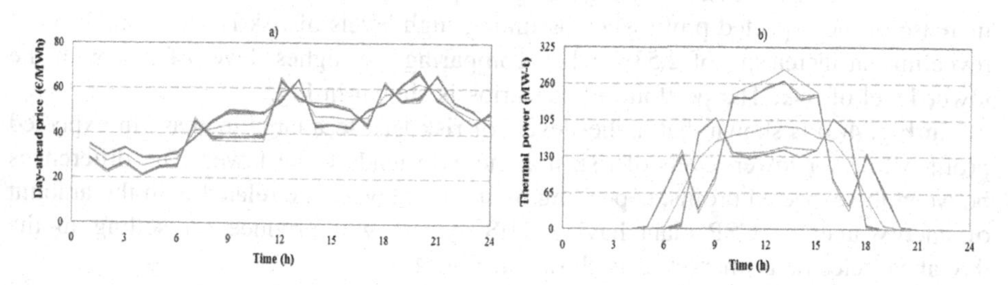


**Fig. 1.** (a) Electricity market price scenarios, (b) Thermal power scenarios.

The optimizing characteristics for processing the problem presented in the case study in what regards the number of continuous variables, binary variables and constraints are shown in Table 2.

**Table 2.** Optimization characteristics of the case study.

| | Continuous variables | Binary variables | Constraints |
|---|---|---|---|
| Case study | 54 002 | 24 000 | 162 000 |

The number of scenarios highly affects the computation burden of the problem to be solved. The application of a scenario reduction algorithm, forward reduction algorithm detailed in [10], was used to assess the implication of the number of scenarios in the expected profit and in the computation time. The scenario reduction algorithm allow to reduce the number of scenarios of the original set maintaining the statistical characteristics. The results are shown in Fig. 3 (Fig. 2).

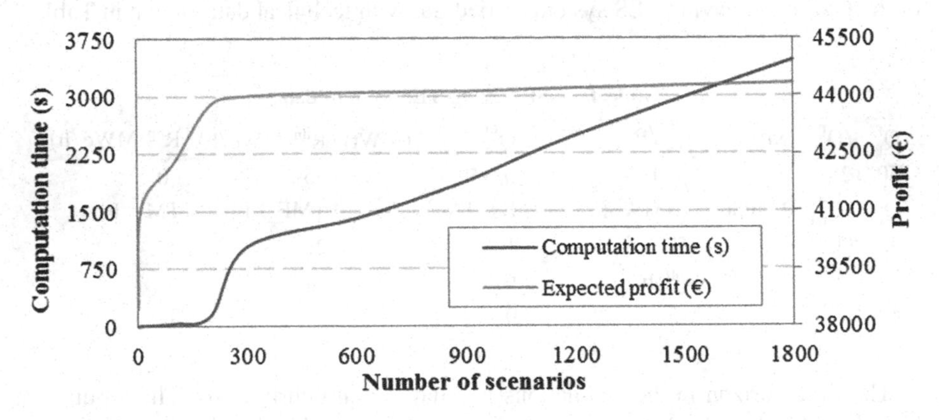


**Fig. 2.** Expected profit, computation time related with the number of scenarios.

Figure 3 shows that for lower number of scenarios the increase in the computational time it is residual, for moderate to higher number of scenarios the increase in the computational time becomes higher and higher assuming almost a linear evolution. On the other and, the profit tends to stabilize at moderate number of scenarios.

The results of the case study regarding the profit deviation, expected profit and the increase of the expected profit when assuming high levels of risk is shown in Table 3 revealing an increasing of 2.53% when comparing the highest level of risk with the lower level of risk. The portfolio of scenarios is shown in Fig. 4.

In Fig. 4(a) is shown that higher levels of risk lead to a large increase in expected profit, while for lower levels of risk this increase tends to be lower. The differences between the expected profits, depending on the risk levels, are related with the amount of energy that the CSP plant having TES system will produce for selling in the day-ahead electricity market as is shown in Fig. 4(b).

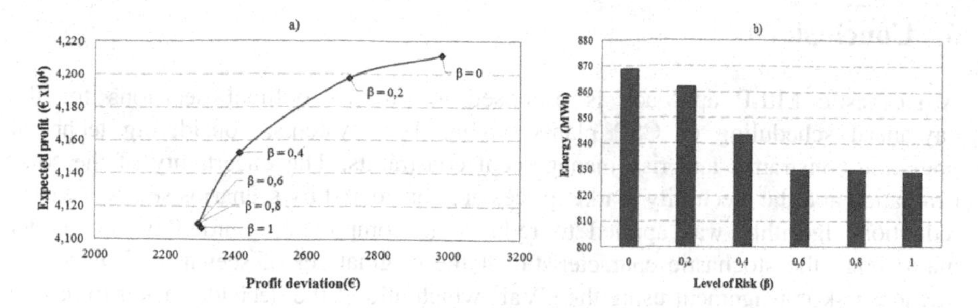


**Fig. 3.** (a) Portfolio of scenarios, (b) Energy to sell in the electricity market.

**Table 3.** Profit assessment for different levels of risk.

| β | Profit deviation (€) | Expected profit (€) | % Increase |
|---|---|---|---|
| 1.0 | 2 301 | 41 072 | – |
| 0.8 | 2 293 | 41 097 | 0.06 |
| 0.6 | 2 292 | 41 097 | 0.06 |
| 0.4 | 2 410 | 41 525 | 1.10 |
| 0.2 | 2 718 | 41 975 | 2.20 |
| 0 | 2 978 | 42 110 | 2.53 |

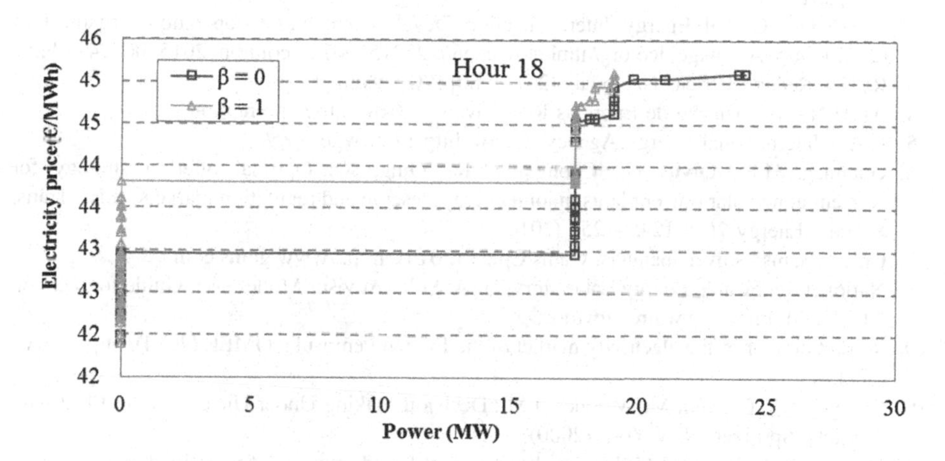


**Fig. 4.** Offering curves for hour 18.

The offering curves for a conservative power producer ($\beta = 1$) and a non-conservative power producer ($\beta = 0$) are shown in Fig. 4.

In Fig. 4 is shown that a conservative power producer delivers less power in the day-ahead electricity market to guarantee his revenues, while the non-conservative one delivers more power in the day-ahead electricity market willing to take more risk to increase his revenues.

## 6 Conclusions

A stochastic MILP approach is proposed to provide optimal decisions for the day-ahead scheduling of CSP plants having TES systems, considering technical operation constraints and risk management constraints. The uncertainty of the solar irradiation and the electricity market prices are considered by scenarios set. A scenario reduction algorithm was applied to reduce the computation burden of the model maintaining the stochastic characteristic of the original set of scenarios. The model includes risk management using the CVaR which allows the decision maker to have a portfolio of expected profits for different risk levels. The case studies are in favor of the approach to support decisions in day-ahead markets. This proposed approach shown to be feasible and accurate for optimal scheduling of the CSP plants with TES.

**Acknowledgement.** This work was supported by R&D Nester, Centro de Investigação em Energia REN – State Grid.

## References

1. Energy 2020 - a strategy for competitive, sustainable and secure energy (2011). http://europa.eu
2. GEIDCO - Global Energy Interconnection Development and Cooperation Organization (2016). http://www.geidco.org/html/qqnycoen/col2015100724/column_2015100724_1.html
3. REN – Redes Elétricas Nacionais (2016). http://www.ren.pt/
4. APREN – Associação de Energias Renováveis (2016). http://apren.pt/pt/
5. IEA – International Energy Agency (2016). https://www.iea.org/
6. Guannan, H.E., Qixin, C., Chongqing, K., Qing, X.: Optimal offering strategy for concentrating solar power plants in joint energy, reserve and regulation markets. IEEE Trans. Sustain. Energy **7**(3), 1245–1254 (2016)
7. Cplex, Gams, Solver manuals. Gams/Cplex (2014). http://www.gams.com
8. National Renewable Energy Laboratory, USA, Solar Advisor Model User Guide for Version 2.0 (2008). http://www.nrel.gov/docs
9. Market oprator of the electricity market of the Iberian Peninsula, OMEL (2014). http://www.omel.es
10. Conejo, A.J., Carrión, M., Morales, J.M.: Decision Making Under Uncertainty in Electricity Markets. Springer, New York (2000)
11. Vasallo, M.J., Bravo, J.M.: A novel two-model based approach for optimal scheduling in CSP plants. Elsevier, ScienceDirect, Solar Energy **126**, 73–79 (2016). Elsevier, ScienceDirect
12. Pousinho, H.M.I., Contreras, J., Pinson, P., Mendes, V.M.F.: Robust optimization for self-scheduling and bidding strategies of hybrid CSP-fossil power plants
13. Vasallo, M.J., Bravo, J.M.: A MPC approach for optimal generation scheduling in CSP plants. Appl. Energy **165**, 357–370 (2016). ELSEVIER, ScienceDirect
14. Madaeni, S.H., Sioshansi, R., Denholm, P.: Estimating the capacity value of concentrating solar power plants with thermal energy storage: a case study of the southwestern united states. Powersystems, **28**(2) (2013). IEEE

15. Xu, Z., Kariuki, S.K., Chowdhury, S., Chowdhury, S.P.: Investigation of thermal storage options for concentrating solar power plants. In: 2012 47th International Universities Power Engineering Conference (UPEC). IEEE (2012)
16. Llamas, J., Bullejos, D., Barranco, V., Ruiz de adna, M.: World location as associated factor for optimal operation model of parabolic trough concentrating solar thermal power plants. IEEE (2016). 978-1-5090-2320-2
17. He, G., Chen, Q., Kang, C., Xia, Q.: optimal offering strategy for concentrating solar power plants in joint energy, reserve and regulation markets. Sustain. Energy, **7**(3), (2016). IEE
18. Chen, R., Sun, H., Guo, Q., Li, Z., Deng, T., Wu, W., Zhang, B.: Reducing generation uncertainty by integrating CSP with wind power: an adaptative robust optimization-based analysis. Sustain. Energy, **6**(2), (2015)
19. Pousinho, H.M.I., Silva, H., Mendes, V.M.F., Collares-Pereira, M., Pereira, C.: Cabrita: self-scheduling for energy and spining reserve of wind/CSP plants by a MILP approach. Energy **78**, 524–534 (2014). ELSEVIER
20. Ejaz, W., Naeem, M., Shahid, A., Anpalagan, A., Jo, M.: Efficient energy management for the internet of things in smart cities. IEEE Commun. Mag. 84–91 (2017)
21. Vukmirovic, S., Erdeljan, A., Kulic, F., Lukovic, S.: Software architecture for Smart Metering systems with Virtual Power Plant, pp. 448–451. IEEE (2010)

# Solar Thermoelectric System with Biomass Back-up

José Teixeira Gonçalves[1(✉)], Cristina Inês Camus[2], and Stanimir Stoyanov Valtchev[1]

[1] Dep. Eng. Electrotecnica, NOVA University of Lisbon –
Faculty of Sciences and Technology, 2829-516 Caparica, Portugal
jt.goncalves@campus.fct.unl.pt, ssv@fct.unl.pt
[2] Dep. Eletric. Energy and Automation,
Politechnic Institute of Lisbon – Superior Institute of Engineering, Lisbon, Portugal
ccamus@deea.isel.pt

**Abstract.** With the objective of having a solar thermoelectric system, running for 24 h a day along the different seasons of the year it is necessary to dimension the adequate storage and back-up systems. The choice of the back-up source of energy depends on how sustainable the power plant should be. In this study, the choice was the use of biomass in order to have a 100% renewable power plant. The selected site was the Alentejo region (Portugal). The local Direct Normal Irradiation (DNI) data was used to simulate with the System Advisor Model program (SAM) considering a solar system with north field and molten salt storage. The system needs no back-up during three months in a year. The use of biomass pellets is a viable alternative because it makes the power plant 100% renewable and dispatchable without loss of energy due to over-dimension of the expensive solar field and molten storage system.

**Keywords:** Solar · Thermoelectric · Backup · Biomass · Pellets

## 1 Introduction

The need to increase the use renewable energy continues in the environmental policy of every country. Climate change keeps being an important issue and all research and investment go into the direction of the Carbon-free technologies.

One of the technologies that has denote an increase in sunny countries is the concentrated solar thermal (CSP) that converts the direct solar irradiation (DNI) into thermal energy through directional mirrors and a receiver. This thermal energy can be stored in molten salt tanks and then converted into electrical energy. The CSP technologies vary with the form of DNI concentration at the receiver. There are four types of CSP: Cylindrical-parabolic collectors, CCP; parabolic dish Stirling; linear Fresnel and central tower receiver [1, 2]. This technology has a high cost, and still needs feed-in tariffs in order to be economically viable.

The central tower solar thermal technology has reached commercial maturity and is expanding. In development since 1980, after the pioneering experience of the Solar I and Solar II in the US and Almeria Platform in Spain during the period 2000-2010 [3].


L.M. Camarinha-Matos et al. (Eds.): DoCEIS 2017, IFIP AICT 499, pp. 358–369, 2017.
DOI: 10.1007/978-3-319-56077-9_35

Currently, the largest power plant in the world with this type of technology is the Ivanpah with 377 MW (California, U.S.A.) [4].

The storage technology is very important to able the power plant to generate for longer periods [2], as it is the case of Gemasolar (a solar tower with circular field) that was the first in the world to run 24 h a day, thanks to the addition of storage and back-up system [5]. The 24 h a day operation is just possible during a few summer periods with only the energy of the DNI and storage. In winter periods the system needs to add another technology to extend over the period, thereby making a hybrid system, but this addition is often made with oil products or natural gas. This makes the power plant less clean.

With this vision comes the main issue. How to make the CSP system that is completely renewable and clean, work 24 h a day both in summer and winter periods?

The objective of this work is to study the use of biomass as a back-up for a CSP system central tower with north field and storage to operate 24 h a day.

## 2 Relationship to Smart Systems

This work demonstrates that a hybrid system such as a thermoelectric solar tower with storage and biomass back-up system can operate 24 h a day, both in summer and in winter and respect the clean and renewable way. Thermoelectric solar systems work automatically, from the solar concentrators drive process (following the sun orientation) [6] thermal fluid process (controlling the valves, the heat transfer fluid, HTF flows, thermal storage, etc.) [7] and the electrical system (controlling the power delivered to the grid, the power consumed in the system, etc.), it is noted that the system is fully automatic and intelligent, so it is important that the back-up system the pellets to be implemented is automatic [8]. The implementation of this automatic back-up system with pellets has several advantages:

- There is no need of human power to drive the biomass system.
- The system works in real time, it allows operation without interruption.
- The power supply is 24 h a day, in winter and in summer.
- This allows the two systems to be operated automatically in only an operating room or other type of control.

The main contribution to the smart system is that the back-up biomass system works automatically, i.e., from the processing control when the thermoelectric solar system requires thermal power, the system actuates automatically (smart) to inlet the pellet biomass to the boiler which triggers combustion, transferring the thermal energy released from combustion to the overall system. All these processes are automated and help in the system reliability to have a safe and non-stop operation.

## 3 State of Art

There are basically four types of CSP, the cylindrical-parabolic collectors, CCP; parabolic dish Stirling; linear Fresnel and central tower receiver.

The elements that differ in each CSP technology are mainly the solar field and receiver. Then they can have heat storage and heat exchanger (depending on the type of CSP and heat transfer fluid, HTF) and finally to convert thermal energy into electric energy a power unit is required.

The system studied in this work was the cavity tower receptor. A remarkable advantage of this technology (central tower solar thermoelectric) compared to other CSP technologies, such as linear Fresnel parabolic or disc is that the receiver operates at a higher thermal efficiency. This is due to the fact that all heliostats concentrate the DNI to a point (surface receptor) [9], thereby raising the operating temperature [10] (which aids in hybridization). The rapid growth of this technology is due, in large part, to this advantage. This type of system has also reached the stage of commercial maturity and is expanding.

## 3.1 Description and Operation of the Thermoelectric Solar System with Biomass Backup

The following key elements will be needed for the system to function: heliostats, tower and receiver, storage tank, power block (heat exchanger, turbine steam, electrical generator, condenser, water pumps, etc.) and boiler for pellets. The proper integration of all elements allows the correct operation of the system. Figure 1 shows the proposal of the system to be dimensioned, it shows all the elements already mentioned and we can observe that the structure of the field north the tower, allows to add the system of storage of the biomass to pellets to the south of the tower.

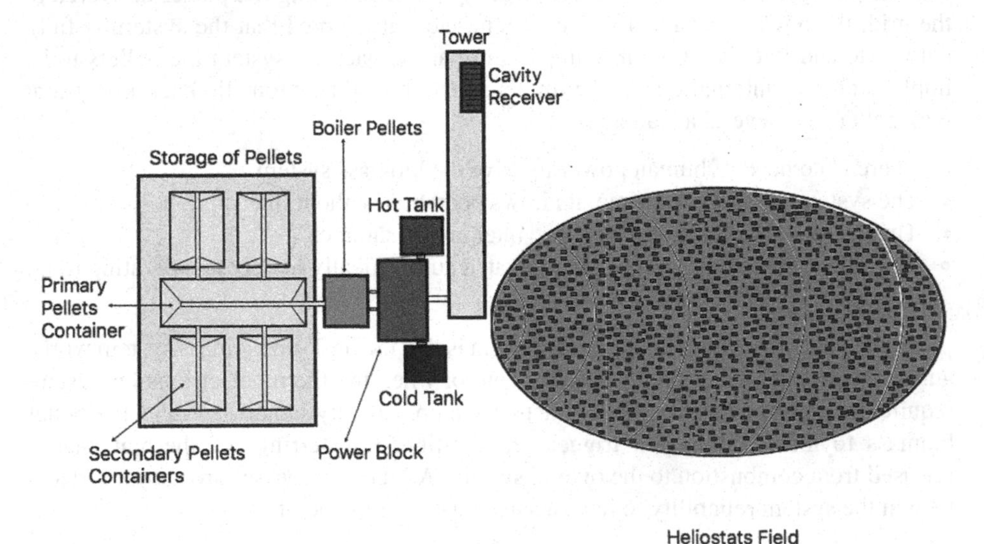


**Fig. 1.** Constitution of the complete structure of the proposed thermoelectric solar system with biomass backup.

### 3.1.1 Boiler of Pellets

The pellet boiler is basically a furnace that operates automatically, it is made of tubes that serve as a heat exchanger, the combustion chamber, the pellets storage, the rook wiper carrier, air pump, etc. The boiler operating principle is shown in Fig. 2.

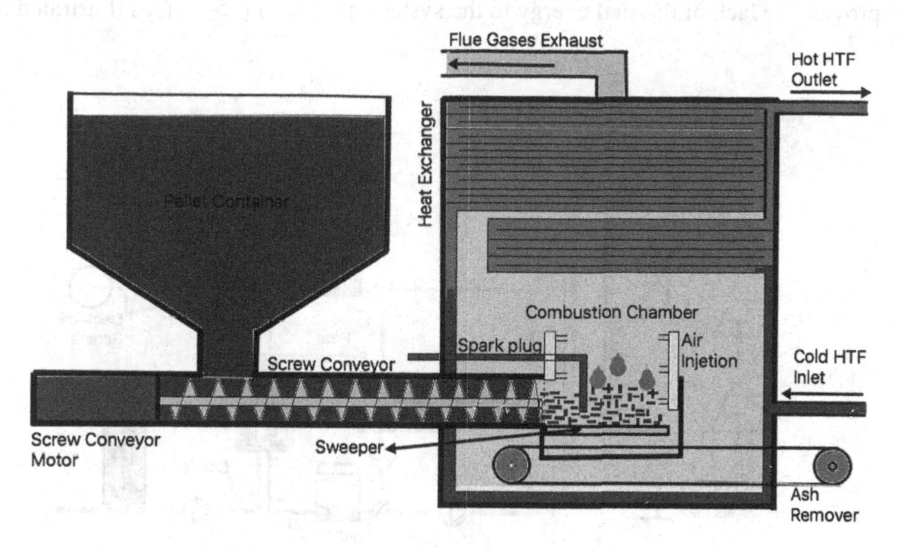


**Fig. 2.** Operating principle of the pellets boiler.

### 3.1.2 Pellets

The fuel pellets are completely natural as they are made primarily from wood chips and sawdust, but can also be made with corn cobs, wheat bran, rice husks, leaves, grasses, and other products.

The pellets are a type of densified biomass and therefore they have better physical and combustion characteristics compared to the raw biomass. Due to the size and shape of the cylinder that the pellets have, it becomes possible to automate the boiler running on this fuel, becoming competitive with gas, Diesel and fuel-oil boilers, since these boilers are also the systems of automatic ignition, automatic transport of the pellets to the combustion chamber, the ash removal mechanism automatic cleaning of the heat exchanger and others. The production of pellets adds the possibility of using different waste biomass into standard fuel systems [8].

In Europe and the United States of America biomass is largely used either on a domestic scale or on an industrial scale such as large thermoelectric power stations [11].

One can convert a coal plant to function with biomass, thereby reducing carbon dioxide emissions, or to create an original biomass power plant [11, 12].

The biomass plants can operate only with pellets, such as the Les Awirs plant (Belgium), which was converted from coal to work 100% with wood pellets, and consumes about 350,000 tons of pellets per year at 80 MW power level [13, 14].

### 3.1.3 Operating System

The operating principle of this system which occurs automatically, can be described in two periods: The first period consists of the system operation with Direct Normal Irradiation (DNI) and storage and the second period the controlling of biomass combustion to provide the lack of thermal energy to the system (period 2) [15–17], as illustrated in Fig. 3.

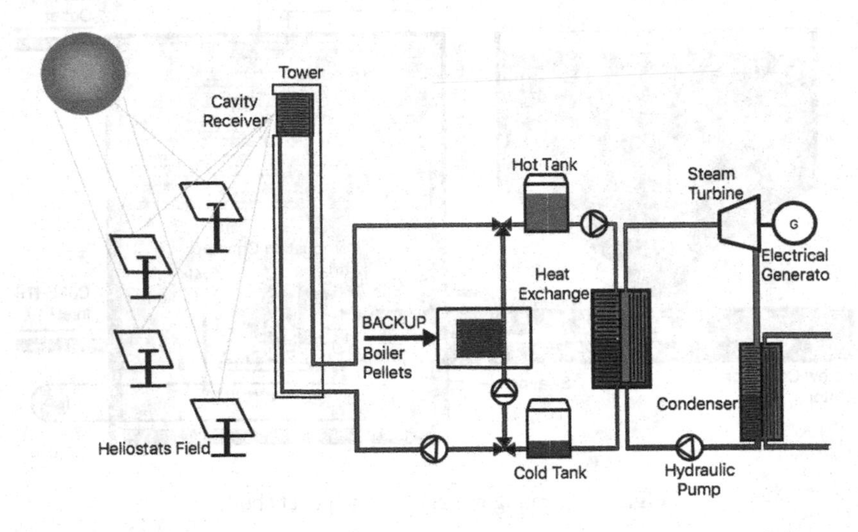


**Fig. 3.** Operating principle of the solar thermoelectric tower with back-up with biomass.

**Period 1**: The Direct Normal Irradiation, DNI, reaches the surface of the heliostat, which acts as a mirror, deflecting or reflecting the sun's rays to the receiver. This causes an increased solar concentration which in turn increases the temperature (thermal energy). This temperature is transferred to a heat transfer fluid (in the case molten salt), known as heat transfer fluid, HTF. If the system has two circuits (one primary and one secondary), the molten salt (HTF) only circulates through the primary circuit.

The primary circuit has two storage tanks: A cold and a hot tank, wherein the HTF flows with the aid of hydraulic pumps. The HTF leaving the cold tank, travels through the tube to the top of the tower reaching the receptor where its temperature rises from approximately 290°C to 575°C. Then it is transferred to the hot tank, where it can be stored and, subsequently, the fluid leaves the hot tank and passes through a heat exchanger to be conducted again to the cold tank where the same process continues (a cycle).

The heat exchanger has the function to transfer the thermal energy from the primary circuit to the secondary circuit. In the secondary circuit the working fluid (water) rises its temperature changing from a liquid state into a vapor state. The vapor flows through a steam turbine, which converts kinetic energy into mechanical rotary energy. This vapor passes then through a condenser to become saturated liquid and continues its secondary circuit of the cycle. The rotating mechanical energy is converted into electricity by an electric generator coupled to the turbine. The electricity is then directed to a transformer, which increases the voltage lowering the current, so it can be transported by electric cables minimizing transmission losses of energy.

**Period 2**: In the period when there is neither DNI nor thermal energy stored in the hot tank, comes into operation the biomass back-up system to transfer the needed thermal energy to the HTF of the primary circuit. The stored pellets are transported by an automatic system in the boiler, into the furnace (above a grid), where its ignition is done by an automatic system (an electric resistance). Through this process the biomass is combusted in the combustion chamber (furnace). The heat released by the combustion process is then transferred to the HTF through the walls, which can be called a heat exchanger. Consequently, the HTF exits the boiler with sufficient thermal energy and goes to the system where it is transferred to the secondary circuit through the heat exchanger and then converted to electricity.

# 4 Research Contribution and Innovation

Some existing solar systems work with back-up of various forms of energy, such as Gemasolar 20 MW (Spain) that adds a fossil fuel back-up (natural gas) for 15% of the annual energy generated by the plant [18, 19]. There are no CSP plants currently running with biomass back-up. This fact makes this work innovative on the following points:

- An automatic pellet boiler biomass (combustion) system works together with a thermoelectric solar system (north field) in a hybrid way ensuring that the system operates 24 h a day, both in summer and in winter.
- A method to simulate with the System Advisor Model (SAM), a hybrid thermoelectric solar system and biomass, as far as the SAM software does not have a specific option for hybridization throughout the year.

# 5 Methodology and Simulation

The simulation was performed on System Advisor Model, SAM (the selected version was the "Version 01/14/2014"), with six (6) blocks to enter the parameters for processing, namely the City and climatic resources, Heliostat field, solar tower and receiver, power block (Rankine Cycle thermal dispatch control and electric generator), thermal storage and system costs. The parameters used in these blocks are described in Table 1. Some points should be described in greater detail, such as the location and climatic resort, the thermal dispatch control system and the pellets' system. As a reference, it is possible to analyze some examples of simulations of CSP performed in SAM in [18].

**1º- Location and Climatic Resources**
In this block, the location where we want to install the plant was set up. The site was near a village in the Alentejo region, with the coordinates described in Table 1. The data from the DNI were extracted from the Photovoltaic platform Geographic Information System (PVGIS), thus having the monthly average data, considered these data as daily

for each month and there was thus obtained in an Excel table with 8760 values of annual DNI. It was then possible to create a new database, transferring some values of a model ("Évora database" as stored in the program) in EPW format to Excel and complete with the extracted DNI from PVGIS. Later this database, in Excel, was converted to the executable TMY3 format to enter as an input to the SAM software.

**Table 1.** Parameters used in system simulation.

| Block | Parameters | Values |
|---|---|---|
| Location and resource | *DNI (kWh/m2) year* | 2319,4 |
| | *Latitude* | 38,165 |
| | *Longitude* | -7,195 |
| Heliostats field | *Heliostat area (m2)* | 91 |
| | *Number of heliostats* | 2067 |
| Tower and receiver | *Receiver type* | Cavity |
| | *External pipe diameter (mm)* | 40 |
| | *Pipe wall thickness (mm)* | 1,25 |
| | *Tower height (m)* | 127,11 |
| | *Tube material (stainless steel)* | AISI 316 |
| Power cycle | *Drawing gross capacity (MW)* | 20 |
| | *Conversion efficiency (%)* | 41,2 |
| Thermal storage | *Type condensation* | Ar |
| | *Type of storage* | 2 tanks |
| | *Hours of storage (hrs)* | 0-11 |
| | *Storage volume (m3)* | 2514,98 |
| | *Initial hot HTF Tem. ($^{o}C$)* | 574 |
| | *Initial cold HTF Tem. ($^{o}C$)* | 290 |
| Direct capital costs | *Site improvements ($/m2)* | 15 |
| | *Heliostats field ($/m2)* | 170 |
| | *Balance of plant ($/kW)* | 350 |
| | *Power block ($/kW)* | 1200 |
| | *Storage ($/kW ht)* | 27 |
| | *Boiler pellets ($/kW e)* | 540 |
| | *Fixed tower cost ($)* | 3000000 |
| | *Receiver reference cost ($)* | 110000000 |
| | *Contingencies (%)* | 7 |
| Indirect capital costs | *EPC and ower costs (%)* | 11 |
| | *Total land cost ($/acre)* | 10000 |
| | *Sales tax of (%)* | 5 |
| | *Direct cost of sales (%)* | 80 |
| | *Cost of O & M ($/kW)* | 65 |

## 2º- Thermal Dispatch Control System

The thermal clearance control system is an automatic control of thermal energy that is routed to the power block. This control is registered to a table where the lines refer to the months of the year and 24 columns refer to the daily hours (Table 2).

**Table 2.** Strategic model for the control of thermal dispatch system.

| | 12am | 1am | 2am | 3am | 4am | 5am | 6am | 7am | 8am | 9am | 10am | 11am | 12pm | 1pm | 2pm | 3pm | 4pm | 5pm | 6pm | 7pm | 8pm | 9pm | 10pm | 11pm |
|---|---|---|---|---|---|---|---|---|---|---|---|---|---|---|---|---|---|---|---|---|---|---|---|---|
| Jan | 3 | 3 | 3 | 3 | 3 | 3 | 3 | 3 | 3 | 3 | 3 | 1 | 1 | 1 | 1 | 1 | 3 | 3 | 3 | 3 | 3 | 3 | 3 | 3 |
| Feb | 3 | 3 | 3 | 3 | 3 | 3 | 3 | 3 | 3 | 1 | 1 | 1 | 1 | 1 | 1 | 1 | 1 | 1 | 3 | 3 | 3 | 3 | 3 | 3 |
| Mar | 3 | 3 | 3 | 3 | 3 | 3 | 3 | 3 | 1 | 1 | 1 | 1 | 1 | 1 | 1 | 1 | 1 | 1 | 3 | 3 | 3 | 3 | 3 | 3 |
| Apr | 3 | 3 | 3 | 3 | 3 | 3 | 3 | 3 | 1 | 1 | 1 | 1 | 1 | 1 | 1 | 1 | 1 | 1 | 1 | 3 | 3 | 3 | 3 | 3 |
| May | 3 | 3 | 3 | 3 | 3 | 3 | 3 | 2 | 1 | 1 | 1 | 1 | 1 | 1 | 1 | 1 | 1 | 1 | 2 | 2 | 2 | 2 | 2 | 3 |
| Jun | 2 | 2 | 2 | 2 | 2 | 2 | 2 | 2 | 1 | 1 | 1 | 1 | 1 | 1 | 1 | 1 | 1 | 2 | 2 | 2 | 2 | 2 | 2 | 2 |
| Jul | 2 | 2 | 2 | 2 | 2 | 2 | 1 | 1 | 1 | 1 | 1 | 1 | 1 | 1 | 1 | 1 | 1 | 1 | 1 | 1 | 1 | 1 | 2 | 2 |
| Aug | 2 | 2 | 2 | 2 | 2 | 2 | 2 | 1 | 1 | 1 | 1 | 1 | 1 | 1 | 1 | 1 | 1 | 1 | 1 | 1 | 2 | 2 | 2 | 2 |
| Sep | 2 | 3 | 3 | 3 | 3 | 3 | 3 | 3 | 1 | 1 | 1 | 1 | 1 | 1 | 1 | 1 | 1 | 1 | 1 | 2 | 2 | 2 | 2 | 2 |
| Oct | 3 | 3 | 3 | 3 | 3 | 3 | 3 | 3 | 3 | 1 | 1 | 1 | 1 | 1 | 1 | 1 | 1 | 1 | 2 | 2 | 3 | 3 | 3 | 3 |
| Nov | 3 | 3 | 3 | 3 | 3 | 3 | 3 | 3 | 3 | 3 | 1 | 1 | 1 | 1 | 1 | 1 | 1 | 1 | 3 | 3 | 3 | 3 | 3 | 3 |
| Dec | 3 | 3 | 3 | 3 | 3 | 3 | 3 | 3 | 3 | 3 | 1 | 1 | 1 | 1 | 1 | 3 | 3 | 3 | 3 | 3 | 3 | 3 | 3 | 3 |

| 1 | 2 | 3 |
|---|---|---|
| 100% with DNI | 50% with storage | 50% with back-up |

In this paper it is proposed a thermal model order strategy so that only 50% of the rated power is generated when there is no DNI, which means that without DNI the power supply is of only 10 MW instead of 20 MW. This will cause the stored energy to last during a double period.

When the source is biomass it will also only be supplied 50% of plant capacity, which means that during the back-up period, the power supply is 10 MW. This will minimize the cost of biomass and cause the system to work 24 h a day. Table 2 shows the order made for each month of the year.

Thus, one can make an assessment of how many hours the biomass will be running for each month (having an average 13.78 h) and using the pellets in the system calculations.

## 3º- The Biomass System

It was considered that the boiler has a standardized performance value of $\eta_{boi} = 80.6\%$.

The lower heating value (LHV) of the pellets is about 17000 kJ/kg = 4060.382 kcal/kg. To calculate the amount of pellets consumed by the system it is sufficient to apply the following two steps:

1. Calculate the energy of the biomass ($E_{bio}$) pellets, which burns to produce the heat required for the system:

$$E_{bio} = \frac{E_{ther.m}}{\eta_{boi}} \tag{1}$$

Where:

$\eta_{boi}$ = Boiler performance.
$E_{ther.m}$ = Monthly thermal energy, calculated with the following expression.

$$E_{ther.m} = P_{syst} \times t_{moy} \times 860\ kcal \quad (2)$$

$t_{moy}$ = Monthly time given in hours.
$P_{syst}$ = Thermal power in the system.

1. Calculate the amount of pellets burning in the boiler ($C_{pel.m}$).

$$C_{pel.m} = \frac{E_{bio}}{LHV} \quad (3)$$

# 6 Results and Discussion

## 6.1 Annual Output of System with Backup of Biomass Pellets

Figure 4 shows several energy conversion processes and system losses that occur. This figure shows the total thermal energy falling on the solar field to power into the grid.

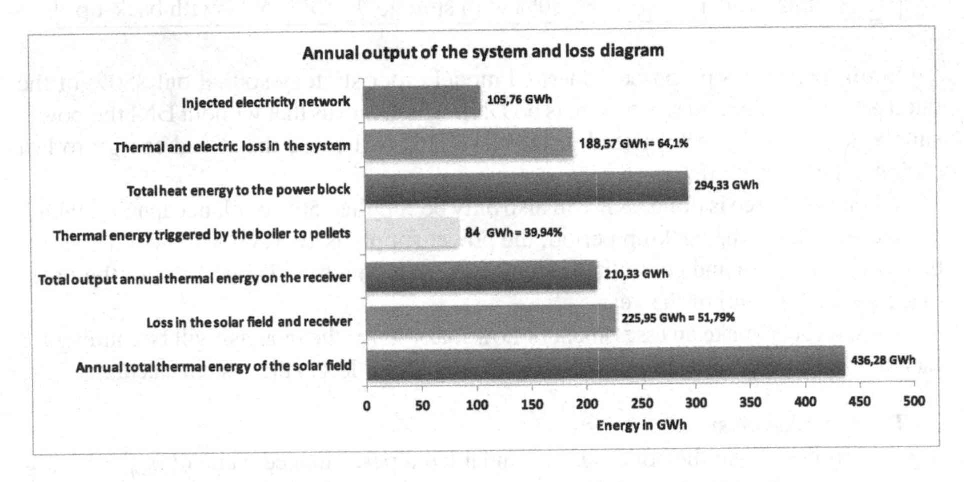


**Fig. 4.** Summary of the annual production of the system and loss diagram.

The total annual thermal incident energy in the solar field is 436.28 GWh, this energy is thus forwarded to the receiver, causing a loss in the solar field and receiver of 225.95 GWh which corresponds to 51.79%. With this loss the total annual thermal energy output of the receiver is 210.33 GWh, which is directed to the power block. To the thermal energy of 210.33 GWh in the HTF is added a thermal energy from the back-up system of 84 GWh corresponding 39.94%. Therefore, the total annual thermal energy power block is 294.33 GWh. This thermal energy is thus converted (after suffering thermal and electrical losses of 188.57) into electrical energy 105.67 GWh injected into the network. With these data, it can be said that the overall system performance is 19.57%.

The energy injected into the network varies each month of the year (see Fig. 5 left), and in July and august it is injected more energy in the network, about 10.77 GWh and 10.04 GWh respectively. The month with less injected energy is the month of January with only 7.27 GWh. This energy generation profile was attained due to the back-up system. Figure 5 (right), shows the amount of pellets each month that the system needs for the production of thermal energy necessary for the system to function in the absence of DNI and storage.

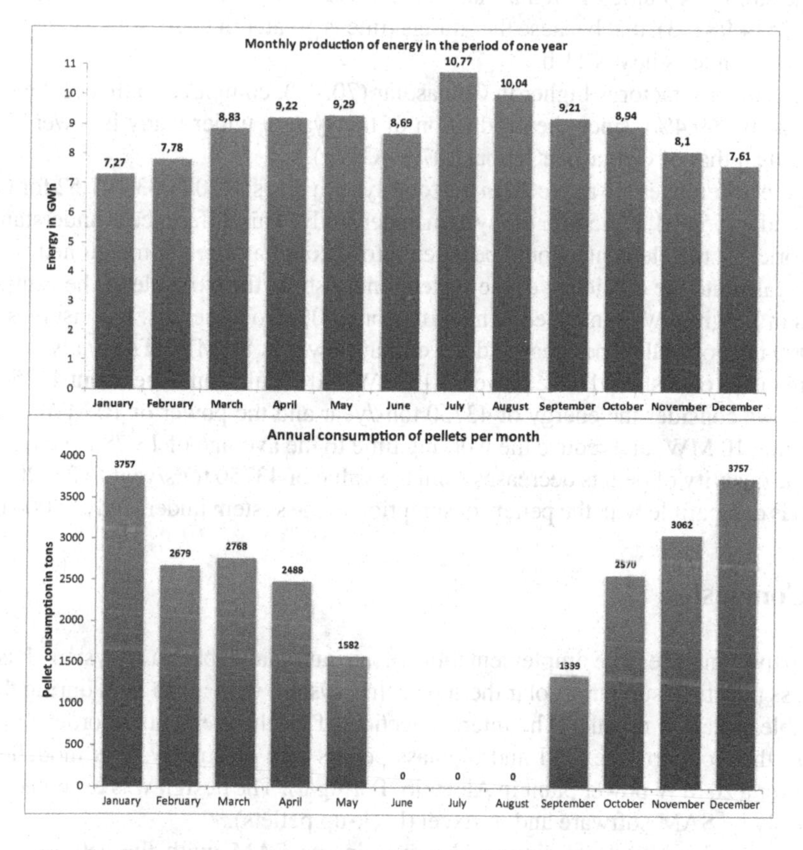


**Fig. 5.** Energy production per month and monthly consumption of pellets.

Note that the months with the highest consumption of biomass are January and December and in June, July and August no biomass is consumed, as during these months there is enough DNI to make the system work 24 h a day.

The total amount of pellets per year that the system will need is about 24 thousand tons, which is less than the amount of pellets each of the three selected producers export (about 180 from 270 thousand tons) per year. These show that only a producer would be enough to power the system.

### 6.2 Validation of the Results

If we analyze the obtained results and compare with the values of central Gemasolar [18], which also have a power of 20 MW and back-up system, we verify that many parameters are compatible such as:

The thermal power generated by the receiver of the system under study is 116.505 MWt and Gemasolar is 120.752 MWt, this small difference is due to differences in receptor structure, availability of DNI and because the Gemasolar applies a higher multiplier of 2.5 than 2.4 in the study system.

The storage volume of Gemasolar (3476.76 m3) is greater than the system under study (2514.98 m3), that because the storage time is greater in Gemasolar, and 15 h after the system under study is 11 h.

The capacity factor is higher in Gemasolar (70.4%), compared to that of the system under study (69.4%), since the production of the system under study is lower (105 76 GWh), than that of Gemasolar (about 107.34 GWh).

These differences are reflected in the total system costs: $ 209,003,197.52 for Gemasolar, and $ 155,901,171.55 for the system under study. This difference is understandable if the price of the elements would be taken into account as already mentioned.

To calculate the efficiency of the system under study the example of the central Les Awirs in Belgium was analyzed, which runs on 100% wood pellets. It consumes about 350,000 tons of pellets per year and the output power is 80 MW. The analyzed above system in this case should have power of 10 MW and should consume about 43750 tons/year. If we consider the energy of 43750 tons/year and the power of 10 MW, i.e. 24 h producing 10 MW, and reduce the working time to the average of 13.78 h a day, in that case the quantity of pellets decreases from the value of 43750 tons/year to 25120 t/year, which is compatible with the pellet consumption in the system under study 24000 t/year.

## 7 Conclusion

This paper analyses the implementation of an automatic back-up system based in biomass pellets to support a solar thermoelectric system with north field during the less available radiation months. The interconnection of all the elements in order to form a system that converts the DNI and biomass pellets into electricity, was modelled and tested for a 20 MW power plant in Alentejo (Portugal). The design was executed simultaneously by SAM software and in Excel (back-up pellets).

A comparison made of simulated values in the SAM, with the actual values of Gemasolar (obtained from a simulation made in the laboratory of renewable energy in the United States of America, USA), demonstrates a reasonable approximation of these simulated values, both in technical parameters and in economic parameters.

The annual figures were all generated in the SAM and Excel, which demonstrates a certain energy balance, entering thus the system about DNI = 2319.4 kWh/m$^2$ and 24000 tons of pellets per year, converted to 105,76 GWh of electricity per year, which is a feasible value for this type of system. In terms of back-up, the biomass demonstrates to be viable, increasing the energy produced and thus the capacity factor from that of 49.1% to 69.4%, but the use of biomass also increases the overall yield of 16.99% to 19.57%.

## References

1. García-Segura, A., Fernández-García, A., Ariza, M.J., Sutter, F., Valenzuela, L.: Durability studies of solar reflectors: a review. Renew. Sustain. Energy Rev. **62**, 453–467 (2016)
2. Khan, J., Arsalan, M.H.: Solar power technologies for sustainable electricity generation–a review. Renew. Sustain. Energy Rev. **55**, 414–425 (2016)
3. Boudaoud, S., Khellaf, A., Mohammedi, K.: Solar tower plant implementation in northern algeria: technico economic assessment. In: 2013 5th International Conference on Modeling, Simulation and Applied Optimization (ICMSAO), pp. 1–6. IEEE, April 2013
4. IVANPANH Solar Electric Generating System. http://www.ivanpahsolar.com
5. Torresol Energy. http://www.torresolenergy.com/TORRESOL/gemasolar-plant/en
6. Kumar, M., Krishna, D.J.: Analytical and numerical study to assess the solar flux on a heliostat based central receiver. In: 2016 International Conference on Energy Efficient Technologies for Sustainability (ICEETS), pp. 762–767. IEEE, April 2016
7. Tehrani, S.S.M., Taylor, R.A.: Off-design simulation and performance of molten salt cavity receivers in solar tower plants under realistic operational modes and control strategies. Appl. Energy **179**, 698–715 (2016)
8. Monteiro, E., Mantha, V., Rouboa, A.: The feasibility of biomass pellets production in Portugal. Energy Sources Part B: Econ. Plan. Policy **8**(1), 28–34 (2013)
9. Wagner, M.J.: Simulation and Predictive Performance Modeling of Utility-Scale Central Receiver System Power Plants. University of Wisconsin, Madison (2008)
10. Boudaoud, S., Khellaf, A., Mohammedi, K., Behar, O.: Thermal performance prediction and sensitivity analysis for future deployment of molten salt cavity receiver solar power plants in Algeria. Energy Convers. Manag. **89**, 655–664 (2015)
11. Johnston, C.M., van Kooten, G.C.: Economics of co-firing coal and biomass: an application to Western Canada. Energy Econ. **48**, 7–17 (2015)
12. REVISTA DA MADEIRA - EDIÇÃO N°137 - OUTUBRO DE 2013. http://www.remade.com.br/br/revistadamadeira_materia.php?num=1712&subject=Pellets&title=Demandadepelletsebiomassanomercadoeuropeu
13. Marchal, D.: Current developments on Belgian pellet market. Wels (Austria) (2007). http://valbiom.be/files/gallery/wels20071200481419.pdf
14. CENTRALE DES AWIRS 100% biomasse. http://corporate.engie-electrabel.be/wp-content/uploads/2016/03/12038_lesawirs_folder_fr_lr.pdf
15. Boudaoud, S., Khellaf, A., Mohammedi, K., Behar, O.: Thermal performance prediction and sensitivity analysis for future deployment of molten salt cavity receiver solar power plants in Algeria. Energy Convers. Manag. **89**, 655–664 (2015)
16. Gonçalves, J.T.: Estudo técnico-económico de um sistema solar termoelétrico com back-up a biomassa (dissertation) (2015)
17. Hussain, C.M.I., Norton, B., Duffy, A.: Technological assessment of di erent solar-biomass systems for hybrid power generation in Europe. Renew. Sustain. Energy Rev. (2016). doi: 10.1016/j.rser.2016.08.016(2016)
18. System Advisor Model (SAM) Case Study. Concentrating Solar Power Systems. https://sam.nrel.gov/case-studies
19. Amadei, C.A., Allesina, G., Tartarini, P., Yuting, W.: Simulation of GEMASOLAR-based solar tower plants for the Chinese energy market: influence of plant downsizing and location change. Renewable Energy **55**, 366–373 (2013)

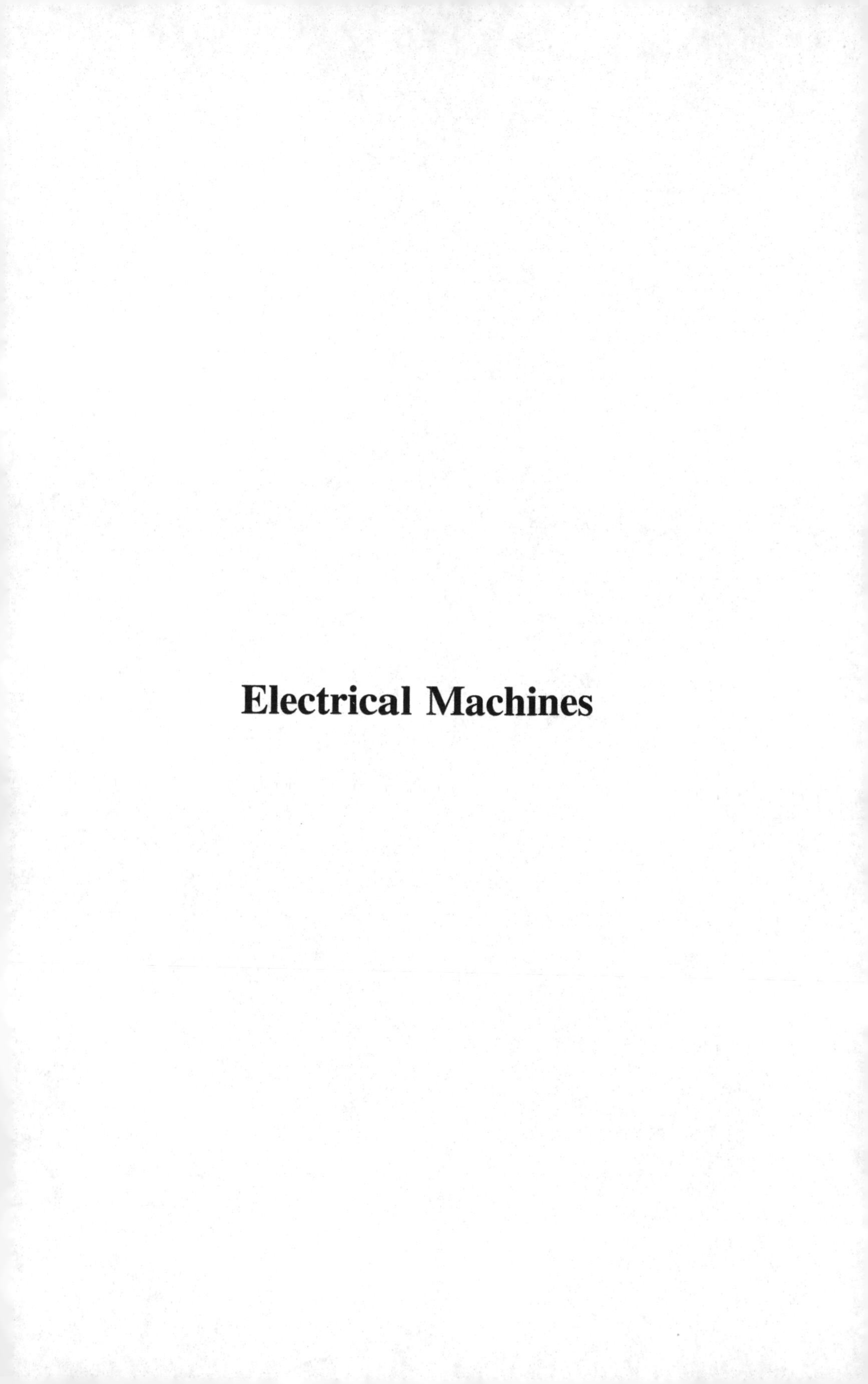

# Electrical Machines

# A Generalized Geometric Programming Sub-problem of Transformer Design Optimization

Tamás Orosz[1,2(✉)], Tamás Nagy[2], and Zoltán Ádám Tamus[1]

[1] Budapest University of Technology and Economics, Budapest, Hungary
orosz.tamas@vet.bme.hu
[2] Robert Bosch kft, Budapest, Hungary

**Abstract.** The first step in transformer design optimization is to solve a non-linear optimization task. Here, not only the physical and technological requirements, but the economic aspects are also considered. Large number of optimization algorithms have been developed to solve this task. These methods result the optimal electrical parameters and the shape of the core and winding geometry. Most of them model the windings by their copper filling factors. Therefore the transformer designer's next task, to find out the detailed winding arrangement, which fits to the optimization results. However, in the case of large power transformers, the calculation of some parameters like: winding gradients, short-circuit stresses etc., needs the knowledge of the exact wire dimensions and winding arrangement. Therefore, an other optimization task should be solved. This paper shows how this sub-problem can be formulated and solved as a generalized geometric program.

**Keywords:** Mathematical programming · Geometric programming · Transformer optimization · Electrical machine design

## 1 Introduction

It is well known that the transformer design optimization is a complex, nonlinear optimization task, due to the nature of the interaction of several physical fields encountered during the design of an electrical machine [1]. Therefore, in practice, this design process is split into more sub-optimization tasks and design stages. The first design stage is the preliminary or tendering design stage, where not only the physical and technological requirements, but the economical aspects are also considered [2]. Large number of optimization algorithms have been published in the literature to solve this task [3]. These methods are using simplified transformer models to calculate the optimal electrical parameters and the shape of the core and windings with adequate precision in a very short time. These approaches generally model the windings by a copper filling factor to obtain the optimal key design parameters. This simplification is widely used in the industry, and gives accurate results for the electrical parameters [4]. Moreover, FEM calculations also use this replacement [5].


L.M. Camarinha-Matos et al. (Eds.): DoCEIS 2017, IFIP AICT 499, pp. 373–381, 2017.
DOI: 10.1007/978-3-319-56077-9_36

In the case of large power transformers, the calculation of some parameters like winding gradients, short-circuit stresses etc., requires deeper knowledge of the exact wire dimensions and winding arrangement. Therefore, the transformer designer's next task is to find out the detailed winding arrangement which fits the optimized winding shape and their electrical parameters. This is another optimization task that should be performed Fig. 1. This design stage is often time consuming and usually needs huge expertise and manual interaction from the designer.

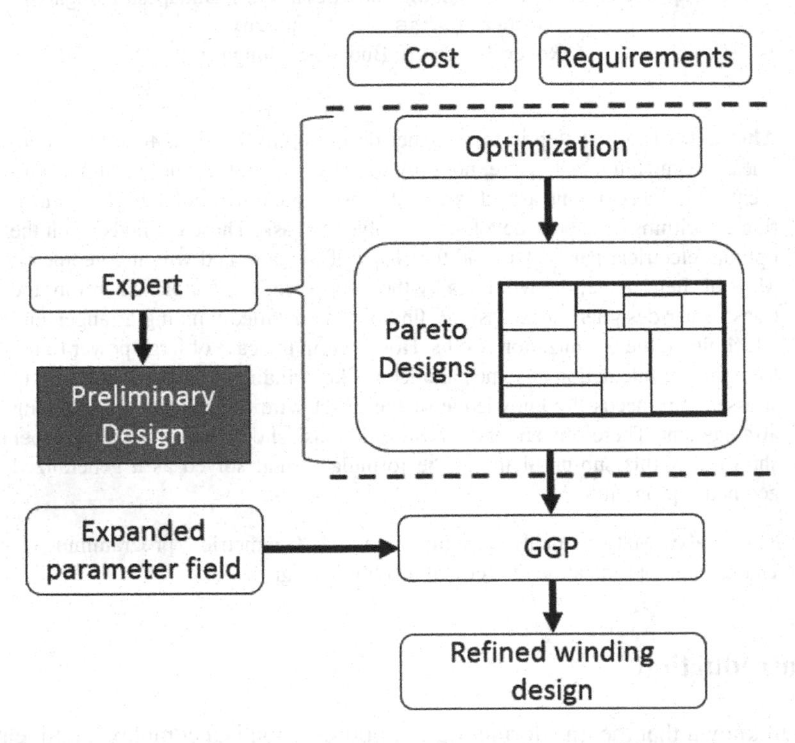


**Fig. 1.** The place of the generalized geometric programming sub-problem in electric machine design process.

This paper demonstrates how this design step can be solved with introducing a novel sub-problem. This sub-problem assumes that the geometrical and electrical parameters of the winding are known from a copper filling factor based transformer optimization. Its application is denonstrated on a case-type winding. It can be generalized for every other type of windings and it can be integrated into the metaheuristic based solution to replace copper filling factor based approximation [6]. A generalized geometric programing can be formulated and solved, which guarantees that the global optimum is found in no time [7].

## 2 Relationship to Smart Systems

The smart system environments will arise new challenges regarding to total cost of ownership, which is the base of the hereby presented optimization process. "The increasing proliferation of renewable energy resources and new size-able loads like electric vehicle (EV) charging stations has posed many technical and operational challenges to distribution grids" [8]. Therefore the conventional passive system design have to be replaced by an active concept resulting in a change of the emphasis of cost efficient design.

## 3 Generalized Geometric Programming

A geometric program is a type of the non-linear mathematical optimization problem characterized by the objective and constraint functions given in the following special form [7, 9]:

$$\begin{aligned} &\min f_0(x) \\ \text{s.t. } &f_i(x) \le 1,\ i = 1, \dots, m \\ &g_j(x) = 1,\ j = 1, \dots, n, \end{aligned} \tag{1}$$

where $x = (x_1, x_2, \dots, x_n)$ is a vector containing the optimization variables, $f_0, \dots, f_m$ are the posynomial functions, and $g_0, \dots, g_m$ are the monomial functions. All the elements of $x$ must be positive. The monomial function $g(x)$ can be expressed as

$$g_j(x) = c_g \cdot x_1^{\alpha_1} \cdot x_1^{\alpha_2} \cdot \dots \cdot x_n^{\alpha_n}, \tag{2}$$

where $c_g$ 2 R, _i 2 R„ and $c_g > 0$. This definition of monomial term is similar, but differs from the standard definition of the monomial used in algebra, where the exponents must be non-negative integers and the coefficient c is 1.

The posynomial function is the linear combination of monomials

$$f_i(x) = \sum_{k=1}^{K} c_k \cdot x_1^{\alpha_{1k}} \cdot x_1^{\alpha_{2k}} \dots x_b^{\alpha_{nk}}. \tag{3}$$

We can introduce the generalized posynomials such that a function $f$ of positive variables $x_1, \dots, x_n$ is a generalized posynomial if it can be formed from posynomials using the operations of addition, multiplication, positive (fractional) power, and maximum.

Geometric programing (GP) problems, in general, are not convex optimization problems, and the main trick in the solution of a GP is a transformation to a convex optimization task. This transformation, which is a simple logarithmic change in the variables, is automatically performed by the solver. The GP modeler only knows the formal rules defined in 1.

There are many major advantages of GP modeling. First, the GP formalism guarantees that the solver finds the globally optimal solution, and if the problem is infeasible, this provides that no feasible point exists. Second, the major advantage of this formulation is the great efficiency of this special class of optimization problems. Finally, these interior-pointbased GP solvers are also very robust [6, 7, 9].

# 4 The Optimization Sub-process

## 4.1 Input Parameters

The main purpose of the optimization sub-process is to determine additional input parameters for the preliminary design optimization. These methods, which employs copper filling factors, are suitable to determine the leakage magnetic field distribution inside the core working window. Even a finite element methods are also applicable for greater accuracy [5]. This calculation yields the maximum value of the radial and axial magnetic flux inside the winding area. These important parameters will be our additional input parameter next to the winding dimensions. Than we can prescribe an optimization problem that minimize the load loss in the desired winding area. In our example we focus on disc type winding arrangement consisting of single conductors.

The following input values are come from the previous sub-optimization process:

- $t_w$ radial thickness of the winding,
- $h_w$ axial height of the winding,
- $r_m$ mean radius of the winding,
- $n$ number of turns,
- $B_{rad}$ radial flux density,
- $B_{ax}$ axial flux density,
- *ff* copper filling factor.

The input parameters taken from the requirement specification:

- $f$ *frequency,*
- $I$ *phase current,*
- $t_{rad}$ *minimum value of radial insulation thickness,*
- $t_{ax}$ *minimum value of axial insulation thickness.*
- *The goal of the main optimization process is to obtain the following values:*
- $n_{ax}$ *axial number of turns,*
- *nrad radial number of turns,*
- *nc number of conductors in a turn,*
- *Acu copper area in a turn,*
- *Vcu copper volume,*
- $w$ *width of a condutor,*
- $h$ *height of a condutor.*

The optimization algorithm requires support parameters describing the physical constraints:

- *P0 heat flow density,*
- $\rho$ *specific resistance,*

- $\alpha$ *ratio of total and exposed surface area,*
- *kc heat conductivity of the insulation paper,*
- *Ploss total load loss,*
- *βhor, βver horizontal, vertical heat flow ratio;*
- *θshor, θsver horizontal, vertical temperature rise above cooling oil,*
- *θphor, θpver vertical temperature rise above cooling oil.*

## 4.2 Winding Geometry

The following posynomial inequality and monomial constraints describe the winding arrangement, this is a disc winding with normal conductors in the examined case:

$$n = n_{ax} \cdot n_{rad}, \qquad n_{ax} \cdot h + n_{ax} \cdot t_{ax} \leq h_w \tag{4}$$

$$t_{hor} \cdot n_c \cdot n_{rad} + w \cdot n_c \cdot n_{rad} \leq t_w, \qquad A_{cu} = n_{rad} \cdot n_c \cdot w \cdot h \tag{5}$$

$$V_{cu} = 2 \cdot \pi \cdot r_m \cdot n \cdot A_{cu}, \qquad ff \leq \frac{n \cdot A_{cu}}{h_w \cdot t_w} \tag{6}$$

$$w \leq w_{max}, \qquad h \leq h_{max} \tag{7}$$

$$n_c \leq 1, \qquad n_{rad} \leq 1. \tag{8}$$

## 4.3 Loss Calculations

The Ploss is minimized in this optimization task, which consist of the ohmic losses and the eddy losses. The eddy loss calculation takes into account the axial and radial components of the magnetic flux. It is assumed that the magnetic field is not modified by the field of the eddy currents [5]:

$$P_{loss} = p_{dc} + p_{ax} + p_{rad}, \qquad p_{dc} = \rho \cdot \frac{2 \cdot \pi \cdot r_m}{A_{C_u}} \cdot I^2 \tag{9}$$

$$p_{ax} = \frac{1}{24\rho} \cdot 2 \cdot \pi \cdot f \cdot w \cdot B_{ax}, \qquad p_{rad} = \frac{1}{24\rho} \cdot 2 \cdot \pi \cdot f \cdot h \cdot B_{rad} \tag{10}$$

## 4.4 Thermal Model

The applied calculation method of the temperature gradient uses an electrical analogy of heat-exchange, to solve the 2D-Poisson equation of the steady-state heat-flow [10]. It implies the decomposition of the whole winding into simplified components (cooling blocks), where the heat flow possible only in the axial or the radial directions (Fig. 2) [11–13]. Here, the axial and the radial component of the heat-flow density is owing through the effective thermal resistance of the insulation and the oil

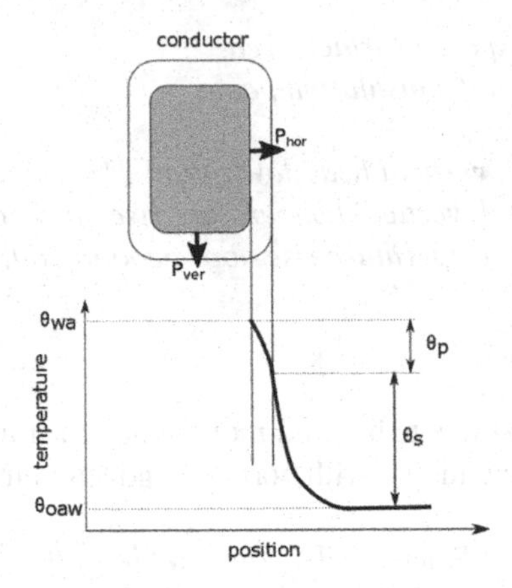


**Fig. 2.** Temperature notations around the winding-oil contact at approximately the half of winding height

boundary layer (Fig. 2). The heat-flow density value defined by the following monomial expression:

$$P_o = \frac{P_{loss}}{N_{ax} \cdot 2 \cdot \pi \cdot r_m} \tag{11}$$

The effective insulation thickness in the axial (vertical) and the radial direction are the following:

$$t_{ver} = t_c, \qquad t_{hor} = \frac{\alpha}{3} \cdot (n_{rad} \cdot n_c \cdot t_c) \tag{12}$$

To calculate the ratio of the radial and axial heat-flow density, between the cooling surfaces the loss ratio parameters are introduced by [11] (β parameters). However, the calculation method of these parameters, which presented in [11] not fulfills the GGP formalism. Therefore a monomial function is fitted in βhor and βver (Fig. 3). Using this formulation, the horizontal and vertical heat-flows and the temperature gradients through the insulation can be formulated by the following monomial expressions:

$$p_{hor} = 0.5 \cdot \frac{p_0 \cdot \beta_{hor}}{\alpha \cdot w_t}, \qquad p_{ver} = 0.5 \cdot \frac{p_0 \cdot \beta_{ver}}{\alpha \cdot h} \tag{13}$$

$$\theta_{p,hor} = \left(\frac{t_C}{2 \cdot k_C}\right) \cdot p_{hor}, \qquad \theta_{p,ver} = \left(\frac{\alpha \cdot N_{rad} \cdot N_C \cdot t_C}{2 \cdot k_C}\right) \cdot \frac{p_{ver}}{3} \tag{14}$$

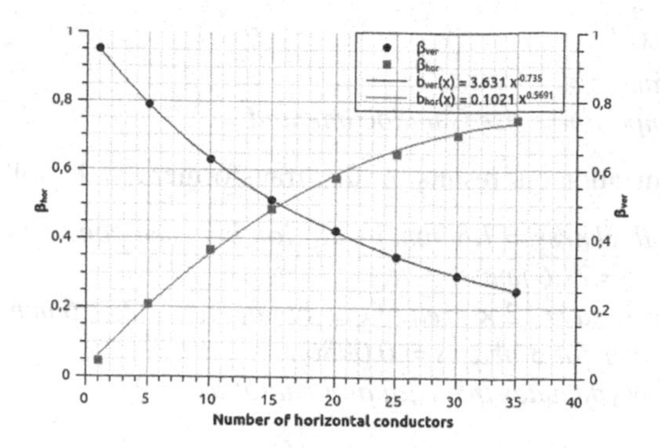


**Fig. 3.** Posynomial form function for the horizontal and the vertical heat ow ratio calculation.

Temperature gradient through the oil boundary layer, in the case of ON or OF cooling:

$$\theta_s = C \cdot a \cdot \sqrt{p_{hor}}, \qquad \theta_s = C \cdot \sqrt{p_{ver}} \tag{15}$$

Where, the auxiliary parameter a = 1 in the case of zig-zag cooling. Regardless of the used formula, the incorporated constant C depends on the established cooling mechanism. Potential interpretations are shown in [11, 13].

The gradient is defined by the following generalized posynomial:

$$min(\theta_{p,hor} + \theta_{s,hor}, \theta_{p,ver} + \theta_{s,ver}) \leq \theta \tag{16}$$

The min() function is not a valid generalized posynomial inequality, but it can be handled by a replacement [9].

## 5 Results

The winding sub-optimization method has been tested on an inner winding of a three-phase, 10 MVA, 33/6.9 kV, star/star Volts/turn = 46.87, Z = 7.34% transformer. The transformer data is presented in [14], and the optimization is based on this real life example transformer. This example gives us the following input parameters: tw = 35 mm, hw = 1100 mm, n = 89, rm = 495/2 mm, I = 836 A, f = 50 Hz. As we consider disc type winding, all oil cooling ducts are considered to be 3 mm wide. The minimal thickness of the insulation of the single conductor is 0.5 mm, this value was chosen taking into account of the voltage level. The transformer specification and relevant design details of the source [14] are:

- *LV conductor: thickness = 2.3 mm, insulation between conductors = 0.5 mm, area of one conductor = 22:684 mm*$^2$
- *Total conductor area = 12 × area of one conductor = 272.21 mm*$^2$
- *LV winding volume for 3 phases = 0.1079 m*$^3$
- $P_{eddy}$ *= 489.55 W (only radial flux considered)*

- $P_{dc} = 21.885\ kW$
- *copper filling factor 60%*
- *the winding temperature rise is undocumented*

After the optimization our results for this transformer are the following:

- *LV conductor: thickness* = *1.8 mm, insulation between conductors* = *0.5 mm, area of one conductor* = *21.6* $mm^2$
- *Total conductor area* = *12* × *area of one conductor* = *237:6* $mm^2$
- *LV winding volume for 3 phases* = *0.0986* $m^3$
- $P_{eddy} = 850.27\ W$ *(includes the axial and radial ux too)*
- $P_{dc} = 21.6\ kW$
- *copper filling factor 55%*
- *the winding temperature rise above oil is 14.1 K*

## 6 Conclusions

Finding a cost efficient transformer design is a complex, non-linear optimization task. Therefore, the transformer design process is separated into several, independent sub-problems and design stages. This paper deals with the preliminary design stage, where numerous algorithms has been applied to find the key design parameters with minimal total owning cost. However, most of these methods are modeling the windings by their copper filling factors - a simplification widely used in the industry with good accuracy - this approach is not appropriate to take into account the numerous possible winding arrangement. The transformer optimization sub-problem is presented and solved in this paper, with the method of generalized geometric programming. The results are shown a good agreement with the realized transformer winding's parameters. The presented model can be generalized to any type of winding arrangements and various type of conductors, for example case of continuously transposed cables.

## References

1. Del Vecchio, R.M., Poulin, B., Feghali, P.T., Shah, D.M., Ahuja, R.: Transformer Design Principles: With Applications to Core-Form Power Transformers. CRC Press, Boca Raton (2010)
2. Orosz, T., Sőrés, P., Raisz, D., Tamus, Á.Z.: Analysis of the green power transition on optimal power transformer designs. Period. Polytech. Electr. Eng. Comput. Sci. **59**(3), 125–131 (2015)
3. Olivares-Galvan, J., Georgilakis, P., Fofana, I., Magdaleno-Adame, S., Campero-Littlewood, E., Esparza-Gonzalez, M.: A bibliographic analysis of transformer literature 2001–2010. In: 8th Mediterranean Conference on IET Power Generation, Transmission, Distribution and Energy Conversion (MEDPOWER 2012), pp. 1-6 (2012)
4. Georgilakis, P.S.: Spotlight on Modern Transformer Design. Springer, Heidelberg (2009)
5. Andersen, O.W.: Transformer leakage flux program based on the finite element method. IEEE Trans. Power Appar. Syst. **92**(2), 682–689 (1973)
6. Orosz, T., Sleisz, Á., Tamus, Z.Á.: Metaheuristic optimization preliminary design process of core-form autotransformers. IEEE Trans. Magn. **52**(4), 1–10 (2016)

7. Boyd, S., Vandenberghe, L.: Convex Optimization. Cambridge University Press, New York (2004)
8. Andresen, M., Costa, L.F., Buticchi, G., Liserre, M.: Smart transformer reliability and efficiency through modularity. In: 2016 IEEE 8th International Power Electronics and Motion Control Conference (IPEMC-ECCE Asia), pp. 3241–3248, May 2016
9. Boyd, S., Kim, S.-J., Vandenberghe, L., Hassibi, A.: A tutorial on geometric programming. Optim. Eng. **8**(1), 67–127 (2007)
10. Orosz, T., Kleizer, G., Iváancsy, T., Tamus, Z.Á.: Comparison of methods for calculation of core-form power transformer's core temperature rise. Period. Polytech. Electr. Eng. Comput. Sci. **60**(2), 88–95 (2016)
11. Dankó, G., Imre, L.: A method for determining the steady-state temperature distribution in the windings discs of oil-cooled transformers. Period. Polytech. Electr. Eng. **20**(4), 89–103 (1976)
12. Ryan, H.M.: High Voltage Engineering and Testing, 3rd Edition, no. 32. IET (2013)
13. Ryder, S., et al.: A simple method for calculating winding temperature gradient in power transformers. IEEE Trans. Power Deliv. **17**(4), 977–982 (2002)
14. Kulkarni, S.V., Khaparde, S.: Transformer Engineering: Design and Practice, vol. 25. CRC Press, Boca Raton (2004)

# Noise, Vibration and Harshness on a Permanent Magnet Synchronous Motor for a Remote Laboratory

Jaime Pando-Acedo(✉), Enrique Romero-Cadaval, Consuelo Gragera-Peña, and María Isabel Milanés-Montero

Universidad de Extremadura, Avda. de Elvas s/n, 06071 Badajoz, Spain
jpandoac@peandes.es, {eromero,cgragera,milanes}@unex.es

**Abstract.** The study of Noise, Vibration and Harshness (NVH) is becoming a key element when it comes to design and maintain a system that have rotatory elements. The understanding of the source of the vibrations can lead to a way of mitigating them, thus ensuring a better operation and expanding the life of the components. In this work, an analysis of the vibrations' frequency spectra of an electric drive is developed, extracting some conclusions from it and deducing a source. The possibility of operating this system remotely makes it a perfect experiment for a remote lab, operating through Internet, being a novelty for this kind of studies.

**Keywords:** Harshness · Noise · NVH · PMSM · Remote · Vibrations

## 1 Introduction

Rotating machines have vibrations associated with its functioning, though it does not mean any abnormal operation. If we particularize for the case of an electric drive, the vibrations' sources may come from the mechanical part of the system (construction, misalignments, unbalance load, design flaws, etc) or the electric part (harmonics, unbalanced voltages, voltage gaps, etc). Noise, vibration and harshness (NVH) studies' aim is to identify those sources and propose solutions [1–4].

This work uses the advantages of a rapid prototyping platform and the ability to remotely control an electric motor in order to develop the bases for a remote controlled laboratory. This will allow users to monitor and experiment with the machine through the internet, for research and educational purposes [5, 6]. The analysis of the vibration spectra plays a key role in the NVH studies, which are especially important for the design and control of electric vehicles. This work is proposed as a first step towards the implementation of active control techniques, like active damping, suitable for the cancellation of vibrations from different sources.

This paper is schematized as follows. For a first section, the NVH techniques available will be discussed and a suitable one for this project will be selected. Then, the control employed for the system, which will use the Field Oriented Control [7] and Pulse Modulating Techniques, will be described. The acquisition of the data and processing will follow, finishing with an analysis of the results obtained and drawing some conclusions.


L.M. Camarinha-Matos et al. (Eds.): DoCEIS 2017, IFIP AICT 499, pp. 382–389, 2017.
DOI: 10.1007/978-3-319-56077-9_37

## 2 Relationship to Smart Systems

In the days we live in, information can be accessed almost anywhere. From the pc, laptop, smartphone, tablet or even wearable devices; all you need is an internet connection. In this sense, this work aims the possibility of remotely supervising or even control a research laboratory.

In Fig. 1, a scheme for this kind of application is shown. The user has an application window from which he can monitor and control the system via the internet. The system connects to the internet using the TCP/IP protocol, and it consist of the control platform, an IGBT inverter and the electric drive. The control platform receives the orders from the user and generate the control signals for the inverter, as well as collects data from the inverter itself and the motor. A camera could be used for supervision of the hardware platform in anytime.

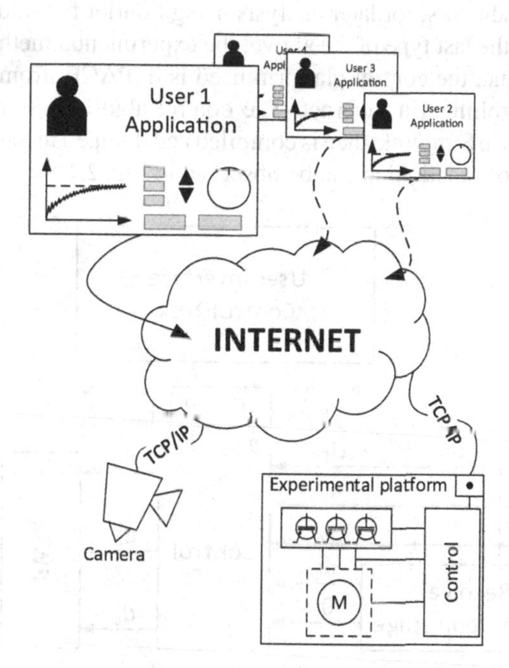


**Fig. 1.** Scheme of a remote laboratory, allowing researchers and students to access and remotely control the system via the internet using protocol TCP/IP.

The remote laboratory represents a good implementation of a smart system, since it can be applied to industry besides research and education. This means that once it is fully developed, new UIs for controlling the system can be programmed using a web interface, for instance.

## 3 NVH and Control Algorithm

There are three groups in which NVH techniques can be classified into: numeric, experimental and analytic. Each of them have its advantages and disadvantages, and one should decide which method suits better for the application. For instance, analytic methods are useful for a first understanding of the problem and modeling. However, due to the increase in microprocessors' speed, numeric methods are getting more powerful and widely used every day. In the particular case of NVH, there are several numeric methods used, like the Finite Elements Method (FEM) or the Boundary Elements Method (BEM) [8].

On the other hand, although it cannot give solutions in design phase because they can only be used with prototypes, experimental methods are useful when it comes to ease of realization, especially if you also use rapid prototyping. The data can be simply obtained using transductors, for later analysis using Fourier for studying the spectra.

This work aims the last type of the above, the experimental method. For conducting this kind of technique, the control platform used is dSPACE, from MATLAB, which allows remote controlling via Ethernet. The control algorithm is first elaborated and tested in simulations in Simulink, then is compiled and dumped in the dSPACE platform. The whole scheme of functioning can be observed in Fig. 2.

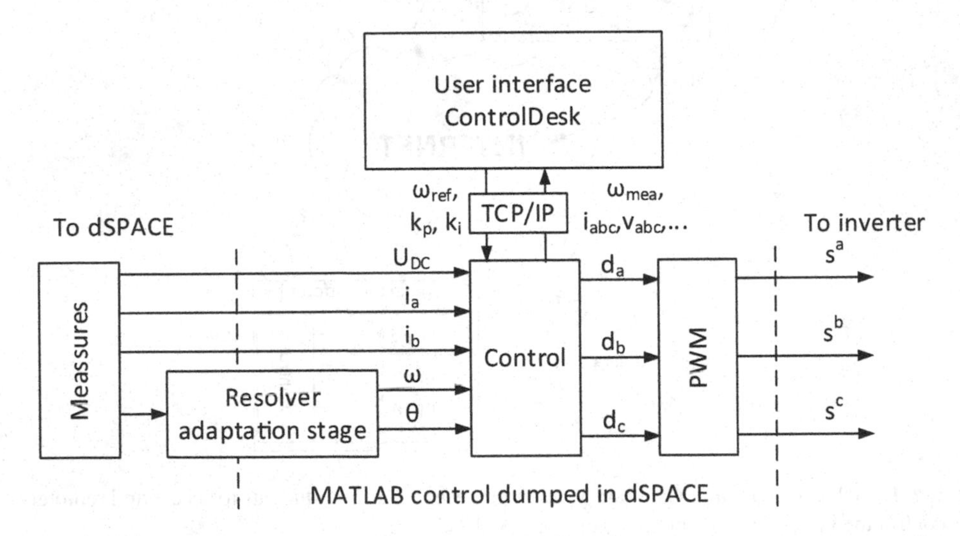


**Fig. 2.** Control block diagram. The block "Measures" takes the voltages, currents and speed values from sensors. "Resolver adaptation stage" block takes the information from the resolver and outputs speed and position. "Control" inputs all values from sensors and outputs the duty cycles for the PWM block. $U_{DC}$, $i_a$, $i_b$, $\omega$ and $\theta$ are the bus voltage, motor currents, speed and position, respectively. $\Omega_{ref}$, $k_p$, and $k_i$ are the speed reference, and controller gains selected by user. The variables $d_{abc}$ are the duty cycles and $s^{abc}$ the switching signals.

The control strategy, a FOC strategy, is depicted in Fig. 3. The measured currents in stationary reference frame are inputs to the system, as well as the motor speed and DC bus voltage. The controller uses these inputs for generating the duty cycles that ensures that the references are met.

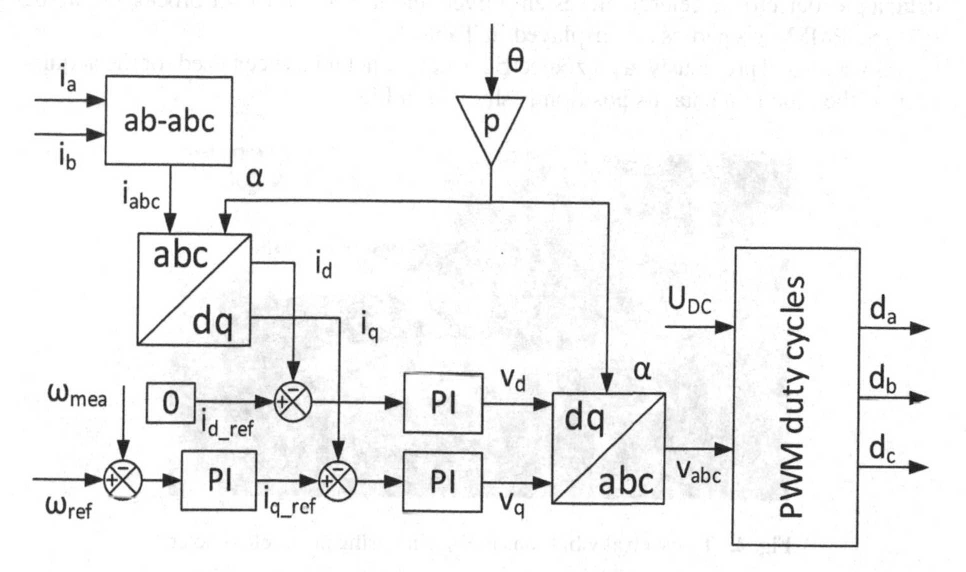


**Fig. 3.** Scheme of the FOC control used for the inverter-PMSM system. $\theta$ is the rotor position, $p$ are the pole pairs and $\alpha$ the electrical angle.

# 4 Experimental NVH Analysis

In this section the physical system will be described, and the later on the process for filtering the data obtained will be explained. For the last part, the results will be presented and some conclusions drawn.

## 4.1 Test Bench

In order to carry out the NVH study, a test bench consisting of a PMSM, an IGBT inverter, the dSPACE platform and a PC running MATLAB Simulink is employed.

**Table 1.** PMSM properties.

| Variable | Value | Units |
|---|---|---|
| Pole pairs (p) | 4 | - |
| Stator windings resistance (R) | 0.2 | Ω |
| Rotor inertia (J) | 3.6 | $Kg{\cdot}m^2{\cdot}10^{-3}$ |
| d-axis inductance ($L_d$) | $1.871{\cdot}10^{-3}$ | H |
| q-axis inductance ($L_q$) | $1.616{\cdot}10^{-3}$ | H |
| Magnet flux linkage | 0.1323 | V·s·rad |

The electric drive is equipped with a resolver for speed measuring, for the obtaining of the current and voltage values, transducers are used. For the acquisition of vibration data, a piezoelectric accelerometer is employed and it stored for later processing in the PC. The PMSM properties are displayed in Table 1.

As was stated previously, a piezoelectric accelerometer has been used for the acquisition of the vibration data, its position is showed in Fig. 4.

**Fig. 4.** Transversal vibrations measuring using an accelerometer.

### 4.2 ControlDesk

The application interface presented to the user is elaborated using ControlDesk. This software allows the user to both monitor and control the dSPACE platform, been able to change control variables without the need of recompiling. One of the biggest advantages of this rapid prototyping tool is that it lets you change parameters of the control while executing it, thus making the tuning much faster.

### 4.3 Analysis of the Results

Once the vibration data is obtained and processed, one is interested in studying the low frequency amplitudes, since these are the one associated to vibrations. The high frequency amplitudes may have different causes, like electromagnetic noise, switching-frequency noise, etc. Therefore, a filter has been employed for removing the latter.

The filter proposed is a moving average, which formula is presented in (1). Lowering the window size $M$ of the average lowers the cut-off frequency, eliminating the high frequency content.

$$P_n = \frac{\sum_{n-M}^{n} P_i}{M} \tag{1}$$

Since the frequency content we are interested in is located below 250 Hz, different values of the window size have been tried out until the high frequency content is eliminated. The raw data is shown in Fig. 5(a), with its frequency spectra (b), and then the filtered data and its frequency spectra (Fig. 5(c) and (d)).

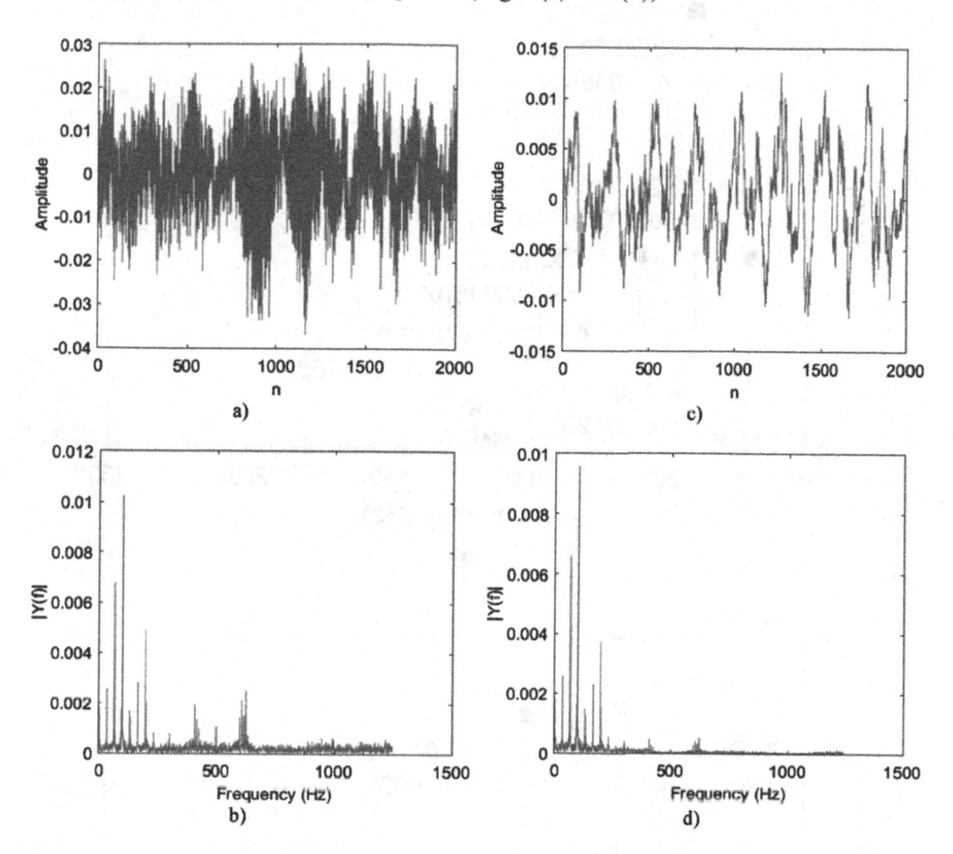


**Fig. 5.** Vibration data. (a) Raw data. (b) Fourier analysis of the raw data. (c) Moving average with window size of 5 points. (d) Fourier analysis of the filtered data.

The speed equation of the PMSM is shown in (2), where $n_m$ the motor rpm, $f$ is the frequency of the inverter's voltage and $p$ are the pole pairs.

$$n_m = f\frac{60}{p}; f = n_m\frac{p}{60} \tag{2}$$

The PMSM has 4 pole pairs (Table 1), so if the motor is running at 300 RPM, the main frequency would be 20 Hz, the second harmonic 40 Hz, and so on. The data obtained shows this behavior exactly (Fig. 6).

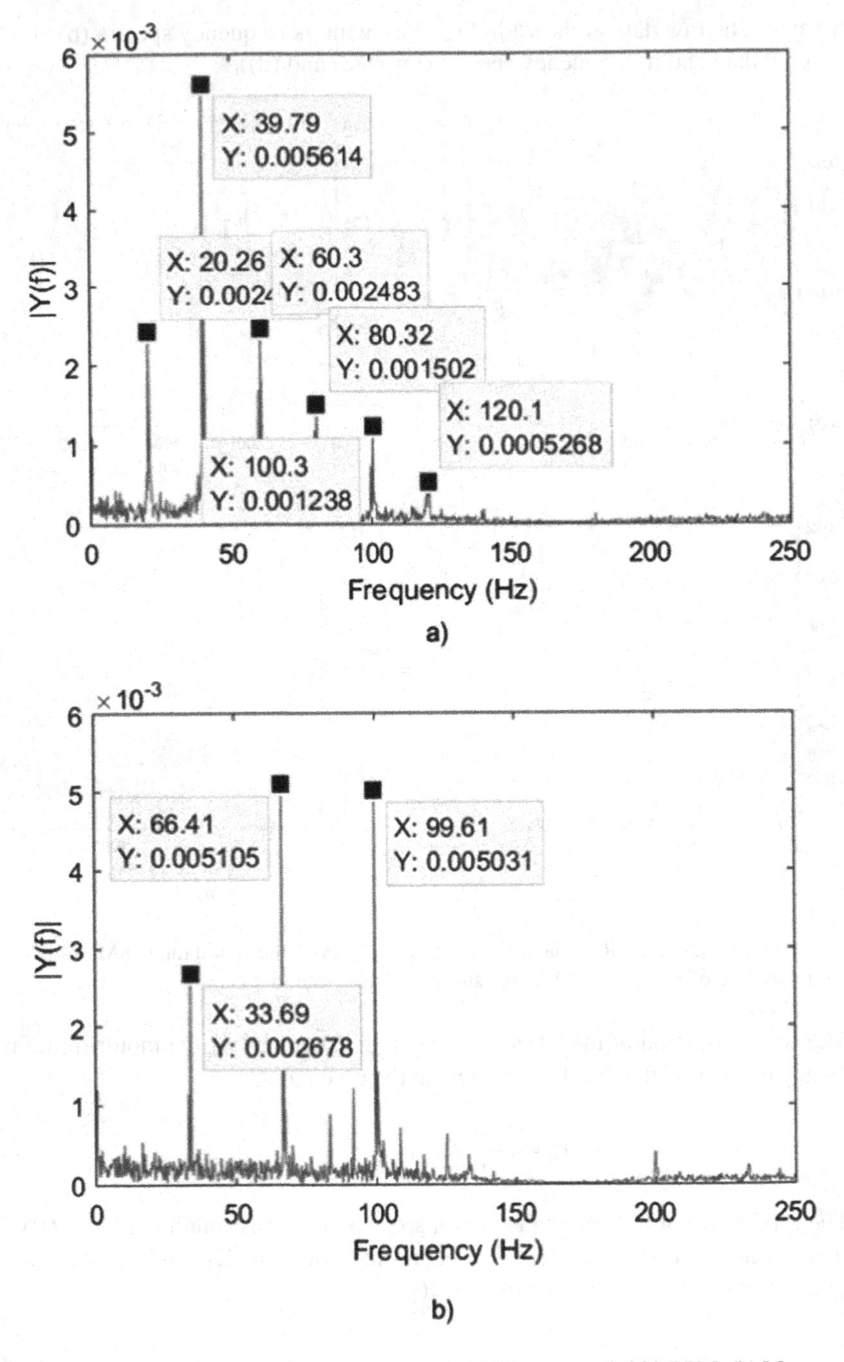


**Fig. 6.** Peaks separating according to motor speed. (a) Motor speed: 300 RPM. (b) Motor speed: 500 RPM.

The results obtained show a relationship with the motor speed that is associated with misalignments in the rotor axis. This causes that the drive has to overcome an extra load at a determined angle, producing the vibration spectra observed.

## 5 Conclusions

Experimental analysis for a NVH study has been the method selected and the data obtained processed. The control employed for this work uses a the FOC technique and a PWM algorithm for controlling the inverter which feeds the PMSM. The platform used, dSPACE, allows the user to remotely control the system via TCP/IP protocol and the user interface, ControlDesk, to take advantage of the rapid prototyping. The processing of the data is necessary for the elimination of the high frequency noise.

Using Fourier analysis, a cause for the vibration spectra is deduced: rotor misalignment. For future works, the authors would like to go deeper into the frequency spectra trying to stablish more causes for the vibrations as well as associating the vibrations' amplitude values with the motor speed.

**Acknowledgements.** This work was supported by the Junta de Extremadura, under grant GR15177.

## References

1. Putri, A.K., Rick, S., Franck, D., Hameyer, K.: Application of sinusoidal field pole in a permanent-magnet synchronous machine to improve the NVH behavior considering the MTPA and MTPV operation area. IEEE Trans. Ind. Appl. **52**, 2280–2288 (2016)
2. Chauvicourt, F., Faria, C., Dziechciarz, A., Martis, C.: Influence of rotor geometry on NVH behavior of synchronous reluctance machine. In: Tenth International Conference on Ecological Vehicles and Renewable Energies (EVER) (2015)
3. Tong, W., Hou, Z.: Vertical vibration analysis on electric vehicle with suspended in-wheel motor drives. In: Electric Vehicle Symposium and Exhibition (EVS27) (2013)
4. Pyrhönen, O., Eskelinen, P.: Advanced measurement of rotor vibration in electric drives. In: IEEE Aerospace and Electronic Systems Magazine, vol. 13 (1998)
5. Langmann, R., Jacques, H.: Generic remote labs in automation engineering. In: IEEE Global Engineering Education Conference (EDUCON) (2016)
6. Kruse, D., Frerich, S., Petermann, M.: Remote labs in ELLI. In: Global Engineering Education Conference (EDUCON) (2016)
7. Miñambres-Marcos, V., Guerrero-Martínez, M.A., Romero-Cadaval, E., Gutiérrez, J.: Issues and improvements of hardware/software co-design sensorless implementation in a permanent magnet synchronous motor using Veristand. In: IEEE International Symposium on Sensorless Control for Electrical Drives and Predictive Control of Electrical Drives and Power Electronics (SLED/PRECEDE), Munich (2013)
8. dos Santos, F., Anthonis, J., Van der Auweraer, H.: Multiphysics thermal and NVH modeling: integrated simulation of a switched reluctance motor drivetrain for an electric vehicle. In: Electric Vehicle Conference (IEVC). LMS International, Leuven (2012)

# Levitating Bearings Using Superconductor Technology Under Smart Systems Scope

Martim V. Carvalho[1], António J. Arsenio[1], Carlos Cardeira[1,2], Paulo J. Costa Branco[1,2], and Rui Melício[2,3(✉)]

[1] Instituto Superior Técnico, Universidade de Lisboa, Lisbon, Portugal
[2] IDMEC, Instituto Superior Técnico, Universidade de Lisboa, Lisbon, Portugal
ruimelicio@gmail.com
[3] Departamento de Física, Escola de Ciências e Tecnologia, Universidade de Évora, Évora, Portugal

**Abstract.** This paper presents a study about the cooling and leakage aspects of superconductor magnetic bearings towards their integration on a smart systems scope to keep the stability of the cooling system. The use of superconductor magnetic bearings in electric power generation, namely in wind power has been increased due to their weight, friction less, and volume reduction advantages. The novelty of this paper is the use of Zero-Field Cooling instead of the common Field Cooling technique. On the other hand, this new kind of bearings needs a constant flow of liquid nitrogen to keep the superconductivity properties. A prototype was modelled, simulated and implemented for experimental validation.

**Keywords:** Superconductor magnetic bearing · Superconductors · Smart grids · Magnetic levitation · Zero field cooling

## 1 Introduction

Technological Innovation for Smart Systems is essential for smart grids. Wind power systems manifest variations in the output power due to the variations wind speeds, thus introducing a new factor of uncertainty and risk on the electrical grid posing challenges in terms of power system security, power system stability, power quality and malfunctions. The significant breakthroughs in science and technology allow the opportunity to link cyber space and physical components, using technological Innovation for Smart Systems. This bridge leads to Cyber-Physical Systems [1].

Large wind power generators with superconductors bearings are in constant development due to their advantages such as weight, friction less, volume reduction and the increased efficiency when compared with conventional bearings technologies [2, 3]. The use of Superconductor Magnetic Bearings (SMB) requires their constant cooling with liquid nitrogen. Smart grids allow the optimization of the available resources, namely in keeping and monitoring the liquid nitrogen critical levels.

Classical rotating bearings are a frequent component of rotational systems. They are widely used in wind energy production, hydraulic power plants and thermo power plants.


L.M. Camarinha-Matos et al. (Eds.): DoCEIS 2017, IFIP AICT 499, pp. 390–397, 2017.
DOI: 10.1007/978-3-319-56077-9_38

The SMBs studied in this work will likely replace the existing classical bearings in a near future. Despite the SMB increased efficiency and life time as well as the frictionless advantages, they have other drawbacks. Actually they need a cryogenic system to maintain the working low temperatures that are needed (by now) to keep the superconductivity properties.

Advanced researches have been carried out to build SMBs. Prototypes have been developed for many applications like the construction of large-scale flywheels [4] or for bearings in the textile industries [5], or for wind energy conversion systems [6].

The main contribution of this paper is the use of the zero field cooling (ZFC) technique instead of the classical field cooling (FC) technique. The approach in this work uses has shown to be more efficient, presenting less Joule losses [7, 8]. Moreover the Nitrogen consumption is analyzed as it is an important feature of SMB, regarding the economic impacts of its use.

## 2 Relationship to Smart Systems

A smart grid is a new type of power grid which highly integrates modern advanced information techniques, communication techniques, computer science and techniques with physical grids [9]. One of the objectives of smart grid is the optimization of resources allocation.

Cyber Physical Systems (CPS) is an important issue for smart grids and smart sensors. The coordination and interaction of cyber layers and physical components of wind systems require handling a number of challenges for essential feasibility and competiveness of future. Successfully integrating wind power into the existing electric power grid is a complex challenge that relies on distributed, interconnected cyber-physical systems, which are proliferating within the engineering industry [1].

One of the objectives of smart grid is the optimization of resources allocation. The use of SMBs require a critical allocation of smart systems, namely smart sensors to support the continuous availability of liquid nitrogen or to measure guidance and levitation forces, inside the SMB. These smart sensors can transfer information to a cloud architecture system, benefiting of existing Internet Cloud Services [2].

## 3 SMB Cooling Prototype

An SMB based on NdFeB permanent magnets and YBCO superconductor bulks, also referred to as high temperature superconductors was designed. The SMB viability was made in [10, 11] and the conception and experimental evaluation the SMB was achieved in [3], as shown in Fig. 1.

In [10] the magnetic fields as well as the levitation and guidance forces were simulated in a finite element analysis tool.

The levitation forces, as shown in [10] are given by:

$$|F_{Lev}| = \frac{A}{\mu}\frac{|B_t|^2}{2} \tag{1}$$

where $A$ is the surface parallel to the $xy$ plan, $B_t$ is the magnetic flux density tangential component and $\mu$ is the magnetic permeability of the YBCO superconductor bulks.

The SMB was simulated in a finite element analysis tool, as shown in Fig. 1. The arrows show the levitation forces $F_{Lev}$ which were shown to be able to sustain the rotor weight [10].

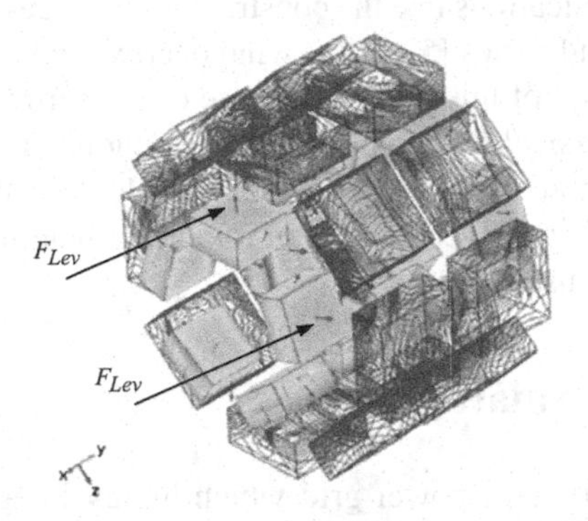


**Fig. 1.** SMB simulation of magnetization directions and contours.

The geometric placement of the permanent magnets and high temperature superconductors is carefully performed to keep symmetry along the main axis and to minimize the air gap of the prototype between the rotating part (rotor) and the static part (stator). Moreover, studies made during the development of this work involve changes in these geometric placements, either in the rotor as in the stator. Additionally, more finite element modelling was designed for simulation and viability study of the bearing, calculating the estimated levitation and guidance forces involved. The designed prototype model is shown in Fig. 2. Experimental validation was achieved by building a structure in conformity with the previously simulated geometry and comparing the simulation results with the ones obtained by measuring the existing forces in the real prototype. The results allowed the conclusion that it is possible to build a superconductor magnetic bearing using the zero field cooling technique, providing an important insight on how the system behaves.

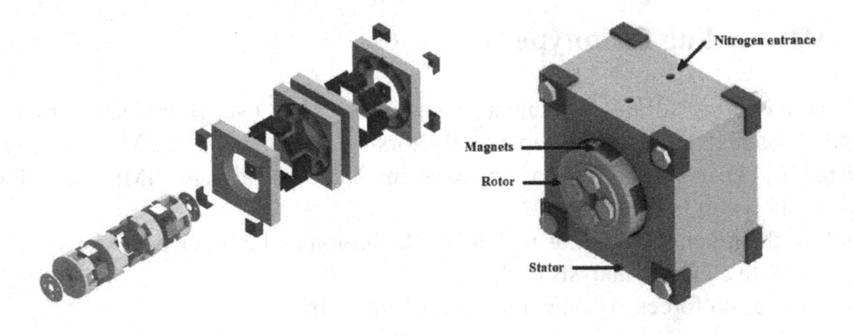


**Fig. 2.** Prototype model.

To keep the YBCO superconductor bulks cooled, channels were created inside the stator for Liquid nitrogen cooling, as shown in Fig. 3.

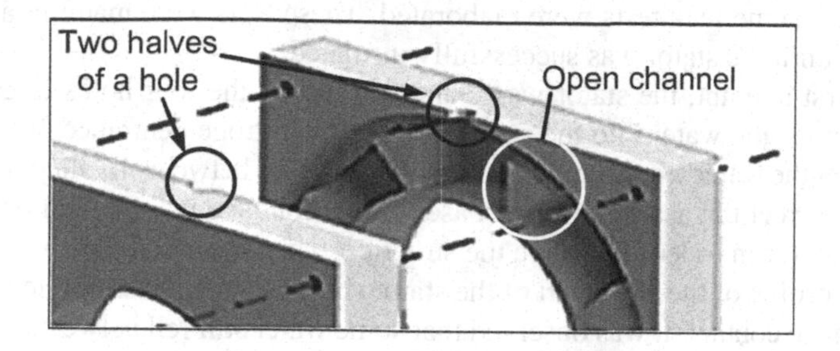


**Fig. 3.** Prototype model: channels for $LN_2$ cooling.

As the results were conclusive, a real prototype was implemented. Polyethylene and PLA were the materials used for the stator and rotor. To confirm the choice of the materials, namely the stator and the fasteners parts, an experience was developed to study the behavior of the structure at working conditions. These conditions are characterized by working at low temperatures, in the order of 77 K (liquid nitrogen temperature) and under clamping forces that close the stator slices together. With the intention of not harming all the structure, only half of the stator part was used in these first tests. Hence, the half of the stator part was totally submerged in liquid nitrogen, without any HTS bulks inside, until the system achieved a state of stability. For this purpose, some cables were attached to the structure so that it could be pulled. An example of this process is shown in Fig. 4.

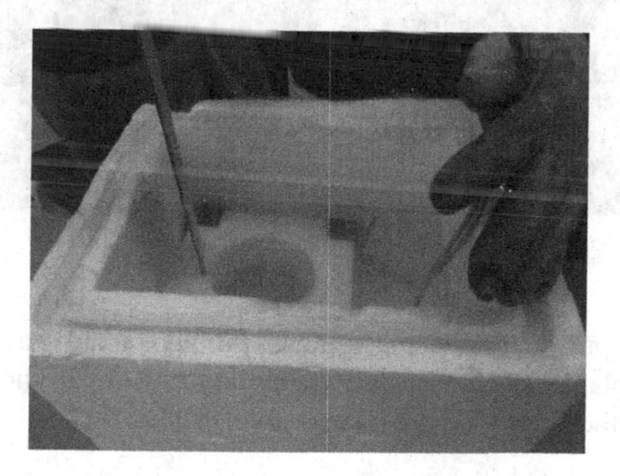

**Fig. 4.** Totally submerged stator in $LN_2$.

- Submerged tests

In this first test, the used materials presented a good resistance within the working temperature range during the first experiences, without breaking or exhibiting any type of weakness.

• Leak tests

To ensure the structure, namely the stator, would enclose the liquid nitrogen in an efficient way, some leak tests were elaborated. These tests were made in a first phase with water, until the stator was successfully insulated.

In the first attempt, the stator was clamped only by the four holes in each corner. After pouring some water into the stator through the nitrogen entrance channel, it was observed that the water would come out through breaches between the slices of the stator. In order to correct this issue, a rubber based insulation tape was applied between each slice of the stator in order to insulate the structure. This insulation tape was cut accordingly to the profile of the inner part of the stator slices, so that it was applied only where the slices make contact. It was observed that some water still fell between the breaches, although in much less quantity. Subsequently it was decided to provide a better distributed and more uniform clamping to the structure, closer to the inner part of the stator. In a first phase, this was achieved by using clamps with PVC plaques to distribute the load. An example of how the structure was clamped is shown in Fig. 5.

**Fig. 5.** Clamps with PVC plaques.

As this solution proved to solve the tightness problem, it led to a change in the design of the structure explained, where as a way to avoid the clamps, 8 holes of 6.5 mm were designed closer to the inner part of the stator to distribute the loads uniformly.

• Nitrogen pouring

Liquid nitrogen is unstable when exposed to room temperature. Since its boiling point is at 77 K, it immediately evaporates. For this reason, when it is poured into the stator it often starts to boil, spilling if not handled with caution. Hence, the pouring of nitrogen is executed slowly in short movements, periodically switching channels of nitrogen entrances. This process usually takes about 10 min to 15 min, until the structure is full of stabilized liquid nitrogen. At this point we know that the temperature within the stator bulks is 77 K.

To complete the leak test, liquid nitrogen was poured inside the stator and the results were satisfactory, as the stator remained well insulated without any nitrogen spills.

- Nitrogen usage

In order to estimate the rate at which the liquid nitrogen would evaporate from inside the stator, a graph of time vs. weight was elaborated. This was achieved by pouring liquid nitrogen inside the stator until the structure was full. To read the weight values a weighting scale was used. With the structure standing on the weighing-scale, the time was measured until the weight stopped falling. With the purpose of not doing any damage to the weighing scale with the liquid nitrogen, a box of Styrofoam was used to protect the scale. The proper tare of this box was made to allow the precise reading of the structure weight. The set-up established to read the weigh values is shown in Fig. 6.

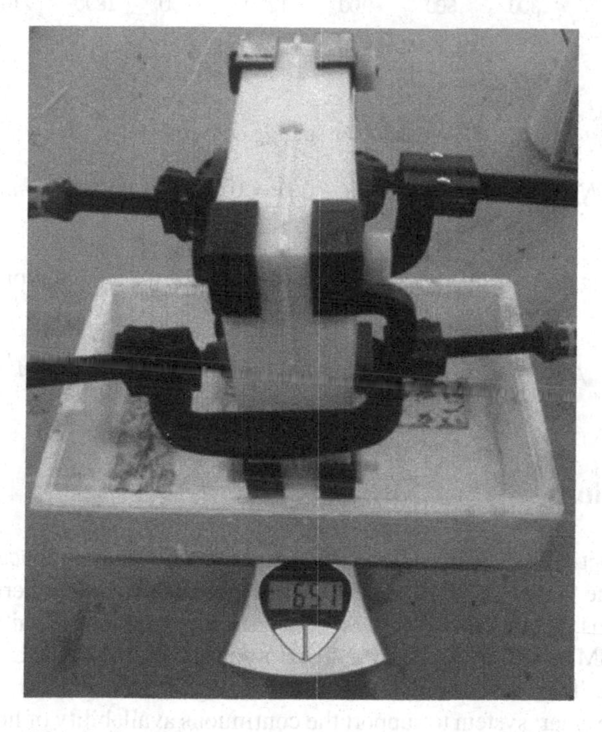

**Fig. 6.** Set-up used for the nitrogen usage information.

The curve of weight vs. time is shown in Fig. 7. It is easy to see that the rate at which the liquid nitrogen evaporates in the first 600 s can be considered linear, and therefore possible to calculate. In the calculation of the nitrogen usage rate, the first value considered was the weight measured after one minute, because after pouring the liquid nitrogen in the stator, it takes some seconds to stabilize. The evaporation rate in *g/min* of the liquid nitrogen in the stator was calculated and is given by:

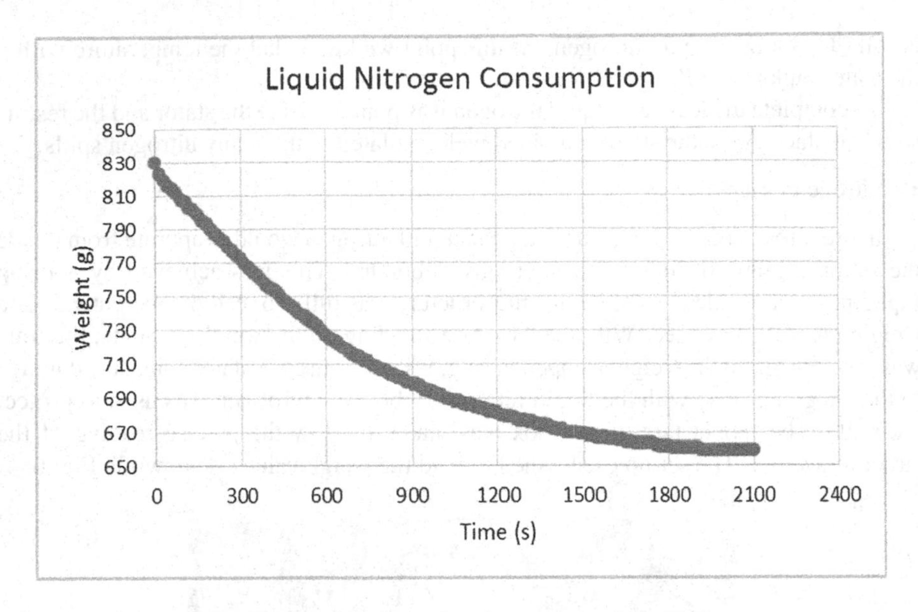


**Fig. 7.** Superconductor bearing consumption of $LN_2$.

$$LN_{er} = (814 - 727)/(600 - 60) = 0.161\ \mathrm{g/s} = 9.67\ g/min$$

Knowing that the density value of $LN_2$ is 0.807 *g/mL*, the evaporation rate in *mL/min* is given by:

$$LN_{er} = 0.161/0.807 = 0.200\,mL/s = 11.979\,mL/min$$

## 4 Conclusions

One of the objectives of smart grid is the optimization of resources allocation. As shown in this paper, the Superconductor Magnetic Bearings usage requires a critical allocation of smart systems, namely smart sensors to support the continuous availability of liquid nitrogen. The SMB viability was shown, a prototype was build and the LN2 consumption was measured. This data is important for smart grid management, and requires a comprehensive smart system to support the continuous availability of liquid nitrogen or to measure and guidance and levitation forces inside the Superconductor Magnetic Bearing.

Further work will deal with the integration of this data into smart systems, in an embedded architecture which embodies advanced automation systems to provide control and monitoring over the continuous availability of liquid nitrogen or to measure and guidance and levitation forces inside the Superconductor Magnetic Bearing.

**Acknowledgments.** This work was supported by Fundação para a Ciência e a Tecnologia, through IDMEC, under LAETA, project UID/EMS/50022/2013 and under project PTDC/EEEI-EEL/4693/2014 – HTSISTELEC. The authors would like to thank Mitera and Fablab EDP for their help and availability to do the complex machinery involved in this work.

## References

1. Viveiros, C., Melício, R., Igreja, J.M., Mendes, V.M.F.: Control and supervision of wind energy conversion systems. In: Camarinha-Matos, Luis M., Falcão, A.J., Vafaei, N., Najdi, S. (eds.) DoCEIS 2016. IAICT, vol. 470, pp. 353–368. Springer, Cham (2016). doi: 10.1007/978-3-319-31165-4_34
2. Seixas, M., Melício, R., Mendes, V.M.F., Collares-Pereira, M., Santos, M.P.: Simulation of offshore wind system with three-level converters: HVDC power transmission in cloud scope. In: Camarinha-Matos, L.M., Baldissera, T.A., Di Orio, G., Marques, F. (eds.) DoCEIS 2015. IAICT, vol. 450, pp. 440–447. Springer, Cham (2015). doi:10.1007/978-3-319-16766-4_47
3. Arsénio, A.J., Carvalho, M.V., Cardeira, C., Costa Branco, P.J., Melício, R.: Conception of a YBCO superconducting zfc-magnetic bearing virtual prototype. In IEEE International Power Electronics and Motion Control Conference, pp. 1226–1230 (2016)
4. Ogata, M., Matsue, H., Yamashita, T., Hasegawa, H., Nagashima, K., Maeda, T., Matsuoka, T., Mukoyama, S., Shimizu, H., Horiuchi, S.: Test equipment for a flywheel energy storage system using a magnetic bearing composed of superconducting coils and superconducting bulks. Supercond. Sci. Technol. **29**(5), 1–7 (2016)
5. Sparing, M., Hossain, M., Berger, D., Berger, A., Abdkader, A., Fuchs, G., Cherif, C., Schultz, L.: Superconducting magnetic bearing as twist element in textile machines. IEEE Trans. Appl. Supercond. **25**(3), 1–4 (2015)
6. Lloberas, J., Sumper, A., Sanmarti, M., Granados, X.: A review of high temperature superconductors for offshore wind power synchronous generators. Renew. Sustain. Energy Rev. **38**, 404–414 (2014)
7. da Costa Branco, P.J., Dente, J.A.: Design and experiment of a new Maglev design using zero-field-cooled YBCO superconductors. IEEE Trans. Ind. Electron. **59**(11), 4120–4127 (2012)
8. Fernandes, J., Montes, I., Sousa, R., Cardeira, C., Costa Branco, P.J.: Superconductor joule losses in the zero-field-cooled Maglev vehicle. IEEE Trans. Appl. Supercond. **26**(3), 1–7 (2016)
9. Zhang, R., Du, Y., Yuhong, L.: New challenges to power system planning and operation of smart grid development in China. In: International Conference on Power System Technology (POWERCON), pp. 1–8 (2010)
10. Arsénio, A.J., Carvalho, M.V., Cardeira, C., Costa Branco, P.J., Melício, R.: Viability of a frictionless bearing with permanent magnets and HTS bulks. In IEEE International Power Electronics and Motion Control Conference, pp. 1231–1236 (2016)
11. Arsénio, A.J., Carvalho, M.V., Cardeira, C., Costa Branco, P.J., Melício, R.: Experimental set-up and efficiency evaluation of zero-field-cooled (ZFC) YBCO magnetic bearings. In: IEEE Transactions on Applied Superconductivity, vol. 99, pp. 1–5 (2017) 10.1109/TASC. 2017.2662605

# An Overview on Preisach and Jiles-Atherton Hysteresis Models for Soft Magnetic Materials

Pedro Melo[1(✉)] and Rui Esteves Araújo[2]

[1] School of Engineering, Polytechnic of Porto, Porto, Portugal
pma@isep.ipp.pt

[2] INESC TEC and Faculty of Engineering University of Porto, Porto, Portugal
raraujo@fe.up.pt

**Abstract.** The design of efficient and high power density electrical machines needs an accurate characterization of magnetic phenomena. Core losses estimation is usually addressed by empirical models, where its lack of accuracy is well known. Hysteresis models are able to take an insight into the magnetic physical mechanisms. Compared to the empirical models, they contribute to a higher accuracy in modeling electromagnetic systems, including core losses estimation. At a macroscopic level, two models are often used: the Preisach and the Jiles–Atherton (J–A) models. This paper presents their basic formulation, as well the main limitations and scope of application. This is a first step to investigate the possible application of hysteresis models, in order to reach accurate core losses estimation in switched reluctance machines.

**Keywords:** Magnetic hysteresis · Preisach model · Jiles-Atherton model · Core losses · Electrical machines

## 1 Introduction

Core losses estimation in ferromagnetic materials has been motivating an increasing research interest, since the first Steinmetz empirical formulation [1]. For electric machines, an accurate characterization and estimation of core losses is a fundamental issue, in order to improve the machine design, efficiency and control. Empirical models, based on losses separation, are unable to characterize the complex mechanisms that rule core losses phenomena [2]. Their parameter identification is based on curve-fitting methods, using manufacturer iron sheet data, experimental data or finite element analysis (FEA) [3]. FEA is the main tool for field distribution analysis in electric machines, where hysteresis is considered only after a stable solution is reached (i.e., post-processing calculation) [4]. Empirical models are still engineer's first choice, due to their simplicity and fast processing. However, for non-sinusoidal flux density machines, like permanent magnet synchronous machine (PMSM) and switched reluctance machine (SRM), conventional models reveal a significant lack of accuracy for core losses estimation. Therefore, different approaches must be addressed.


L.M. Camarinha-Matos et al. (Eds.): DoCEIS 2017, IFIP AICT 499, pp. 398–405, 2017.
DOI: 10.1007/978-3-319-56077-9_39

Hysteresis models provide a deeper insight into the magnetic mechanism, allowing a better characterization of the magnetic system behavior, including core losses. Moreover, it is possible to integrate them with FEA. In literature, two different approaches for modeling hysteresis can be found [5]: physical models and mathematical (phenomenological) models. The first ones are based on magnetization physics, at micromagnetic scale. In the second ones, hysteresis is described with a mathematical relation between the macro scale magnetization (M) and the magnetic field strenght (H). Phenomenological models have smaller computation times, while the physical ones give a much better insight into the magnetic physics. The Preisach and Jiles–Atherton (J–A) models are considered two classic macroscopic approaches for hysteresis modeling. The former is a phenomenological model, with higher accuracy than the later. The Preisach model can represent major and minor loops for different materials, due to its phenomenological nature. The J-A model includes some physical features, but it cannot represent minor loops. On the other hand, it has a smaller parameter number, which contributes to more efficient computational and memory requirements [6]. Both models need a large amount of measurement data [7] and show limitations in characterizing the magnetization phenomena. Therefore, several extensions to the original models were proposed. Also, since they have some complementary features, there have been attempts to integrate both models. However, this is beyond the paper scope.

This paper intends to make a description of the Preisach and J-A hysteresis models. The fundamental goal is to give a first contribution to the main research objective of the PhD thesis: to develop a methodology for characterizing core losses in SRM. The paper is structured as follows: Sect. 2 presents the PhD work relationship to smart systems. In Sect. 3, the original Preisach model is presented, where its main features are highlighted. Section 4 has a similar structure for the J-A model. Section 5 includes the conclusions and the next steps in this work.

## 2 Relationship to Smart Systems

Most often, smart systems must face several demanding requirements. In order to do it, they must integrate different technologies, where sensing, actuation and control levels are fundamental. Their performance relies on a multitude of energy conversion devices, like sensors and actuators. Many devices include ferromagnetic materials. Therefore, an accurate characterization of these materials, including their losses, is fundamental, under different views (e.g., higher efficiency actuators, sensors with smaller error and accurate models, for a better control performance). Moreover, this may be extended to the design level, where machines, sensors and controllers can be simulated in a more realistic way, optimizing their performance before the prototype stage. A good example of a smart system is an autonomous electric vehicle. Its propulsion is based on electric motors, so very accurate models must be employed, in order to achieve the instantaneous drive performance requirements. Also, the vehicle efficiency is a fundamental issue. The energy management system has a key role, where loss characterization for different operation levels is fundamental.

## 3 Preisach Model

In the Preisach model, the physical mechanisms of hysteresis are not considered, only a general mathematical description of its behavior (i.e., a phenomenological representation). Therefore, it can be used independently of the hysteresis physical nature under analysis. It is a quasi-static (rate-independent) scalar model, where H(t) is unidirectional. It assumes that the material has a large set of small dipoles, where each one has a rectangular hysteresis loop (hysterons). Mathematically, it may be expressed by[1] [8]:

$$M(t) = \int_{-\infty}^{+\infty} d\alpha \int_{-\infty}^{\alpha} p(\alpha, \beta) f_{\alpha,\beta}(H(t)) d\beta. \tag{1}$$

Where $\alpha, \beta$ are, respectively, the transition "up" and "down" inputs of the generic hysteron ($f_{\alpha,\beta}$) – Fig. 1-(a). Its magnetization can be $-m_s$ or $+m_s$, depending on the magnetic field strength (H(t)) and its history, where $m_s$ is the elementary magnetization, equal for all hysterons.

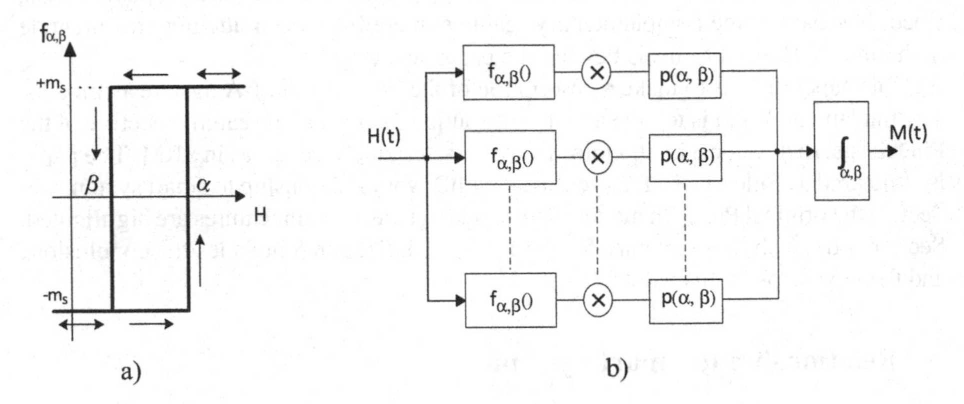


**Fig. 1.** (a) Preisach typical hysteron and (b) Model schematic representation

The hysteresis loop results from the superposition of these elementary rectangular loops, affected by the Preisach function, $p(\alpha, \beta)$, which is a histeron weight function – Fig. 1(b). Hysterons include both symmetric and asymmetric loops, allowing to model minor loops (by its nature, they are asymmetric ones) [8]. Each hysteron must verify a set of rules, reflecting the phenomenological nature of the model:

- If $H_h > H_{sat}$ => $f_{\alpha,\beta} = +m_s$; consequence: $\alpha \leq H_{sat}$
- If $H_h < -H_{sat}$ => $f_{\alpha,\beta} = -m_s$; consequence: $\beta \geq -H_{sat}$
- $\alpha \geq \beta$

[1] (1) is defined for magnetic process. Since this is a phenomenological model, in the general formulation, M(t) and H(t) are replaced by, respectively, the generic functions y(t) and u(t).

Where "=" is related to reversible thermodynamics phenomena (i.e., stored energy, able to be recovered) and ">" reflects the irreversible phenomena (i.e., energy dissipative effects). This rule is reflected in the integral limits of (1). Moreover, since these limits include the entire Preisach plan $(\alpha, \beta)$, for $\alpha < \beta$, $p(\alpha, \beta)$ must be 0. At the macro-magnetic scale, if $H > H_{sat}$, then $M = M_{sat}$; also, for $H < -H_{sat}$, one has $M = -M_{sat}$. $M_{sat}$ is the material saturated magnetization and $H_{sat}$ is the corresponding magnetic field strength. The mathematical basis of Preisach model can be highlighted under a geometrical analysis. For: $-H_{sat} \leq H \leq H_{sat}$, the area inside the rectangular triangle, in Fig. 2-(a), defines the hysteron domain, necessary to model the hysteresis loops.

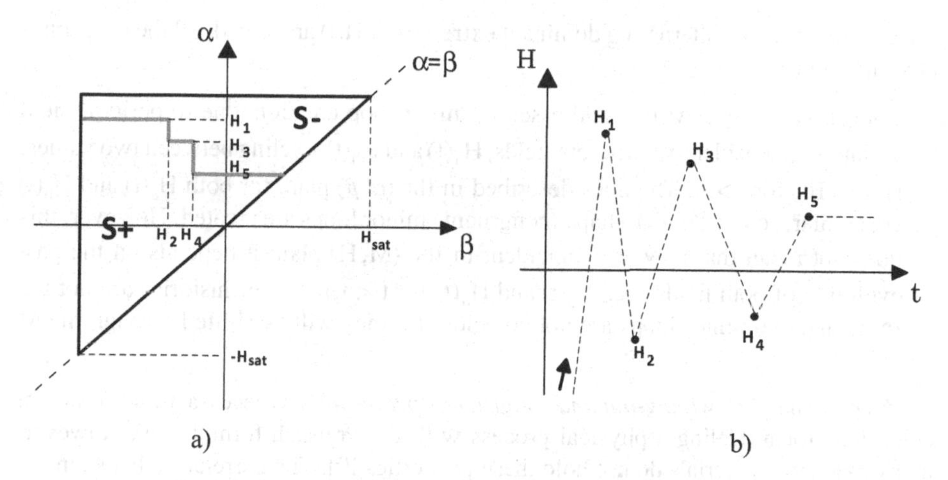


**Fig. 2.** (a) H(t) reversal extremes and (b) H(t) extremes representation in the Preisach plan

The staircase line depicted in Fig. 2-(a) represents the path in the Preisach plan, performed by the input H(t), in Fig. 2-(b). Initially, it has a very large negative value. Equation (1) can then be replaced by:

$$M(t) = m_s \iint_{S+} p(\alpha, \beta) d\alpha d\beta - m_s \iint_{S-} p(\alpha, \beta) d\alpha d\beta. \tag{2}$$

M(t) depends on the positive (S+) and negative (S−) areas, which divide the triangle in Fig. 2-(b) – $f_{\alpha,\beta} = +m_s$, for $(\alpha, \beta) \in S+$ and $f_{\alpha,\beta} = -m_s$, for $(\alpha, \beta) \in S-$. S+ and S− are defined by the staircase line. It should be noted that the past extreme values of H(t) dictate the line´s shape: in fact, they are the coordinates of its vertices. Therefore, the staircase line models the hysteresis memory effect. Its construction is based on two simple rules:

- For monotonically increasing input, a horizontal segment of the staircase line moves upwards, until an extreme value of H(t) is reached.
- For monotonically decreasing input, a vertical segment line moves from right to left, until an extreme of H(t) is reached.

The mathematical structure of the Preisach model has two fundamental properties [8]:

- *Wiping-Out Property*: it is related to the memory process. The sequence of reversal points in H(t) (i.e., local maximum and minimum values) can be redundant: there is a subset with the fundamental reversal values, able to define the state of the system. The following conditions characterize this subset: $H_{max\_1} > H_{max\_2} > \ldots > H_{max\_n}; H_{min\_1} < H_{min\_2} < \ldots < H_{min\_n}$

  Where: $t(H_{min\_1}) < t(H_{max\_1}) < t(H_{min\_2}) < t(H_{max\_2}) < \ldots < t(H_{min\_n}) < t(H_{max\_n})$

In short, only the alternating dominant extremes of H(t) are stored, all the other ones are wiped out.

- *Congruency Property*: it is addressed to minor loop creation, due to periodic field variations. Consider two different fields, $H_a(t)$ and $H_b(t)$, cycling between two values, $H_1$ and $H_2$, for $t > t_c$. The lines described in the $(\alpha, \beta)$ plan, for both $H_a(t)$ and $H_b(t)$ are similar, so two similar-shape (congruent) minor loops are created. However, this does not mean that they are coincident in the (M, H) plan: it depends on the past evolution of both fields (i.e., $H_a(t)$ and $H_b(t)$ for $t < t_c$). If their histories are not the same, the two minor loops are not coincident – they will be shifted through the M axis.

As proven in [8], *wiping-out and congruency properties* are necessary and sufficient conditions for modeling a physical process with the Preisach formulation. However, many magnetic materials do not hold these properties [9]. The correlation between the parameters and the magnetization process is not always clear [10]. Also, a complex parameter identification is usually demanded, with a heavy computation burden.

## 4 Jiles-Atherton (J-A) Model

The J-A model is a scalar pseudo-physics based model, where some nonlinear phenomena are considered (e.g., magnetic moment interaction, Barkhausen jumps due to domain wall motion, domain wall bending and domain rotation). The original J-A model takes the anhysteretic magnetization curve (only domain rotation is considered, without including losses due to domain wall motion), in addition to reversible and irreversible magnetization phenomena [11] (i.e., at any instant, the source energy is considered to be divided into stored magnetostatic energy and energy dissipated due to domain wall movement (hysteresis loss)). The analytical anhysteretic magnetization ($M_{an}$) was first obtained, based on Langevin theory. It was developed for paramagnetic materials and extended to ferro and ferrimagnetic materials [12]. The original Langevin function is given by:

$$M_{an}(H_a) = M_{sat}\left[coth\left(\frac{H_a}{a}\right) - \frac{a}{H_a}\right]. \tag{3}$$

Where $H_a$ is the applied field strength and $a = \frac{k_B T}{\mu 0 m}$ ($k_B$ is the Boltzmann constant, $T$ is the material absolute temperature, $\mu_0$ is the free space permeability and $m$ is the magnitude of each elementary permanent magnetic dipole in the material).

Langevin approach assumes that the material individual magnetic moments do not interact with each other, being sensible only to the applied field and thermal agitation. However, for ferromagnetic materials, $H_a$ in (3) is no longer suitable. In fact, there is a strong interaction between the elementary moments, particularly below the Curie temperature [12]. In 1907, Weiss introduced the concept of "molecular field" ($H_m$), in order to express this moment interaction. He postulated that $H_m$ (an internal fictitious field) is directly proportional to M, which, in turn, adds to $H_a$. The resultant field ($H_r$) is given by:

$$H_r = H_a + H_m = H_a + \gamma M(H_a). \tag{4}$$

It follows that $H_a$ is replaced by $H_r = H_a + \gamma M_{an}(H_a)$, in (3). This equation describes the saturation phenomenon, but not the hysteresis one: irreversible magnetic phenomena are not included (e.g., Barkhausen jumps), only reversible ones (e.g., magnetic domain rotation).

The hysteresis equation is deduced from an energy equilibrium condition [11], between the magnetic energy supplied by an external source ($M_s$), the change in the stored magnetostatic energy (A) and the hysteresis loss (B), related to the irreversible magnetization $M_{irr}$. For an initially demagnetized material ($M_s = M_{an}$), this energy relation is given by:

$$\mu_0 \int M_{an} \cdot dH_r = \underbrace{\mu_0 \int M \cdot dH_r}_{(A)} + \underbrace{\mu_0 k(1-c) \int \delta \frac{dM_{irr}}{dH_r} \cdot dH_r}_{(B)}. \tag{5}$$

Where $k$ is the pinning parameter[2], related to the energy dissipated due to hysteresis. The parameter $\delta$ makes sure that (B) is always a dissipative energy (i.e., $\delta \frac{dM_{irr}}{dH_r} > o$). The coefficient $c$ is a measure of the reversible change in magnetization ($0 < c < 1$; if $c = 0$, then magnetization change is irreversible; if $c = 1$, then magnetization change is reversible). From (5), for static or quasi-static equilibrium conditions, energy losses occur only due to changes in $M_{irr}$. Considering the irreversible and reversible ($M_{rev}$) components of magnetization (M), one has:

$$M = M_{rev} + M_{irr}. \tag{6}$$

After differentiating (5) and considering (6), it follows that:

[2] Due to pinning domain wall effect, as a consequence of material imperfections.

$$M_{an} = M_{rev} + M_{irr} + k(1-c)\delta \frac{dM_{irr}}{dH_r}. \tag{7}$$

If only irreversible phenomena exist, $c = 0$ and $M_{rev} = 0$. Under this scenario, one has (from (7)):

$$M_{irr} = M_{an} - k\delta \frac{dM_{irr}}{dH_r}. \tag{8}$$

Substituting (8) in (7), $M_{rev}$ is given by:

$$M_{rev} = c\left(M_{an} - M_{irr}\right). \tag{9}$$

Thus, in the absence of irreversible phenomena, the magnetization curve $M(H_a)$ is coincident with the anhysteretic curve ($M_{an}(H_a)$). Considering (6), (8), (9) and (4), the rate of change of magnetization with the applied field is given by:

$$\frac{dM}{dH_a} = \frac{ck\delta \cdot dM_{an}/dH_a + \left(M_{an} - M\right)}{k\delta - \gamma\left(M_{an} - M\right)}. \tag{10}$$

Where (10) is ruled by the displacement of the magnetization from the anhysteretic curve. This equation models the magnetic hysteresis behavior for the J-A model. Together with (3), the model has five physical parameters: $a$, $\gamma$, $c$, $k$ and $M_{sat}$.

A critical analysis to J-A model physical basis is presented in [13]: it points out that (5) is a balance of co-energies, which does not reflect the first thermodynamics law in every instant (in fact, only for a complete hysteretic cycle). Moreover, other non-physical considerations were also employed (e.g., the need for the "δ" parameter). Some authors do not consider the J-A model as a physical one [14]. In this paper it is designated as a "pseudo-physics" model.

## 5 Conclusions

Traditional methods for core losses prediction in soft magnetic materials and electrical machines are based on curve-fitting empirical models. New machine designs and non-sinusoidal excitation demand for improved accurate approaches.

Hysteresis models allow a deeper insight in characterizing the magnetization phenomena. They can be divided into physical and phenomenological models. Two classic macroscopic approaches are the Preisach and J–A models. Compared with curve-fitting methods, they give more accurate results, since they are able to represent the material's behavior (under certain limitations). Also, they can be integrated with FEA. Although Preisach model is a phenomenological one, it gives more accurate results than the J-A model. Moreover, it can be applied to other materials than soft-magnetic, because the Preisach function does not consider their physical properties. The static J-A model has a higher physical basis and is more efficient in terms of computational time and memory requirements. Parameter identification is simpler, but it cannot represent minor loops. With the Preisach model this is possible.

Both the original models have several limitations, mainly due to the non-inclusion of several magnetization physical phenomena. Several extended models have been proposed, attempting to overcome those limitations. They show complementary advantages, since modeling magnetization phenomena is a very complex task. In short, this issue remains as a challenging open research field.

This PhD intends to characterize accurately the core losses in SRM. As a future stage, it is intended to address energy–based models, where it is expected to reach a deeper insight in soft magnetic materials energy balance. Parameter estimation effort, computation burden and integration with FEA must also be addressed.

## References

1. Steinmetz, C.P.: On the law of hysteresis. Trans. Am. Inst. Electr. Eng. **9**(1), 1–64 (1892)
2. Du, R., Robertson, P.: Dynamic Jiles-Atherton model for determining the magnetic power loss at high frequency in permanent magnet machines. IEEE Trans. Magn. **51**(6), 1–10 (2015)
3. Rasilo, P., Belahcen, A., Arkkio, A.: Experimental determination and numerical evaluation of core losses in a 150-kVA wound-field synchronous machine. IET Electr. Power Appl. **7**(2), 97–105 (2013)
4. Demenko, I.D., Hameyer, K., Pietrowski, W., Krzysztof Zawirski, A., Mazgaj, W., Warzecha, A.: Influence of simplifications of the rotational magnetization model on calculation accuracy. COMPEL: Int. J. Comput. Math. Electr. Electr. Eng. **34**(3), 740–753 (2015)
5. Alatawneh, N., Pillay, P.: Modeling of the interleaved hysteresis loop in the measurements of rotational core losses. J. Magn. Magn. Mater. **397**, 157–163 (2016)
6. Hussain, S., Lowther, D.A.: Establishing a relation between Preisach and Jiles-Atherton models. IEEE Trans. Magn. **51**(3), 1–4 (2015)
7. Motoasca, S., Scutaru, G., Gerigan, C.: Improved analytical method for hysteresis modelling of soft magnetic materials In: 2015 International Aegean Conference on Electrical Machines & Power Electronics (ACEMP), 2015 International Conference on Optimization of Electrical & Electronic Equipment (OPTIM) & 2015 International Symposium on Advanced Electromechanical Motion Systems (ELECTROMOTION), pp. 558–563. IEEE (2015)
8. Mayergoyz, I.D.: Mathematical Models of Hysteresis and their Applications. Academic Press, Cambridge (2003)
9. Torre, E.D.: Magnetic Hysteresis. Wiley, Hoboken (2000)
10. Bertotti, G.: Hysteresis in Magnetism: for Physicists, Materials Scientists, and Engineers. Academic press, Cambridge (1998)
11. Jiles, D., Atherton, D.: Theory of ferromagnetic hysteresis. J. Magn. Magn. Mater. **61**(1–2), 48–60 (1986)
12. Cullity, B.D., Graham, C.D.: Introduction to Magnetic Materials. Wiley, Hoboken (2011)
13. Zirka, S.E., Moroz, Y.I., Harrison, R.G., Chwastek, K.: On physical aspects of the Jiles-Atherton hysteresis models. J. Appl. Phys. **112**(4), 043916 (2012)
14. Cardelli, E.: Advances in magnetic hysteresis modeling. Handb. Magn. Mater. **24**, 323–409 (2015)

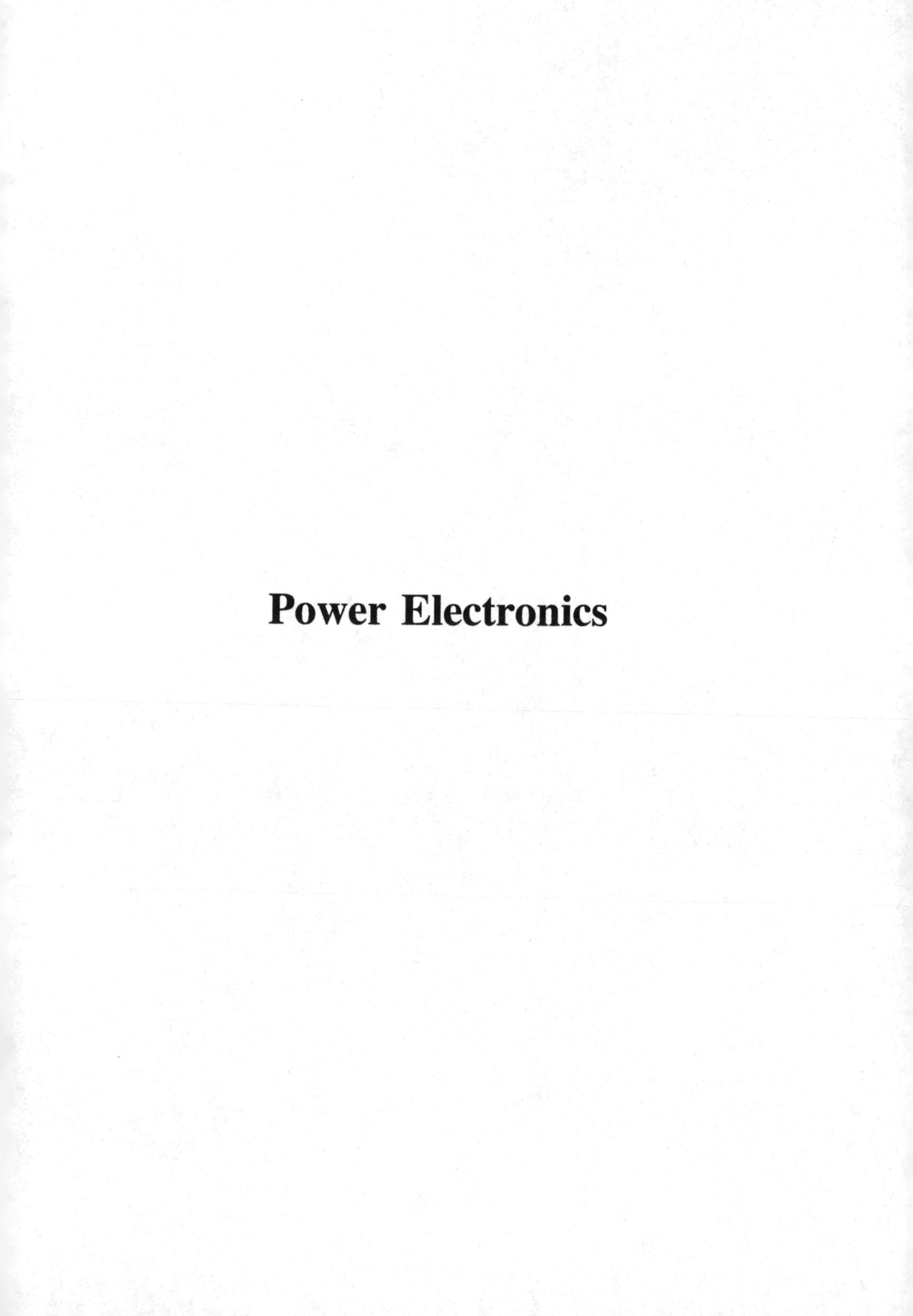

# Power Electronics

# Comparative Analysis of qZS-Based Bidirectional DC-DC Converter for Storage Energy Application

Oleksandr Matiushkin[1], Oleksandr Husev[1], Kostiantyn Tytelmaier[1], Kaspars Kroics[2,3(✉)], Oleksandr Veligorskyi[1], and Janis Zakis[2]

[1] Department of Biomedical Radioelectronics Apparatus and Systems, Chernihiv National University of Technology, Chernihiv, Ukraine
oleksandr.husev@ieee.org, kostya.tytelmaier@gmail.com

[2] Institute of Industrial Electronics and Electrical Engineering, Riga Technical University, Riga, Latvia
janis.zakis@ieee.org

[3] Institute of Physical Energetics, Riga, Latvia
kaspars.kroics@gmail.com

**Abstract.** This paper presents a comparative analysis of the bidirectional qZS-based dc-dc converter for storage application with a traditional solution based on the boost dc-dc converter. The analysis estimates the energy stored in the capacitors and inductors, blocking voltage across semiconductors, and conduction losses. The comparison is based on the mathematical and simulation analysis.

**Keywords:** DC-DC converter · quasi-Z-Source · Storage batteries

## 1 Introduction

Renewable energy sources are increasingly important today. European Commission has set a target to raise the share of renewable energy to 20% before 2020 [1–7]. The topic of power electronics converters for renewable applications is extremely popular. Growing number of renewable energy installations leads to an increase in the storage systems installations. The technologies devised require bidirectional power interface with minimum losses between two dc busses or a dc bus and an energy storage device (battery, supercapacitor, etc.). For these purposes, Bidirectional DC-DC Converters (BDCs) should be used. In applications with lower step-up or step-down ratio, a common solution is a non-isolated type [8–14]. Many research papers focus on optimal solutions of the bidirectional energy transfer for applications where galvanic isolation is not required.

Recent solutions based on the Impedance-Source (IS) networks have been extended to various fields of application. IS-based converters have much higher boost capability than traditional boost converters. Also, they have no forbidden switching states. As a result, the proven advantages of IS-based converters consist in wide-range input voltage regulation along with enhanced reliability [15–19]. At the same time, it has been clearly demonstrated that the proposed solutions have drawbacks in dc-ac applications in terms of power density and efficiency [15].


L.M. Camarinha-Matos et al. (Eds.): DoCEIS 2017, IFIP AICT 499, pp. 409–418, 2017.
DOI: 10.1007/978-3-319-56077-9_40

Though many studies focus on the IS-based converter, no comparative analyses of the bidirectional dc-dc converter based on any IS network have been reported. Due to the enhanced boost functionality, it can be an attractive and competitive solution.

This paper provides a comparative analysis of the proposed bidirectional quasi-Z-Source converter based on the criteria of passive volume, voltage stress on semiconductors and conduction losses.

## 2 Relation to Smart Systems

The concept of the "smart grid" is very popular and well known and can be considered as a part of the future "smart system". It includes an electrical grid that has a variety of operational and energy measures, including smart meters, smart appliances, renewable energy resources and storage systems. Electronic power conditioning and control of the production and distribution of electricity are important aspects of the smart grid. The issue of communication is a very important element of the smart system and smart grid in particular, many researchers focus on that topic. At the same time, new standards of communication technologies such as 5G or IPV6 and their further development may satisfy a very high level of requirements and true obstacles lie in the "muscles" of the future smart system. It is evident that power electronics converters are present in any key module of the smart energy distributed systems. They are providing energy conversion and transfer and can be considered like "muscles" of the system. Such parameters like power density, weight, efficiency, reliability and cost of power electronics facilities define the feasibility of the abovementioned modern concept. Research and development of the power electronics facilities is a contribution to the future smart system.

## 3 Calculation Approach

This section focuses on the calculation of the passive elements of the circuit. To calculate the voltage drop and currents that flow through electrical components, the steady state analysis is used. Switches are assumed as ideal (Fig. 1).

### 3.1 qZS

The qZS converter in a boost mode is controlled by a single switch $T_2$ ($T_1$ and $T_3$ are OFF). To reduce the conduction losses in the diode, $T_1$ and $T_3$ are opened when free-wheeling diodes of these switches are conducting.

A period of switching consists of two time intervals: when the switch $T_2$ is opened (time interval $T_0$) and when the transistor is not conducting (time interval $T_1$).

Assuming that, energy flows from the battery, and the output source can be replaced by a load, or simply, by a resistor.

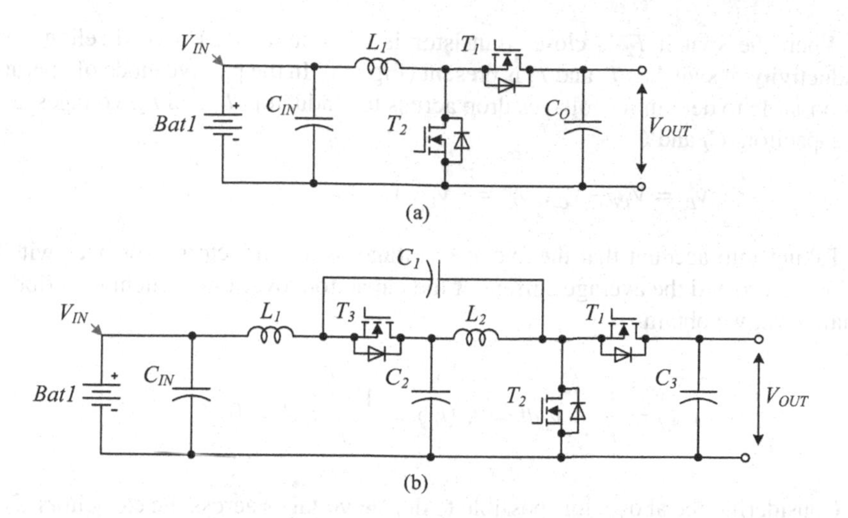


**Fig. 1.** Traditional converter (a) and qZS converter (b).

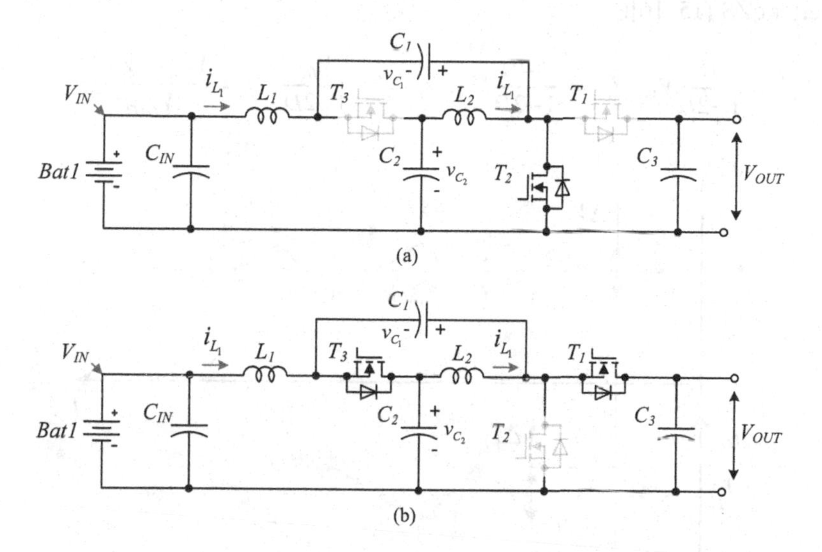


**Fig. 2.** qZS in the active mode (a) and passive mode (b) of operation.

When the switch $T_2$ is opened (transistor is conducting), $T_1$ and $T_3$ are OFF (Fig. 2a). In this case, it possible to determine the voltage drop across the inductors $L_1$ and $L_2$, the output current (current of discharging the capacitor $C_3$). Since voltage drop across the passive components equals the voltage source, it can be expressed for a circuit that consists of the inductor $L_1$, the capacitor $C_2$, and the input voltage source (battery):

$$v_{L_1} = L_1 \frac{di_{IN}}{dt},\ v_{L_2} = v_{C_2},\ \frac{v_{OUT}}{R_L} = -C_3 \frac{dv_{OUT}}{dt},\ v_{L_1} - v_{C_1} = V_{IN}. \tag{1}$$

When the switch $T_2$ is close (transistor is not conducting), freewheeling diodes conductivity of switches $T_1$ and $T_3$ is present (Fig. 2b). In the passive mode of operation, it is possible to determine voltages drop across the inductors $L_1$ and $L_2$, voltages across the capacitors $C_1$ and $C_2$:

$$v_{L_1} = V_{IN} - v_{C_2},\ v_{L_2} = -v_{C_1},\ v_{C_1} + v_{C_2} = v_{OUT}. \quad (2)$$

Taking into account that the average voltage of the inductors over one switching period is zero and the average current of the capacitors over one switching period also equals zero, we obtain

$$\langle v_L \rangle = \frac{1}{T}\int_0^T v_L dt = 0,\ \langle i_C \rangle = \frac{1}{T}\int_0^T i_C dt = 0. \quad (3)$$

Considering the above, it is possible to derive voltages across the capacitors $C_1$ and $C_2$ and the output voltage, through input source voltage and duty cycle for the boost converter qZS [15, 16]:

$$V_{C_1} = \frac{D}{1-2D}V_{IN}, V_{C_2} = \frac{1-D}{1-2D}V_{IN}, V_{OUT} = \frac{V_{IN}}{1-2D}, D = \frac{V_{OUT}-V_{IN}}{2V_{OUT}}. \quad (4)$$

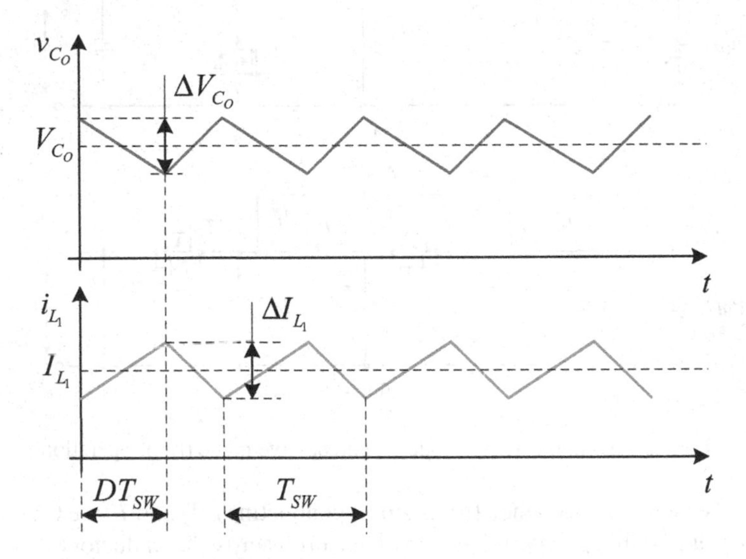


**Fig. 3.** Voltage across the capacitor and current in the inductor

During the period, currents through the inductor vary linearly (Fig. 3). Thus, the input current can be linearized as well. It is well known that the input ripple factor is the ratio of the input current ripple to a steady component of the current:

$$K_{IN} = \frac{\Delta i_{IN}}{\langle i_{IN} \rangle}. \quad (5)$$

Taking into account that capacitor voltage change during the period is negligible, the ripple current along with the steady state component can be expressed as:

$$\Delta i_{IN} = \frac{V_{L_1}}{L_1} TD = \frac{V_{IN} + V_{C_1}}{L_1} TD, \ \langle i_{IN} \rangle = I_{IN} = \frac{P}{V_{IN}}. \quad (6)$$

Thus, based on Eqs. (4)–(6), the values of the inductances $L_1$ and $L_2$ are completely identical and can be estimated as:

$$L_1 = \frac{(V_{OUT}^2 - V_{IN}^2)V_{IN}}{4V_{OUT}K_{IN}Pf}, \ L_2 = \frac{(V_{OUT}^2 - V_{IN}^2)V_{IN}}{4V_{OUT}K_{IN}Pf}. \quad (7)$$

The output ripple factor is determined as the ratio of the output voltage ripple to a steady component of the output voltage:

$$K_{OUT} = \frac{\Delta V_{OUT}}{\langle v_{OUT} \rangle}. \quad (8)$$

Similar to the input current, during the period, output voltage varies linearly (Fig. 3); thus, the output voltage can be linearized:

$$\Delta V_{OUT} = \frac{TDV_{OUT}}{C_3 R_L}, \ \langle v_{OUT} \rangle = V_{OUT}, \ R_L = \frac{V_{OUT}^2}{P}. \quad (9)$$

Thus, based on Eqs. (4), (8) and (9), and considering that a load resistor can be expressed through power, required output capacitance can be obtained as:

$$C_3 = \frac{P(V_{OUT} - V_{IN})}{2V_{OUT}^3 K_{OUT} f}. \quad (10)$$

Similarly, capacitance of the capacitors $C_1$ and $C_2$ can be found where the ripple factors are equal ($K_{C1} = K_{C2} = K_{OUT}$):

$$C_1 = \frac{P}{K_{OUT} V_{OUT} V_{IN} f}, \ C_2 = \frac{(V_{OUT} - V_{IN})P}{K_{OUT}(V_{OUT} + V_{IN})V_{OUT} V_{IN} f}. \quad (11)$$

To further simplify analysis, it is assumed that both passive elements are equal and that the voltage of the inductors and the capacitors corresponds to $V_{IN} = V_{OUT}/2$:

$$L = \frac{3 \times V_{OUT}^2}{32 \times K_{IN} Pf}, \ C = \frac{P}{4V_{OUT}^2 K_{OUT} f}. \quad (12)$$

### 3.2 Traditional Boost Converter

A traditional converter is calculated by the same approach [8]. Table 1 shows the results of calculation for both converters.

**Table 1.** Comparison of passive elements

| Topology | Inductors | | | Capacitors | | |
|---|---|---|---|---|---|---|
| | № | Average current | Value | № | Average voltage | Value |
| qZS bidirectional converter | $L_1$, $L_2$ | $\frac{P_{OUT}}{V_{IN}}$ | $L\frac{8(V_{OUT}^2 - V_{IN}^2)V_{IN}}{3V_{OUT}^3}$ | $C_1$ | $\frac{V_{OUT} - V_{IN}}{2}$ | $C\frac{4V_{OUT}}{V_{IN}}$ |
| | | | | $C_2$ | $\frac{V_{OUT} + V_{IN}}{2}$ | $C\frac{4V_{OUT}(V_{OUT} - V_{IN})}{V_{IN}(V_{OUT} + V_{IN})}$ |
| | | | | $C_3$ | $V_{OUT}$ | $C\frac{2(V_{OUT} - V_{IN})}{V_{OUT}}$ |
| Conventional bidirectional converter | $L_1$ | $I_{IN}$ | $L\frac{8V_{IN}^2(V_{OUT} - V_{IN})}{V_{OUT}^3}$ | $C_1$ | $V_{OUT}$ | $C\frac{2(V_{OUT} - V_{IN})}{V_{OUT}}$ |

# 4 Simulation Results

Figure 4 shows the ripple of the input current and the output voltage for two converters. It can be seen that the steady state current and steady state voltage are the same in the

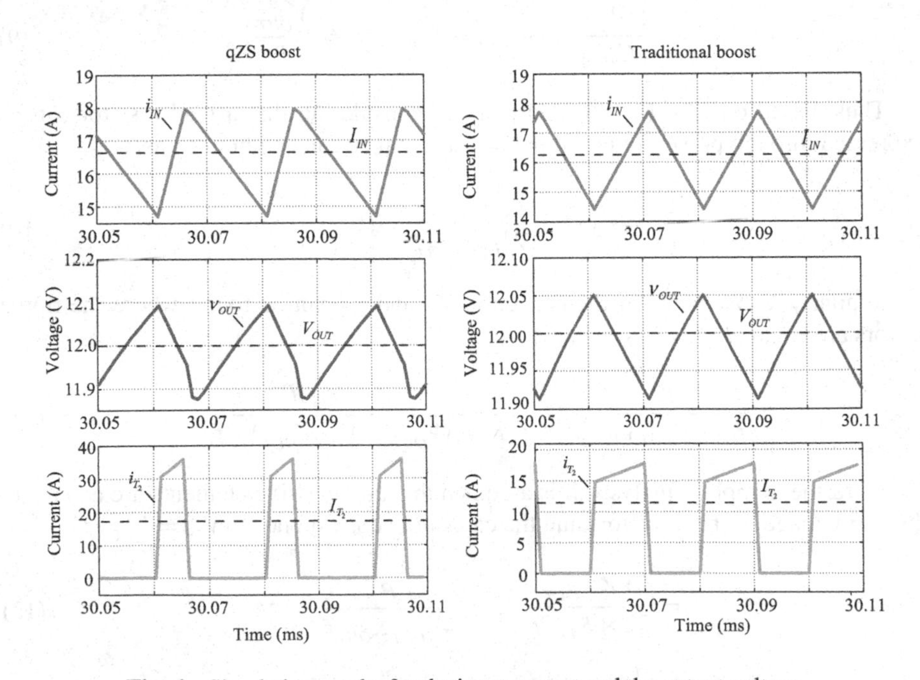


**Fig. 4.** Simulation results for the input current and the output voltage.

two cases. But it is also shown that the ripple of the output voltage qZS converter is not of a linear character like that of a conventional converter. But absolute peak values correspond to the expected and coincide with the calculation part. Thus, it can be implemented in the experiment.

In the simulation (for both converters), the following parameters were taken into account: $P = 100$ W, $f = 50$ kHz, $V_{IN} = 6$ V, $V_{OUT} = 12$ V, $K_{OUT} = 1.2\%$, $K_{IN} = 20\%$.

## 5 Comparative Analysis

The theoretical equations (Table 1) and simulation results presented in previous sections can be represented graphically. Figure 5 shows the curves for passive elements estimated in relative units as a function of the input voltage.

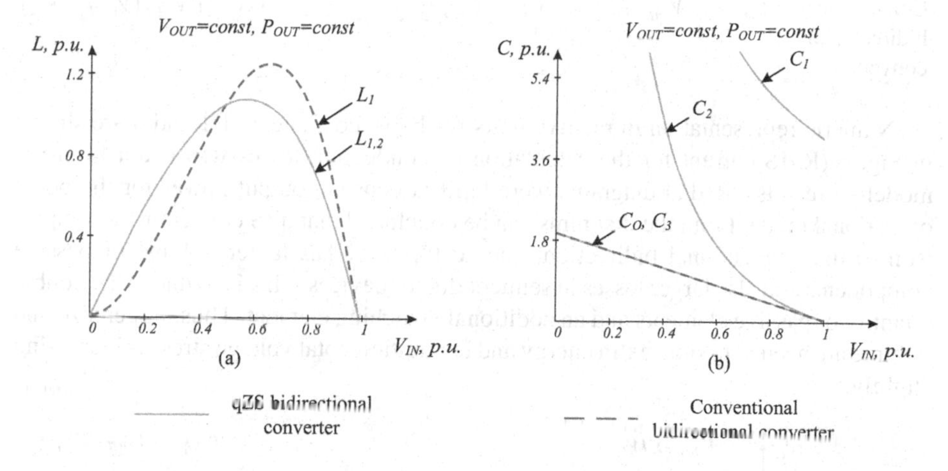


**Fig. 5.** Comparison of passive elements: inductances (a) and capacitances (b).

To estimate the contribution of power switches and passive elements to the volume of the converter, the following criteria were taken into consideration. First of all, voltage stress on semiconductors and conduction losses in semiconductors were considered:

$$T_W = \sum_{i=1}^{N_T} V_{Ti},\ P_{CL} = \sum_{i=1}^{N_T} I_{RMSi}^2 \times R_{DSoni}. \tag{13}$$

Energy stored in the inductors and maximum energy stored in the capacitors is proportional to their volume:

$$Vol_L \cong E_{LW} = \sum_{i=1}^{N_L} \frac{L_i I_{AVi}^2}{2},\ Vol_C \cong E_{CW} = \sum_{i=1}^{N_C} \frac{C_i V_{AVi}^2}{2}. \tag{14}$$

Based on the above formulas (13) and (14), results for each considered topology are presented in Table 2. The acquired expressions is valid only for the CCM operation mode.

**Table 2.** Summarized results of the analysis

| Topology | Voltage stress | | Inductors energy | | Capacitors energy | |
|---|---|---|---|---|---|---|
| | № | Voltage | № | Energy, p.u. | № | Energy, p.u. |
| qZS bidirectional converter | $T_1, T_2$ | $V_{OUT}$ | $L_1, L_2$ | $L\frac{8(V_{OUT}^2 - V_{IN}^2)P^2}{3V_{IN}V_{OUT}^3}$ | $C_1$ | $C\frac{V_{OUT}(V_{IN} - V_{OUT})^2}{V_{IN}}$ |
| | $T_3$ | $V_{OUT} - \frac{V_{IN}}{2}$ | | | $C_2$ | $C\frac{V_{OUT}^3(V_{OUT} - V_{IN})}{2V_{IN}(V_{OUT} + V_{IN})}$ |
| | | | | | $C_3$ | $CV_{OUT}(V_{OUT} - V_{IN})$ |
| Conventional bidirectional converter | $T_1, T_2$ | $V_{OUT}$ | $L_1$ | $L\frac{4(V_{OUT} - V_{IN})P^2}{V_{OUT}^3}$ | $C_O$ | $CV_{OUT}(V_{OUT} - V_{IN})$ |

Numeric representation in relative units for boost coefficients 1.1 and 4 are shown in Fig. 6 (RMS current for the calculation of conduction losses was taken from the modeling results). Radial diagrams were built at constant output power for the boost operational mode. From the diagrams can be concluded that qZS converter by comparison to the conventional bidirectional dc-dc topology has larger volume of passive components and also larger losses in semiconductor devices. That is so due to the double quantity of passive elements and an additional switching element. Thus, larger size and volume are needed to store extra energy and in addition, total voltage stress is increasing slightly.

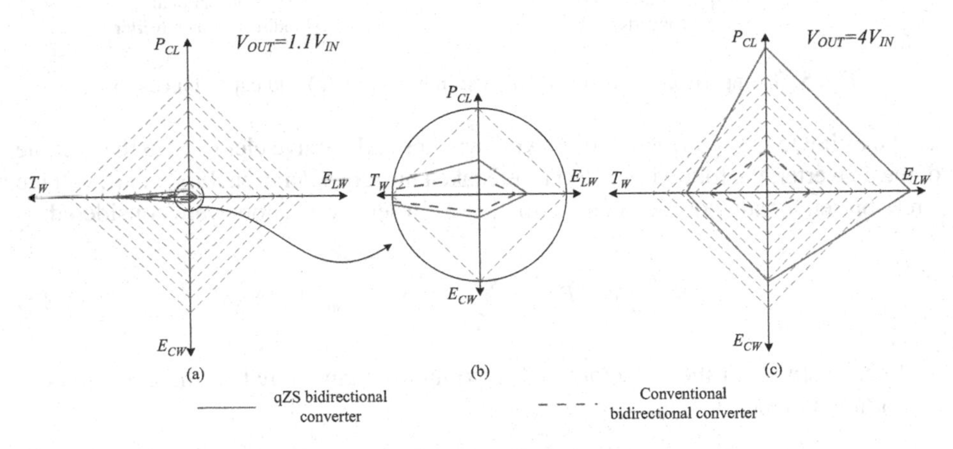


**Fig. 6.** Comparative analysis at constant output power: for VO = 1.1VIN (a) and (b); for VOUT = 4VIN (c).

Figure 6c shows that at high boost coefficients, the qZS converter shows even worse results than at low coefficients (Figs. 6a and b). Also, the losses in the qZS converter is higher. Though the duty cycle of the conducting transistor is twice reduced, the peak value of current is twice larger, and as a result, the RMS value of the current is larger.

## 6 Conclusions

This paper presents the comparative analysis of qZS and conventional bidirectional dc-dc topologies. In the analysis, four parameters were taken into account: energy stored in the inductive and capacitive passive elements (which are reflected in their volumes), voltage stress on the semiconductor devices and conduction losses.

The comparison shows that the characteristics of the qZS topology are worse than those for the conventional solution in almost all cases in the boost mode of operation. Only when the boost coefficient dropped close to 1.1, the topologies showed similar results. The reason lies in the larger number of passive components and semiconductors. As a result, the cost, volume and complexity are higher, which restricts the use of that topology in real applications. The conventional solution seems to be preferable in the application discussed.

**Acknowledgments.** This research work has been supported by Latvian National Research Programme “LATENERGI”, Latvian Council of Science (Grant 673/2014) and co-supported by Ukrainian Ministry of Education and Science (Grant № 0116U004695 and № 0116U006960).

## References

1. Renewables 2016 Global Status Report. Renewable Energy Policy Network for the 21st Century, REN21, France (2016)
2. Solar Energy White Paper, Where we are Now and What’s Ahead, Natcore Technology, New Orleans (2012)
3. The Future of Solar Energy, An Interdisciplinary MIT Study. Massachusetts Institute of Technology (2015)
4. IEA PVPS International Energy Agency Photovoltaic Power Systems Programme. IEA PVPS Report: 2014 Snapshot of Global PV Markets, March 2015. http://www.iea-pvps.org/fileadmin/dam/public/report/technical/PVPS_report_-_A_Snapshot_of_Global_PV_-_1992-2014.pdf
5. Carbone, R.: Energy Storage in the Emerging Era of Smart Grids. InTech, Rijeka (2011)
6. Howey, D.A., Mahdi Alavi, S.M.: Rechargeable Battery Energy Storage Systems Design. Wiley, Hoboken (2015)
7. Fernao Pires, V., Romero-Cadaval, E., Vinnikov, D., Roasto, I., Martins, J.F.: Power converter interface for electrochemical energy storage systems – a review. Energy Convers. Manag. **86**, 453–475 (2014)
8. Basic calculation of a boost converter’s power stage, Application report SLVA372C – November 2009 – revised January 2014
9. Liu, S.-I., Liu, J., Zhang, J.: Research on output voltage ripple of boost DC/DC converters. Int. MultiConference Eng. Comput. Scientists **2**, 1–5 (2008)
10. Chao, K.-H., Tseng, M.-C., Huang, C.-H., Liu, Y.-G.: Design and implementation of a bidirectional DC-DC converter for stand-alone photovoltaic systems. Int. J. Comput. Consum. Control **2**(3), 44–55 (2013)

11. Tytelmaier, K., Husev, O., Veligorskyi, O., Yershov, R.: A Review of non-isolated bidirectional DC-DC converters for energy storage systems. In: II International Young Scientists Forum on Applied Physics and Engineering, pp. 1–7 (2016)
12. Ardi, H., Ajami, A., Kardan, F., Nikpour, S.: Analysis and implementation of a non-isolated bidirectional DC-DC converter with high voltage gain. IEEE Trans. Power Electron. **63**(8), 4878–4888 (2016)
13. Mohammadi, M.R., Farzanehfard, H.: A new family of zero-voltage-transition nonisolated bidirectional converters with simple auxiliary circuit. IEEE Trans. Industr. Electron. **63**(3), 1519–1527 (2015)
14. Eichhorn, T.: Boost converter efficiency through accurate calculation. In: Power Electronics Technology (2008)
15. Siwakoti, Y.P., Peng, F.Z., Blaabjerg, F., Loh, P.C., Town, G.E.: Impedance-source network for electric power conversion Part I: a topological review. IEEE Trans. Power Electron. **30**(2), 699–716 (2015)
16. Zakis, J., Vinnikov, D., Roasto, I., Ribickis, L.: quasi-Z-Source inverter based bi-directional DC/DC converter: analysis of experimental result. In: 7th International Conference-Workshop Compatibility and Power Electronics, pp. 394–399 (2011)
17. Vinnikov, D., Roasto, I., Zakis, J.: New bi-directional DC/DC converter for supercapacitor interfacing in high-power application. In: 14th International Power Electronics and Motion Control Conference, pp. 38–43 (2010)
18. Husev, O., Blaabjerg, F., Roncero-Clemente, C., Romero-Cadaval, E., Vinnikov, D., Siwakoti, Y.P., Strzelecki, R.: Comparison of impedance-source networks for two and multilevel buck-boost inverter applications. IEEE Trans. Power Electron. **31**(11), 7564–7579 (2016)
19. Panfilov, D., Husev, O., Blaabjerg, F., Zakis, J., Khandakji, K.: Comparison of three-phase three-level voltage source inverter with intermediate DC-DC boost converter and quasi-Z-Source inverter. IET Power Electronics **9**(6), 1238–1248 (2016)
20. Liu, Y., Abu-Rub, H., Ge, B., Blaabjerg, F., Ellabban, O., Loh, P.C.: Impedance Source Power Electronics Converters. Wiley, IEEE Press, Hoboken (2016)

# Single-Phase Wireless Power Transfer System Controlled by Magnetic Core Reactors at Transmitter and Receiver

Luis Romba[1], Stanimir S. Valtchev[1(✉)], and Rui Melício[2,3(✉)]

[1] UNINOVA-CTS and Faculdade de Ciências e Tecnologia, Universidade Nova de Lisboa, Lisbon, Portugal
luis.rjorge@netvisao.pt, ssv@fct.unl.pt

[2] IDMEC, Instituto Superior Técnico, Universidade de Lisboa, Lisbon, Portugal
ruimelicio@gmail.com

[3] Departamento de Física, Escola de Ciências e Tecnologia, Universidade de Évora, Évora, Portugal

**Abstract.** The applications of wireless power transmission have become widely increasing over the last decade, mainly in the battery charging systems for electric vehicles. This paper focuses on the single-phase wireless power transfer prototype controlled by magnetic core reactors in either side of the system: that of the transmitter, and that of the receiver. The described wireless power transfer system prototype employs a strong magnetic coupling technol ogy to improve the power transmission efficiency. In the same time, a magnetic core reactor is used to control the "tuning" between the transmitter and the receiver frequencies, allowing for that increase of the system efficiency. Finally, practical results of the implemented prototype are presented.

**Keywords:** Wireless power transfer · Strongly magnetic coupling · Magnetic core reactor

## 1 Introduction

The Wireless Power Transfer (WPT) is a system that allows the transfer of electric power from a point in space to another point in space without any physical support, i.e., without wires or galvanic contact [1]. For relatively short and midrange distances, i.e., from a few millimeters to some meters, the transmission is performed mainly by magnetic field. The WPT systems used at short distances have a wide range of purposes. The applications start from battery charging of domestic appliances, mobile phones and laptops, even of Implantable Medical Devices (IMD), e.g. pacemakers [2, 3]. The performance of these devices is largely dependent on the high efficiency of the power transfer. The best way to optimize the power transfer efficiency is to increase the (magnetic) coupling between the two coils. This is one of the rules of thumb that should be applied to the WPT systems [4].

Due to the increased demand for the green energy systems, the WPT systems for midrange distances have had a significant development in the research related to battery charging applications for hybrid and electric vehicles. Hence, in the practice the


L.M. Camarinha-Matos et al. (Eds.): DoCEIS 2017, IFIP AICT 499, pp. 419–428, 2017.
DOI: 10.1007/978-3-319-56077-9_41

midrange battery charger uses the strongly coupled magnetic resonance technology [5]. The strongly coupled magnetic resonance technology is characterized by two pairs of control loops, one for the transmitter the other for the receiver (Fig. 1). The internal impedance of the Source Generator can be annotated later as $R_s$, and the impedance of the Load can be marked as further on as $R_l$. Those two impedances (real, at the resonant frequency) are excluded further on, from the $R\ L\ C$ resonators, leaving the resonator circuits independent of the generator and the load.

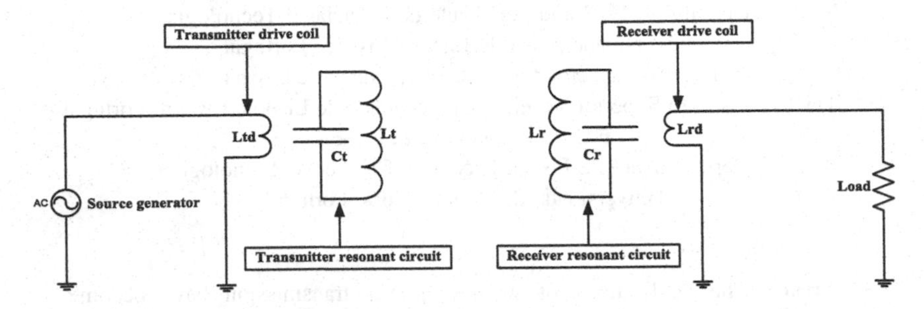


**Fig. 1.** Simplified equivalent circuit of the WPT system.

The equivalent circuit of the WPT using strongly coupled magnetic resonance technology is shown in Fig. 1.

The transmission efficiency, when a strongly coupled magnetic resonance technology is used, drops down with any directional misalignment. The maximum efficiency occurs when a correct impedance matching between the source and load will be achieved [6].

The impedance adjustment can be obtained in two ways. One way is by tuning the operating frequency, and the other is by keeping a fixed operating frequency but tuning the impedance matching between the transmitter and the receiver systems [6, 7]. The described WPT system includes two magnetic core reactors (MCR), one on each side of the system, i.e., the transmitter and the receiver, as shown in Fig. 2.

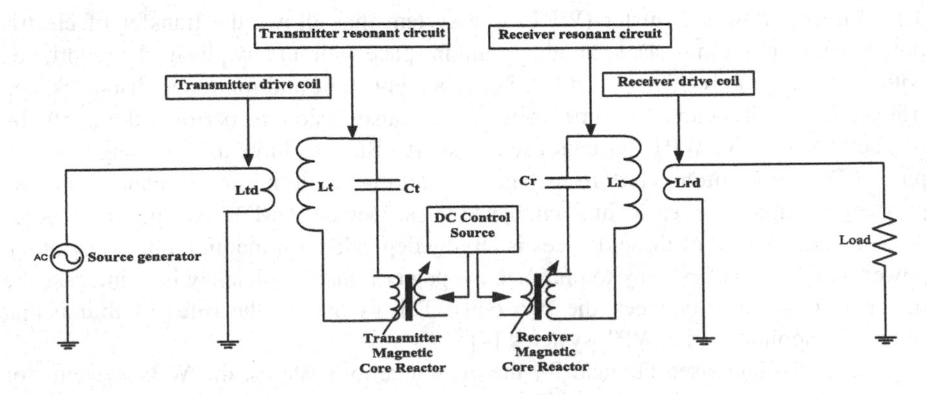


**Fig. 2.** WPT with magnetic core reactor in both sides.

The MCR is in fact, a part of the well-known in the past “magnetic amplifier” [8]. It is formed by two circuits, one is the control or primary circuit where a DC current is applied for controlling the secondary circuit, where the circulation of AC current is regulated [8]. When the control current (DC) changes, it causes a variation of the magnetic flux. This change in the magnetic flux produces a variation of the magnetic permeability of the circuit. If the circuit is working in the linear part of the curve of the hysteresis cycle, the variation of the magnetic flux is approximately constant [9]. However, if the operating point is above the knee of the curve of the hysteresis loop, the variation is always much more significant and the control is more efficient. This enables the frequency adjustment to be carried out either at the transmitter or the receiver of the WPT system, by changing the value of the inductance connected in series with the corresponding resonant coil, causing a change in the total circuit inductance [8, 9].

## 2 Relationship to Smart Systems

In contemporary societies the combination of nanoelectronics, embedded systems, cybernetics and intelligent systems, has and will continue to have a fundamental importance. That combination (smart system) is present in almost all modern products and services [10], being the decisive set of elements and key enabling technologies in many innovations and solutions for societal challenges, being also a very significant factor in the competitiveness, employment and prosperity of countries and regions [10].

The smart systems combine cognitive functions with sensing, actuation, data communication and energy management in an integrated way. The enabling principles of this combination are now the nanoelectronics, micro-electro-mechanics, magnetism, photonics, chemistry and radiation [11]. The smart systems are integrated with the (natural, built and social) environment, networks for power and data, other smart systems and the human beings [11].

The mobility sector is facing major challenges, e.g. the $CO_2$ emissions reduction, together with the implied improvement of air quality especially in large cities and the reduction of traffic congestion still using the existing infrastructure wherever possible.

The development is made possible by electronics, with new components and systems, as well as by introduction of new methods and tools. It has come to support the concepts that will provide new vehicles, transport systems and infrastructures with intelligence and flexibility to meet the challenges of industry in this sector [12]. The innovation, development and research related to the smart mobility has its emphasis on the capabilities in the areas of detection, communication, geo-positioning, computing and decision making and control based on Electronics, Components and Systems (ECS) [12]. ECS functions will lead to resource-efficient transportation as they enable partly or fully electrified as well as advanced conventional vehicles that are clean, $CO_2$-optimized and smartly connected to renewable energy sources [12].

Two examples of the development of specific applications in smart mobility are: smart systems for electro-mobility and smart systems for automated driving. In the first are included electric drive train, battery and battery management and charging technologies and the second is the included automated driving without the driver in the loop, on highways and in urban regions, but allowing the driver to intervene [11].

## 3 Modelling

The magnetic core circuit of a WPT system, when a correct alignment is achieved, is shown in Fig. 3. The coils are wound on the middle leg of each "E" part of the cores, upper "E" part corresponding to the primary, and the bottom part to the secondary. According to the structure of the magnetic core and the dimension of the air gap, the reluctance values are indicated as follows: $\Re_1, \Re_2, \Re_3\Re_4, \Re_5$ for the primary branches, and $\Re_6, \Re_7, \Re_8\Re_9, \Re_{10}$ are the secondary parts of the magnetic core reluctance. The $\Re_{ag1}, \Re_{ag2}, \Re_{ag3}$, are the reluctance elements that correspond to the air gap, and $\Re_{l1}, \Re_{l2}, \Re_{l3}\Re_{l4}$ are the magnetic leakage reluctance values of the primary and the secondary. The simplified equivalent magnetic circuit is shown in Fig. 4, where seven loops or sub-circuits are marked.

Based on this circuit (Fig. 3), the equation of the total reluctance can be deduced as:

$$\Re_{Total} = 2\frac{\ell_c}{\mu_0\mu_r S_c} + \frac{3\left(\frac{\ell_{ag}}{\mu_0 S_{ag}}\right)\left(\frac{\ell_\ell}{\mu_0 S_\ell}\right)^2 + 12\left(\frac{\ell_\ell}{\mu_0 S_\ell}\right)\left(\frac{\ell_{ag}}{\mu_0 S_{ag}}\right)\left(\frac{\ell_c}{\mu_0\mu_r S_c}\right) + 4\left(\frac{\ell_c}{\mu_0\mu_r S_c}\right)\left(\frac{\ell_\ell}{\mu_0 S_\ell}\right)^2}{2\left(\frac{\ell_\ell}{\mu_0 S_\ell}\right)^2 + 16\left(\frac{\ell_\ell}{\mu_0 S_\ell}\right)\left(\frac{\ell_c}{\mu_0\mu_r S_c}\right) + 6\left(\frac{\ell_\ell}{\mu_0 S_\ell}\right)\left(\frac{\ell_{ag}}{\mu_0 S_{ag}}\right) + 24\left(\frac{\ell_c}{\mu_0\mu_r S_c}\right)\left(\frac{\ell_{ag}}{\mu_0 S_{ag}}\right)} \tag{1}$$

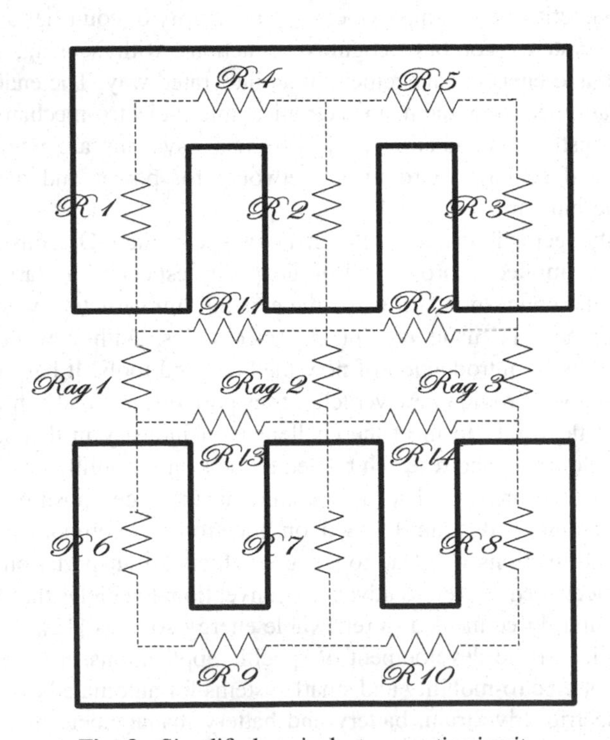


**Fig. 3.** Simplified equivalent magnetic circuit.

The result is shown in Fig. 4.

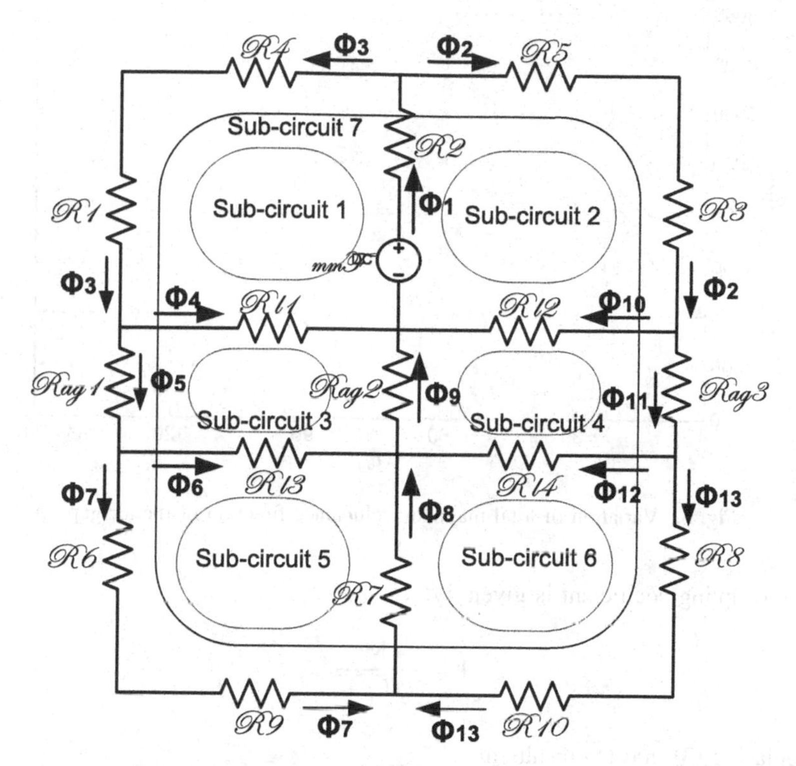


**Fig. 4.** Equivalent magnetic circuit with sub-circuits.

The variation of $\Re_{Total}$, as function of $\ell_{ag}$, fixed all the others parameters is shown in Fig. 5.

The mutual inductance, $M_{ps} = M_{sp} = M$ is given by:

$$M = \frac{N_p.N_s}{\Re_{Total}} \tag{2}$$

The result of the replacing of the (1) in (2) is given by:

$$M = \frac{N_p.N_s}{2\frac{\ell_c}{\mu_0\mu_r S_c} + \frac{3\left(\frac{\ell_{ag}}{\mu_0 S_{ag}}\right)\left(\frac{\ell_\ell}{\mu_0 S_\ell}\right)^2 + 12\left(\frac{\ell_\ell}{\mu_0 S_\ell}\right)\left(\frac{\ell_{ag}}{\mu_0 S_{ag}}\right)\left(\frac{\ell_c}{\mu_0\mu_r S_c}\right) + 4\left(\frac{\ell_c}{\mu_0\mu_r S_c}\right)\left(\frac{\ell_\ell}{\mu_0 S_\ell}\right)^2}{2\left(\frac{\ell_\ell}{\mu_0 S_\ell}\right)^2 + 16\left(\frac{\ell_\ell}{\mu_0 S_\ell}\right)\left(\frac{\ell_c}{\mu_0\mu_r S_c}\right) + 6\left(\frac{\ell_\ell}{\mu_0 S_\ell}\right)\left(\frac{\ell_{ag}}{\mu_0 S_{ag}}\right) + 24\left(\frac{\ell_c}{\mu_0\mu_r S_c}\right)\left(\frac{\ell_{ag}}{\mu_0 S_{ag}}\right)}} \tag{3}$$

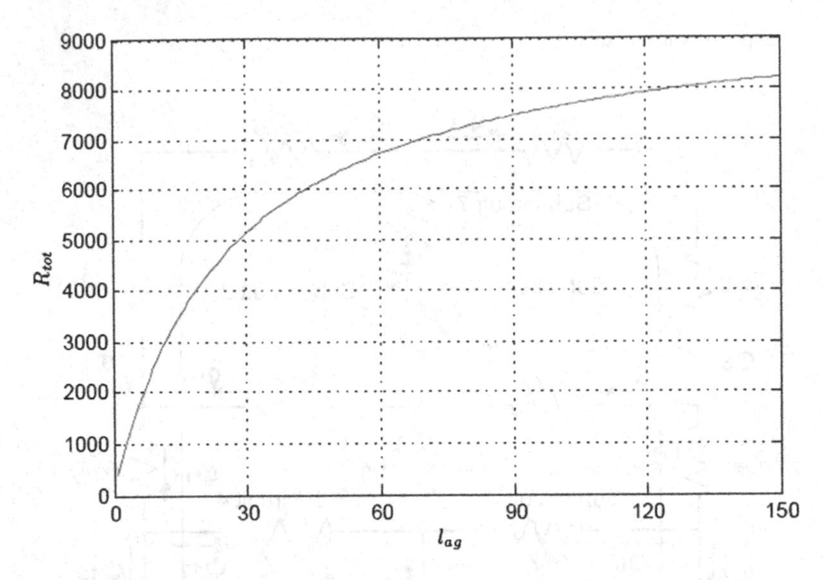


**Fig. 5.** Variation of total magnetic reluctance function of the air gap.

The coupling coefficient is given by:

$$k = \frac{M}{\sqrt{L_p.L_s}} \tag{4}$$

Replacing (3) into (4) results in:

$$k = \frac{N_p.N_s}{\sqrt{L_p.L_s}\left[\frac{2\ell_c}{\mu_0\mu_r S_c} + \frac{3\left(\frac{\ell_{ag}}{\mu_0 S_{ag}}\right)\left(\frac{\ell_\ell}{\mu_0 S_\ell}\right)^2 + 12\left(\frac{\ell_\ell}{\mu_0 S_\ell}\right)\left(\frac{\ell_{ag}}{\mu_0 S_{ag}}\right)\left(\frac{\ell_c}{\mu_0\mu_r S_c}\right) + 4\left(\frac{\ell_c}{\mu_0\mu_r S_c}\right)\left(\frac{\ell_\ell}{\mu_0 S_\ell}\right)^2}{2\left(\frac{\ell_\ell}{\mu_0 S_\ell}\right)^2 + 16\left(\frac{\ell_\ell}{\mu_0 S_\ell}\right)\left(\frac{\ell_c}{\mu_0\mu_r S_c}\right) + 6\left(\frac{\ell_\ell}{\mu_0 S_\ell}\right)\left(\frac{\ell_{ag}}{\mu_0 S_{ag}}\right) + 24\left(\frac{\ell_c}{\mu_0\mu_r S_c}\right)\left(\frac{\ell_{ag}}{\mu_0 S_{ag}}\right)}\right]} \tag{5}$$

## 4 Case Study

The global physical model presented in this paper was performed in Power Electronics Laboratory, Faculty of Science and Technology, New University of Lisbon and is shown in Fig. 6. The implemented schematic diagram is based on Fig. 2.

The transmitter coils set, the receiver coils set and the two magnetic core reactors, one for each circuit [13, 14] are shown in Fig. 7.

The tests were conducted step by step. In the first step a frequency of the function generator was selected (Fig. 2). This choice is a hardware simulation of the reality

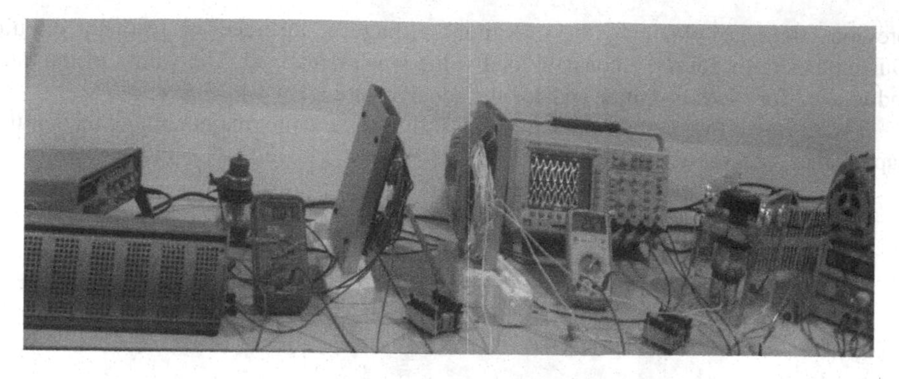

**Fig. 6.** Global physical model implemented.

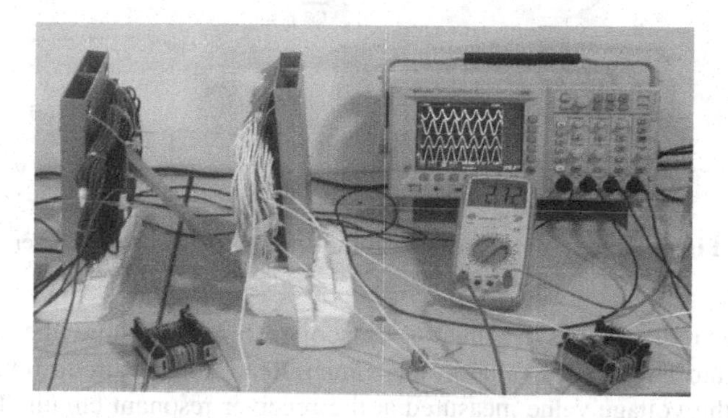

**Fig. 7.** Partial view implemented circuit with the two MCRs.

where the transmitter frequency is defined as the transmitter control has decided. The second step was to tune the transmitter resonant circuit by applying DC current to the transmitter magnetic core reactor until obtaining the maximum voltage that is observed at the resonant circuit. This step corresponds to the set point optimization. The third step was to tune the receiver resonant circuit following an identical procedure to the

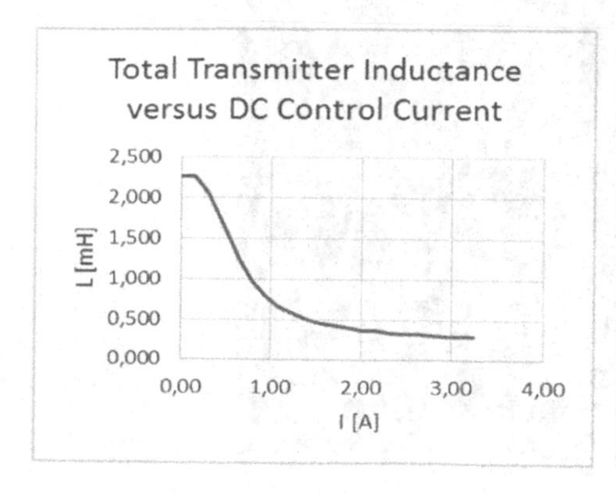


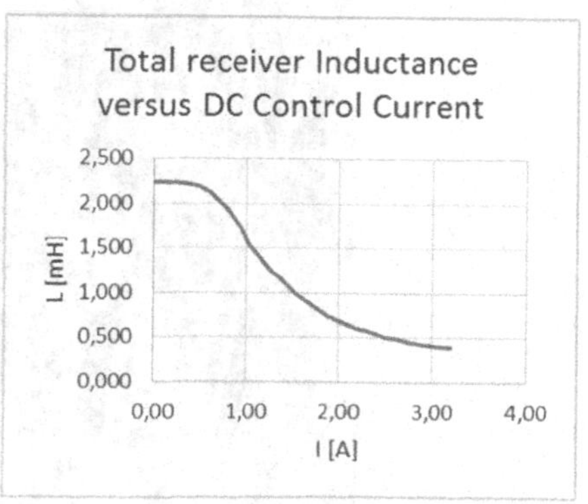


**Fig. 8.** The total inductance values to the transmitter and receiver.

previous step, i.e. obtaining the maximum voltage at the receiver resonant circuit. During those procedures, a constant load value was maintained. The values of the total inductance for the transmitter and for the receiver are shown in Fig. 8.

The resonant frequency values for each inductance value, maintaining constant the capacitance, are shown in Fig. 9.

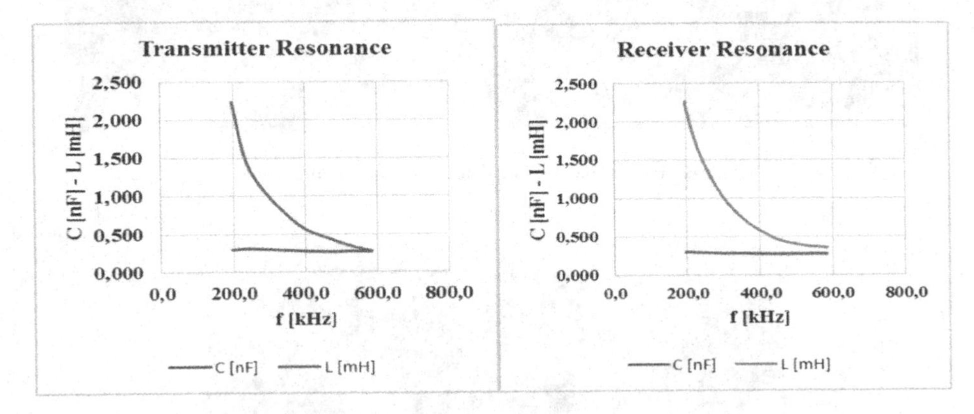


**Fig. 9.** The resonance values to the transmitter and the receiver.

The measured voltages by the oscilloscope (Fig. 10) show the input voltage (channel 1), and the voltage value in the transmitter resonant circuit (channel 2). The channel 3 is the voltage value measured at the receiver resonant circuit. The channel 4 is the voltage value applied to the load.

The voltage values in each test point of the circuit (channels 1, 2, 3, 4) are shown in Fig. 10.

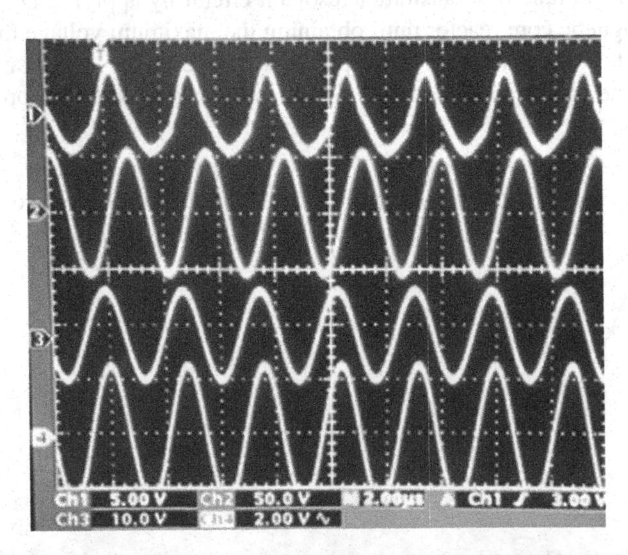


**Fig. 10.** The voltage values in each test point of the circuit (channels 1, 2, 3, 4).

## 5 Conclusions

In this paper a different approach is presented, applying the magnetic reluctance elements to determine the mutual inductance and the coupling coefficient. This process makes possible to obtain, with reasonable precision, the most important values in order to design a project of a wireless power transfer through a simple distances measurement. In the case study, it is shown a WPT implemented physical model in which a frequency control is presented using two magnetic core reactors. Through the curves shown in Figs. 8 and 9, it is possible to predict the DC current that will be necessary for controlling the magnetic core reactors, maintaining constant the resonant capacitor values, and in the same time, to tune the transmitter and the receiver at the same frequency. The stronger magnetic coupling, as presented in the circuits, shown in the Figs. 6 and 7, allows to obtain a sine wave at the output even if the input signal is not a pure sinusoidal wave. As it is shown in Fig. 10, it is made possible by the two resonant circuits mounted at each side: the transmitter (distorted, channel 1) and the receiver (channel 4).

## References

1. Ke, L., Yan, G., Yan, S., Wang, Z., Liu, D.: Improvement of the coupling factor of Litz-wire coil pair with ferrite substrate for transcutaneous energy transfer system. Prog. Electromagnet. Res. **39**, 41–52 (2014)
2. Park, S.I.: Enhancement of wireless power transmission into biological tissues using a high surface impedance ground plane. Prog. Electromagnet. Res. **135**, 123–136 (2013)
3. Jolani, F., Mehta, J., Yu, Y., Chen, Z.: Design of wireless power transfer systems using magnetic resonance coupling for implantable medical devices. Prog. Electromagnet. Res. Lett. **40**, 141–151 (2013)
4. Benjamin, L.C., Hoburg, J.F., Stancil, D.D., Goldstein, S.C.: Magnetic resonant coupling as a potential means for wireless power transfer to multiple small receivers. IEEE Trans. Power Electron. **24**(7), 1819–1825 (2009)
5. Qiu, C., Chau, K.T., Ching, T.W., Liu, C.: Overview of wireless charging technologies for electric vehicles. J. Asian Electr. Veh. **12**(1), 1679–1685 (2014)
6. Austrin, L., Engdahl, G.: Modeling of a three-phase application of a magnetic amplifier. In: 24th International Congress of the Aeronautical Sciences, pp. 1–8, Yokohama, Japan (2004)
7. Trinkaus, G.: Magnetic Amplifiers Another Lost Technology. High Voltage Press, USA (2000). Based on original work from U.S. Navy 1951
8. Xu, D.Q., Joós, G., Lévesque, M., Maier, M.: Integrated V2G, G2V, and renewable energy sources coordination over a converged fiber-wireless broadband access network. IEEE Trans. Smart Grid **4**(3), 1381–1390 (2013)
9. Fabbri, G., Tarquini, G., Pasquali, L., Anniballi, L., Odoardi, S., Teodori, S., Santini, E.: Impact of V2G/G2 V technologies on distributed generation systems. In: 23th IEEE International Symposium on Industrial Electronics, pp. 1–4, Istanbul, Turkey (2014)
10. Thales, L.G.: ARTEMIS-IA Pre-Brokerage Event, Brussels, Belgium (2014)
11. Neul, N.: Smart Systems Integration in the Context of ECSEL, Munich, Germany (2016)

12. Multi Annual Strategic Research and Innovation Agenda for ECSEL Joint Undertaking (2016)
13. Romba, L.F.R., Valtchev, S.S., Melício, R.: Improving magnetic coupling for battery charging through 3D magnetic flux. In: IEEE International Power Electronics and Motion Control Conference, pp. 291–297, Varna, Bulgaria (2016)
14. Romba, L.F.R., Valtchev, S.S., Melício, R.: Three-phase magnetic field system for wireless energy transfer. In: IEEE International Symposium on Power Electronics, Electrical Drives and Motion, pp. 73–78, Capri, Italy (2016)

# Soft-Switching Current-FED Flyback Converter with Natural Clamping for Low Voltage Battery Energy Storage Applications

Roman Kosenko[1,2(✉)] and Dmitri Vinnikov[1]

[1] Tallinn University of Technology, Tallinn, Estonia
roman.kosenko@ttu.ee

[2] Chernihiv National University of Technology, Chernihiv, Ukraine

**Abstract.** This paper introduces a new galvanically isolated current-fed step-up dc-dc converter intended for high voltage gain applications. The converter have fully-controllable voltage doubler rectifier, with control signals synchronous to that of the inverter switches. The proposed converter can regulate output voltage within wide range of the power and input voltage variations. Proposed converter does not require snubbers or resonant switches and with proposed control sequence ensures switches operation under soft-switching conditions in all transient states. Soft-switching in semiconductors allows achieving high efficiency. Moreover, the input side current is continuous. The operating principle for the energy transfer from the current-fed to voltage-fed side is described, design guidelines along with experimental verification of the proposed converter are shown in this paper. The converter proposed can be used as a front-end converter for grid connected battery storage.

**Keywords:** Current-fed converter · Soft-switching · Flyback converter · ZVS · ZCS · Switching sequence · Clampless · Energy storage

## 1 Introduction

New trends in European Union green energy policy encourage use of distributed power generation sources [1]. For more efficient use of alternative energy sources in residential applications it is preferable to include battery energy storage (BES) to those systems [2–5]. Battery energy storage systems (BESS) are used as buffer to store exceeded energy so that later it can help to suppress grid shortages or even inject power to the grid in case of high electric energy demand. This helps to minimize the impact of renewable power generation sources on electric grid stability [6].

BESSs in residential installations have power rating up to 8 kW. Battery chemistry for battery energy storage also varies significantly. In recent years lithium iron phosphate (LiFePO4) batteries are preferable for residential usage as they have one of the highest safety and reliability characteristics among lithium batteries. Another important feature is that they can preserve nominal capacity even after few thousands of discharge-charge cycles [7–9].


L.M. Camarinha-Matos et al. (Eds.): DoCEIS 2017, IFIP AICT 499, pp. 429–436, 2017.
DOI: 10.1007/978-3-319-56077-9_42

For connecting the BESS to the AC-grid the additional inverter stage is used. The intermediate DC-link of the system is traditionally formed by different voltage-fed (VF) DC-DC converters [10, 11]. One of the most reliable and widely used solution is dual active bridge (DAB) converter as they have soft switching capability and thus high operational performance [12]. Soft-switched current-fed (CF) converters with soft-switching capability is a topic of interest in recent years. Among them there are converters with additional clamping circuits [13, 14], with resonant-assisted transition [15], and others additional circuits [16, 17]. Additional advantage of current-source converters is continuous input current. This places them as preferable choice for battery powered applications [10, 11].

Fully-controllable switches at the rectifier side of isolated CF converters brings new challenges and benefits [18–26]. This paper introduces CF flyback DC-DC converter with novel regulation1 algorithm. This algorithm has duration of transient switching intervals that is constant in a whole operation range and so ensures soft switching across that range. High efficiency can be achieved due to minimizer dynamic losses even at relatively high operational frequencies.

## 2 Relationship to Smart Systems

Nowadays smart systems are used in all major sectors of human life such as healthcare, safety, automotive and environment. Typical smart system consist of diverse components that are usually physically organized in different physical nodes. And with the rapidly increasing interest in the concept of the Internet of things all those nodes and even entire systems have to be operating and connected to each-other all the time without any interruptions. To ensure this the most critical smart-systems accommodates backup battery-based power supplies. Battery energy storage in household is used to accumulate energy from renewable energy sources when it is not used or from the grid when it has a lowest price. After battery is charged it can be used when it is required by user or when grid is under heavy load and requires additional power. This means that battery energy storage together with smart learning system with energy demand tracking as a part of smart house can increase stability of local power supply system and decrease yearly average grid power usage. So a major research direction of power electronics in smart systems is the development of high-efficiency robust power converters for battery-storage.

## 3 General Description

Topology of the discussed converter is shown in Fig. 1. It can be considered as derivation of the CF dual-inductor push-pull topology [25] proposed earlier. It consists of the CF inverter with four-quadrant (4Q) switches at the input side and the controllable rectifier at the output VF side. Isolating transformer introduces additional step-up needed to interface low-voltage sources to the high voltage DC-link. Converter provides low level of ripple CF terminal that is faced to battery. By controlling the duration of the reactive energy circulation interval regulation of output voltage is achieved.

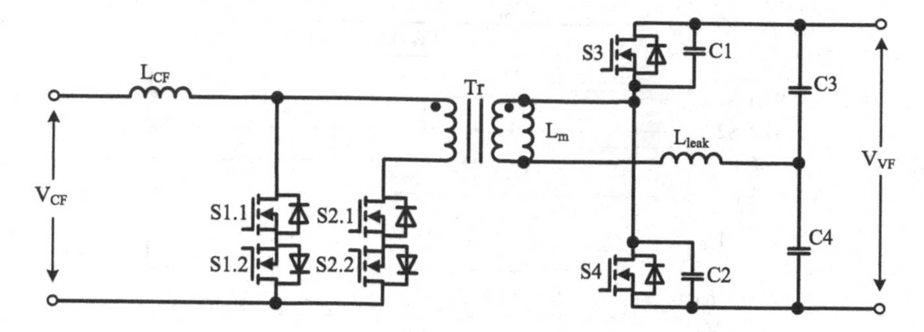


**Fig. 1.** Proposed converter topology.

Full soft-switching CF flyback converter topology can be used as interface power converter for connecting battery to the DC-grid or as the first conversion stage in AC-grid connected battery storage. In such applications the proposed topology and proposed control algorithm have the following distinguish features:

- High voltage gain that allows to use transformers with lower turns ratio to connect low voltage 12 V and 24 V batteries to a standard 400 V DC-grid.
- Current-fed terminal that ensures that the battery is operating with continuous input current. This allows increasing battery elements lifetime.
- Natural clamping of input inductor is achieved with the special control algorithm without any additional active or passive elements.
- Soft-switching is achieved in all switching elements through the wide operation range in both energy transfer directions.
- Reduced number of semiconductor devices as compared to the full-bridge boost converter (FSS-IFBBC) [24], and the same number of switches that in the dual inductor push-pull converter [25] and CF push-pull converter [26] that are using same control algorithm.
- High transformer utilization factor due to the double-ended operation with one part of the switching period (t0–t7) proposed topology operates similar to the flyback converter and during the other part (t7–t12) it operates similar to the forward converter.
- Transformer flux runaway protection capability without use of RCD snubbers or additional transformer windings.

During operation the transistors S1.2 and S2.2 are controlled synchronously to S1.1 and S2.1 and are turned off (acting like diodes) for transient intervals.

## 4 Converter Operation

This section describes converter operation with power flow from $V_{CF}$ to $V_{VF}$ terminal. Generalized control algorithm and simulated waveforms presented in Fig. 2. Switching period with the proposed control algorithm consists of twelve time intervals that are described in details further.

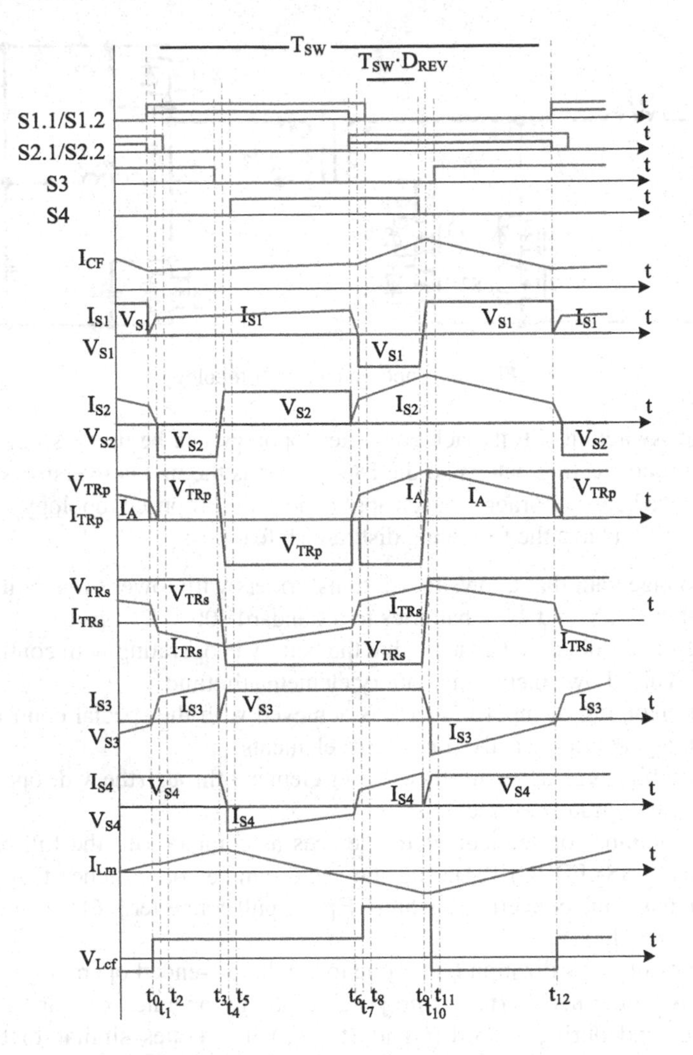


**Fig. 2.** Simulated operation waveforms.

**Interval 1** ($t_0 < t < t_1$). Interval starts with turn off of switch S2.2. Current is been redirected to the body diode (BD) of S2.2. S1 with assistance of the transformer leakage inductance $L_{leak}$ is turned on under soft-switching conditions. Leakage inductance acts like a snubber by slowing down current rise in S1 to separate voltage fall and current rise and thereby to minimize switching losses. The inductor $L_{CF}$ is storing energy. Switch S3 is kept on forcing the inverter side transformer current to transfer from S2 to S1.

**Interval 2** ($t_1 < t < t_2$). Interval starts after S2 current reaches zero and BD of S2.2 is turned off naturally. Inductor $L_{CF}$ is storing energy from CF terminal. Transformer-magnetizing inductance starts accumulating energy from C3 through S3. The duration of this interval is calculated so that the current through S1 will have sufficient time to drop to zero at minimal input voltage at nominal power.

**Interval 3** ($t_2 < t < t_3$). Interval starts with turning off switch S2.1 under ZCS conditions. Interval is needed to separate switching transitions in CF and VF sides and in most cases it can be omitted.

**Interval 4** ($t_3 < t < t_4$). Switch S3 is turned off with soft-switching assisted by snubber capacitor C1. Capacitor C1 is charging from zero to VVF and C2 is discharged from VVF to zero. The transformer output The voltage polarity at the transformer output is changed.

**Interval 5** ($t_4 < t < t_5$). Interval begins when the transformer current starts flowing through BD of S4. This interval is a part of power delivery mode from CF to VF terminal. Energy accumulated in the transformer magnetizing inductance $L_M$ is delivered to the output terminal through the BD of S4. Interval duration is Duration of this interval is calculated so that it is equal to the snubber capacitor recharge time in point of the minimal operational power. In this case soft-switching will be achieved in the whole operation range while the conduction losses across BD will be minimized.

**Interval 6** ($t_5 < t < t_6$). During this interval S4 turned on at ZVS conditions. Energy accumulated in $L_M$ is delivered to the VF terminal through switch S4.

**Interval 7** ($t_6 < t < t_7$). Switch transients on this interval are analogical to those **on interval 1**. 4Q switch S2 is turned on with soft-switching assisted by $L_{leak}$.

**Interval 8** ($t_7 < t < t_8$). Interval is analogical to **interval 2**. At the beginning of interval current through S2 reaches zero by those forcing BD of S2.1 to turn off naturally. Reverse energy transfer interval is started. LCF accumulating energy from VF (C4 through S3 and Tr) and CF terminals.

**Interval 9** ($t_8 < t < t_9$). During this interval S2.1 turned off under ZCS conditions. Processes in converter are analogical to **interval 8**. Duration of intervals 8 and 9 defines the converter gain factor.

Simplified converter voltage gain factor can be derived analogically to [25] and is expressed as follows:

$$F_{FRW} = \frac{V_{VF}}{2nV_{CF}} = \frac{2}{1 - 4 \cdot D_{REV}}, \quad (1)$$

where n is the transformer turns ratio.

**Interval 10** ($t_9 < t < t_{10}$) and **Interval 11** ($t_{10} < t < t_{11}$). Are analogical to intervals 4 and 5 accordingly. Starting from instant $t_{10}$ energy is delivered to transferred from the input to the output terminal through the transformer and body diode of S4 like in classical forward DC-DC converter.

**Interval 12** ($t_{11} < t < t_{12}$). At the instant $t_{11}$ switch S4 is turned on under ZVS conditions. Current is transferred from BD to switch thereby decreasing conduction losses.

## 5 Experimental Results

For the experimental study converter was designed to operate with a 4-cell $LiFePO_4$ battery was selected. Experimental prototype design parameters are shown in Table 1. Semiconductor devices parameters used are shown in Table 2.

**Table 1.** Converter parameters

| Parameter | Decsignator | Value |
|---|---|---|
| Nominal power, W | P | 100 |
| CF voltage, V | $V_{CF}$ | 10–15 |
| VF voltage, V | $V_{VF}$ | 400 |
| Operation frequency, kHz | $f_{SW}$ | 100 |
| Transformer turns ratio (Np/Ns) | N | 1:6 |
| Magnetizing inductance, mH | $L_{TX\_m}$ | 3.4 |
| Transformer secondary leakage ind., uH | $L_{TX\_leak}$ | 8 |
| Inductance of input inductors, uH | $L_{A, LB}$ | 44 |
| Capacitance of output filter, uF | $C_3, C_4$ | 2.2 |
| Capacitance of snubber, pF | $C_1, C_2$ | 265 |

**Table 2.** P semiconductor devices parameters

| Component | Device | Parameters |
|---|---|---|
| CF MOSFETs | Infineon BSC035N10NS5 | $V_{DS} = 100$ V; $R_{DS(on)} = 3.5$ mΩ $I_D = 100$A, $t_{rr} = 62$ ns, $C_{OSS} = 770$ pF |
| VF MOSFETs | Infineon IPB60R190C6 | $V_{DS} = 650$ V; $_{DS(on)} = 190$ mΩ $I_D = 59$ A, $t_{rr} = 430$ ns, COSS = 85pF |

Experimental data is shown in Table 3. The difference in theoretical and experimental voltage gains can be explained by the assumptions and simplifications used in theoretical equation.

**Table 3.** Experimental results forward operation

| VCF, V | DREV | Theoretical gain | Experim. gain | Losses, W | η, % |
|---|---|---|---|---|---|
| 10 | 0.104 | 3.42 | 3.33 | 7.89 | 92.11 |
| 11 | 0.086 | 3.05 | 3.03 | 6.29 | 93.71 |
| 12 | 0.069 | 2.77 | 2.78 | 5.39 | 94.61 |
| 13 | 0.051 | 2.51 | 2.56 | 4.53 | 95.47 |
| 14 | 0.033 | 2.31 | 2.38 | 3.81 | 96.19 |
| 15 | 0.017 | 2.15 | 2.22 | 3.66 | 96.34 |

## 6 Conclusions

A novel current-fed flyback converter was proposed and analyzed. One of the main features of proposed converter is that all semiconductors are operating under soft-switching conditions at all switching transients. Also proposed converter due to utilization of the circulating energy features higher DC voltage gain than in traditional current-fed converters. Moreover, due to the special control algorithm it futures natural clamping of input inductor convertor does not require any additional clamping circuits. Peak power stage efficiency during experimental verification reached 96.34%. Converter switching frequency can be increased without significant efficiency drop thanks to the soft-switching operation. This allowing to decrease size and price of magnetic components and so the price of BESS in general.

**Acknowledgments.** This work has been supported by the Estonian Research Council grant PUT744 and by the Estonian Centre of Excellence in Zero Energy and Resource Efficient Smart Buildings and Districts, ZEBE, grant 2014-2020.4.01.15-0016 funded by the European Regional Development Fund.

## References

1. Blaabjerg, F., Ionel, D.M.: Renewable energy devices and systems – state-of-the-art technology, research and development, challenges and future trends. Electr. Power Compon. Syst. **43**(12), 1319–1328 (2015)
2. Heymans, C., Walker, S.B., Young, S.B., Fowler, M.: Economic analysis of second use electric vehicle batteries for residential energy storage and load-levelling. Energy Policy **71**, 22–30 (2014)
3. Weniger, J., Tjaden, T., Quaschning, V.: Sizing of residential PV battery systems. Energy Procedia **46**, 78–87 (2014)
4. Chiang, S.J., Chang, K.T., Yen, C.Y.: Residential photovoltaic energy storage system. IEEE Trans. Ind. Electron. **45**(3), 385–394 (1998)
5. Chen, S.X., Gooi, H.B., Wang, M.Q.: Sizing of energy storage for microgrids. IEEE Trans. Smart Grid **3**(1), 142–151 (2012)
6. Wang, Z., Gu, C., Li, F., Bale, P., Sun, H.: Active demand response using shared energy storage for household energy management. IEEE Trans. Smart Grid **4**(4), 1888–1897 (2013)
7. Dunn, B., Kamath, H., Tarascon, J.: Electrical energy storage for the grid: a battery of choices. Science **334**, 928–935 (2011)
8. Stan, A.I., Świerczyński, M., Stroe, D.I., Teodorescu, R., Andreasen, S.J.: Lithium ion battery chemistries from renewable energy storage to automotive and back-up power applications—an overview. In: Proceedings of OPTIM 2014, pp. 713–720 (2014)
9. Whittingham, M.S.: History, evolution, and future status of energy storage. Proc. IEEE **100**, 1518–1534 (2012)
10. Pires, F., Romero-Cadaval, E., Vinnikov, D., Roasto, I., Martins, J.F.: Power converter interfaces for electrochemical energy storage systems – a review. Energy Convers. Manage. **86**, 453–475 (2014)
11. Kong, X., Choi, L.T., Khambadkone, A.M.: Analysis and control of isolated current-fed full bridge converter in fuel cell system. In: Proceedings of IECON 2004, vol. 3, pp. 2825–2830 (2004)
12. Tan, N.M.L., Abe, T., Akagi, H.: Design and performance of a bidirectional isolated DC–DC converter for a battery energy storage system. IEEE Trans. Power Electron. **27**(3), 1237–1248 (2012)
13. Prasana, U.R., Rathore, A.K.: Extended range ZVS active-clamped current-fed full-bridge isolated DC/DC converter for fuel cell applications: analysis, design, and experimental results. IEEE Trans. Ind. Electron. **60**(7), 2661–2672 (2013)
14. Andersen, R.L., Barbi, I.: A ZVS-PWM three-phase current-fed push-pull DC–DC converter. IEEE Trans. Ind. Electron. **60**(3), 838–847 (2013)

15. Chen, R.Y., et al.: Study and implementation of a current-fed full-bridge boost DC–DC converter with zero-current switching for high-voltage applications. IEEE Trans. Ind. Appl. **44**(4), 1218–1226 (2008)
16. Wang, H., et al.: A ZCS current-fed full-bridge PWM converter with self-adaptable soft-switching snubber energy. IEEE Trans. Power Electron. **24**(8), 1977–1991 (2009)
17. Chakraborty, D., Rathore, A.K., Breaz, E., Gao, F.: Parasitics assisted soft-switching and naturally commutated current-fed bidirectional push-pull voltage doubler. In: Proceedings of 2015 IEEE Industry Applications Society Annual Meeting, Addison, TX, pp. 1–8 (2015)
18. Xuewei, P., Rathore, A.K.: Naturally clamped zero-current commutated soft-switching current-fed push-pull DC/DC converter: analysis, design, and experimental results. IEEE Trans. Power Electron. **30**(3), 1318–1327 (2015)
19. Prasanna, U.R., Rathore, A.K., Mazumder, S.K.: Novel zero-current-switching current-fed half-bridge isolated DC/DC converter for fuel-cell-based applications. IEEE Trans. Ind. Appl. **49**(4), 1658–1668 (2013)
20. Chub, A., Kosenko, R., Blinov, A.: Zero-voltage switching galvanically isolated current-fed full-bridge DC-DC converter. In: Proceedings of CPE-POWERENG 2016, Vol. 5 (2016)
21. Nayanasiri, D.R., et al.: A switching control strategy for single- and dual-inductor current-fed push-pull converters. IEEE Trans. Power Electron. **30**(7), 3761–3771 (2015)
22. Blinov, A., Vinnikov, D., Ivakhno, V.: Full soft-switching high stepup dc-dc converter for photovoltaic applications. In: Proceedings of EPE 2014 - ECCE Europe, pp. 1–7, November 2014
23. Kosenko, R., Husev, O. Chub, A.: Full soft-switching high step-up current-fed DC-DC converters with reduced conduction losses. In: Proceedings of POWERENG 2015, pp. 170–175, 11–13 May 2015
24. Chub, A., et al.: Full soft-switching bidirectional current-fed DC-DC converter. In: Proceedings of RTUCON 2015, pp. 1–6, 14 October 2015
25. Kosenko, R., et al.: Full soft-switching bidirectional isolated current-fed dual inductor push-pull DC-DC converter for battery energy storage applications. In: 57th International Scientific Conference on Power and Electrical Engineering of Riga Technical University (RTUCON 2016), Riga and Cesis, Latvia, 13–14 October 2016
26. Kosenko, R., Chub, A., Blinov, A.: Full-soft-switching high step-up bidirectional isolated current-fed push-pull DC DC converter for battery energy storage applications. In: The 42nd Annual Conference of IEEE Industrial Electronics Society (IECON 2016), Florence, Italy, 24–27 October 2016

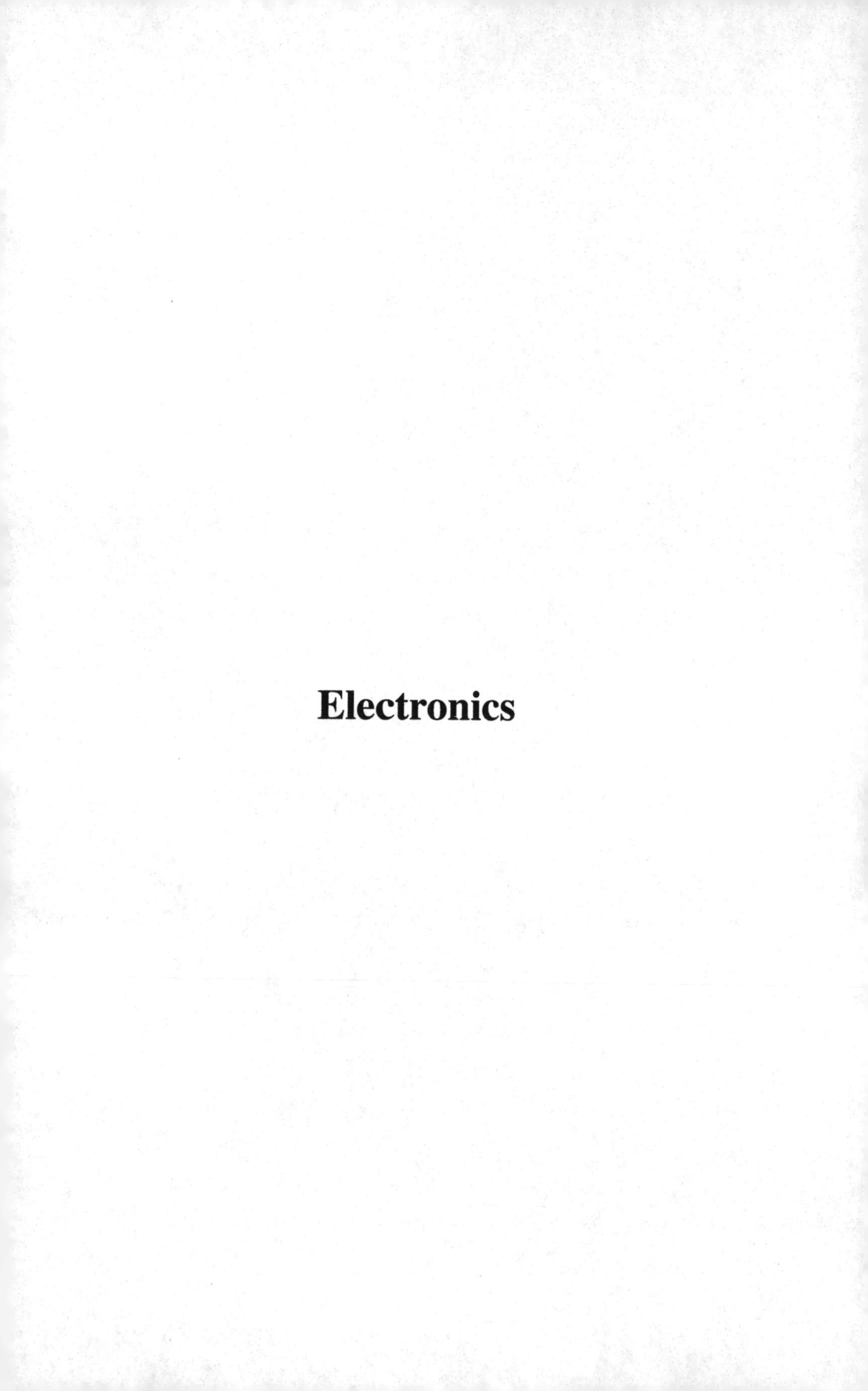

# Electronics

# Design Methodology for an All CMOS Bandgap Voltage Reference Circuit

Ricardo Madeira(✉) and Nuno Paulino

Centre for Technologies and Systems (CTS) – UNINOVA, Department of Electrical Engineering (DEE), Universidade Nova de Lisboa (UNL), 2829-516 Caparica, Portugal
r.madeira@campus.fct.unl.pt, nunop@uninova.pt

**Abstract.** The Internet of Thing (IoT) has given rise to the integration of smart systems in the industrial, healthcare, and social environments. These smart systems are often implemented by system-on-chip (SoC) solutions that require power management units, sensors, signal processors, and wireless interfaces. Hence, an independent voltage reference circuit is crucial for obtaining accurate measurements from the sensors and for the proper operation of the SoC. Making, bandgap circuits indispensable in these types of applications. Typically, bandgap circuits are implemented using bipolar transistors to generate two voltages with opposite temperature coefficients (Complementary To Absolute Temperature – CTAT; Proportional do Absolute Temperature - PTAT) which are added resulting in a temperature independent voltage reference. The disadvantage of using bipolar devices is that the power supply voltage must be larger than the ON base-emitter voltage, resulting in voltages larger than 0.7 V. The low power demands of IoT and of technology scaling, have forced lower values for the power supply voltage and thus new bandgap circuits using only CMOS transistors have gathered increased interest. In these, the MOS transistors operate in the weak inversion region where its current has an exponential relation with its gate-to-source voltage, as in the bipolar devices. Making it possible to generate both the PTAT and the CTAT voltages and thus produce a temperature independent voltage reference. This paper describes a design methodology of an all CMOS bandgap voltage reference circuit, in which one of the transistors works in the moderate region and the other in the weak inversion, to achieve the lowest possible voltage variation with temperature. The circuit produces a voltage reference of 0.45 V from a minimum power supply voltage of 0.6 V, with a total variation of 1.54 mV, over a temperature range of −40 to 100°C, resulting in a temperature coefficient of 24 ppm/°C, and a power supply rejection ratio (PSRR) of −40 dB.

**Keywords:** Low-voltage CMOS bandgap voltage reference · Analog design methodology · Weak inversion operation

## 1 Introduction

Reference voltage generators are required in almost any Internet of Thing (IoT) System on Chip (SoC). To ensure the proper behavior of these systems, these reference voltages


L.M. Camarinha-Matos et al. (Eds.): DoCEIS 2017, IFIP AICT 499, pp. 439–446, 2017.
DOI: 10.1007/978-3-319-56077-9_43

should be stable, independently of variations in: environment temperature, supply voltage, and fabrication process. The bandgap reference (BGR) is the most commonly voltage reference generator used [1]. Typically, a bandgap produces a constant reference voltage by summing two temperature dependent voltages – one Proportional to Absolute Temperature (PTAT) and the other Complementary to Absolute Temperature (CTAT). Most of the bandgaps uses the $V_{BE}$ of a bipolar transistor to generate both CTAT and PTAT voltages [1–3]. However, due to the technology scaling and the need to reduce power consumption, the supply voltage has been greatly reduced. At supply voltages bellow 1 V, bipolar transistors are not suitable since they need a minimum $V_{BE}$ value of 0.7 V to be turned ON. Therefore, to keep pace with this decrease in the power supply several all CMOS bandgap reference voltage circuits have been proposed [4–8]. In these new topologies, the PTAT and CTAT voltages are generated by using a MOS transistor in the weak inversion (or subthreshold region). In this region the transistor current depends exponentially on the gate-to-source voltage ($V_{gs}$), similar to the bipolar transistor. This paper describes a complete design methodology and simulation test of an all CMOS bandgap, in order to achieve the best performance.

## 2 Relationship to Smart Systems

The smart systems together with IoT play an important role on the new generation of information technology and it is predicted to have a major impact in the human life in the near future. This relatively recent technology aims to enable things to be connected anytime, anywhere, with anything and anyone ideally using any path/network or service. Such systems must be small, low power, and durable, hence they are often implemented by SoC solutions, that contain power management units, sensors, signal processors, and wireless transceivers. All this circuits and sensor require constant reference voltages so that they can work properly and obtain accurate and precise measurements. Moreover, since smart systems can be placed in any kind of environment, these reference voltages must be stable and robust even in harsh environments where the temperature variation can be high.

## 3 Proposed Circuit and Design Methodology

Figure 1 shows the simplified schematic of the all CMOS bandgap voltage reference circuit [5], where the operational amplifier was implemented using a simple two stage miller compensated amplifier. The diode connected transistors $M_{5,8}$ operate in the weak inversion region where their current varies exponentially with $V_{gs_{5,8}}$. Thus, $V_{gs_{5,8}}$ is given by (1) where $I_{D0}$ is the technology current, $n$ is the substrate value, $V_T$ is the thermal voltage, $L$ and $W$ are the transistor length and width, and $V_{th}$ is the threshold voltage. For the 130 nm CMOS technology the value of $I_{D0}$ and $n$ for an NMOS transistor were determined by simulation and are 0.9 μA, and 1.17, respectively.

$$V_{gs} = nV_T \ln\left(\frac{I_D L}{I_{D0} W}\right) + V_{th} \quad (1)$$

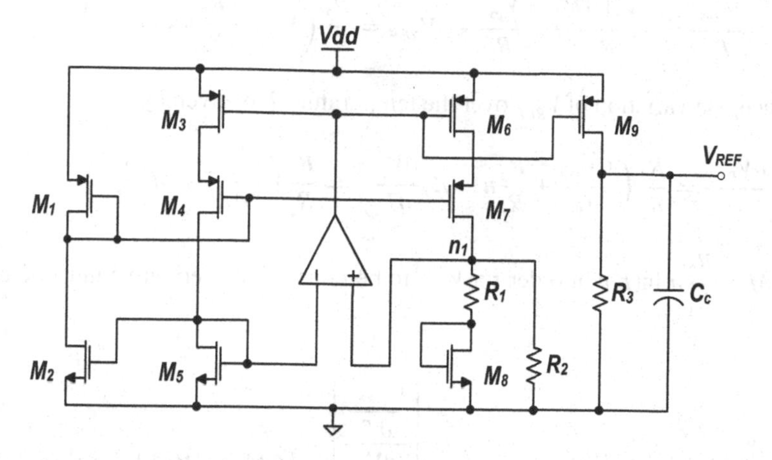


**Fig. 1.** Simplified schematic of the proposed bandgap circuit.

The transistors $M_{5,8}$ along with the resistances $R_{1,2}$ are responsible for the generation of a PTAT and CTAT current. The values of these currents can be derived by applying KCL to node $n_1$ (2), where it is assumed that the amplifier gain is high so that $V^- = V^+$, and that the current $I_D$ is equal in all the four branches of the circuit and so it can be derived through Ohm's law (3).

$$\frac{V_{gs_5} - V_{gs_8}}{R_1} + \frac{V_{gs_5}}{R_2} - I_n = 0 \quad (2)$$

$$I_D = \frac{V_{REF}}{R_3} \quad (3)$$

Considering $L_5 = L_8 = L$, $V_{th5} = V_{th8}, I_{D0_5} = I_{D0_8} = I_{D0}, n_5 = n_8 = n, I_{D5} \approx I_{D8} \approx I_D$ and $W_5 = KW_8$, then $V_{gs_5} - V_{gs_8}$ from (2) can be replaced by (1), resulting in

$$\begin{aligned} V_{gs_5} - V_{gs_8} &= n_5 V_T \ln\left(\frac{I_{D5}L_5}{I_{D0_5}W_5}\right) + V_{th_5} - n_8 V_T \ln\left(\frac{I_{D8}L_8}{I_{D0_8}W_8}\right) - V_{th_8} \\ &= nV_T \ln\left(\frac{\dfrac{I_{D5}L_5}{I_{D0_5}W_5}}{\dfrac{I_{D8}L_8}{I_{D0_8}W_8}}\right) = nV_T \ln\left(\frac{\dfrac{I_D L}{I_{D0}W_5}}{\dfrac{I_D L}{I_{D0}KW_5}}\right) = nV_T \ln(K) \end{aligned} \quad (4)$$

Replacing (4) and (3) in (2) results

$$\frac{nV_T \ln(K)}{R_1} + \frac{V_{gs_5}}{R_2} - \frac{V_{REF}}{R_3} = 0$$
$$\leftrightarrow \frac{V_{REF}}{R_3} = \frac{nV_T \ln(K)}{R_1} + \frac{V_{gs_5}}{R_2} \leftrightarrow V_{REF} = \frac{R_3}{R_2}\left(V_{gs_5} + \frac{R_2}{R_1} nV_T \ln(K)\right) \tag{5}$$

Hence, the variation of $V_{REF}$ over the temperature $T$ is given by

$$\frac{\partial V_{REF}}{\partial T} = \frac{R_3}{R_2}\left(\frac{\partial V_{gs_5}}{\partial T} + \frac{R_2}{R_1} n \ln(K) \frac{\partial V_T}{\partial T}\right) = \frac{R_3}{R_2}\left(\frac{\partial V_{gs_5}}{\partial T} + M \frac{\partial V_T}{\partial T}\right) \tag{6}$$

where $M = \frac{R_2}{R_1} n \ln(K)$. In order for $V_{REF}$ to be constant its derivative must be equal to zero,

$$\frac{R_3}{R_2}\left(\frac{\partial V_{gs_5}}{\partial T} + M \frac{\partial V_T}{\partial T}\right) = 0 \leftrightarrow M = \left|\frac{\frac{\partial V_{gs_5}}{\partial T}}{\frac{\partial V_T}{\partial T}}\right| = \left|\frac{634.5 \times 10^{-6}}{86.3 \times 10^{-6}}\right| = 7.35 \tag{7}$$

where $\partial V_T / \partial T = k/q = 86.3 \mu V/°C$, in which $k$ is the Boltzmann constant ($k = 1.38 \times 10^{-23}$ J/K) and $q$ is the elementary charge ($q = 1.60 \times 10^{-19} C$). By simulating a NMOS diode connected transistor biased by a current (Fig. 2). source it possible to determine the $V_{gs}$ variation with temperature (Fig. 3). The current was set to 500 nA and $L = 10$ µm, the temperature was swept from −40 to 100°C for different $W$ values (1 µm, 5 µm, 10 µm, and 20 µm). As expected from (1) as $W$ decreases $V_{gs}$ increases. Also, $V_{gs}$ decreases with the increase in the temperature, i.e. it has a negative temperature coefficient like the bipolar transistors. For the chosen current and $L$ values, and knowing that $V_{th} = 0.220$ V (extracted by simulation), then $W$ must be higher than 5 µm to be in

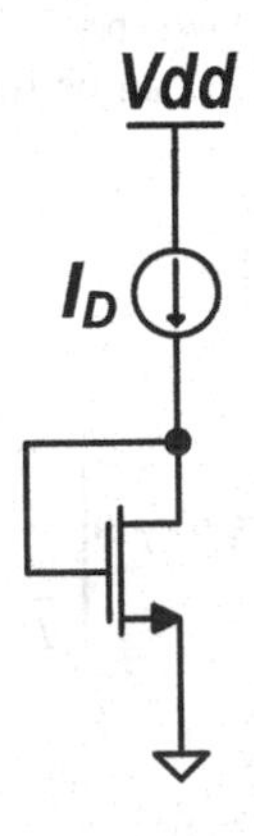


**Fig. 2.** Simplified schematic of the test bench to determine $\partial V_{gs} / \partial T$.

the moderate region ($V_{gs} \approx V_{th}$) or weak inversion ($V_{gs} < V_{th}$) using even higher $W$ values. Table 1 shows the slope, i.e. $\partial V_{gs}/\partial T$, for each $W$ value from Fig. 3. Using the value of Table 1 obtained through simulation for $W_5 = 5\ \mu m$ results in $M = 7.35$ (7).

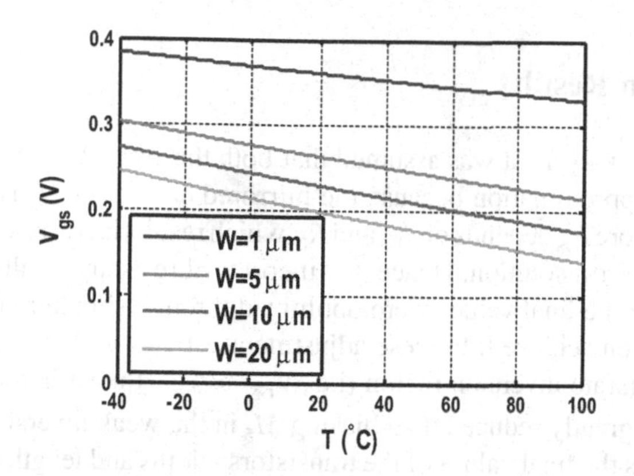


**Fig. 3.** Simulation results of the $V_{gs}$ as a function of the temperature for different $W$ using the test bench of Fig. 2 with $I_D = 500$ nA and $L = 10\ \mu m$.

**Table 1.** Simulation results of $\partial V_{gs}/\partial T$ of an NMOS transistor with $L = 10\ \mu m$ and $I_D = 500$ nA.

| $W$(μm) | $\frac{\partial V_{gs}}{\partial T}$ (μV/°C) |
|---|---|
| 1 | −395.6 |
| 5 | −634.5 |
| 10 | −697.7 |
| 20 | −750.4 |

The sizing of the resistors $R_{1,2,3}$ was performed using the same method used in [2]. Where the total resistance value was set according with the area constraints which resulted in a total resistance ($R_{total}$) value of 4 MΩ. Then, at room temperature ($T = 300$ K), $V_{gs}(W = 5\ \mu m) = 258$ mV, $V_T = 0.026$ V, $V_{REF} = 0.45$ V, $K = 4$ (for employing common centroid techniques), then

$$V_{REF} = \frac{R_3}{R_2}\left(V_{gs_5} + MV_T\right) \leftrightarrow 0.45 = \frac{R_3}{R_2}(0.258 + 7.35 \times 0.026) \quad (8)$$

$$M = \frac{R_2}{R_1} n \ln(K) \leftrightarrow 7.35 = \frac{R_2}{R_1} \times 1.17 \times \ln(4) \quad (9)$$

$$R_{total} = R_1 + R_2 + R_3 = 4 \times 10^6 \Omega \quad (10)$$

Solving (8), (9), and (10) results in

$$R_1 = 397\,\text{K}\Omega$$
$$R_2 = 1.80\,\text{M}\Omega \quad (11)$$
$$R_3 = 1.80\,\text{M}\Omega$$

## 4 Simulation Results

In the theoretical analysis, it was assumed that both the current in $M_5$ and $M_8$ are equal. This is a rough approximation because the mirrored current of $M_5$ is divided between $M_8$ and $R_2$. Therefore $I_{D8}$ depends on $I_{D5}$ and $R_2$, which results in a transcendental equation without a closed-form solution. Hence, the theoretical resistance values were used as a starting point and the final values were optimized through simulation until a minimum $\partial V_{REF}/\partial T$ had been achieved. In these adjustments, it was verified that by keeping $M_5$ closer to the moderate inversion region (i.e., $V_{GS_5} \approx V_{th}$), the variation of $V_{REF}$ over the temperature was greatly reduced than biasing $M_5$ in the weak inversion.

Table 2 shows the final values of the transistors widths and lengths, the resistors and capacitors values. Where the capacitance $C_C$ is implemented by a MIM capacitor that can be placed on top of the bandgap resistors. Hence, its area was chosen to be equal to the total resistance area, resulting in a 20 pF capacitor

**Table 2.** Transistors, resistors, and capacitor final sizing values.

| **# Transistor** | $(W/L)$(μm) |
|---|---|
| $M_1$ | 0.64/10 |
| $M_3, M_6, M_9$ | 3/0.75 |
| $M_4, M_7$ | 3/3 |
| $M_2, M_5$ | 5/10 |
| $M_8$ | $4 \times 5/10$ |
| **# Resistor** | MΩ |
| $R_1$ | 0.260 |
| $R_2$ | 2.455 |
| $R_3$ | 1.281 |
| **# Capacitor** | $pF$ |
| $C_c$ | 20 |

Figure 4 shows the simulation results of $V_{REF}$ as a function of the temperature for $V_{DD} = 0.9\,\text{V}$ The maximum and minimum value of $V_{REF}$ are $451.69\,\text{mV}$ and $450.16\,\text{mV}$, respectively. Resulting in a total variation of $1.54\,\text{mV}$ and, a temperature coefficient ($TC = (\Delta V_{REF} \cdot 1 \times 10^6)/(V_{REF \cdot 27^\circ C}\Delta T)$) of 24 ppm/°C. Figure 5 shows the simulation results of $V_{REF}$ when sweeping the supply voltage from 0 to 1.2 V at 27°C. The smallest $V_{DD}$ voltage for the bandgap to provide a stable reference voltage is 0.6 V. Lastly, Fig. 6 shows the Power Supply Rejection Ratio performance for $V_{DD} = 0.9\,\text{V}$ at 27°C. The highest PSRR value of −40 dB occurs for frequencies bellow 20 kHz.

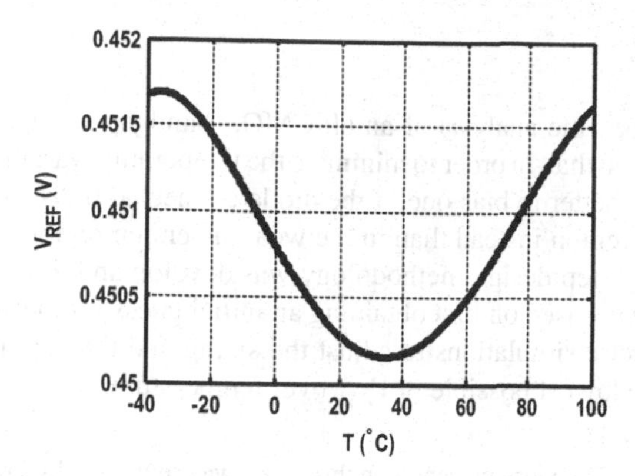


Fig. 4. Simulation results of $V_{REF}$ as a function of the temperature for $V_{DD} = 0.9$ V.

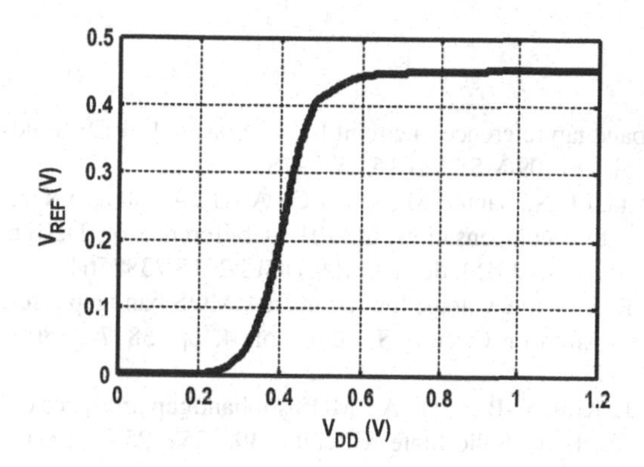


Fig. 5. Simulation results for $V_{REF}$ as a function of $V_{DD}$ at 27°C.

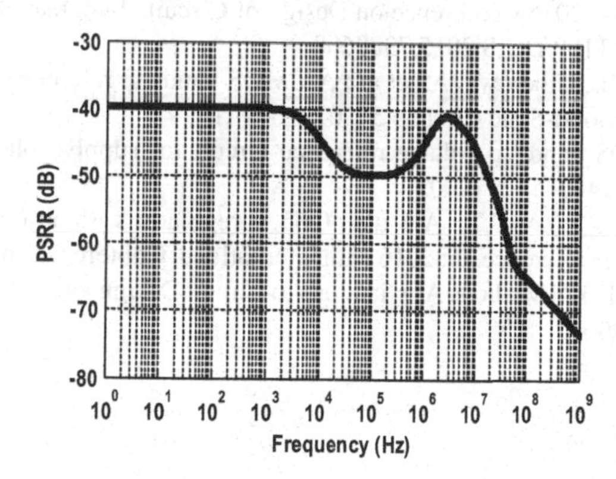


Fig. 6. Simulation results of the power supply rejection ratio (PSRR) of the bandgap circuit for $V_{DD} = 0.9$ V and $T = 27\,°C$.

## 5 Conclusion

This paper described the analysis of an all CMOS bandgap voltage reference circuit. This analysis showed that, in order to minimize the temperature variation of the reference voltage value, it is better to bias one of the diode connected transistors of the bandgap in the moderation region instead than in the weak inversion region. From this analysis a complete step-by-step design methodology was develop and described in this paper. This methodology is based on first obtaining an initial guess for the sizes of the devices and the using Spectre simulations to adjust the sizing and thus optimize the circuit in order to obtain the lowest possible of $V_{REF}$ over temperature.

**Acknowledgments.** The work presented in this paper was supported by Proteus project funded by the European Union's H2020 Programme under grant agreement no. 644852.

## References

1. Razavi, B.: The bandgap reference [a circuit for all seasons]. IEEE Solid-State Circuits Mag. **8**, 9–12 (2016). doi:10.1109/MSSC.2016.2577978
2. Pereira, M.S., Costa, J.E.N., Santos, M., Vaz, J.C.: A 1.1 μA voltage reference circuit with high PSRR and temperature compensation. In: 2015 Conference on Design Circuits Integrated Systems (DCIS), pp. 1–4, (2015). doi:10.1109/DCIS.2015.7388564
3. Jiang, Y., Lee, E.K.F.: A low voltage low 1/f noise CMOS bandgap reference. In: 2005 IEEE International Symposium on Circuits System, vol. 4, pp. 3877–3880 (2005). doi:10.1109/ISCAS.2005.1465477
4. Doyle, J., Lee, Y.J., Kim, Y.-B., et al.: A CMOS subbandgap reference circuit with 1-V power supply voltage. IEEE J. Solid-State Circuits **39**, 252–255 (2004). doi:10.1109/JSSC.2003.820882
5. Quendera, F., Paulino, N.: A low voltage low power temperature sensor using a 2nd order delta-sigma modulator. In: 2015 Conference on Design of Circuits Integrated Systems (DCIS), pp. 1–6 (2015). doi:10.1109/DCIS.2015.7388608
6. Banba, H., Shiga, H., Umezawa, A., et al.: A CMOS bandgap reference circuit with sub-1-V operation. IEEE J. Solid-State Circuits **34**, 670–674 (1999)
7. Ytterdal, T.: CMOS bandgap voltage reference circuit for supply voltages down to 0.6 V. Electron. Lett. **39**, 1427–1428 (2003)
8. Lin, F.T., Tsai, J.H., Liao, Y.T.: A 3 μW, 0.65 V regulator with an embedded temperature compensated voltage reference. In: 2016 13th International Conference on Synthesis, Modeling and Analysis Simulation Methods Applications to Circuit Design, pp. 1–4 (2016). doi:10.1109/SMACD.2016.7520647

# Reconfigurable Photonic Logic Architecture: An Overview

Vitor Silva[1,2(✉)], Manuel Barata[1,2], and Manuela Vieira[1,2,3]

[1] Electronics Telecommunication and Computer Department, ISEL Instituto Superior de Engenharia de LisboaIPL, Instituto Politécnico de Lisboa, 1959-007 Lisbon, Portugal
vsilva@deetc.isel.ipl.pt

[2] CTS-UNINOVA, Quinta da Torre, Monte Caparica, 2829-516 Caparica, Portugal

[3] DEE-FCT-UNL, Quinta da Torre, Monte Caparica, 2829-516 Caparica, Portugal

**Abstract.** The photosensor studied in this document, is an amorphous silicon structure deposited on transparent glass allowing illumination on both sides. It responds to wavelengths from near infrared range to the ultra-violet. The front illumination surface is used for inputting light signal. The dynamic characteristics of the photosensor are altered by using optical bias on either surface of the sensor, thus the same input results in different outputs. Experimental studies made with the photosensor evaluate its applicability as a multiplexer, demultiplexer and logical operations device. A memory effect was observed. A programmable pattern emission system, and a validation and recovery system were built to illuminate the sensor with input light signals and to analyze the resulting output.

**Keywords:** Optoelectronics · Photosensor · a-SiC technology · Pin · Amorphous silicon · Logical functions · Digital light signal

## 1 Introduction

Research work of the a-Si:H, a-SiC:H pi'npin photosensor [1] has shown that it has a nonlinear property of changing its bandwidth not only with the electrical bias but also a change of its bandwidth occurs with a light shining over the sensor, in either surface, despite the applied electrical bias [2]. Several applications have been done or proposed and the possibility of using this sensor for communication is studied even though the throughput is presently 12000 bps per channel.

This suggests that the channel selection which is made by digital switching over electrical channels to be de-multiplexed can, with this approach, be selected at the sensor with light selectors thus bringing to the photon side of the interface the capability of choosing the desired channel. Using Manchester code instead of the previous work with NRZ (No Return to Zero) code improves channel discrimination at the reception. The sensor is also looked upon its capability of producing a single electrical output with several input channels. This photocurrent translation is studied to identify logical operations as the AND, OR, NOT and XOR, and other more complex as the


L.M. Camarinha-Matos et al. (Eds.): DoCEIS 2017, IFIP AICT 499, pp. 447–462, 2017.
DOI: 10.1007/978-3-319-56077-9_44

adder. A memory effect has also been identified and studied herein. Experiments with random values show the dispersion of the signals at the reception and several strategies to reproduce the input channels are proposed. Several programs had to be built and are used as tools to illuminate the sensor and analyze the sensor output. Observation of the device's behavior arose several questions that span the proposed work: How many channels can be transmitted by WDM, in visible light, with the sensor by selecting the desired channel with ultra-violet light?; What wavelengths ranges are best suited for the de-multiplexing function?; Which are the logical functions that can be translated by this sensor?; How to build complex logical functions with the device?; How to de-multiplex light channels using light selectors with one sensor?; How to build a memory due to charges stored by light?. The answer to these questions is presented in this document, along with other study paths.

## 2 Relationship to Smart System

A smart system uses a feedback loop of data, providing evidence for decision-making. The system can monitor, measure, analyze, communicate and act, based on the sensors' information [3]. Smart systems can be classified in several levels: those that collect usage and performance data to be used for a next version of a smarter system, those that collect, process and present data for humans to make a better decision, and those that collect data to take action without any human intervention.

A smart system is based on data, and that is its core. Data can be easily acquired and stored but it will also easily increase in quantity. As a resource, its ownership is also a point to be taken into consideration. Data must be analyzed, fused with other data sources, interpreted and sourced into mathematical models that will hopefully provide a highly reliable decision making tool for society and especially for those whose jobs are involved in taking decisions.

A system needs feedback, and that aspect is fundamental in any smart system, not only to increase its performance but also to improve its robustness. Data of different sources may come from the same physical reality which brings our attention to redundancy. This redundancy, which can also be constituted by different systems, is also an advantage due to possible error conditions or flaws.

The device presented in this paper is a sensor that is sensitive to light and as such it can be inserted into a smart system data loop. One of the observed characteristics of the sensor is that it holds memory of the visible light that has impinged onto its surface, even when not connected to a circuit. This could be applied as a sensor to be located in a closed box or case; an environment where no electricity is present. If the box or case is opened under visible light and then shut again, it would be possible to know that at least one opening occurred by reading the sensor under a controlled no light condition.

## 3 Device Structure and Operation

The sensor is a two stacked pin structure (p(a-SiC:H)- i'(a-SiC:H)-n(a-SiC:H))-(p (a-SiC:H)-i(a-Si:H)-n(a-Si:H)). Thicknesses and optical gap optimized for light absorption in the red and blue ranges; i- (1000 nm; 1.8 eV), i'- (200 nm; 2.1 eV) [4].

This structure is 1 $cm^2$ in area, built with silicon carbon [5], can be seen in Fig. 1, where the wavelength arrows indicate absorption depths during operation. The wavelengths mostly used are: $\lambda_V = 400$ nm, $\lambda_B = 470$ nm, $\lambda_G = 524$ nm and $\lambda_R = 626$ nm.

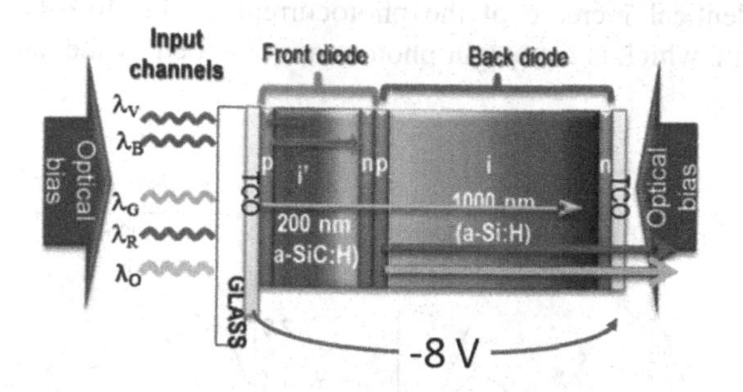


**Fig. 1.** Sensor structure and operation (Color figure online)

The experimental setup use LEDs as light sources: as digital signals and as bias lighting. The digital signals are usually: violet (400 nm), blue (470 nm), green (524 nm) and red (626 nm), and shone over the entire front surface of the device. The background lighting is ultra-violet (390 nm) and is set either at the back or at the front side of the sensor, and lighting is applied in a continuous non pulsating flux. Compared to the optical bias intensity, the intensity of the signal sources is lower.

Using a monochromator with 10 nm steps, from 300 to 800 nm, spectral response curves were produced, and the resulting photocurrent stored in a file with the corresponding input wavelength. The device operation point was set at −8 V because a previous study suggests that, that electrical bias is the most suited to allow for a steady state optical bias and channels with visible light impulses, while outputting a photocurrent that changes with the input wavelengths [6]. The spectral response for a violet (400 nm) optical bias at the front and back surfaces of the sensor is presented in Fig. 2.

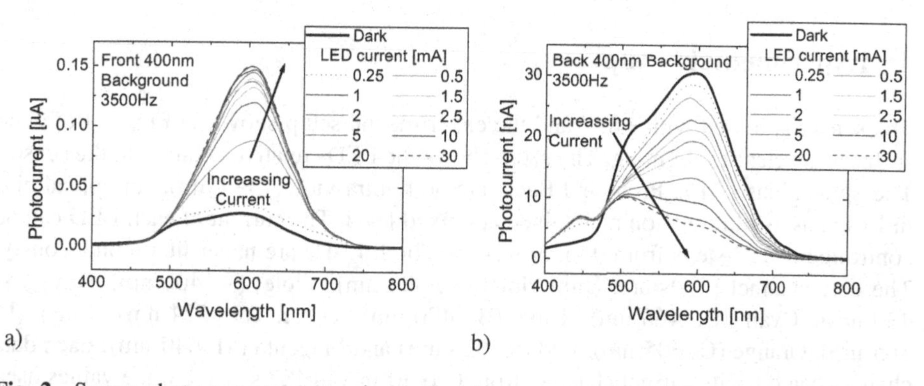


**Fig. 2.** Sensor photocurrent output using optical bias wavelength of 400 nm at the (a) front and (b) back sides. (Color figure online)

The experimental results of Fig. 2(a), show that the photocurrent increases in the 470 to 700 nm bandwidth when the sensor is lit on the front surface. To compare both graphs in of Fig. 2, the thick black curve represents the dark level and is the same on both figures. With increasing LED current through the LED that shines over the back surface, the photocurrent decreases gradually in the 470–700 nm bandwidths, and there is an almost identical increase of the photocurrent in the 400–470 nm range. The photocurrent gain, which is the output photocurrent divided by the dark current level, is shown in Fig. 3.

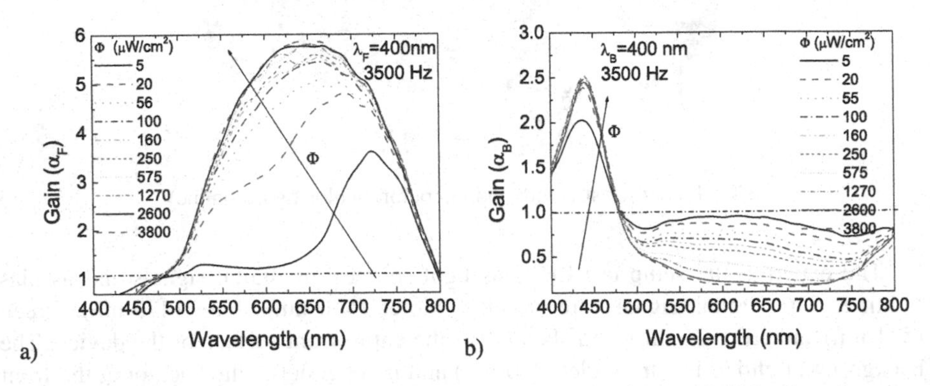


**Fig. 3.** Photocurrent gain due to bias lighting upon the (a) front and (b) back surfaces.

The spectral gain shown in Fig. 3(a), is reduced within the short wavelengths (<470 nm) and increased over the long wavelengths (>470 nm), behaving as a selective filter centered in 650 nm with bias illumination at the front. With bias lighting shining over the back surface, Fig. 3(b), the opposite behavior is noticed; the short wavelengths gain increases while the long wavelengths gain decreases. This resembles a filter centered at 440 nm. The sensor acts as a selective filter, where the gain of the short and long pass wavelengths is controlled by the optical bias. The gains of both filters suffer almost no changes when the LED currents are above 10 mA.

## 4 Experimental Setup

The sensor is biased with −8 V and the experimental setup shown in Fig. 4. Both end surfaces are clear for lighting. The setup shows the LEDs relative position to the sensor. The optical bias LEDs, Front and Back, are both ultra-violet (390 nm). An optical bias lit LED has its illumination maintained at a fixed level. The current of each LED can be controlled in 127 steps from 0 to 30 mA, and both LEDs are never lit simultaneously. The data channel LEDs are: Ultra-violet (U: 390 nm), Violet (V: 400 nm), Navy (N: 430 nm), Cyan (C: 460 nm), Blue (B: 470 nm), Green (G: 524 nm), Lime (L: 565 nm), Orange (O: 605 nm), Red (R: 626 nm) and Magenta (M: 640 nm). Each data channel can have its current changed from 0 to 30 mA in 127 steps, but the values used are usually 0 to 5 mA range with different synchronized patterns.

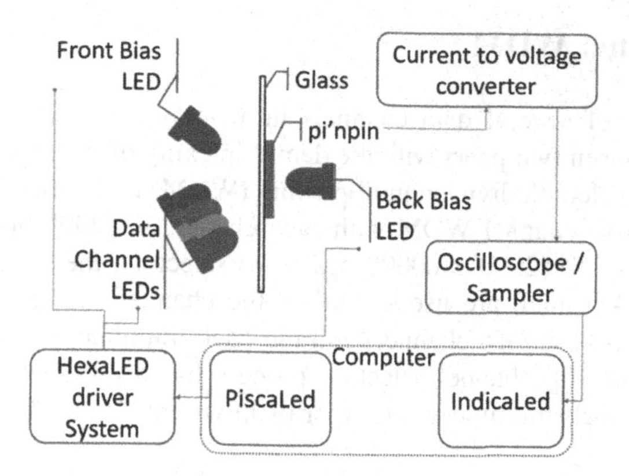


**Fig. 4.** Sensor and LED relative positions and equipment (Color figure online)

All LEDs are software controlled by the PiscaLed [7, 8] system that drives them physically by communicating with the HexaLed driver [9]. Data values supplied by the PiscaLed are shaped into configurable frames and encoded or not in Manchester coding; the frame can be raw, with the data bits presented to the channel LEDs, in a frame with a preamble, start of frame, data values and end of frame. The preamble is a sequence of several [010101] bits on all channels, used at the reception to determine the maximum intensity value, used for normalization and synchronization. The start of frame begins with a slight difference in the preamble [0110] followed by data. The sequence of [010100] at the end of the frame precedes the switching off of all channels. To analyze the results several programs developed within this thesis were used namely the IndicaLed [9–12].

Each data bit can be coded in non return to zero (NRZ) or Manchester [13]. When a large number of equal data bits are transmitted the resulting NRZ output is either no illumination (bits = 0) or continuous (bits = 1); resulting in no photocurrent changes. With Manchester coding there is always a change in photocurrent due to the pulsed pattern. There are more solutions to overcome the NRZ pattern for example by the introduction of a scrambler or other type of encoding [14].

The photocurrent output of the sensor is named differently according to its optical illumination and signal source: dark, back or front. The Dark photocurrent is defined as the output current when the sensor is in darkness. The Front signal is the photocurrent output with optical bias applied to the front surface of the sensor and pulsing data LEDs shining, and consequently the Back photocurrent output is the electrical signal produced with the optical bias at the back surface and the pulsating data LEDs lit.

The photocurrent output is converted to a voltage by the current to voltage converter and is connected to an oscilloscope. The voltage signal is sampled at the oscilloscope and sent to the computer, as shown in Fig. 4. The Data stored consists on the data LEDs input and converted output photocurrent, so that the analysis can be validated by having both the expected result and the obtained output. The signals are stored as received, Dark, Front and Back signals read with their minimum value set to 0.0 V, and normalized between 0.0 and 1.0 for analysis.

## 5 Five Channel WDM

The multiplexing of several data channels into one channel is an efficient way to communicate between two peers with the demultiplexing of the single channel into the original ones. Wavelength division multiplexing (WDM) uses different wavelengths for each channel. A five channel WDM with each channel at 12000 bps allows for a raw transmission rate of 5*12000 = 60000 bps. Two selectors, the front and back illumination of the background, are used to select the channels. As a summary the front illumination would sieve two channels and the back illumination would sieve another two, and the remaining channel selection process has a different approach [15]. To study the five channels multiplexer an experiment was made and the result is displayed in Fig. 5.

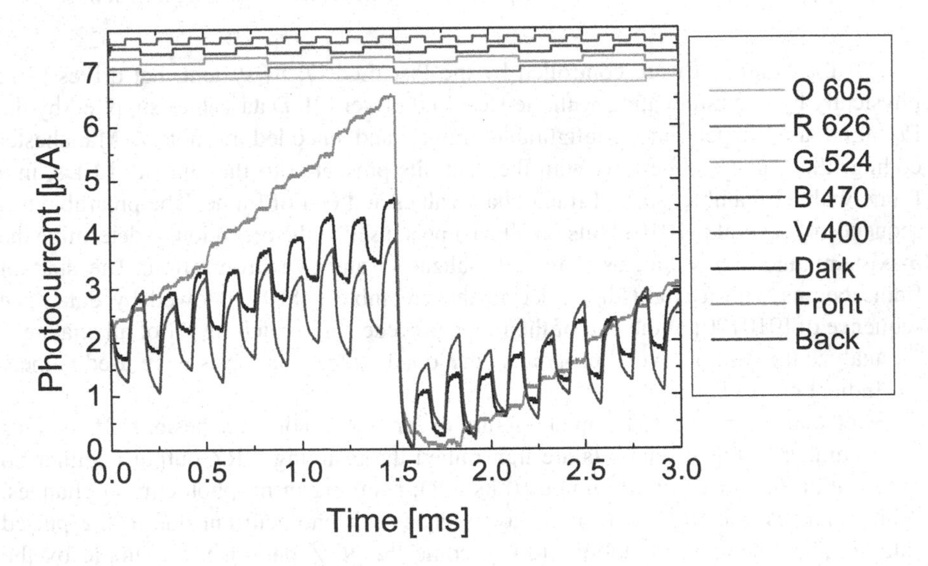


**Fig. 5.** A five channel multiplexer output. (Color figure online)

Shown in Fig. 5 is the output of the pi'npin when it is bathed by the signal wavelengths (dark) and simultaneously by the front illumination of the background (front) and the background at the back of the device (back). The digital input signals applied to the LEDs, at the top of the figure, contain the input data to each channel. These digital signals help guiding the eyes to the front signal and showing that all 32 combinations are there. The back signal follows the violet input channel (400 nm) very strongly and distinguishes between dubious classifications of the front signal. The blue (470 nm) channel can also be recovered by the back signal. After the extraction of both violet and blue channels, the red = 626 nm, orange = 605 nm and green = 524 nm can be easily identified.

The orange wavelength can be extracted because the gain of that wavelength ($\alpha_{Ofront} = 4.53$, $\alpha_{Oback} = 0.54$) is different from the other gains. Using the clustering approach [16], applied to the front and back signals of Fig. 5 results in what is shown in Fig. 6.

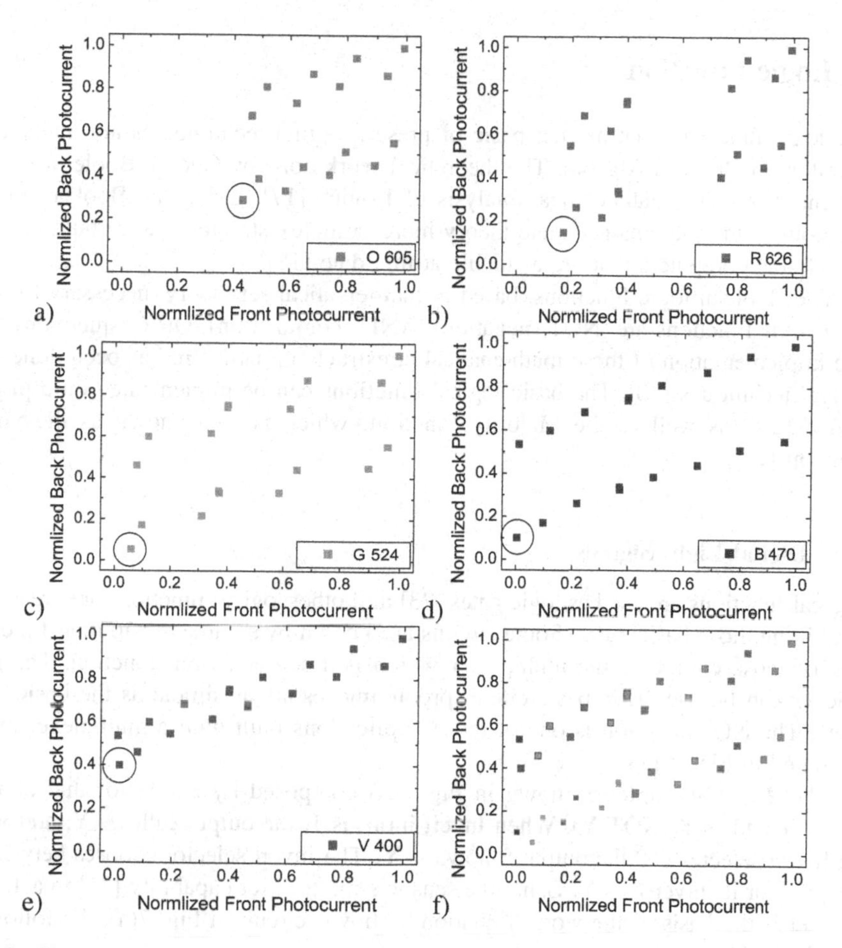


**Fig. 6.** Clusters of (a) O = 605 nm (b) R = 626 nm (c) G = 524 nm (d) B = 470 nm (e) V = 400 nm and (f) all ORGBV. (Color figure online)

The pi'npin sensor responds to the orange O 605 nm signal depending on the other signals that are also shining at the sensor. With five different wavelengths, $2^5 = 32$ possible combinations lead to the 16 areas in Fig. 6(a) which represent the combinations that have the presence of the O 605 wavelength. The area marked with a circle belongs to the influence of O 605 when no other signal wavelength is active. The clusters for the other wavelengths are also shown in Fig. 6(b) to (e). The graph of Fig. 6(f) is the over position of all individual wavelengths. There are only 31 areas on this graph because de 00000 bit is not plotted, as each individual graph only indicates where the wavelength is active.

Additional channels may be introduced using only the front and back selectors, but due to the cross gains of one channel over another difficult the extraction of the individual channels [16].

## 6 Logic Functions

The logic functions that are the basic of present digital technologies are mathematic operators of Boolean Algebra. The theoretical work done by George Boole in 1847 is presented in “The mathematical analysis of Logic” [17]. Using the Boolean logical functions with mathematical state theory more complex structures were built creating the reliable computers that are available at this date [18].

The Boolean logic functions, based in mathematical set theory, necessary to build other logic functions are: NOT (negation), AND (conjunction), OR (disjunction) [19]. The implementation of these mathematical constructs in hardware has been done with several techniques [20]. The basic logical functions can be implemented on a pi’npin device [21], as well as the Majority function, which is also known as the voting function [22].

### 6.1 Digital Light Signals

Logical functions are used as logic gates [23] and other logical functions are created by combining those basic gates. Some circuits are known by a name due to their function. One of those circuits is the multiplexer which is a combinatorial function. The multiplexer can be used as a basic circuit producing results as simple as the basic logic gates. The NOT function is one of those applications built with a multiplexer and is presented in Fig. 7 [24].

The 2 × 1 multiplexer shown in Fig. 7(a) composed by a selector that chooses inputs Y and ~Y (NOT Y). When Invert input is 1, the output follows Y, and when the Invert selector is 0 the output follows ~Y. The Invert selector chooses between: a Y signal, or its inverse ~Y. Using the sensor’s multiplexer capability [25] to act as an inverter is the basis of the work. Equation 1 shows, circuit in Fig. 7(a), the following transformations:

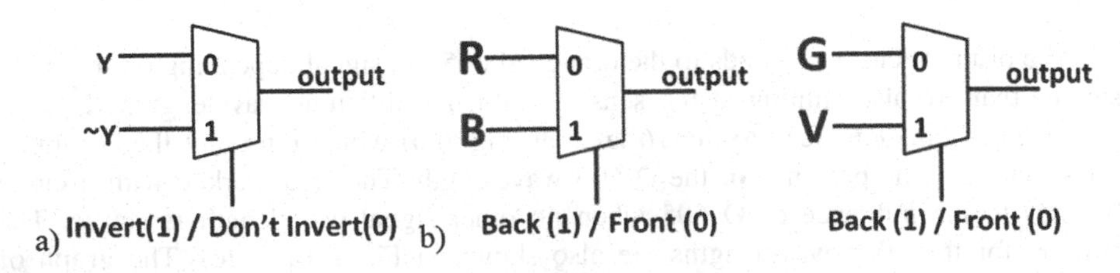


**Fig. 7.** A multiplexer is used (a) as an invert function, and (b) 2 examples of the NOT function using the digital light signal pair (Red, Blue) and (Green, Violet) pair. (Color figure online)

$$output = Y \cdot \sim Invert + \sim Y \cdot Invert = Y \oplus Invert \qquad (1)$$

Equation 1, simplified as the exclusive OR (XOR) relation between Y and Invert inputs. If Invert is 0 the output follows Y input, otherwise it follows ~Y [26].

A digital light signal is defined as: a signal pair with two components where one is the inverse of the other, and the wavelengths of one is from the long and the other from the short filter ranges [27].

Any long, short combination as a signal pair can be used, for example: (Red, Blue) pair and (Green, Violet), both shown in Fig. 7(b).

The two different digital light signal pairs are presented in Fig. 8 with the same data sequence.

By subjecting both digital light signal pairs shown in Fig. 8 to the Front and Back bias lighting, it is clear that the Front signal follows the Red and Green input, and Back signal follows Blue and Violet signals.

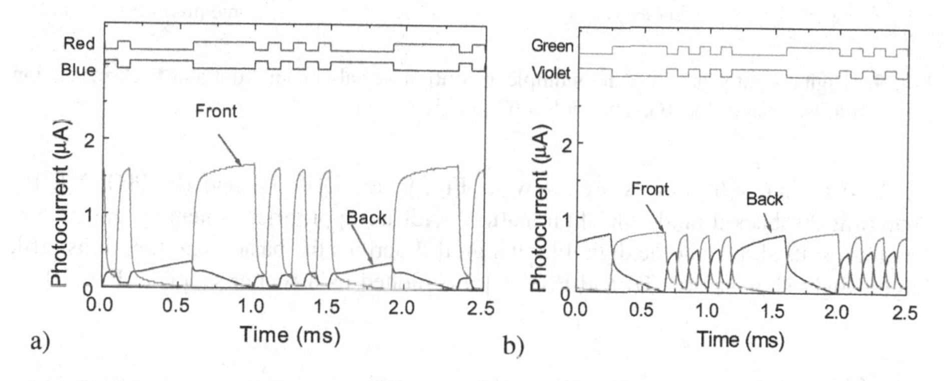


**Fig. 8.** Examples of digital light signals: (a) Red-Blue pair, (b) Green-Violet pair. (Color figure online)

A visible digital light signal D is the multiplexed signal of two wavelength channels; from the long and from the short wavelength ranges. Represented as D[L, S] where L and S represent the channel color [26], long and short-wavelength range, respectively.

**NOT Gate.** The NOT gate shown in Fig. 7(b) is built by a long and a short wavelength channels (for example red and blue or green and violet) is controlled by the violet irradiation. The construction of this gate urged the definition of the digital light signal. Thus the digital light signal is the base of all logical functions built with this device. Figure 8 shows examples of the NOT function, with the Back ultra-violet light.

Transmitting a False or a True value for a long period of time would degrade the recovery of the output signals, as only the ac component is used. To improve the photocurrent output it is necessary that a True or False bit value changes in time during its bit length. This can be solved by using the Manchester coding [13] which guarantees that there is a polarity change in the middle of each True or False bit time length.

The photocurrent to be useful must be time dependent, so the logic signal, even if does not have its state changed, must vary in time. This obliges the digital signals to be synchronized and differential, thus a digital light signal must be Manchester coded [20, 21].

Two examples of digital light signals with Manchester coding are shown in Fig. 9.

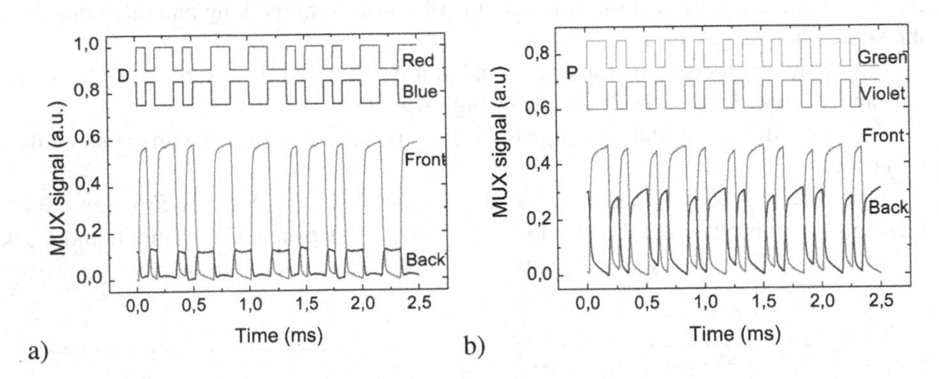


**Fig. 9.** Digital light signals and its multiplexer output signals under front and back irradiation (a) D(Red, Blue) and (b) P(Green, Violet) (Color figure online)

Two digital light signals are show in Fig. 9: (a) D[R, B] and (b) P[G, V]. By changing the background side illumination with an appropriate intensity and wavelength, results show that the digital light signal describes the basic logic functions [26]. Digital light signals D[R, B] and P[G, V] are plotted as a scatter graph in Fig. 10.

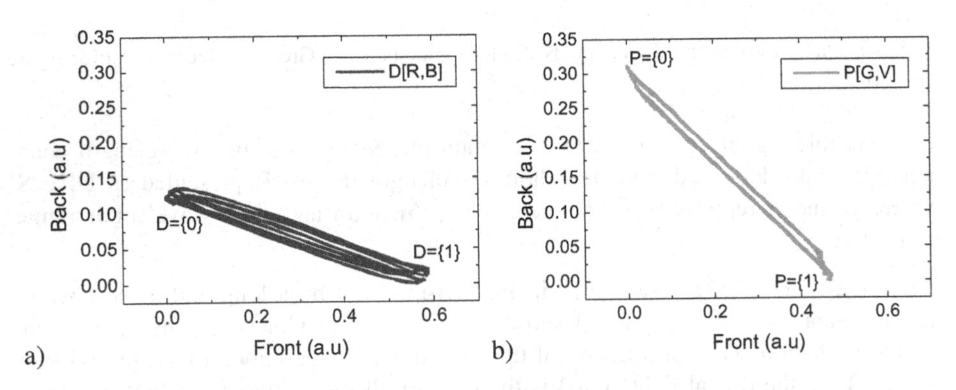


**Fig. 10.** Scatter graphs of **Fig. 8** (a) D[R, B] and (b) P[G,V] (Color figure online)

Shown in Fig. 10 are the scatter graphs of the signals presented in Fig. 9. It is clear that only two values exist in each digital light signal D[R, B] = {{0}, {1}}, and P[G, V] = {{0}, {1}}. The scatter graph also shows that the front and back signal are inversely correlated, due to the negative slope of the graphs [28] as expected due to each digital light signal being a composition of a short and long wavelength [29].

A digital light signal can be written as Eq. 2.

$$D[R,B] = [R(t) \cdot i_R \quad B(t) \cdot i_B] \times \begin{bmatrix} \alpha^{V}_{R,Front} & \alpha^{V}_{R,Back} \\ \alpha^{V}_{B,Front} & \alpha^{V}_{B,Back} \end{bmatrix} = [L_R(t) \quad S_B(t)] \tag{2}$$

R(t) is the digital sequence for example = {0,1,1,0,0,1,0,1} and B(t) the inverse of that sequence e.g. = {1,0,0,1,1,0,1,0}. The dark photocurrent for each wavelength is $i_R$ for the Red and $i_B$ for the Blue.

## 6.2 Interaction of Two Digital Light Signals

Logic operators besides the NOT have expressions with two variables. Boolean logic has the associative and commutative properties of algebra that are applied when more than two variables are involved. The number of input variables and the operation that produces a single logical output is called a gate. In this section two light variables are studied as the basic building block of a logical gate. In order to effectively classify the output of the logical gate the sampling of the signal will also be addressed.

Two different digital light signals D[R, B] (Fig. 9(a) and P[G, V] (Fig. 9b) illuminated the device and their combination is presented in Fig. 11 [27].

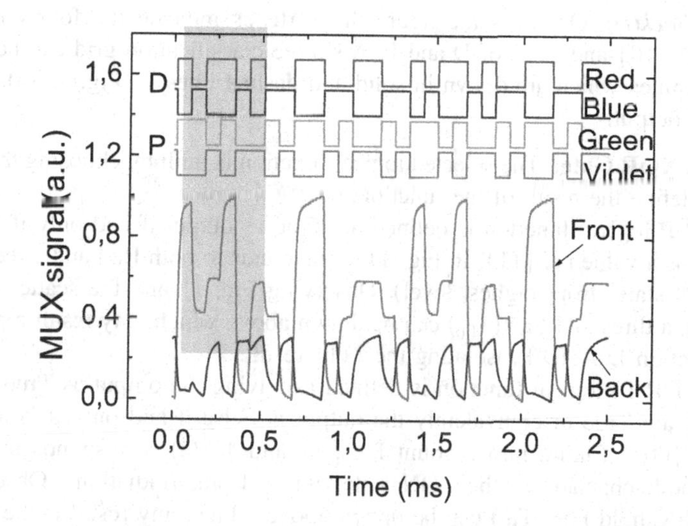


**Fig. 11.** Result of the interaction of two digital light signals D[Red, Blue] and P[Green, Violet] (Color figure online)

The interaction of two digital light signal pairs red-blue and green-violet is shown in Fig. 11. This signal is shown in Fig. 12 (a) as a scatter graph where the Front signal is set as the horizontal axis and the Back signal is set as the vertical axis.

The pattern selected in Fig. 11, shown in detail in Fig. 12 (b), holds all possible combinations of the D and P signal and will be analyzed further in the text.

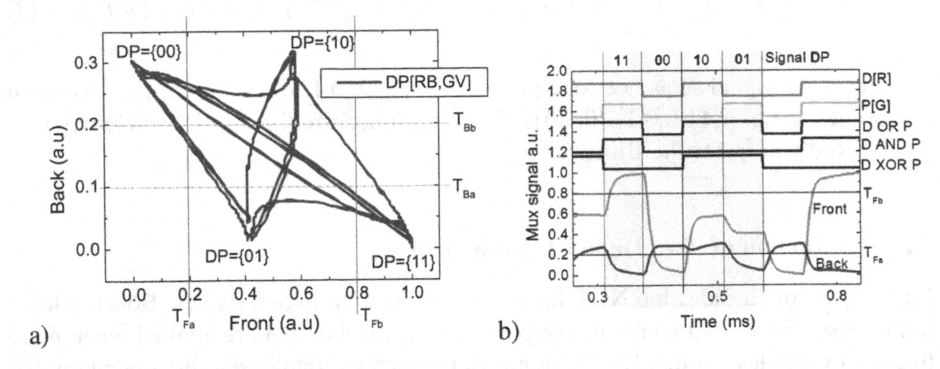


**Fig. 12.** (a) Scatter graph of the resulting Back and Front bias signals of Fig. 11 and (b) detail of Fig. 11, contains all 4 combinations of two digital signals, front and back are the interaction of two signals D,P, and the OR, AND, XOR signals represent the expected output. (Color figure online)

Figure 12 shows signals Front and Back as a scatter graph, the sequence path where at each point, the back and front values meet at the same time instant i.e. *functionOf (Front(t), Back(t))*. On the same figure the vertexes indicate the four combinations {00}, {01}, {10} and {11} of D and P. A simple classification grid can be used with threshold values which are drawn in solid and dashed lines on Fig. 12 (a), to classify the logical output.

**AND, OR, XOR Gate.** The selected logical function is built by choosing the threshold lines that define the result of the function.

D AND P logical function is defined as: True as output if and only if both inputs have True as a value i.e. {11}. In Fig. 11 corresponds to both Red and Green channels in their ON state (front highest level). Observing Fig. 11 and the scatter function of Fig. 12 (a), a threshold line ($T_{Fb}$) can be drawn above which, any result represents the output function D AND P, isolating the {11} result.

The (D OR P) logical function is defined as having the output as True if either or both {DP} are True or equivalently the output is False if and only if both {DP} are False, i.e. {00}. Taking into account Figs. 11 and 12 (a) corresponds to either red, green or both channels in their ON state (under front irradiation). Observing both figures a threshold line ($T_{Fa}$) can be drawn above which, any result is the function D OR P, i.e. {{01}, {10}, {11}}.

The Exclusive OR function, XOR, between two variables {DP} is True when only one of them holds the True value, i.e. {{01}, {10}}. It excludes the {11} from the OR. It can also be defined as the difference operation, which is True, when the variables hold a value which is different from the other. Observing the scatter graph of Fig. 12 (a) it is possible to confine the {{01}, {10}} values between the two threshold lines $T_{Fa}$ and $T_{Fb}$.

## 7 Memory Effect

The p'inpin device has been used throughout this work as a stateless component which means that its output is only dependent of its inputs regardless of the outcome history of past inputs or outputs. This however does not hold when the outputs are observed as a sequence of inputs. The identification of the input sequence, using a classification function with the output of the pi'npin, shows that past sequences do influence the output of the device which does in fact hold a memory of the illumination beforehand. Although this is a difficult task for signal identification it is necessary to understand this phenomenon to ease the identification task and also to use it as an advantage by constructing a volatile memory with the pi'npin device.

**Paradigm Change**

The observations made to indicate that there is a memory function that characterizes the pi'npin device started when the readings of the dark signal did not remain constant between experiments. Initially it was assumed that there was an influence of noise from the surrounding equipment and other devices. Several modifications were gradually made to the experimental setup to reduce the influence of the noise sources.

To understand the start condition of the pi'npin device and its influence over the output, changes were made to the software of the user interface and that of the microcontroller that drives the LEDs, to allow static readings [24, 25]. The selection of a start condition, programmable delays and the possibility of short sequences, all time controlled and with adequate triggering for the oscilloscope, allowed the study that is presented in this section.

In all experiments till this point the optical bias was used as a steady lighting, namely in the digital light logical functions and WDM [18, 21, 26]. This optical bias paradigm has to be changed, and seen as two light Control signals; the Back and Front Control signals, to use the sensor as a volatile memory [30]. On this section, the Front and Back lighting are no longer steady state, but impulses.

**Conceptual Approach**

The experimental setup with the relative positioning of the LEDs to the sensor is shown in Fig. 13 (a) along with the equipment. The LED currents and their timing patterns are controlled by PiscaLed system [24, 25].

There is a sequential operation with three phases to use the sensor as a volatile memory: Control, Hibernation and Data phase. Control phase is when only the Front Control and or the Back Control signals lit the sensor. The Hibernation phase has no illumination over the sensor. Data phase when the sensor is illuminated with Data Sense signal. Phases are represented in Fig. 13 as switches (a), and as a timeline (b).

***Data Sense Signal Phase Duration:*** to study the decay of the stored value through time, an experiment was set in which the Data Sense signal used in the previous experiments is repeated cyclically with a 0.5 ms interval. This is shown in Fig. 14, where the signals at the top guide the eyes. Presented in Fig. 14, are the results of the applied Back and Front Control signals, each with duration of 3 ms followed by the Hibernation phase lasting 3 ms. The Data Sense signal sequence is equivalent to 16 cycles at a frequency of 6000 Hz. These 16 cycles are then repeated.

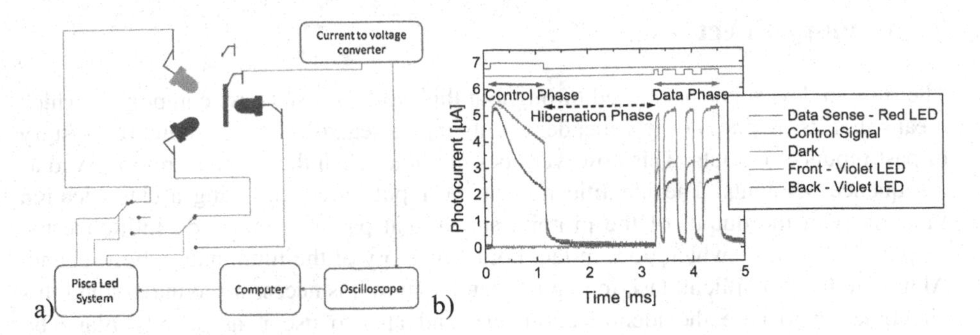


**Fig. 13.** (a) Experimental setup showing D (Data), H (Hibernation) and C (Control) phases as switches. (b) Outputs represented on same plot by three different conditions: dark, front control and back control.

Results show that there is amplitude decay in time of the output Data Sense signal despite the Control signal is either Front or Back. The decay rate, were it linear, is roughly twice higher in the Front Control signal (−83 μA/s) than in the Back Control signal (−46 μA/s) which means that they instead of crossing will maintain an amplitude equivalent to the Dark reference signal.

Furthermore, the decayed amplitude of the Data Sense due to the Front Control will eventually be less than the highest amplitude of the Data Sense due to the Back Control. Observing Fig. 14 that would occur around 25 ms; this determines the volatile memory hold time due to the data sense signal profile.

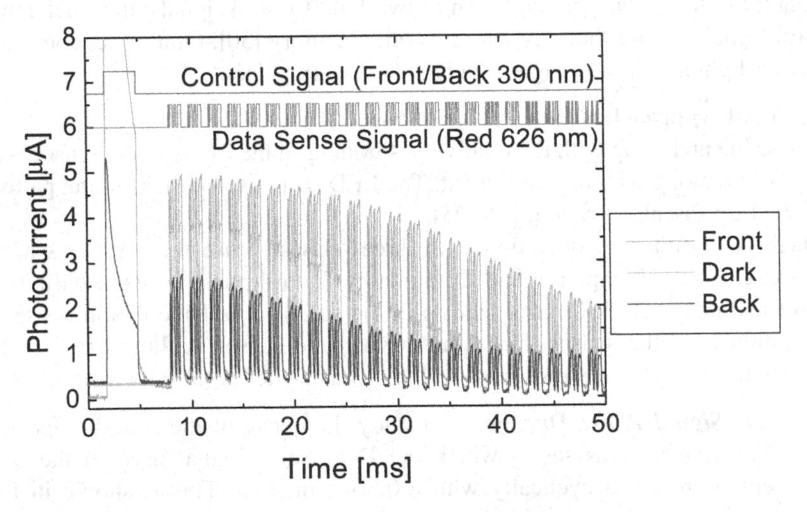


**Fig. 14.** Back and front control signals followed by data sense with a repeated cycle pattern.

## 8 Conclusion

The p'inpin device has proven to be versatile and with several applications in which it was used: multiplexer, logical gates and a volatile memory. This flexibility is its main advantage.

By experimental observation the simultaneous incidence of the front and back optical illumination is equivalent to a front bias illumination. This characteristic could be used by having the back surface always illuminated and using the front bias when convenient which could speed up the acquisition process.

The sensor illumination has always flooded the whole surface. Studying the use of masks to allow only certain parts of the sensor being used by spatially distributing the input wavelengths over the surface can produce curious results.

The photocurrent varies linearly with the flux intensity of the input signals. By using analogue signals a "light transistor" can be built using the background illumination as a constant amplification setting.

The possibility to change the pi'npin heterostructures by depositing more pins and enabling their junctions to be accessible by optical transparent contacts will definitely improve the research of this simple photosensitive device. The ideal device would be one that could also emit light; with optical inputs and outputs it would effectively be considered as an optical gate. The demultiplexer would then emit in separate channels, the logical functions would work as gates that could be connected as circuits including the memory component.

Due to the low throughput of 12000 bps of each channel the pi'npin sensor can be used in audio and signalling applications.

**Acknowledgments.** Work supported by FCT, CTS multi annual funding, through PIDDAC Programs PTDC/EEA-ELC/111854/2009, PTDC/EEA-ELC/120539/2010.

## References

1. Vieira, M., Fantoni, A., Louro, P., Fernandes, M., Schwarz, R., Lavareda, G., Carvalho, C. N.: Self-biasing effect in colour sensitive photodiodes based on double p-i-n a-SiC: H heterojunctions. Vacuum **82**(12), 1512–1516 (2008)
2. Vieira, M., Fernandes, M., Louro, P., Vieira, M.A., Barata, M., Fantoni, A.: Multilayered a-SiC:H device for wavelength-division (de) multiplexing applications in the visible spectrum. In: MRS Proceedings, vol. 1066, pp. 1066–A08-1, February 2011
3. The Royal Academy of Engineering. Smart infrastructure: the future
4. Vieira, M., Louro, P., Fernandes, M., Vieira, M.A., Torre, Q., Caparica, M.: Three transducers embedded into one single SiC photodetector: LSP direct image sensor, optical amplifier and demux device. In: Betta, G.F.D. (ed.) Advances in Photodiodes, pp. 403–426. Intech, Rijeka (2011)
5. De Cesare, G., Irrera, F., Lemmi, F., Palma, F., Tucci, M.: a-Si:H/a-SiC: H heterostructure for bias-controlled photodetectors. MRS Proc. **336**, 885 (2011)
6. Vieira, M.A.: Three Transducers for One Photodetector: essays for optical communications, FCT-UNL Universidade Nova de Lisboa (2012)

7. Silva, V.: PiscaLed - User's Manual, GIAMOS - ISEL/IPL (2013)
8. Silva, V.: PiscaLed - Technical Manual, GIAMOS - ISEL/IPL (2013)
9. Silva, V.: HexaLed- User's Manual, GIAMOS - ISEL/IPL (2016)
10. Silva, V.: IndicaLed, GIAMOS - ISEL/IPL (2015)
11. Silva, V.: Papagaio, GIAMOS - ISEL/IPL, p. 6 (2015)
12. Silva, V.: Descodifica, GIAMOS - ISEL/IPL (2013)
13. ATMEL, manchester coding basics: application note (2009)
14. Choi, D.: Report: parallel scrambling techniques for digital multiplexers. AT&T Tech. J. **65**(5), 123–136 (1986)
15. Silva, V., Barata, M., Louro, P., Vieira, M.A., Vieira, M.: Five channel WDM communication using a single a:SiC-H double pin photo device. Appl. Surf. Sci. **380**, 318–325 (2015)
16. Silva, V.: Reconfigurable photonic logic architecture, UNL - Universidade Nova de Lisboa (2015)
17. Boole, G.: The Mathematical Analysis of Logic. Philosophical Library, New York (1847)
18. Nishimura, H.: Boolean valued decomposition theory of states. Publ. Res. Inst. Math. Sci. **21**(5), 1051–1058 (1985)
19. Tanenbaum, A., Austin, T.: Structured Computer Organization, 6th edn. Pearson Education International, London (2013)
20. Millman, J., Grabel, A.: Microelectronics, 2nd edn. McGraw-Hill Publishing Company, New York (1987). ISBN 978-0070423305
21. Silva, V., Vieira, M.A., Louro, P., Vieira, M., Barata, M.: Optoelectronic digital capture device based on Si/C multilayer heterostructures. In: Camarinha-Matos, L.M., Tomic, S., Graça, P. (eds.) DoCEIS 2013. IAICT, vol. 394, pp. 555–562. Springer, Heidelberg (2013). doi:10.1007/978-3-642-37291-9_60
22. Moore, J.B., Tan, K.T.: Majority Logic Coding and its Multinomial Representation (1975). http://users.cecs.anu.edu.au/~john/papers/JOURSUB/06.pdf
23. Stallings, W.: Computer Organization and Architecture Designing for Performance, 8th edn. Prentice Hall, Upper Saddle River (2010)
24. Silva, V., Vieira, M.A., Louro, P., Barata, M., Vieira, M.: Simple and complex logical functions in a SiC tandem device. In: Camarinha-Matos, L.M., Barrento, N.S., Mendonça, R. (eds.) DoCEIS 2014. IAICT, vol. 423, pp. 592–601. Springer, Heidelberg (2014). doi:10.1007/978-3-642-54734-8_66
25. Vieira, M., Vieira, M.A., Silva, V., Louro, P., Costa, J.: SiC monolithically integrated wavelength selector with 4 channels. MRS Proc. **1536**, 79–84 (2013)
26. Silva, V., Vieira, M.A., Vieira, M., Louro, P., Fantoni, A., Barata, M.: Logic functions based on optical bias controlled SiC tandem devices. Phys. Status Solidi **11**(2), 211–216 (2014)
27. Silva, V., Vieira, M., Louro, P., Barata, M.: Logical functions in a tandem SiC device. Microelectron. Eng. **126**, 79–83 (2014)
28. Navidi, W.: Statistics - for Engineers and Scientists, 3rd edn. McGraw-Hill Professional, New York (2010)
29. Silva, V., Vieira, M.A., Louro, P., Barata, M., Vieira, M.: AND, OR, NOT logical functions in a SiC tandem device. Procedia Technol. **17**, 557–565 (2014)
30. Silva, V., Barata, M., Vieira, M.A., Louro, P., Vieira, M.: Light memory operation based on a double pin SiC device. In: Camarinha-Matos, L.M., Baldissera, T.A., Di Orio, G., Marques, F. (eds.) DoCEIS 2015. IAICT, vol. 450, pp. 265–272. Springer, Cham (2015). doi:10.1007/978-3-319-16766-4_29

# Microneedle Based ECG – Glucose Painless MEMS Sensor with Analog Front End for Portable Devices

Miguel Lima Teixeira[1,2](✉), Camilo Velez[2], Dian Li[3], and João Goes[1]

[1] Centre for Technologies and Systems (CTS) – UNINOVA, Department of Electrical Engineering (DEE), Faculty of Sciences and Technology (FCT NOVA), NOVA University of Lisbon, 2829-516 Caparica, Monte da Caparica, Portugal
mdl.teixeira@campus.fct.unl.pt

[2] Electrical and Computer Engineering Department (ECE), University of Florida, Gainesville, FL 32611, USA

[3] Massachusetts Eye and Ear, Schepens Eye Research Institute, Boston, MA 02114, USA

**Abstract.** A portable microelectromechanical system (MEMS) for mobile phones, or other portable devices, that measures body electrical signals, as well as, extracts transdermal biological fluid for invivo analysis is proposed. This system integrates two sensing methods: three points finger electrocardiography (ECG) and glucose monitoring, through one electrode with a microneedle-array. This work presents the: (1) device modeling and microneedle-array' fabrication method, (2) signal processing and biasing circuitry' design and simulation, (3) Analog Front End (AFE) for measured signals, and (4) Glucose sensor characterization. Design parameters and geometries are obtained by solving the capillarity model inside the microneedles and running optimization numeric methods. The AFE consists in a differential band pass filter that provides amplification, filtering, and noise rejection. This work presents clear technological innovation, for its miniaturization and integration of known biological signals' measurement methods in a portable Smart System, which points in the direction of Internet of Things' goals.

**Keywords:** Electrocardiogram · ECG · Glucose · Sensor · Microneedles · MEMS · Electrodes · Mobile phones · Cellphones · Electronics · Biological signals · Healthcare · Portable

## 1 Introduction

Many organs in human body, such as heart, brain, and muscles, produce measurable electrical signals [1] and their detection is necessary to collect clinical information that permits health diagnosis that many times prevent body failures. In order to monitor health conditions such as Diabetes, it is also necessary to evaluate enzyme's concentration in body fluid [2].


L.M. Camarinha-Matos et al. (Eds.): DoCEIS 2017, IFIP AICT 499, pp. 463–478, 2017.
DOI: 10.1007/978-3-319-56077-9_45

According to the World Health Organization, cardiovascular diseases are the main cause of death in the world and 422 million people worldwide had diabetes in 2015, causing an average of 1.5 million deaths per year, as in 2012 (http://www.who.int/diabetes/en/). Portable sensors that can measure biological signals related with those illnesses and prevent body failure, are desired to be integrated in our normal life. Further more, the healthcare' current and future trend is to give patients the ability to monitor their biological signals, making health diagnosis and at certain times preventing serious health conditions, during their normal day life, reducing significantly the hospital costs. Mobile phones are excellent candidates to be carriers of such portable sensors, due to their omnipresence in our daily lives and because general population keeps them near themselves, as much as possible, for communication purposes. This work intends to combine those facts proposing a MEMS system that integrates both sensing methods (ECG - Glucose) in one electrode, using microneedles.

One of the most important techniques to measure electrical signals in clinic and biomedical application is the ECG (electrocardiogram), which records the electrical activities of the heart over time [1, 3–7]. The standard commercial ECG electrodes utilize an electrolytic gel to form conductive interface between skin and electrode. Using modern microsystem design techniques, it is possible to eliminate the use of gel by creating and using microneedles as electrodes [1, 4, 8]. At the same time, those microneedles can be used to extract interstitial fluid (ISF) to perform enzyme based diagnostic test, as shown in [9–12]. Due to careful microneedle array design the sensor is painless, because the length of microneedles is less than 500 μm and so they don't reach the pain sensitive dermis' area, which will be explained in Sect. 3.2.1. These concepts were applied in this device.

This work is divided in three main areas: microneedle fabrication, microfluidics for glucose sensor, and sensors' and AFE' electrical characterization. The microneedles for ECG electrodes and glucose electrodes share the same fabrication process. In this approach the microfluidic and electric models can be separated, because the ISF is static inside the channel during the glucose measurement. The link between both electrical sensors is the reference signal. A circuit for amplifying, eliminating noise and filtering is proposed for both systems, to connect them with a microcontroller.

The PhD study in which this work is being developed has the following research question and hypothesis: Research question: What would be a suitable way to do a low power system on chip (100 s of μW) that performs advanced computations for low frequency signals (below 10 kHz) as the brain does, as for example pattern recognition - namely Electrocardiogram (ECG) complete Arrhythmia detection, or fully implanted closed loop Brain Machine Interface (BMI) with brain signals' complex interpretation; without the use of a general purpose CPU? Hypothesis: One possible way is to develop a system on chip inspired in the brain that uses neurotechnology and neuromorphic circuits, creating an amplitude to time conversion with an Integrate and Fire circuit, which gives a stream of pulses. That system can then use pulse processing to do computations with the output pulses, for example with an automaton - without using conventional digital signal processing (DSP). If such kind of system is designed and prototyped then it may do these advanced computations with power consumption below 100 μW. Because its pulse timing structure only has information for relevant input amplitudes defined by the user, reducing the redundancy and lowering the hardware specifications.

The motivation of this work is presented next: With this power consumption, the system could be implanted in the Brain as a battery less standalone device to compensate for loss brain function, or could be used as a small ECG recording and arrhythmia detection device that is in contact with the patient chest. Currently these kinds of devices don't exist for ECG, or have very simple functions in the case of BMI, due to the limitation in computation power of such systems on chip.

## 2 Contribution to Smart Systems

This work presents clear technological innovation in a portable cellphone with smart sensors that measure body signals of the upmost importance, as for its miniaturization and integration of known biological signals' measurement methods in a portable Smart System, which points in the direction of Internet of Things' goals.

Moreover, the present work shows a clear contribution to Smart Systems as it paves the path of integrating healthcare and fitness solutions directly in mobile devices, that are one of the most used nodes to access Internet of things (IoT). In an IoT era, these are important developments to improve humans' daily life.

## 3 Device Design

The system was divided in four major modules to facilitate the design process: the microfluidic model that governs the microneedle fabrication, the glucose sensor, the ECG electrode, and the AFE for both physical measurements. Every module is independent from the others in terms of modeling, simulations and development predictions. The microfluidic model and the glucose sensor can be independent, because the method described in this document to monitor the glucose is for zero flow measurements. In the other hand, the ECG and the glucose sensor can be independent, because they will only share the reference electrode and the voltage in this electrode can be set as a design parameter for both modules.

### 3.1 Capillarity Modeling

Every microneedle that will be inserted inside the dermis of the human skin will experiment a capillarity drawn action. As explained in [9, 13, 14], the ISF inside the dermis area of the skin will have a different pressure as compared with the atmosphere. Difference in pressures and the interactions of the fluid with the needle walls will create forces (van der Waals, dipole-dipole interaction, London dispersion–fluctuating induced pole, and hydrogen bonds) that will contribute to changes on the surface tension of the fluid at the interface. Changes on the surface tension will generate the so-called capillarity drawn. Equation (1) shows the pressure difference in terms of liquid-gas surface tension $(\gamma_{lg})$ the contact angle between the liquid (θ), the volume and the solid ($A_{sl}$) and liquid area ($A_{lg}$).

$$P = \gamma_{lg}\left(\cos\theta\left(\frac{dA_{sl}}{dV}\right) - \frac{dA_{lg}}{dV}\right). \tag{1}$$

Figure 5.4, page 96, of reference [14] presents the capillary rise in a vertically standing cylindrical microchannel. As shown in that figure, the liquid will flow inside the needle until it reaches the distance h. A simplified model for h was demonstrated in [13] and can be expressed as:

$$h = (2 \cdot \gamma_{lg} \cdot \cos(\theta))/(\rho \cdot g \cdot a). \tag{2}$$

Where ρ is the fluid density, g the gravity constant, and a the needle' radius. This maximum drawn distance h will be used later in this paper to optimize the design geometry.

## 3.2 Glucose Sensor

### 3.2.1 Operating Principles

The relationship between glucose concentrations in ISF and blood was studied in the past. In reference [15] the author compared the venous, capillary blood glucose and ISF glucose concentration in one subject during 3 days. A strong correlation was found which could lead to the useful method of measuring the glucose concentration from ISF instead of directly from blood. As the ISF can be extracted from the first layers of the skin, the measurement of glucose concentration from ISF can be painless. This is accomplished by making the microneedles for extraction of ISF smaller than 500 μm long so they don't touch nerve cells when inserted in the skin. This is not possible in the extraction of blood that is always a process with some degree of pain. Thus, a small microneedle based glucose sensor can be designed. This is a major advantage of ISF glucose sensors as they can be painless, if designed properly. Comparing to usual commercially available blood glucose meters, for which the sample extraction is painful and can cause sensibility and irritation in the area of skin where the sample was taken, possibly making the user to have to switch the measurement location, between measurements.

Figure 1 presents the glucose sensor' cross section view. The interstitial fluid (ISF) collected by the microneedle first flows over glucose oxidase (GOX), which is immobilized in the channel, Fig. 1. GOX will catalyze the glucose oxidation process to generate gluconic acid and hydrogen peroxide, Eq. (3). Then hydrogen peroxide when in contact with the working electrode will be decomposed into water and electrons, Eq. (4). Such electrons increase the ISF's electrical conductivity. A 0.7 V potential difference is added between the working electrode and the reference electrode that will induce a current. According to the whole reaction's equation the concentration of glucose in ISF is proportional to this current amplitude [10].

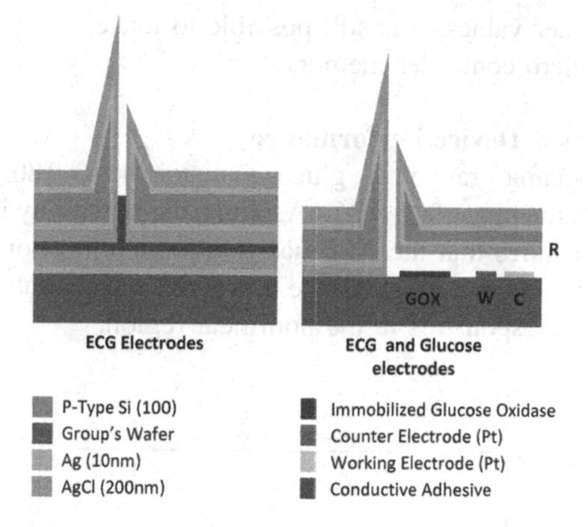


**Fig. 1.** Microneedle cross section for the ECG electrodes (left) and ECG and glucose sensor electrode (right), W: working electrode, C: counter electrode, R: reference electrode, GOX: enzyme.

$$Glucose + O_2 \rightarrow Gluconicacid + H_2O_2. \tag{3}$$

$$H_2O_2 \rightarrow O_2 + 2H^+ + 2e^-. \tag{4}$$

### 3.2.2 First-Order Model

The ISF inside of the channel can be treated as a linear resistor when the glucose concentration is small and a nonlinear one when the concentration is large. The lumped element model is simply a voltage source connected with a resistance in parallel. The governing equation of the reaction is Michaelis–Menten kinetics equation is:

$$i = \frac{I_{\max}[S]}{K_m + [S]}. \tag{5}$$

Where $i$ is current that is generated. $I_{\max}$ is the maximum current achieved by the system. $[S]$ is concentration of substrate. $K_m$ is Michaelis constant, which is the substrate concentration when current is half of $I_{\max}$.

The linearity between current and concentration depends on the assumption that the oxygen concentration in the ISF is enough. In reality, its concentration is limited. So when glucose concentration is more than 200 mg/dl, it is nonlinear. The commercial glucose measurement devices based on this principle also present a nonlinear region above 200 mg/dl.

From Fig. 3 of [16], that is based in the original data and curves of Daly et al. [17], it can be seen that the fluctuation of blood sugar in normal humans during a day with three meals is less than 125 mg/dl. The proposed glucose sensor in this paper will work properly, because the linear region of the sensor is larger than the ordinary human glucose concentration region: 3 mg/dl to 125 mg/dl [16, 17]. Moreover, when the blood sugar is at higher values, it is still possible to make a measurement by using a table stored in the micro controller memory.

### 3.2.3 Glucose Sensor Device Performance

The prospective dynamic range of glucose meter is: 0–180 mg/dl, Fig. 2. The resulting current is in the range 0 to 0.86 nA. Thus, the sensitivity is 0.0048 nA*dl/mg. This small current requires that the AFE has a high amplification gain, only possible with two or more amplifiers stages. Above a glucose concentration of approximately 180 mg/dl, the device response is in the nonlinear region.

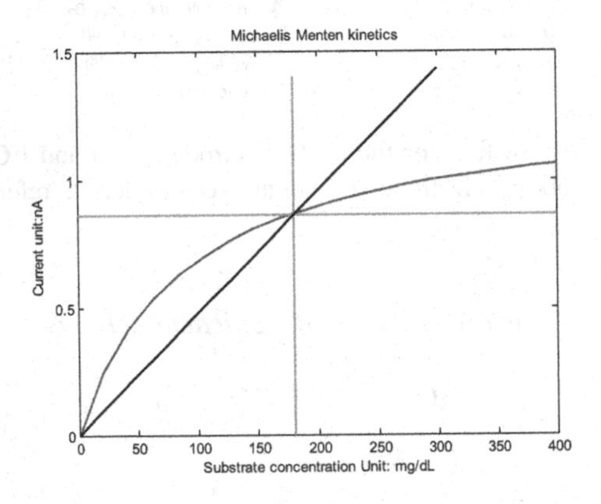


**Fig. 2.** The performance of glucose sensor, current versus substrate (glucose) concentration. The relation between glucose concentration and current is in blue. The perpendicular lines show the prospective dynamic range (Color figure online)

## 3.3 ECG Electrodes

The designed finger ECG electrodes consist in dry electrodes that have a 3D shape. The electrodes are deposited on top of an array of micro needles to decrease the overall electrode' impedance. Each ECG electrode consists in a 2-wafer structure. The top wafer is micro fabricated to have an array of hollow micro needles. After this micro-fabrication, the electrodes' metal (AgCl +Ag) is sputtered in all the sides of the wafer and due to the microneedles geometry it is ensured that there is an electric contact between both sides of the wafer, through the micro needles' holes, Fig. 1. The bottom wafer for the ECG electrode has a metal line that makes a contact between the top wafer and the electronic circuitry. Both wafers are joined together by an Ag conductive glue, in the case of the 2 ECG electrodes, without the glucose sensor.

The extraction of ECG signal from the finger is a novel method. In [3] is presented a proof a principle of finger ECG, using commercial planar dry electrodes and in [5] is presented an heart beat rate monitor system that uses planar dry electrodes for the fingers. The micro needle based electrode for ECG is also a novel approach. Reference [8] shows the microneedle device with a chamber, in the back of the microneedle structure, containing a conductive gel, to improve the conductance of these type of electrodes. It has been shown that dry microneedle electrodes are capable of acquiring signals from ECG, electromyography (EMG) and even electro encephalography (EEG), signals that are lower voltage signals, than ECG and EMG, and so require a very low impedance electrode, [1, 4]. These microneedles's electrodes were designed to record ECG in the human chest. As an ECG measurement device has never been integrated directly in a cell phone.

The first order model for the ECG measurement and the lumped element model for the skin and electrode is presented in [1, 18]. To do an ECG measurement it is required to use two active electrodes, in two fingers, or one finger and hand, and a third reference electrode in the other hand, or in another finger, or in a different zone.

### 3.3.1 ECG Device Performance

An estimate of the ECG electrodes' device performance was made. The electrode area is the area of the whole device, without electronic circuitry, as the electrode metal layer is deposited over all the microneedles in the array: electrode area = 3.8 mm * 2.5 mm with a AgCL layer, 25 nm thick.

The maximum input signal range is –1 mV to 1 mV. The ECG electronic circuitry performs an amplification of 500 times, it has to have a linear response from 0.05 Hz until 500 Hz, that is half of the sampling frequency so it can measure accurately each one of the peaks that make the ECG QRSPT complex, and measure a minimum detectable signal of 5 μV. These and other specifications are presented in the next Tables 1 and 2.

**Table 1.** Overall circuit' specifications.

| Specifications | Value |
|---|---|
| Supply voltage Vdd | 3.6 V |
| Imax (source) [7] | 40 mA |
| Power | 144 mW |

**Table 2.** ECG AFE specifications.

| Specifications | Value |
|---|---|
| Vbias [1] | 1.65 V |
| Input maximum signal range [5] | –1 mV to +1 mV |
| MDS [6] | 5 μV |
| Micro needles height [2] | 20 μm |
| Sampling frequency (fs) [3] | 1 kHz |

To accomplish these circuit requirements it is important that the circuit has low noise. More details about the electronic circuits are given in Sect. 5.

### 3.4 Design Optimization

In order to obtain the geometry values to generate the proper mask for the fabrication process it is necessary to analyze the behavior of the microfluidic module and its interaction with the human skin. Every microneedle will experiment the capillarity drawn for the ISF. It is necessary to design the system with the proper length so the ISF can fill the needle and the micro channel covering the electrodes. This will be the most important parameter to be considering for the mask design. Considering this parameter an optimization problem can be written in order to maximize the fluid volume inside the needle (cylinder volume) and the channel (rectangular channel). Figure 3 shows the distances for the fluidic module.

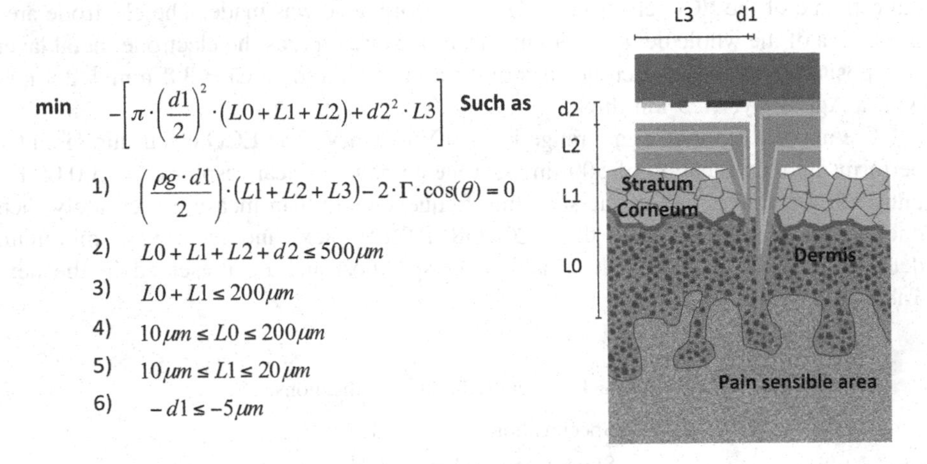


**Fig. 3.** Left: standard format for the optimization problem. Objective function, constraints (from 1 to 3) and boundaries (4 to 6). Right: microneedle inserted in human skin. Geometry variables used in the optimization process. Main dimension constrains related with skin interaction are shown.

The optimization problem can be written in the standard format as in Fig. 3. Constrains for this problem include: (1) the capillarity equation shown in (2) with the ISF $\rho = 1060\ kg/m^3$ [18], the ISF surface tension $\Gamma = 60\ mJ/m^2$ [19], and the contact angle $\theta = 110°$ [20] between ISF and needle walls. (2) Represent the wafer thickness. (3) Is the maximum needle' dimension in order to avoid reaching dermis' pain sensitive area according to [6]. (4) is the dermis typical thickness for human skin. This area contains the ISF. (5) is the stratum corneum thickness for human skin. And (6) is a fabrication limitation by the 20 mm head of the Heidelberg laser stepper available in the University of Florida's Nanoscale Research Facility (NRF) clean room that will be used in this fabrication process. The Matlab® optimization results obtained, after rounding the numbers to fit the fabrication requirements on the available clean room, are: microneedle chamber length L3 = 170 μm and microneedle diameter d1 = 20 μm.

# 4 Device Fabrication

Device fabrication is summarized next:

A. Backside deep reactive ion etching (DRIE) Bosh process (360 μm) using Aluminum (Phosphoric – Acetic - Nitric (PAN) etched) and Photoresist as mask.
B. Backside reactive ion etching (RIE) process (15 μm) using Aluminum (PAN etched).
C. Front side DRIE Bosh process (170 μm) using thick Photoresist as mask.
D. Sharpening etching (HNA solution).
E. Backside metal deposition (Ag) using sputtering and electrochemical deposition (TSV).
F. Front side metal deposition (Ag) using sputtering and electrochemical deposition (through-silicon via (TSV)) [9].
G. FeCl3 chemical oxidation to generate the AgCl and finalize the Ag/AgCl deposition of the TSV on both sides.
H. Metal deposition for electrodes on the Group's Wafer. For electrodes in the glucose sensor.
I. Glucose immobilization.
J. Silicon/Silicon anodic bounding for glucose sensor and conductive adhesive bounding for ECG electrodes.

# 5 Integration and Packaging

## 5.1 Mounting and Packaging

The device is integrated in a System in Package (SiP) that is inside the cellphone. The system is controlled by the ultra-low power microcontroller MSP430FR5739, from Texas Instruments [21, 22]. The biosensor device, as was previously explained, consists in three different electrode structures: two dry ECG sensors and a third wet sensor for ECG and Glucose detection. One of the dry ECG electrodes (active electrode) is integrated in the system in package with the signal processing electronic circuitry. The other dry electrode (reference electrode) is also inside the cellphone, but is connected with a cable to the SiP. The third sensor includes an active ECG electrode and Glucose sensor, this last sensor is disposable and a platform similar to a SD card platform is used to change this disposable sensor. A cable connects this sensor to the electronic circuitry in the SiP. The disposition of these three sensors is shown in Fig. 4.

As it was referred before both ECG dry electrodes and the glucose and ECG sensor consist in two Si wafers. The bottom wafer for the ECG electrode has a metal line that makes a contact between the top wafer and the bottom wafer and then connects to the electronic circuitry. Both wafers are joined together by an Ag conductive glue, in the case of the 2 ECG electrodes, without the glucose sensor. The contact in the bottom wafer is wire bonded to a wafer with the electronic circuitry and the microcontroller.

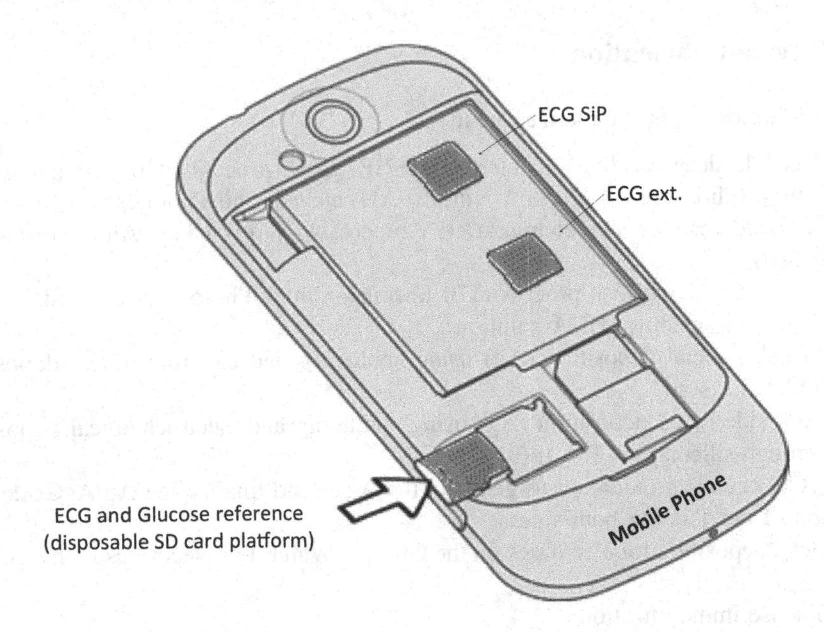


**Fig. 4.** Schematic example of sensors and SiP disposition in the mobile phone

In the glucose and ECG sensor, the two wafers are joined by anodic bonding and the bottom wafer contains metal lines that are connect to a cable connector by wire bonding, this system is all inside the SD card platform.

## 5.2 Signal Processing Electronic Circuitry

The two active ECG electrodes are connected to the electronic signal processing circuitry similar to the one shown in Fig. 2 of [23], these are connected to a differential amplifier that is connected to a high pass filter amplifier that eliminates the DC voltage and then the signal goes through a high gain low pass filter amplifier that connects to and ADC channel of the microcontroller. The overall total gain of this amplifier is 500. Spice simulations were made using BSIM 3 models for MOSIS 350 nm technology. A band pass filter with two amplifier stages and a capacitor feedback configuration was designed, Fig. 5.

This circuit is a band pass filter. The negative feedback loop defined by R2 and C2 gives the high pass frequency of the circuit and R1 C1 plus the low pass filter behavior of the amplifier (any non ideal amplifier due to the devices, transistors and the way they behave in the amplifier stages has intrinsically a Gain roll of frequency that gives a low pass filter behavior) define the low pass frequency. The feedback gain is defined by Cin/Cf = C1/C2, being Cin the input capacitance and Cf the feedback capacitance [24, 25], and in this example it is 100, note for just one amplifier stage. This structure is very common in bio amplifiers and provides good results [25]. R2 and R4 are pseudoresistors to achieve a high resistance value. Each amplifier stage consists in a single

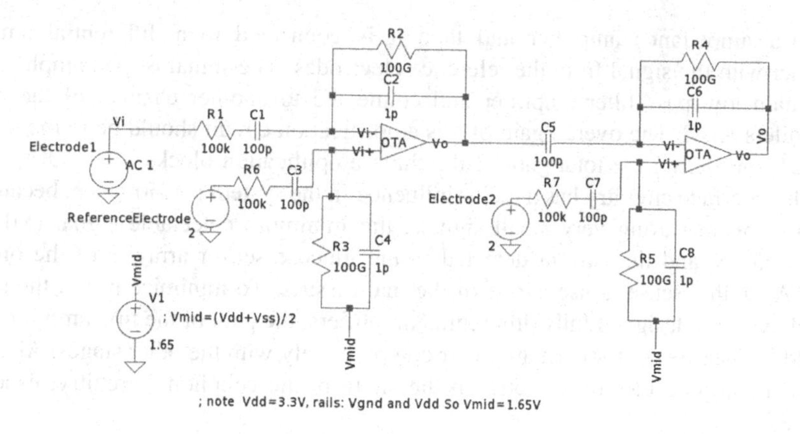


**Fig. 5.** Single ended band pass filter with capacitor feedback for ECG AFE.

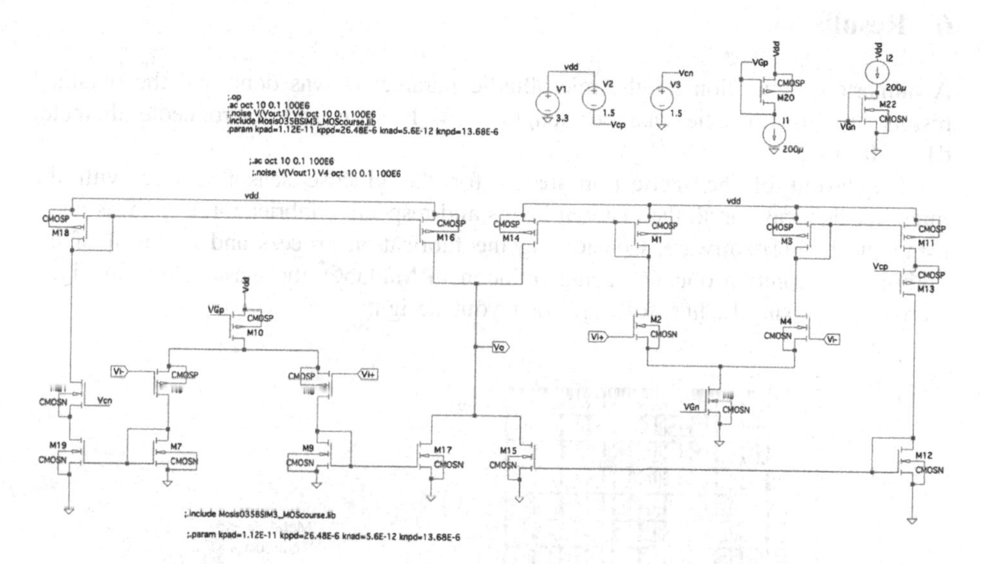

**Fig. 6.** Single ended dual input, PMOS and NMOS input, current mirror OTA.

ended dual input, PMOS and NMOS input, current mirror Output Transconductance Amplifier (OTA) presented in Fig. 6.

The glucose sensors are also connected to electronic circuitry. Figure 6 as the glucose sensor has 100 microneedles the signals from each electrode are connected in parallel (working electrodes' signals are connected in parallel, and reference electrodes' signals are connected in parallel). This way the measured current is approximately 100 times the current of a single microneedle, so in the order of 100 nA. This is considering that the reaction in all microneedles is approximately at same stage and in the maximum current reaction point, of 1 nA per microneedle. The overall signal of the working electrodes and the signal from the reference electrode are then connected to the signal processing circuitry. The current from the working electrodes is converted to a voltage in a transimpedance amplifier and then it is connected to a differential amplifier together with the signal from the reference electrodes. The signal is then amplified in a high gain low pass filter amplifier and connected to another channel of the microcontrollers ADC. The overall gain of this amplification circuit should be in the order of 90 dB, considering the total gain of the three amplification blocks.

The interface circuitry has a major influence in the system performance, because the sensors are measuring very small signals, the minimum detectable signal (MDS) of ECG is 5 μV and the current detected in the glucose sensor array is of the order of 100 nA. In this sense, noise is one of the main issues. To minimize noise: the system must have well designed fully differential amplifiers, the gain of the first amplifier stage should be maximized for each amplifier comparatively with the other stages. Also there should be a good electromagnetic insulation from the cellphone circuitry, especially from RF.

## 6 Results

A numeric optimization for the microfluidic parameters was done and the obtained results are: microneedle chamber length L3 = 170 μm and microneedle diameter d1 = 20 μm.

The layout of the fabrication masks for the glucose sensor device with the microneedle array for all the material layers and respective fabrication steps was made using the Ledit® software. Considering the fabrication process and as result of the optimization solution obtained using fmincon in Matlab®, the masks show in Fig. 7 were created using Ledit® software for layout design.

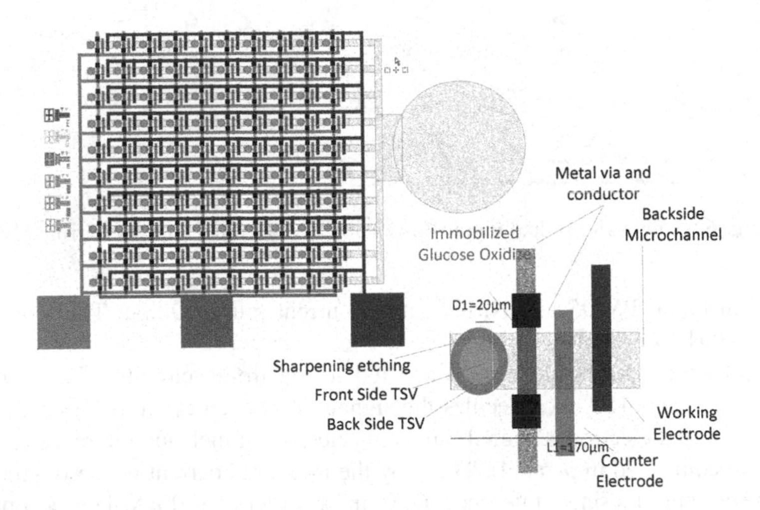


**Fig. 7.** Array of 10 needles by 10 needles design layout, made with Ledit® software, and close-up of a single microneedle' layout with its own glucose sensor electrodes.

Figure 7 presents the final layout of the sensors array with the array of 10 needles by 10 needles, the electrical pads for the glucose sensing and the interconnection channels with a general outlet for generating the pressure difference. Alignment marks for the mask (global alignment, vernier and photoresists development) were also included in this layout. The final system (glucose sensor) area is 3.8 mm × 2.5 mm.

The AFE simulations results met the design specifications. Figure 8 presents the transient analysis with a real ECG noisy signal with typical signal wander due to movement and variation in the electrodes' contact with skin. And Fig. 9 presents the AFE Bode plot from the AC analysis. Table 3 presents the AFE simulation results.

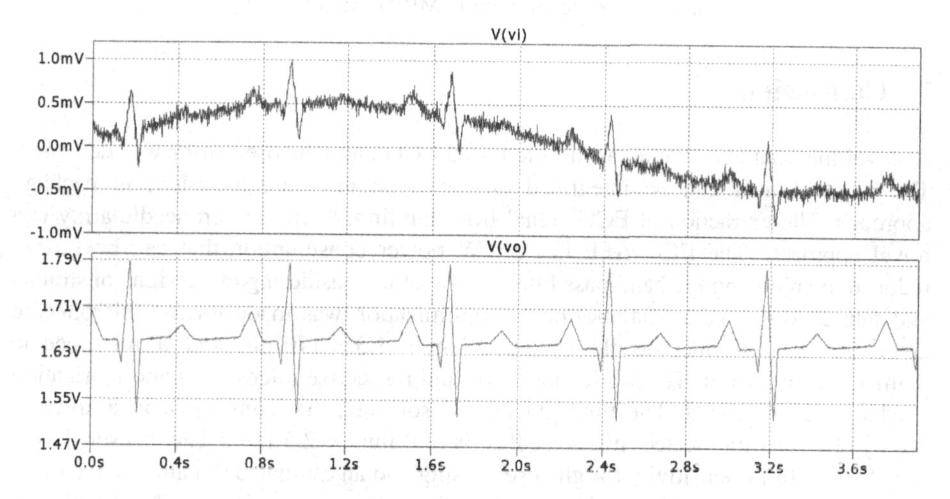


**Fig. 8.** Transient AFE simulation results – top: input signal ECG in blue with noise and signal wander, bottom: AFE signal output in red. (Color figure online)

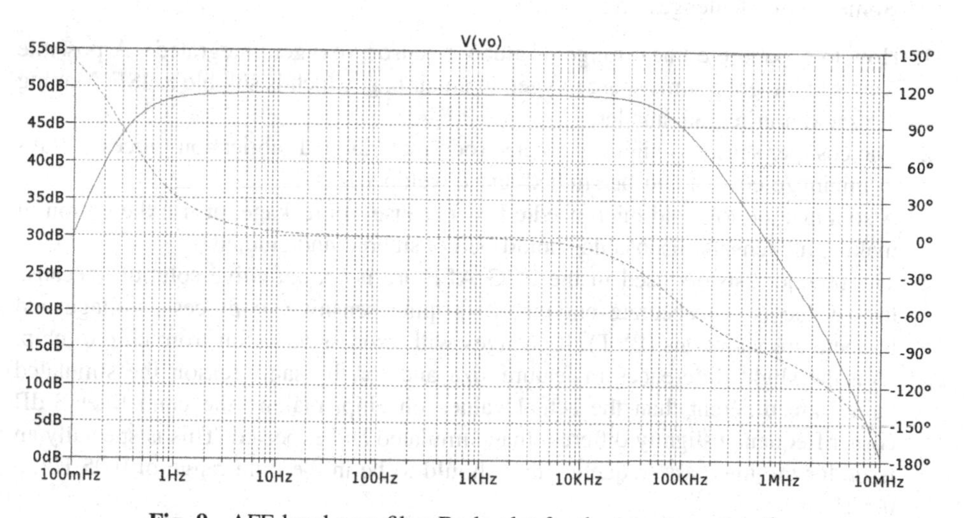


**Fig. 9.** AFE band pass filter Bode plot for the two stages together.

**Table 3.** ECG AFE simulated specifications.

| Specifications | Value |
|---|---|
| Supply voltage Vdd | 3.3 V |
| Imax (source) | 2 mA |
| Power | 6.6 mW |
| Input maximum signal range | −1 mV to +1 mV |
| Vbias and Vcm (common mode) | 1.65 V |
| Gain | 49 dB (282x) |
| 3 dB bandwidth | 77 kHz |
| Common mode rejection ratio (CMRR) | 52 dB (391x) |

# 7 Conclusion

A novel minimal invasive micro needle based ECG and Glucose sensor was designed. The whole sensor can be integrated on smart phone using a system in package approach. The extraction of ECG signal from the finger with a microneedle array is a novel approach. The ECG AFE has 2 mW power consumption that can be further reduced, by reducing the band pass filter bandwidth. Considering the design constraints and fabrication tools, a Matlab® design optimization was made for the microneedle array obtaining a microneedle chamber length L3 = 170 μm and a microneedle diameter d1 = 20 μm. The fabrication steps and respective micro and nano fabrication masks were developed. The final glucose sensor area, that consists in in a array of 10 by 10 microfabricated microneedles is 3.8 mm × 2.5 mm. The sensor has a 0.0048 nA*dl/mg sensitivity for glucose sensing and an estimated dynamic range of 0–180 mg/dl and so it is suited to be used for daily ISF sugar monitoring. The ECG AFE was designed and simulated complying with all the design specifications. It has a Gain of 49 dB and a good CMRR of 52 dB.

Some future challenges are:

1. How to expand the linear range of glucose sensor. (oxygen integrated). A possible way is to integrate some spare oxygen in an air bag, which is soluble to ISF. Making the saturation region smaller.
2. The cost per sensor is high due to electrode material and fabrication process. Ways to minimize cost should be studied and developed.
3. Also a method to avoid microneedle loss because of breakage, or use the option of making all sensors disposable should be considered and studied.
4. The pseudo resistors used in the ECG AFE are not a definitive solution, because their value is not accurate enough, it is quite sensitive to process, voltage and temperature variations (PVT). It changes with process variation from chip to chip, with the slight differences in feature size and for the same reason the simulated values are different than the actual values on chip, making the band filter 3 dB corner frequency slightly different than simulated and expected. This is specially an issue for the low pass frequency that should to be in the tight range of 0.05 Hz to 0.5 Hz.

This work shows a clear contribution to Smart Systems as it paves the path of integrating healthcare and fitness solutions directly in mobile devices, that are one of the most used nodes to access Internet of things (IoT). In an IoT era, these are important developments to improve humans' daily life.

## References


1. O'Mahony, C., Pini, F., Blake, A., Webster, C., O'Brien, J., McCarthy, K.G.: Microneedle-based electrodes with integrated through-silicon via for biopotential recording. Sens. Actuators A Phys. **186**, 130–136 (2012). doi:10.1016/j.sna.2012.04.037
2. Newman, J.D., Turner, A.P.F.: Home blood glucose biosensors: a commercial perspective. Biosens. Bioelectron. **20**, 2435–2453 (2005). doi:10.1016/j.bios.2004.11.012
3. Lourenço, A., Silva, H., Fred, A.: Unveiling the biometric potential of finger-based ECG signals. Comput. Intell. Neurosci. **2011**, 1–8 (2011). doi:10.1155/2011/720971
4. Forvi, E., Bedoni, M., Carabalona, R., Soncini, M., Mazzoleni, P., Rizzo, F., O'Mahony, C., Morasso, C., Cassarà, D.G., Gramatica, F.: Preliminary technological assessment of microneedles-based dry electrodes for biopotential monitoring in clinical examinations. Sens. Actuators A Phys. **180**, 177–186 (2012). doi:10.1016/j.sna.2012.04.019
5. Raju, M.: Heart-Rate and EKG Monitor Using the MSP430FG439 (2007)
6. Sarpeshkar, R.: Ultra Low Power Bioelectronics Fundamentals, Biomedical Applications, and Bio-Inspired Systems (2010). http://dx.doi.org/10.1017/CBO9780511841446
7. Fulford-Jones, T.R.F., Wei, G-Y, Welsh, M.: A portable, low-power, wireless two-lead EKG system. In: 26th Annual International Conference of the IEEE EMBS, pp 2141–2144 (2004)
8. Yu, L.M., Tay, F.E.H., Guo, D.G., Xu, L., Nyan, M.N., Chong F.W., Yap, K.L., Xu, B.: A MEMS-based bioelectrode for ECG measurement. In: 2008 IEEE Sensors, pp. 1068–1071. IEEE (2008)
9. Mukerjee, E.V., Collins, S.D., Isseroff, R.R., Smith, R.L.: Microneedle array for transdermal biological fluid extraction and in situ analysis. Sens. Actuators A Phys. **114**, 267–275 (2004). doi:10.1016/j.sna.2003.11.008
10. Wang, P.M., Cornwell, M., Prausnitz, M.R.: Minimally invasive extraction of dermal interstitial fluid for glucose monitoring using microneedles. Diab. Technol. Ther. **7**, 131–141 (2005). doi:10.1089/dia.2005.7.131
11. Zimmermann, S., Fienbork, D., Stoeber, B., Flounders, A.W., Liepmann, D.: A microneedle-based glucose monitor: fabricated on a wafer-level using in-device enzime inmobilization. In: 12th International Conference on Solid State Sensors, Actuators and Microsystems, pp. 99–102 (2003)
12. Zhao, Y., Li, S., Davidson, A., Yang, B., Wang, Q., Lin, Q.: A MEMS viscometric sensor for continuous glucose monitoring. J. Micromech. Microeng. **17**, 2528–2537 (2007). doi:10.1088/0960-1317/17/12/020
13. Bruss, H.: Theoretical Microfluidics. Oxford University Press, Oxford (2008)
14. Bruus, H.: Theoretical microfluidics, Third fall. Lect Notes Fall 2006 (2006). doi:10.1111/j.1574-6968.2009.01808.x
15. Thennadil, S.N., Rennert, J.L., Wenzel, B.J., Hazen, K.H., Ruchti, T.L., Block, M.B.: Comparison of glucose concentration in interstitial fluid, and capillary and venous blood during rapid changes in blood glucose levels. Diab. Technol. Ther. **3**, 357–365 (2004). doi:10.1089/15209150152607132

16. Suckale, J., Solimena, M.: Pancreas islets in metabolic signaling - focus on the Beta-cell, pp. 7156–7171 (2008)
17. Daly, M.E., Vale, C., Walker, M., Littlefield, A., George, K., Alberti, M.M., Mathers, J.C.: Acute effects on insulin sensitivity and diurnal metabolic profiles of a high-sucrose compared with a high-starch diet. Am. J. Clin. Nutr. **67**, 1186–1196 (1998)
18. Dorgan, S.J., Reilly, R.B.: A model for human skin impedance during surface functional neuromuscular stimulation. IEEE Trans. Rehabil. Eng. **7**, 341–348 (1999)
19. Cutnell, J., Johnson, K.: Physics, 4th edn. Wiley, Hoboken (1998)
20. ASM International. Materials and Coatings for Medical Devices: Cardiovascular (2009)
21. Texas Instruments. Texas Instruments MSP430FR5739 Mixed Signal Microcontroller (2013). http://www.ti.com/product/msp430fr5739
22. Texas Instruments. MSP430FR573x MSP430FR572x Mixed Signal Microcontroller datasheet (2013)
23. Raju, M.: Heart-Rate and EKG Monitor Using the MSP430FG439, pp. 1–12 (2007)
24. Xiao, Z.: Low Power Circuits and Systems for Brain Machine Interface. University of Florida (2013)
25. Wattanapanitch, W., Fee, M., Sarpeshkar, R.: An energy-efficient micropower neural recording amplifier. IEEE Trans. Biomed. Circuits Syst. **1**, 136–147 (2007). doi:10.1109/TBCAS.2007.907868

# Crystalline Silicon PV Module Under Effect of Shading Simulation of the Hot-Spot Condition

Ruben S. Anjos[2], Rui Melício[1,3(✉)], Victor M.F. Mendes[1,2], and Hugo M.I. Pousinho[1,2]

[1] Departamento de Física, Escola de Ciências e Tecnologia, Universidade de Évora, Évora, Portugal
ruimelicio@gmail.com
[2] Instituto Superior de Engenharia de Lisboa, Lisbon, Portugal
[3] Instituto Superior Técnico, IDMEC, Universidade de Lisboa, Lisbon, Portugal

**Abstract.** This paper centers on the silicon crystalline PV module technology subjected to operation conditions with some cells partially or fully shaded. A shaded cell under hot-spot condition operating at reverse bias are dissipating power instead of delivering power. A thermal model allows analyzing the temperature increase of the shaded cells of the module under hot-spot condition with or without protection by a bypass diode. A comparison of the simulation results for a crystalline PV module without shading and with partial or full shading is presented.

**Keywords:** Hot-spot · PV systems · Shading effect · Breakdown voltage · Modelling

## 1 Introduction

The demand for energy, the scarcity of nonrenewable fuels and the need for $CO_2$ emissions decrease have resulted in a consciousness of the significance of energy efficiency and energy savings [1, 2] and DSM software and Smart Grid architecture have been developed in order to endure consumers on energy use. Also, renewable energy sources coming from solar and wind energy sources are attractive to go into exploitation, considering large scale and mini or micro scale conversion systems [1, 3]. A photovoltaic system (PVS) converts solar energy into electric energy. The main device of a PVS is a cell. Cells may be assembled to form modules or arrays. A module is composed by cluster of cells linked in series. An array can be a module or a set of modules linked in parallel, series, total cross tied or bridge-linked to form large PVS. These PVSs may have or not a tracking system in order to achieve higher energy conversion values during sunny days due to the diverse positions to accumulate the sun´s irradiation [4]. The array performance depends on the solar irradiation, operating conditions, array configuration, environmental parameters as temperature and wind speed, and shading. The shading on arrays occurs due to the dust, bird droppings, leaves, snow or shadowing caused by near buildings, near trees or a passing cloud. The shading can be partial or full with respect to a PV module [5]. The existence of shaded cells in a module causes a mismatch on the


L.M. Camarinha-Matos et al. (Eds.): DoCEIS 2017, IFIP AICT 499, pp. 479–487, 2017.
DOI: 10.1007/978-3-319-56077-9_46

equilibrium of photo-generation of electron-hole pairs of the cells. Under this mismatch, a non-shaded cell of the module operates at a generation of carriers higher than those of the shaded ones [5]. Consequently, the shaded cells are into reverse bias [5]. The hot-spot occurs after a cell, or cluster of cells in a PV module, operates at reverse bias, dissipating power and operating at abnormally high temperatures. Cells exposed to higher temperatures will degrade at a higher rate than others. If a cell is shaded and operation at high temperatures occurs for a long time, then irreparable damage is due to happen to the solar cell. If a bypass diode is used, a cell in reverse bias renders useless the rest of the cells under the bypass diode [6]. The hot-spot occurrence is particularly frequent and always happens when a shadow covers the module, thus reducing the photo-generated current [7, 8]. In order to protect shaded cells from breakdown voltage (BV), bypass diodes linked in antiparallel with the cell or module is applied. In the 1990s authors contributed to improve the PVS module design and to determine the maximum solar cells number per bypass diode in order to avoid the hot-spots condition [9, 10]. Based on these knowledges, according to IEC 61215 a hot-spot endurance test became part of the type approval for crystalline silicon modules [8].

The diodes bypass limit the reverse voltage that can be applied to a cell, thus avoiding it from reaching the BV when the cell is shaded [6]. Due to impurities and defects inside the silicon, some cells may display a great reverse current even before reaching the BV. This impurities and defects inside the silicon are commonly modeled by inserting a shunt resistance, whose value depends on the distribution of defects/impurities and concentration inside the cell [6]. If the shunt resistance is low enough the hot-spot condition can occur even before that the PV cell enters the BV even if bypass diodes are used [6]. In this case, due to the hot-spot heating, the PV cell can reach a temperature high enough to cause permanent damage, although this takes longer to occur than when the PV cells operates in breakdown voltage [6].

This paper is considered with the exposition to shading of a silicon crystalline PV module and as a consequence subjected to condition of hot-spot. A thermal model is used in order to allow estimating the temperature of the area subjected to the condition of hot-spot. The paper is systematized as follows. Section 2 presents the relationship to smart systems. Section 3 presents the modeling of the PVS module forward biased, using the five parameters model, i.e., the single-diode, shunt and series resistances circuit; the model of the PV module reverse biased; and the thermal model which allows estimating the hot-spot temperature. Section 4 presents the case studies. In the first case, the I-V characteristic curves are simulated considering the reverse bias of the PV module: simulation results for a silicon crystalline PV module technology without shading are compared with the module partially shaded or fully shaded; without or with bypass diode. In the second case, a thermal model allows analyzing the temperature increase of the shaded cells of the module subjected to condition of hot-spot without or with a bypass diode. Finally, concluding remarks are given in Sect. 5.

## 2 Relationship to Smart Systems

Technological Innovation has entered into the age of smart systems and smart grids, when machines implement human functions of analyzing real data and making decisions [11]. A Smart system is an embedded systems architecture which embodies advanced automation systems to provide control and monitoring over the something's functions, for example a PVS. Hence, the advances in science and technology allow the opportunity to bridge physical components and cyber space, leading to Cyber-Physical Systems [12, 13]. The coordination and interaction of smart systems of PVSs require handling number of challenges to increase the reliability and the service life time of the PV modules. The performance of a PV modules depends on the operating conditions, i.e., depends not only on solar irradiation and temperature, but also on the array configuration and shading effects. The hot-spot on a PV cell produces degradation or a permanent damage in the shaded cells, with a consequent reduction of the provided power [14].

Statistics show degradation rates of the nominal power for thin-film and silicon PVS modules of 0.8%/year [15]. Hence, to raise the service life and the reliability of PVS modules one has to recognize the challenges involved. Successfully integrating PV power into the existing electric grid is a complex challenge that relies on distributed, interconnected smart systems, which are proliferating within the engineering industry [12].

## 3 Modeling

The circuit for the PVS entails in a current controlled generator, a single-diode, a shunt and series resistances. Thus for the forward bias module the I-V characteristic is formulated [5] given by:

$$I = I_S - I_0(e^{\frac{V + IR_S}{mV_T}} - 1) - \frac{V + IR_S}{R_p} \tag{1}$$

where the output module current is $I$, the photo-generated current is $I_S$, the reverse bias saturation diode current is $I_0$, the output voltage is $V$, the shunt and the series resistances are respectively $R_p$ and $R_S$, the ideality factor is $m$, the thermal voltage of the $p$-$n$ junction is $V_T$.

This circuit for the reverse bias module is shown in Fig. 1. Where in Fig. 1 the current through diode $D$ is $I_D$, described by the Shockley diode equation, the reverse breakdown current which is a portion of $I_p$ dissipated in $Rp$ when the module is reverse biased is $I_{bd}$ [5, 6]. For the reverse bias module, the $I_{bd}$ [11] is given by:

$$I_{bd} = \frac{V + IR_S}{R_p} a \, (1 - \frac{V + IR_S}{V_{bd}})^{-b} \tag{2}$$

where the portion of the linear current involved in avalanche breakdown is $a$.

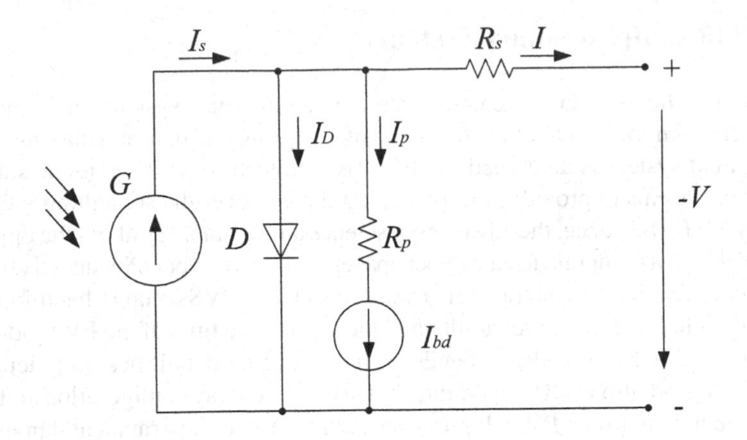


**Fig. 1.** Circuit for the reverse bias module.

For the reverse bias module the I-V characteristic is formulated by an implicitly function [16] given by:

$$I = I_S - I_0(e^{\frac{V+IR_S}{mV_T}} - 1) - \frac{V+IR_S}{R_p}[1 + a(1 - \frac{V+IR_S}{V_{bd}})^{-b}] \tag{3}$$

where the BV is $V_{bd}$, the avalanche breakdown factor is $b$.

The thermal model allows estimating the temperature $T_{HS}$ of the module's area $A_{HS}$ under hot-spot condition [5, 6]. The time instant t at which the PV module goes into shading condition and reverse biased is denoted by $t_{HS}$. Considering the PV module staying into shading condition from $t_{HS}$ until at time $t \geq t_{HS}$, the temperature $T_{HS}\ (t)$ is incremented relatively to the ambient one, $T_{amb}$. The expression for computing $T_{HS}\ (t)$ [6] is given by:

$$T_{HS}(t) = T_{amb} + i + ii \tag{4}$$

where $i$ and $ii$ [6] are respectively given by:

$$i = R_{TH}A\ G[\gamma + (1-\gamma)e^{-(\frac{t-t_{HS}}{R_{TH}C_{TH}})}] \tag{5}$$

$$ii = P_{diss}R_{TH-HS}[1 - e^{-(\frac{t-t_{HS}}{R_{TH-HS}C_{TH-HS}})}] \tag{6}$$

where the whole PV module area is $A$, the solar irradiance is $G$, the relative gap in the irradiation between the non-shaded modules and the shaded module linked in series is $\gamma$, the dissipated power on the $R_p$ resistance of the shaded module is $P_{diss}$, the thermal capacitance and thermal resistance of the module with area $A_{HS}$ under hot-spot condition are $C_{TH\text{-}HS}$ and $R_{TH\text{-}HS}$, respectively, the thermal capacitance and thermal resistance of the remaining module area are $C_{TH}$ and $R_{TH}$. In the thermal model, the parameters $R_{TH\text{-}HS}$, $C_{TH\text{-}HS}$ are constants and depend of the materials which composing the upper layers of the module cells [6]. Considered hot-spot condition on the two cells of the PV module with 36 cells is shown in Fig. 2.

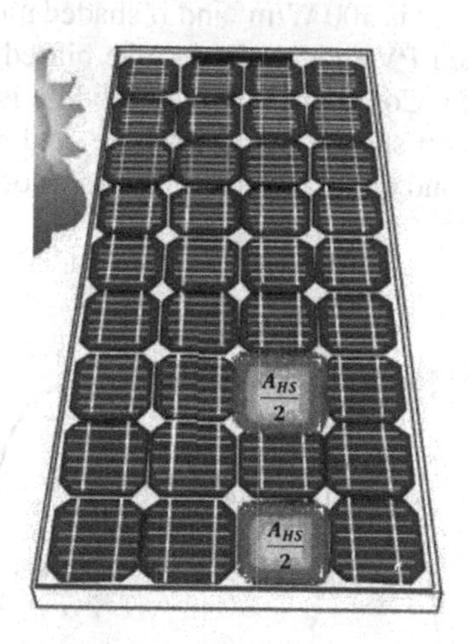


**Fig. 2.** Two cells shaded PV module under hot-spot condition [5].

In Fig. 2, $A_{cell}$ is the area of each cell of the module and $A_{HS}$ is the area of the module under hot-spot condition. The range of values of $A_{HS}$ for the simulation is taken from [6], where is said to be set experimentally in the range between 5% and 10% of $A_{cell}$.

## 4 Simulation Results

The models for the PV module forward biased, for the PV module reverse biased and for the thermal model that allows assessing the temperature of the area under hot-spot condition are implemented in Matlab/Simulink.

The module has 36 cells linked in series, the shaded area $A_{HS}$ is equal to the area of two cells, i.e., taking approximately 5.6% of the area $A$ associated with the capture of energy. The data of the manufacture for the crystalline silicon PV module Isofotón I-53 at STC [17] is displayed in Table 1.

**Table 1.** Isofotón I-53 solar module data

| Technology | $V_m^*$ | $I_m^*$ | $V_{oc}^*$ | $I_{sc}^*$ |
|---|---|---|---|---|
| Crystalline | 17.4 V | 3.05 A | 21.65 V | 3.27 A |

The forward breakover voltage considered for silicon diode is 0.7 V. The equivalent circuit for the reverse bias module shown in Fig. 1 has the values for the $R_P$ and $R_S$ given in [16].

*First Case Study.* The simulation results for a silicon crystalline PV module technology without shading are compared with the module partially shaded (MPS) or fully shaded (MFS) in order to study the current and voltage when the module is reverse biased. Considering the MPS or MFS linked in series with one non-shaded module, if the shaded module is a MPS his irradiance is 500 W/m$^2$ and if shaded module is MFS his irradiance is 0 W/m$^2$. When the shaded PV module is reverse biased we assume $V_{bd} = -10$ V, $a = 1.93$ and $b = 1.10$ in (3). Considered that the module is without bypass diode: the I-V reference curve of the non-shaded module (black) and respective symmetric curve (also in black but in the second quadrant), the I-V curves of the MPS (orange) or MFS (blue) are shown in Fig. 3.

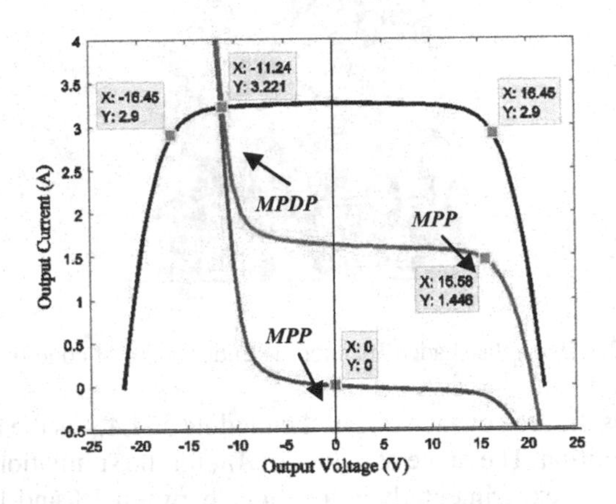


**Fig. 3.** No bypass protection. I-V curves (black), MPS (orange), MFS (blue) (Color figure online).

Figure 3 reveals that at the maximum power dissipation point (MPDP) the MPS and the MFS is reverse biased with a same voltage value of −11.24 V and a current value of 3.22 A. The maximum power point (MPP) when the module is partially shaded is at a voltage value of 15.58 V and a current value of 1.45 A. Considered that the module is with bypass diode: the I-V reference curve of the non-shaded module (black) and respective symmetric curve (also in black but in the second quadrant), the I-V curves of the MPS (orange) or MFS (blue) are shown in Fig. 4.

Figure 4 reveals that at MPDP the shaded module is reverse biased with the value of voltage of −0.70 V and the value of the current of about 3.38 A. A comparison of Figs. 3 and 4 allows observing the effect of the bypass protection. The maximum power dissipated in the shaded module without bypass diode is 36.20 W at MPDP while for the shaded module with bypass diode is 2.28 W at MPDP. For the two case studies the maximum power point (MPP) when the module is partially shaded is at a voltage value of 15.58 V and a current value of 1.45 A, while when the module is fully shaded the voltage value is 0.00 V and a current value is 0.00 A. Hence, the bypass diode does not affect the module at MPP in [17].

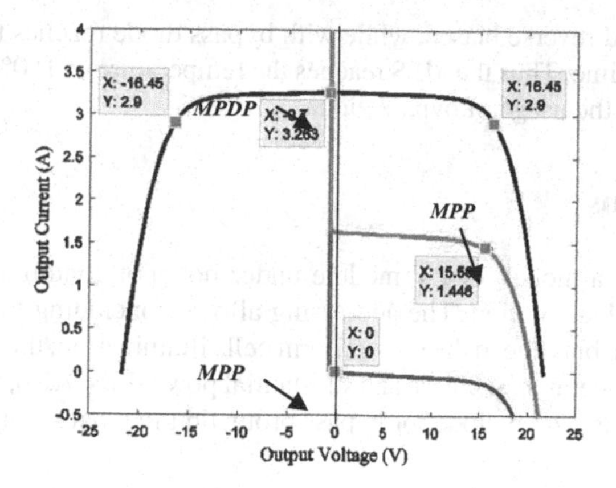


**Fig. 4.** Bypass protection. I-V curves (black), MPS (orange), MFS (blue) (Color figure online).

*Second Case Study.* The simulation results for a silicon crystalline PV module technology without shading are compared with the module partially shaded or fully shaded in order to evaluate the evolution of the module temperature under hot-spot condition. The value of $T_{amb}$ considered is 25°C and the time instant $t_{HS}$ considered is 5 s. Consider the MPS operating at MPDP the curves of the temperature $T_{HS}$ for the area under hot-spot condition in function of the time: without bypass diode (blue) and with bypass diode (green) are shown in Fig. 5.

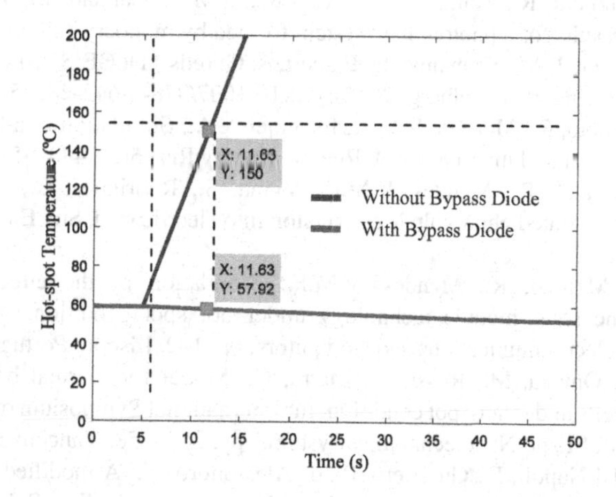


**Fig. 5.** MPS, temperature on the area of hot-spot condition (Color figure online).

The PV module has a $T_{HS}$ of approximately 60°C for a lower than $t_{HS} = 5$ s. Figure 5 reveals that the MPS without bypass diode reaches the critical temperature of 150°C [6] at time of 11.63 s, i.e., 6.63 s after getting shaded and reverse biased, while with bypass diode decreases temperature 57.92°C in same time. The curves of the temperature for the MFS in function of the time reveals that the MFS without bypass diode reaches the critical temperature of 150°C [6] at time of 11.71 s, i.e., 6.71 s after getting shaded and reverse biased, while with bypass diode reaches the temperature of 56.79°C in same time. Thus the MPS reaches the temperature of 150°C faster than MFS not depending on the usage of bypass diode.

## 5 Conclusions

An addressing of a model for PV module under hot-spot condition is study in what regards the model simulation. The addressing allows concluding that when a module operates at reverse bias due to the non-uniform cells illumination, the module is exposed to a quantified temperature increase and substantial power loss. Also, allows quantifying the advantage of using the diode for bypass protection in what regards performance of the module.

**Acknowledgment.** This paper was in part supported by Portuguese Funds, Foundation for Science and Technology, UID-EMS-50022-2013, LAETA project 2015/2020.

## References

1. Fialho, L., Melício, R., Mendes, V.M.F., Estanqeiro, A., Collares-Pereira, M.: PV systems linked to the grid: parameter identification with a heuristic procedure. Sustain. Energy Technol. Assess. **10**, 29–39 (2015)
2. Saraiva, S., Melício, R., Matias, J.C.O., Cabrita, C.M.P., Catalão, J.P.S.: Simulation and experimental results for a photovoltaic system formed by monocrystalline solar modules. In: Camarinha-Matos, L.M., Shahamatnia, E., Nunes, G. (eds.) DoCEIS 2012. IAICT, vol. 372, pp. 329–336. Springer, Heidelberg (2012). doi:10.1007/978-3-642-28255-3_36
3. Fialho, L., Melício, R., Mendes, V.M.F., Estanqeiro, A.: Simulation of a-Si PV system grid connected by boost and inverter. Int. J. Renew. Energy Res. **5**(2), 443–451 (2015)
4. Fialho, L., Melício, R., Mendes, V.M.F., Viana, S., Rodrigues, C., Estanqeiro, A.: A simulation of integrated photovoltaic conversion into electric grid. Sol. Energy **110**, 578–594 (2015)
5. Anjos, R.S., Melício, R., Mendes, V.M.F.: Simulation of the effect of shading on monocrystalline solar module technology under hot spot condition. In: Conference on Electronics, Telecommunications and Computers, pp. 1–2, Lisbon, Portugal (2016)
6. Giaffreda, D., Omaña, M., Rossi, D., Metra, C.: Model for thermal behavior of shaded photovoltaic cells under hot-spot condition. In: International Symposium on Defect and Fault Tolerance in VLSI and Nanotechnology Systems, pp. 252–258, Vancouver, Canada (2011)
7. Daliento, S., Di Napoli, F., Guerriero, P., d` Alessandro, V.: A modified bypass circuit for improved hot spot reliability of solar panels subject to partial shading. Sol. Energy **134**, 211–218 (2016)
8. IEC Standard 61215: Crystalline silicon terrestrial photovoltaic (PV) modules-design qualification and type approval. In: International Electrotechnical Commission (1995)
9. Arnett, J.C., González, C.C.: Photovoltaic module hot-spot durability design and test methods. In: 15th Photovoltaic Specialists Conference, pp. 1099–1105, Kissimmee, USA (1981)
10. Shepard Jr., N.F., Sugimura, R.S.: The integration of bypass diodes with terrestrial photovoltaic modules and arrays. In: 17th Photovoltaic Specialists Conference, pp. 676–681, Orlando, USA (1984)

11. Arsan, T.: Smart systems: from design to implementation of embedded smart systems. In: 13th International Symposium on Smart MicroGrids for Sustainable Energy Sources enabled by Photonics and IoT Sensors, pp. 59–64, Nicosia, Cyprus (2016)
12. Viveiros, C., Melício, R., Igreja, J.M., Mendes, V.M.F.: Control and supervision of wind energy conversion systems. In: Camarinha-Matos, L.M., Falcão, A.J., Vafaei, N., Najdi, S. (eds.) DoCEIS 2016. IAICT, vol. 470, pp. 353–368. Springer, Cham (2016). doi: 10.1007/978-3-319-31165-4_34
13. Fialho, L., Melício, R., Mendes, V.M.F., Collares-Pereira, M.: Simulation of a-Si PV system linked to the grid by DC boost and three-level inverter under cloud scope. In: Camarinha-Matos, L.M., Baldissera, T.A., Di Orio, G., Marques, F. (eds.) DoCEIS 2015. IAICT, vol. 450, pp. 423–430. Springer, Cham (2015). doi:10.1007/978-3-319-16766-4_45
14. Rossi, D., Omaña, M., Giaffreda, D., Metra, C.: Modeling and detection of hotspot in shaded photovoltaic cells. IEEE Trans. Very Large Scale Integr. (VLSI) Syst. **23**(6), 1031–1039 (2015)
15. Jordan, D.C., Kurtz, S.R.: Photovoltaic degradation rates-an analytical review. Prog. Photovoltaics Res. Appl. **21**, 12–29 (2013)
16. Ramaprabha, R., Mathur, B., Santhosh, K., Sathyanarayanan, S.: Matlab based modelling of SPVA characterization under reverse bias condition. In: 3rd International Conference on Emerging Trends in Engineering and Technology, pp. 334–339, Goa, India (2014)
17. Fialho, L., Melício, R., Mendes, V.M.F.: PV system modeling by five parameters and in situ test. In: International Symposium on Power Electronics, Electrical Drives and Motion, pp. 573–578, Ischia, Italy (2014)

# Author Index

Printed in the United States
By Bookmasters